Bioelectronics

Edited by
I. Willner, E. Katz

Further Titles of Interest

H. Baltes, O. Brand, G. K. Fedder, C. Hierold, J. G. Korvink, O. Tabata (Series Eds.)
Advanced Micro & Nanosystems (Book Series)

H. Baltes, O. Brand, G. K. Fedder, C. Hierold, J. G. Korvink, O. Tabata (Volume Eds.)
Vol. 1: Enabling Technology for MEMS and Nanodevices
2004, ISBN 3-527-30746-X

O. Brand, G. K. Fedder (Volume Eds.)
Vol. 2: CMOS-MEMS
2005, ISBN 3-527-31080-0

C. S. S. R. Kumar, J. Hormes, C. Leuschner
Nanofabrication Towards Biomedical Applications Materials and Methods
2005, ISBN 3-527-31115-7

R. C. Advincula, W. J. Brittain, K. C. Caster, J. Rühe (Eds.)
Polymer Brushes
2004, ISBN 3-527-31033-9

M. Köhler, W. Fritzsche
Nanotechnology
An Introduction to Nanostructuring Techniques
2004, ISBN 3-527-30750-8

M. Komiyama, T. Takeuchi, T. Mukawa, H. Asanuma
Molecular Imprinting
From Fundamentals to Applications
2003, ISBN 3-527-30569-6

G. Hodes (Ed.)
Electrochemistry of Nanomaterials
2002, ISBN 3-527-29836-3

W. Menz, J. Mohr, O. Paul
Microsystem Technology
2000, ISBN 3-527-29634-4

Bioelectronics

From Theory to Applications

Edited by
Itamar Willner and Eugenii Katz

WILEY-VCH Verlag GmbH & Co. KGaA

Editors:

Prof. Dr. Itamar Willner
Dr. Eugenii Katz

The Hebrew University of Jerusalem
Institute of Chemistry
Givat Ram, Jerusalem 91904
Israel

Cover illustration: The graphic was provided by Dr. Andrew N. Shipway.

■ This books published by Wiley-VCH are carefully produced. Nevertheless, authors, editors, and publisher do not warrant the information contained in these books, including this book, to be free of errors. Readers are advised to keep in mind that statements, data, illustrations, procedural details or other items may inadvertently be inaccurate.

Library of Congress Card No.: applied for
A catalogue record for this book is available from the British Library.

Bibliographic information published by Die Deutsche Bibliothek
Die Deutsche Bibliothek lists this publication in the Deutsche Nationalbibliografie; detailed bibliographic data is available in the internet at <http://dnb.ddb.de>.

© 2005 WILEY-VCH Verlag GmbH & Co. KGaA Weinheim

All rights reserved (including those of translation into other languages). No part of this book may be reproduced in any form – by photoprinting, microfilm or any other means – nor transmitted or translated into machine language without written permission from the publishers. Registered names, trademarks, etc. used in this book, even when not specifically marked as such, are not to be considered unprotected by law.

Printed in Singapore
Printed on acid-free paper

Composition Laserwords Private Ltd, Chennai, India
Printing and Bookbinding Markono Print Media Pte Ltd, Singapore

ISBN-13: 978-3-527-30690-9
ISBN-10: 3-527-30690-0

Contents

Preface XIII

List of Contributors XV

1 **Bioelectronics – An Introduction** 1
  Itamar Willner and Eugenii Katz
  References 12

2 **Electron Transfer Through Proteins** 15
  Jay R. Winkler, Harry B. Gray, Tatiana R. Prytkova, Igor V. Kurnikov, and David N. Beratan
2.1 Electronic Energy Landscapes 15
2.2 Theory of Electron Tunneling 15
2.3 Tunneling Pathways 17
2.4 Coupling-limited ET Rates and Tests of the Pathway Model 19
2.5 Multiple Tunneling Pathway Models 23
2.6 Interprotein Electron Transfer: Docking and Tunneling 27
2.7 Some New Directions in Electron Transfer Theory and Experiment 28
2.8 Concluding Remarks 31
  References 31

3 **Reconstituted Redox Enzymes on Electrodes: From Fundamental Understanding of Electron Transfer at Functionalized Electrode Interfaces to Biosensor and Biofuel Cell Applications** 35
  Bilha Willner and Itamar Willner
3.1 Introduction 35
3.2 Electrodes Functionalized with Reconstituted Redox Proteins 43
3.2.1 Reconstituted Flavoenzyme-Electrodes Using Molecular or Polymer Relay Systems 43

Bioelectronics. Edited by Itamar Willner and Eugenii Katz
Copyright © 2005 WILEY-VCH Verlag GmbH & Co. KGaA, Weinheim
ISBN: 3-527-30690-0

3.2.2	Electrical Contacting of Flavoenzymes by Reconstitution on Carbon Nanotubes and Conducting Polymer Wires 53
3.2.3	Electrical Contacting of Flavoenzymes by Means of Metallic Nanoparticles 57
3.2.4	Integrated Electrically Contacted Electrodes Composed of Reconstituted Quinoproteins 65
3.2.5	Reconstituted Electrically Contacted Hemoproteins 67
3.2.6	Reconstituted *de novo* Hemoproteins on Electrodes 69
3.3	Electrical Contacting of Redox Proteins by Cross-linking of Cofactor-Enzyme Affinity Complexes on Surfaces 73
3.3.1	Integrated NAD(P)$^+$-Dependent Enzyme-Electrodes 73
3.3.2	Integrated Electrically Contacted Hemoprotein Electrodes 80
3.4	Reconstituted Enzyme-Electrodes for Biofuel Cell Design 83
3.5	Conclusions and Perspectives 91
	References 93

4   Application of Electrically Contacted Enzymes for Biosensors   99
Frieder W. Scheller, Fred Lisdat, and Ulla Wollenberger

4.1	Introduction 99
4.2	Biosensors – Precursors of Bioelectronics 99
4.3	Via Miniaturization to Sensor Arrays – The Biochip 102
4.4	The Route to Electrically Contacted Enzymes in Biosensors 104
4.5	Routine Applications of Enzyme Electrodes 107
4.6	Research Applications of Directly Contacted Proteins 109
4.6.1	Protein Electrodes for the Detection of Oxygen-derived Radicals 109
4.6.2	Cytochrome P 450 – An Enzyme Family Capable of Direct Electrical Communication 117
4.7	Conclusions 123
	References 123

5   Electrochemical DNA Sensors   127
Emil Palecek and Miroslav Fojta

5.1	Introduction 127
5.1.1	Indicator Electrodes 128
5.1.2	Electrochemical Methods 128
5.2	Natural Electroactivity and Labeling of Nucleic Acids 129
5.2.1	Electroactivity of Nucleic Acid Components 129
5.2.2	Analysis of Unlabeled Nucleic Acids 131
5.2.3	Electroactive Labels of Nucleic Acids 136
5.2.4	Signal Amplification 140
5.3	Sensors for DNA and RNA Hybridization 140
5.3.1	DNA Hybridization 142

5.3.2	Electrochemical Detection in DNA Sensors	*143*
5.3.3	Single-surface Techniques	*143*
5.3.4	Double-surface Techniques	*153*
5.3.5	Concluding Remarks to DNA Hybridization Sensors	*158*
5.4	Sensors for DNA Damage	*159*
5.4.1	DNA Damage	*159*
5.4.2	Relations Between DNA Damage and its Electrochemical Features	*162*
5.4.3	DNA-modified Electrodes as Sensors for DNA Damage	*167*
5.4.4	Sensors for DNA Strand Breaks	*168*
5.4.5	Detection of Covalent Damage to DNA Bases	*170*
5.4.6	Genotoxic Substances Interacting with DNA Noncovalently	*173*
5.4.7	Electrochemically Induced DNA Damage	*176*
5.4.8	Analytical Applications of Electrochemical Sensors for DNA Damage	*177*
5.4.9	Concluding Remarks to DNA Damage Sensors	*180*
	References *181*	

6	**Probing Biomaterials on Surfaces at the Single Molecule Level for Bioelectronics** *193*	
	Barry D. Fleming, Shamus J. O'Reilly, and H. Allen O. Hill	
6.1	Methods for Achieving Controlled Adsorption of Biomolecules	*194*
6.2	Methods for Investigating Adsorbed Biomolecules	*195*
6.3	Surfaces Patterned with Biomolecules	*197*
6.4	Attempts at Addressing Single Biomolecules	*201*
6.5	Conclusions *205*	
	References *207*	

7	**Interfacing Biological Molecules with Group IV Semiconductors for Bioelectronic Sensing** *209*	
	Robert J. Hamers	
7.1	Introduction *209*	
7.2	Semiconductor Substrates for Bioelectronics	*210*
7.2.1	Silicon *210*	
7.2.2	Diamond *211*	
7.3	Chemical Functionalization *213*	
7.3.1	Covalent Attachment of Biomolecules to Silicon Surfaces	*213*
7.3.2	Hybridization of DNA at DNA-modified Silicon Surfaces	*215*
7.3.3	Covalent Attachment and Hybridization of DNA at Diamond Surfaces	*217*

7.4	Electrical Characterization of DNA-modified Surfaces	219
7.4.1	Silicon	219
7.4.2	Impedance Spectroscopy of DNA-modified Diamond Surfaces	225
7.5	Extension to Antibody–Antigen Detection	225
7.6	Summary	227
	References	228

8 Biomaterial-nanoparticle Hybrid Systems for Sensing and Electronic Devices 231

Joseph Wang, Eugenii Katz, and Itamar Willner

8.1	Introduction	231
8.2	Biomaterial–nanoparticle Systems for Bioelectrochemical Applications	232
8.2.1	Bioelectrochemical Systems Based on Nanoparticle-enzyme Hybrids	232
8.2.2	Electroanalytical Systems for Sensing of Biorecognition Events Based on Nanoparticles	235
8.3	Application of Redox-functionalized Magnetic Particles for Triggering and Enhancement of Electrocatalytic and Bioelectrocatalytic Processes	250
8.4	Conclusions and Perspectives	259
	References	261

9 DNA-templated Electronics 265

Kinneret Keren, Uri Sivan, and Erez Braun

9.1	Introduction and Background	265
9.2	DNA-templated Electronics	266
9.3	DNA Metallization	268
9.4	Sequence-specific Molecular Lithography	271
9.5	Self-assembly of a DNA-templated Carbon Nanotube Field-effect Transistor	276
9.6	Summary and Perspective	279
	References	284

10 Single Biomolecule Manipulation for Bioelectronics 287

Yoshiharu Ishii and Toshio Yanagida

10.1	Single Molecule Manipulation	287
10.1.1	Glass Microneedle	289
10.1.2	Laser Trap	289
10.1.3	Space and Time Resolution of Nanometry	290
10.1.4	Molecular Glues	291
10.1.5	Comparisons of the Microneedle and Laser Trap Methods	291

10.2	Mechanical Properties of Biomolecules 291	
10.2.1	Protein Polymers 291	
10.2.2	Mechanically Induced Unfolding of Single Protein Molecules 294	
10.2.3	Interacting Molecules 296	
10.3	Manipulation and Molecular Motors 297	
10.3.1	Manipulation of Actin Filaments 298	
10.3.2	Manipulation of a Single Myosin Molecule 300	
10.3.3	Unitary Steps of Myosin 300	
10.3.4	Step Size and Unconventional Myosin 302	
10.3.5	Manipulation of Kinesin 303	
10.4	Different Types of Molecular Motors 304	
10.5	Direct Measurements of the Interaction Forces 304	
10.5.1	Electrostatic Force Between Positively Charged Surfaces 305	
10.5.2	Surface Force Property of Myosin Filaments 305	
	References 306	
11	**Molecular Optobioelectronics** 309	
	Eugenii Katz and Andrew N. Shipway	
11.1	Introduction 309	
11.2	Electronically Transduced Photochemical Switching of Redox-enzyme Biocatalytic Reactions 310	
11.2.1	Electronic Transduction of Biocatalytic Reactions Using Redox Enzymes Modified with Photoisomerizable Units 312	
11.2.2	Electronic Transduction of Biocatalytic Reactions Using Interactions of Redox Enzymes with Photoisomerizable "Command Interfaces" 316	
11.2.3	Electronic Transduction of Biocatalytic Reactions of Redox Enzymes Using Electron Transfer Mediators with Covalently Bound Photoisomerizable Units 322	
11.3	Electronically Transduced Reversible Bioaffinity Interactions at Photoisomerizable Interfaces 323	
11.3.1	Reversible Immunosensors Based on Photoisomerizable Antigens 326	
11.3.2	Biphasic Reversible Switch Based on Bioaffinity Recognition Events Coupled to a Biocatalytic Reaction 330	
11.4	Photocurrent Generation as a Transduction Means for Biocatalytic and Biorecognition Processes 332	
11.4.1	Enzyme-Biocatalyzed Reactions Coupled to Photoinduced Electron Transfer Processes 332	
11.4.2	Biorecognition Events Coupled to Photoinduced Electron Transfer Processes 334	
11.5	Conclusions 335	
	References 336	

12	The Neuron-semiconductor Interface 339
	Peter Fromherz
12.1	Introduction 339
12.2	Ionic–Electronic Interface 340
12.2.1	Planar Core-coat Conductor 343
12.2.2	Cleft of Cell-silicon Junction 346
12.2.3	Conductance of the Cleft 349
12.2.4	Ion Channels in Cell-silicon Junction 358
12.3	Neuron–Silicon Circuits 362
12.3.1	Transistor Recording of Neuronal Activity 362
12.3.2	Capacitive Stimulation of Neuronal Activity 367
12.3.3	Two Neurons on Silicon Chip 372
12.3.4	Toward Defined Neuronal Nets 377
12.4	Brain–Silicon Chips 383
12.4.1	Tissue-sheet Conductor 383
12.4.2	Transistor Recording of Brain Slice 385
12.4.3	Capacitive Stimulation of Brain Slices 388
12.5	Summary and Outlook 392
	References 393

13	S-Layer Proteins in Bioelectronic Applications 395
	Stefan H. Bossmann
13.1	Introduction 395
13.1.1	Upcoming Nanotechnology Applications 396
13.2	S-layer Proteins and Porins 396
13.2.1	The Building Principles of Tailored S-layer Proteins Layers 397
13.2.2	Chemical Modification of S-layers 400
13.2.3	Interaction by Noncovalent Forces 401
13.3	Experimental Methods Developed for Hybrid Bioelectronic Systems 402
13.3.1	Electron Microscopy 402
13.3.2	Combined X-Ray and Neutron Reflectometry 402
13.3.3	Atomic Force Microscopy Using Protein-functionalized AFM-cantilever Tips 403
13.3.4	Scanning Electrochemical Microscopy 404
13.4	Applications of S-layer Proteins at Surfaces 404
13.4.1	S-layer Proteins as Permeability Barriers 404
13.4.2	S-layer Proteins at Lipid Interfaces 405
13.4.3	Introduction of Supramolecular Binding Sites into S-layer Lattices 412
13.5	Molecular Nanotechnology Using S-layers 414
13.5.1	Patterning of S-layer Lattices by Deep Ultraviolet Irradiation (DUV) 414

13.5.2	Synthesis of Semiconductor and Metal Nanoparticles Using S-layer Templates Design of Gold and Platinum Superlattices Using the Crystalline Surfaces Formed by the S-layer Protein of *Bacillus sphaericus* as a Biotemplate *416*	
13.5.3	Generation of S-layer Lattice-supported Platinum Nanoclusters	*418*
13.5.4	Formation and Selective Metallization of Protein Tubes Formed by the S-layer Protein of *Bacillus sphaericus* NCTC 9602 *419*	
13.5.5	S-layer/Cadmium Sulfide Superlattices *421*	
13.6	Immobilization and Electrochemical Conducting of Enzymes in S-layer Lattices *421*	
13.6.1	S-layer and Glucose Oxidase-based Amperometric Biosensors	*421*
13.6.2	S-layer and Glucose Oxidase–based Optical Biosensors	*422*
13.7	Conclusions *423*	
	References *423*	

14 Computing with Nucleic Acids *427*
Milan N. Stojanovic, Darko Stefanovic, Thomas LaBean, and Hao Yan

14.1	Introduction *427*	
14.2	Massively Parallel Approaches *428*	
14.3	The Seeman–Winfree Paradigm: Molecular Self-assembly	*435*
14.4	The Rothemund–Shapiro Paradigm: Simulating State Machines	*439*
14.5	Nucleic Acid Catalysts in Computation *442*	
14.6	Conclusion *453*	
	References *454*	

15 Conclusions and Perspectives *457*
Itamar Willner and Eugenii Katz

Subject Index *463*

Preface

The integration of biomolecules with electronic elements to form functional devices attracts substantial recent research efforts. The entire field was named with the general buzzword, "bioelectronics". Exciting advances in the area include the integration of enzymes, antigen/antibodies, DNA, or bioreceptors with electronic units to yield specific biosensors for clinical diagnosis, detection of pathogens, environmental and food analysis, and homeland security applications. Another general scientific effort is directed to the coupling of neurons with electronic elements to assemble neuroelectronic junctions and neuronal networks that are anticipated to act as "brain computers" and information processing devices. Other merging research efforts include the development of biofuel cells, and biomolecule-based motors and devices. Progress in the rapidly developing area of nanotechnology introduced new concepts and scientific paradigms to bioelectronics. Conjugation of biomolecules and metallic or semiconducting nanoparticles yields hybrid materials with unique electronic and photonic properties that provide fascinating scientific and technological opportunities. New nanostructured sensors, electronic nanocircuitry based on biomolecular templates, nanostructured devices and nanoscale drug delivery systems are a few viable examples where bioelectronics "meet" nanotechnology.

The various topics covered highlight key aspects and the future perspectives of bioelectronics. The book discusses theoretical limitations in the electronic coupling of biomolecules with electronic elements, the chemical strategies to immobilize biomolecules such as proteins or DNA on electronic transducers, and to apply the systems as biosensors. The junction between bioelectronics and nanotechnology is introduced by exemplifying the microscopic imaging of biomolecular assemblies on surfaces at the single molecule level, the use of biomolecules as a mold to synthesize functional nano-objects and devices, and the use of biomolecule-nanoparticle hybrid systems as functional biosensing elements. The assembly of neuronal networks as information processors, and the use of biomolecules as information storage and computing systems are further topics that are discussed in detail.

The different topics addressed in this book will be of interest to the interdisciplinary community active in the area of bioelectronics. It is hoped that the collection of the different chapters will provide chemists, biologists, physicists, material scientists and engineers with a comprehensive perspective of the field. Furthermore, the book is aimed

Bioelectronics. Edited by Itamar Willner and Eugenii Katz
Copyright © 2005 WILEY-VCH Verlag GmbH & Co. KGaA, Weinheim
ISBN: 3-527-30690-0

to attract young scientists and introduce them to the field while providing newcomers with an enormous collection of literature references. We, indeed, hope that the book will spark the imagination of scientists to further develop the topic.

Finally, we would like to thank all scientists that contributed to this effort and made possible the publication of this book.

Jerusalem, January 2005

Itamar Willner
Eugenii Katz

List of Contributors

DAVID N. BERATAN
Departments of Chemistry and
Biochemistry
Duke University
Durham, NC 27708
U.S.A.

STEFAN H. BOSSMANN
Kansas State University
Department of Chemistry
111 Williard Hall
Manhattan, KS 66506-3701
U.S.A.

EREZ BRAUN
Technion – Israel Institute of
Technology
Department of Physics
Haifa 32000
Israel

BARRY D. FLEMING
Inorganic Chemistry Laboratory
University of Oxford
South Parks Road
Oxford, OX1 3QR
United Kingdom

MIROSLAV FOJTA
Institute of Biophysics, ASCR
Kralovopolska 135
65612 Brno
Czech Republic

PETER FROMHERZ
Department of Membrane and
Neurophysics
Max Planck Institute for Biochemistry
82152 Martinsried, Munich
Germany

HARRY B. GRAY
Beckman Institute
Caltech
Pasadena, CA 91125
U.S.A.

ROBERT JOHN HAMERS
Department of Chemistry
University of Wisconsin
Madison, WI 53706
U.S.A.

H. ALLEN O. HILL
University of Oxford
Inorganic Chemistry Laboratory
South Parks Road
Oxford, OX3 8AJ
United Kingdom

YOSHIHARU ISHII
Japan Science and Technology
Agency (JST)
Soft Nanomachines Project, CREST
Osaka University
Nanobiology Bld. 7F
1–3 Yamadaoka, Suita
Osaka 565-0871
Japan

Bioelectronics. Edited by Itamar Willner and Eugenii Katz
Copyright © 2005 WILEY-VCH Verlag GmbH & Co. KGaA, Weinheim
ISBN: 3-527-30690-0

EUGENII KATZ
Institute of Chemistry
The Hebrew University of Jerusalem
91904 Jerusalem
Israel

KINNERET KEREN
Stanford University
Department of Biochemistry
Stanford, CA 94305
U.S.A.

IGOR V. KURNIKOV
Department of Chemistry
Northwestern University
Evanston, IL 60208
U.S.A.

THOMAS LABEAN
Duke University
Computer Science
Durham, NC 27708
U.S.A.

FRED LISDAT
University of Applied Sciences
Bahnhofstrasse
15745 Wildau
Germany

SHAMUS J. O'REILLY
Department of Chemistry
Inorganic Chemistry Laboratory
University of Oxford
South Parks Road
Oxford, OX1 3QR
United Kingdom

EMIL PALECEK
Institute of Biophysics, ASCR
Kralovopolska 135
65612 Brno
Czech Republic

TATIANA R. PRYTKOVA
Department of Chemistry
Duke University
Durham, NC 27708
U.S.A.

FRIEDER W. SCHELLER
University of Potsdam
Department of Analytical Biochemistry
Karl-Liebknecht-Strasse 24–25
14476 Golm
Germany

ANDREW N. SHIPWAY
Institute of Chemistry
The Hebrew University of Jerusalem
91904 Jerusalem
Israel

URI SIVAN
Technion – Israel Institute of Technology
Department of Physics
Haifa 32000
Israel

DARKO STEFANOVIC
Department of Computer Science
University of New Mexico
Albuquerque, NM 87131
U.S.A.

MILAN N. STOJANOVIC
Department of Medicine
Columbia University
New York, NY 10032
U.S.A.

JOSEPH WANG
The Biodesign Institute
Center for Bioelectronics and Biosensors
Arizona State University
Tempe, AZ 85287-6006
U.S.A.

BILHA WILLNER
Institute of Chemistry
The Hebrew University of Jerusalem
Jerusalem 91904
Israel

ITAMAR WILLNER
Institute of Chemistry
The Hebrew University of Jerusalem
91904 Jerusalem
Israel

JAY R. WINKLER
Beckman Institute
Caltech
Pasadena, CA 91125
U.S.A.

ULLA WOLLENBERGER
Department of Analytical Biochemistry
University of Potsdam
Chair of Analytical Biochemistry
Karl-Liebknecht-Strasse 24–25
14476 Golm
Germany

HAO YAN
Department of Chemistry and
Biochemistry
Arizona State University
Tempe, AZ 85287
U.S.A.

TOSHIO YANAGIDA
Graduate School of Frontier
Biosciences
Osaka University
Nanobiology Bld. 7F
1–3 Yamadaoka, Suita
Osaka 565-0871
Japan

1
Bioelectronics – An Introduction

Itamar Willner and Eugenii Katz

The integration of biomolecules with electronic elements to yield functional devices attracts substantial research efforts because of the basic fundamental scientific questions and the potential practical applications of the systems. The research field gained the buzzword "bioelectronics" aimed at highlighting that the world of electronics could be combined with biology and biotechnology [1–3]. Mother Nature has in course of evolution processed the most effective catalysts (enzymes), and biomolecules of optimal recognition and binding capabilities that lead to highly selective and specific biopolymer complexes (antigen–antibody, hormone-receptor, or duplex DNA complexes). Similarly, biology provides the fastest and most complex computing and imaging systems where optical information is processed and stored in the form of three-dimensional memorable images (vision process). The tremendous biochemical and biotechnological progress in tailoring new biomaterials by genetic engineering or bioengineering provides unique and novel means to synthesize new enzymes and protein receptors, and to engineer monoclonal antibodies or aptamers for nonbiological substrates (such as explosives or pesticides) and DNA-based enzymes. All these materials provide a broad platform of functional units for their integration with electronic elements. The latter electronic elements may involve, for example, electrodes, field-effect transistor devices, piezoelectric crystal, magnetoresistance recording media, scanning tunneling microscopy (STM) tips and others. The bioelectronic devices, Figure 1.1, may operate in dual directions: In one configuration, the biological event alters the interfacial properties of the electronic element, thus enabling the readout of the bioreaction by monitoring the performance of the electronic unit such as the readout of the potential, impedance, charge transport, or surface resistance of electrodes or field-effect transistors, or by following the resonance frequencies of piezoelectric crystals. The second configuration of bioelectronic systems uses the electronic units to activate the biomaterials toward desired functions.

The major activities in the field of bioelectronics relate to the development of biosensors that transduce biorecognition or biocatalytic processes in the form of electronic signals [4–6]. Other research efforts are directed at utilizing the biocatalytic electron transfer functions of enzymes to assemble biofuel cells that convert organic fuel substrates into electrical energy [7, 8]. Exciting opportunities exist in the electrical

Bioelectronics. Edited by Itamar Willner and Eugenii Katz
Copyright © 2005 WILEY-VCH Verlag GmbH & Co. KGaA, Weinheim
ISBN: 3-527-30690-0

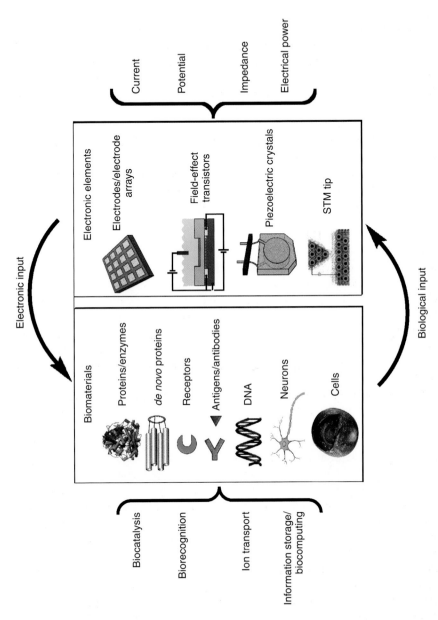

Fig. 1.1 Integrated systems of biomaterials and electronic elements for bioelectronic applications.

interfacing of neuronal networks with semiconductor microstructures. The excitation of ion conductance in neurons may be followed by electron conductance of semiconductor devices, thus opening the way to generating future neuron-semiconductor hybrid systems for dynamic memory and active learning [9]. The recent progress in nanotechnology and specifically in nanobiotechnology adds new dimensions to the area of bioelectronics. Metal and semiconductor nanoparticles, nanorods, nanowires, and carbon nanotubes represent nano-objects with novel electronic properties. Recent studies revealed that the integration of these objects with biomolecules yields new functional systems that may yield miniaturized biosensors, mechanical devices and electronic circuitry [10–12].

A fundamental requirement of any bioelectronic system is the existence of electronic coupling and communication between the biomolecules and the electronic supports. Special methods to immobilize biomolecules on solid supports while preserving their bioactive structures were developed. Ingenious methods to structurally align and orient biomaterials on surfaces in order to optimize electronic communication were reported [13]. Although impressive advances in the functional tailoring of biomolecule electronic units–hybrid systems were accomplished, challenging issues await scientific solutions. The miniaturization of the bioelectronic systems is a requisite for future implantable devices, and these types of applications will certainly introduce the need for biocompatibility of the systems. The miniaturization of the systems will also require the patterned, dense organization of biomolecules on electronic supports. Such organized systems may lead to high throughput parallel biosensing and to devices of operational complexity. The development of methods to address and trigger specific biomolecules in the predesigned arrays is, however, essential. This book attempts to highlight different theoretical and experimental topics that place bioelectronics as a modern interdisciplinary research field in science.

The understanding of charge transport phenomena through biological matrices attracted in the past decades, and continues to evolve, intensive theoretical and experimental work. The seminal contributions of the Marcus theory [14], the superexchange charge transfer theory [15], and the definition of superior tunneling paths in proteins [16] had a tremendous impact on the understanding of biological processes such as the electron transfer in the photosynthetic reaction center, or the charge transport in redox-proteins that are the key reactions for numerous electrochemical and photoelectrochemical biosensing systems. A continuous feed back between elegant experimental work employing structurally engineered proteins and theoretical analysis of the results led to the formulation of a comprehensive paradigm for electron transport in proteins [17]. This topic is addressed in detail in Chapter 2. The charge transport through DNA has recently been a serious scientific debate [18, 19], and contradicting results claiming conductive [20], superconductive [21], semiconductive [22] or insulating [23] properties of DNA were reported. Theories describing charge transport through DNA (electrons or holes) that included hopping mechanisms, tunneling paths, or ion-assisted electron transfer were developed [24, 25]. Charge transport through DNA is anticipated to play a key role in the electrical detection of DNA and in the analysis of base mismatches in nucleic acids, in the use of DNA nanowires as circuitry in devices, and as a means to readout sequence specific DNA structures (DNA computers).

The electrical contacting between biomolecules and electrodes is an essential feature for most bioelectronic systems. Numerous redox enzymes exchange electrons with other biological components such as other redox-proteins, cofactors or molecular substrates. The exchange of electrons between the redox-centers of proteins and electrodes could activate the biocatalytic functions of these proteins, and may provide an important mechanism for numerous amperometric biosensors. Nonetheless, most of the proteins lack direct electron transfer communication with electrodes, and the lack of electrical communication between the biomaterials and the electronic elements presents one of the fundamental difficulties of bioelectronic systems. Although the barriers for charge transport between redox-proteins are easily explained by the Marcus theory and the spatial insulation of the redox-centers of enzymes by the protein matrices, they hinder the construction of electrically communicated biomolecular-electronic hybrid systems. Ingenious methods for the electrical contacting of biomolecular assemblies associated with electronic units were developed in recent years [13]. The structural engineering of proteins with electron relays [26], the immobilization of redox enzymes in conductive polymers or redox-active polymers [5], the steric alignment of proteins on electron relays associated with electrodes [27], or the incorporation of redox-active intercalators in DNA [28] represent a few means to electronically communicate the biomolecules with the electronic elements. These aspects are addressed in several sections of the book (Chapters 3 and 4) and are exemplified here with the electrical communication of redox enzymes with electrodes for the generation of amperometric biosensors and biofuel cells, and with the intercalation of a redox-label into double-stranded DNA for the electrical probing of DNA. The integration of glucose oxidase, which lacks direct electrical communication with electrodes, into a redox-active hydrogel film consisting of tethered Os(II)-polypyridine complex (1) units, and linked to the electrode, facilitates the electrical contact between the enzyme and the conductive support, Figure 1.2(A). The flexible redox-units linked to the polymer electrically wire the redox-center of the enzyme with the electrode by mediated electron transfer. Glucose sensing electrodes based on this charge transport concept are already on the market, and the design of microsized electrically wired enzyme electrodes for invasive continuous monitoring of glucose are close to commercial realization [29]. A different application of electrically contacted enzyme electrodes rests in the design of biofuel cells [7, 8], Figure 1.2(B). Fuel cell systems represent a well-established technology, where electrical power is generated by two complementary oxidation and reduction processes occurring at a catalytic anode and cathode, respectively. While the generation of electrical power by electrically contacted redox enzymes, in a biofuel cell configuration has probably little value in global energy production, the systems might have important merit as implantable devices that generate electrical power from body fluids. For example, a glucose-based biofuel cell utilizing electrically contacted enzyme electrodes could use blood as a fuel for the electrical powering of pace makers, insulin pumps or prosthetic elements.

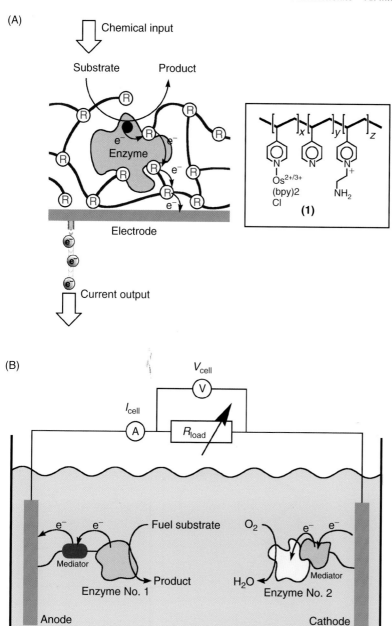

Fig. 1.2 (A) Electrical contacting of a redox-enzyme with an electrode by an electroactive polymer and the application of the system as an amperometric biosensor. (B) A biofuel cell configuration based on electrically contacted enzyme electrodes.

1 Bioelectronics – An Introduction

The electrical contacting between molecular species and electrodes may be stimulated by specific biorecognition events. For example, the intercalation of doxorubicin (**2**) into the double-stranded DNA formed between a primer nucleic acid associated with an electrode and the complementary analyte DNA enables the electrochemical reduction of the intercalator and the subsequent catalytic reduction of O_2 to H_2O_2, Figure 1.3. The latter product induces in the presence of luminol and horseradish peroxidase (HRP) the formation of chemiluminescence as a readout signal for the DNA duplex formation on the electrode [28]. The analysis of DNA by different electrochemical methods is discussed in Chapter 5.

Scanning probe microscopy techniques have introduced exciting opportunities in surface science and specifically in the characterization of biomolecules on surfaces. Scanning tunneling microscopy allows one to probe tunneling currents through proteins, thereby imaging the structure of individual protein molecules. Atomic force microscopy (AFM) not only permits the imaging of single biomolecules on surfaces but also permits the specific affinity interactions between complementary

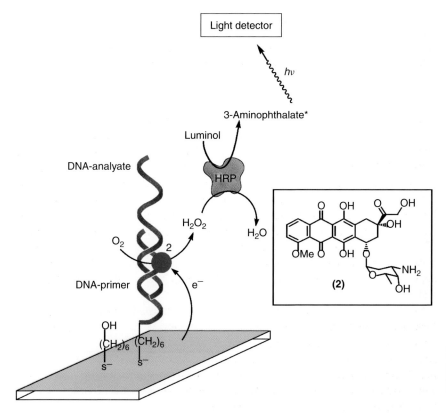

Fig. 1.3 The biochemiluminescent detection of DNA by the intercalation of a redox-active substrate into the double-stranded DNA assembly and its electrochemical activation.

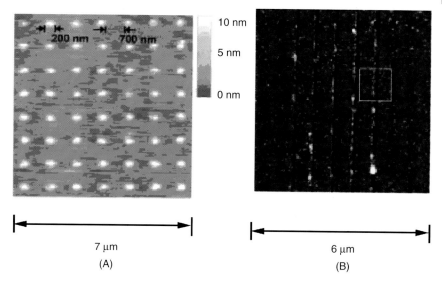

Fig. 1.4 (A) AFM image of a retronectin protein array generated by dip-pen nanolithography. (B) AFM image of a patterned surface consisting of a DNA monolayer treated with a DNase-modified AFM tip that cleaves off the DNA units upon contact with the surface. (Part A is adapted from [33] and Part B is adapted from [32], with permission).

antigen–antibody pairs, or double-stranded DNA complexes to be followed [30, 31]. Scanning probe microscopes also add new dimensions as tools for patterning surfaces with biomolecules. The use of dip pen–lithography for the generation of biomolecular patterns [32], Figure 1.4(A) or the application of enzyme-functionalized AFM tips as a biocatalytic patterning tool [33], Figure 1.4(B), are just two examples demonstrating the potential of these nano-tools to fabricate dense biomolecular arrays. Realizing that bioelectronics involves the intimate coupling of biomolecules to electronic supports, the use of scanning probe microscopy to characterize the structure-function relationships of single biomolecules, and to actuate single biomolecules are inevitable for the future development of the field. Some aspects of scanning probe microscopy for bioelectronic applications and the manipulation of single biomolecules are addressed in Chapters 6 and 10.

Self-organization of biomolecules leads to unique 2D- and 3D-nanostructures that include structurally defined pores or channels. These materials may act as templates for the assembly of other materials, and the generation of systems of hierarchical structural complexity. Figure 1.5 shows a scanning force microscopy image of S-layer protein from *Bacillus sphaericus* on a silicon surface exhibiting square lattice symmetry with a lattice constant of 13.1 nm. Alternatively, the pore or channel structures may be utilized as "microreactors" of predefined dimensions for the synthesis of metallic or semiconductor nano-objects. This topic is addressed in Chapter 13, where the applications of S-layer proteins in bioelectronic systems are discussed.

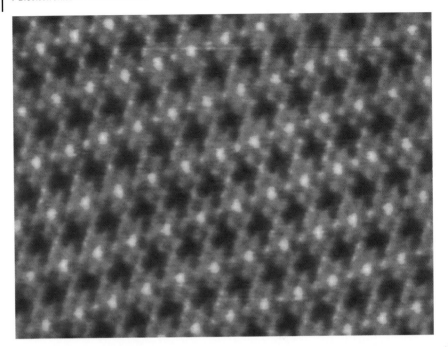

Fig. 1.5 AFM image of an S-layer protein from *Bacillus sphaericus* on a silicon surface. The image size corresponds to 150 × 113 nm.
(Adapted from http://nanotechweb.org/articles/news/2/3/15/1, with permission).

Nanoparticles exhibit unique electronic, optical, catalytic and photoelectrochemical properties [34–36]. The dimensions of nanoparticles are comparable to those of biomolecules such as enzymes, antigens/antibodies or DNA. Not surprisingly, the conjugation of biomolecules with metal and semiconductor nanoparticles yields hybrid systems of new electronic and optoelectronic properties. Indeed, tremendous progress was accomplished in the realization of biomolecule–nanoparticle hybrid systems for various bioelectronic applications [37]. The electrical contacting of redox enzymes with electrodes by means of Au nanoparticles [38], the use of metal nanoparticle–nucleic acid conjugates for the catalytic deposition of metals and inducing electrical conductivity between electrodes [39], the electrochemical analysis of metal ions originating from the chemical dissolution of metallic [40] or semiconductor [41] nanoparticle labels associated with DNA, or the photoelectrochemical assay of enzyme reactions by means of semiconductor nanoparticles [42] represent a few examples that highlight the potential of biomolecule–nanoparticle hybrid systems in biosensor design. Recent advances in the integration of biomolecules with semiconductors and the application of biomolecule–nanoparticle hybrids in bioelectronics are highlighted in Chapters 7 and 8, respectively. Several other applications of biomolecule–nanoparticle or biomolecule–carbon nanotube systems are also discussed in other sections of the book.

Exciting opportunities exist in the applications of biomolecules as templates for the synthesis of metallic or semiconductor nanowires [43]. Such nanowires provide great promise for future nanocircuitry and for the assembly of nanodevices. The possibility of preparing DNA of desired shapes and base sequence, the availability of enzymes acting as biocatalytic tools for manipulating DNA, the binding of metal ions to the phosphate units of DNA chains, the specific intercalation of molecular components into the DNA biopolymers, and the specific DNA–protein interactions, turn DNA into an ideal matrix for its use as a template in the synthesis of nanowires consisting of metals or semiconductors. Indeed, tremendous progress has been accomplished by using DNA as a template for the generation of nanowires and patterned nanowires [44]. This subject is highlighted in Chapter 9, which demonstrates the use of patterned Au

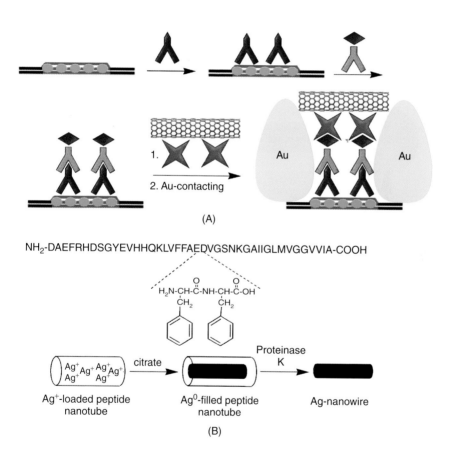

Fig. 1.6 (A) Assembly of a nanotransistor based on a carbon nanotube bridging two Au nanocontacts on a DNA template. The carbon nanotube is positioned on the DNA by the initial binding of RecA protein to the DNA, followed by the association of RecA-antibody and a biotinylated anti-antibody, and the fixation of avidin-coated tube to the assembly. (B) Formation of a Ag wire in the channel of a diphenylamine peptide tube, followed by the enzymatic dissolution of the peptide template.

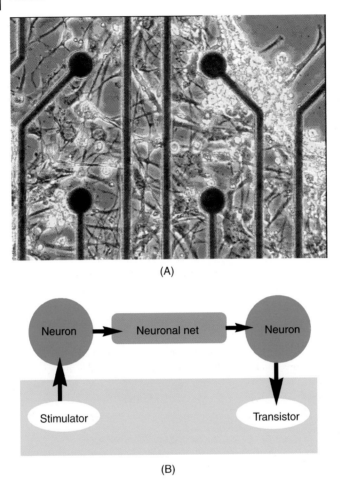

Fig. 1.7 (A) Neurons on top of a multielectrode array (adapted from http://physicsweb.org/article/news/7/4/17#neuronsonelectrode with permission). (B) A neuroelectronic hybrid system consisting of two neurons; the first neuron is activated by a capacitive stimuli, the signal transmission occurs through a neuronal network to a second neuron, where the information is recorded by a transistor.

nanowires on DNA as electrical contacts for the assembly of a nanotransistor. The construction of the biomolecule-base nanotransistor [45], Figure 1.6(A), is based on the assembly of a carbon nanotube between gold contacts formed on a DNA template using biorecognition events as driving motives for the construction of the nanodevice. Recent advances in this area suggest that self-assembled protein tubules or filaments may similarly be employed as templates for the synthesis of nanowire system [46].

For example, Figure 1.6(B) depicts an impressive micrometer-long Ag wire exhibiting a width of ca 20 nm that was generated by the *in situ* reduction of silver ions in the template composed of aromatic short-chain peptides (diphenylalanine β-amyloids), that are considered as key proteins in the development of Alzheimer's disease.

The interfacing of neurons with semiconductors provides a challenging approach to mimicking brain functions by bioelectronic information storage and processing memory devices. Tremendous progress has been reported in the past decade in the directed growth and organization of neurons connected by synapses on electrode arrays associated with semiconductor supports to form neuroelectronic junctions and neuronal networks [47], Figure 1.7(A). The possibility of eliciting neuronal activity by capacitive stimuli induced by the semiconductor, and the imaging of neuronal functions by a transistor element, represent two fundamental write–read functions of bioelectronic computing devices. The further stimulation of one neural cell and the transfer of the neuronal activity through the synapse to a second cell that enables the readout of the information by the transistor unit, Figure 1.7(B), represents an information transfer and processing device. Recent research demonstrates the organization and operation of such functional neuronal arrays as a viable approach to fabricate "brain computers". The neurobiology and physics involved in neuronal dynamics and thinking computation systems are discussed in Chapter 12.

The combinatorial synthesis of nucleic acids paves the way for synthesizing mixtures of numerous DNAs with base sequence encoded information. The possibilities of addressing this mixture of DNA in solution, as well as on surfaces, by hybridization or by biocatalytic transformations, such as replication or scission, allow the manipulation of target(s) DNA in the mixtures. Furthermore, the possibility of amplifying minute amounts of DNA enables us to fish out and identify the biomanipulated DNA. Not surprisingly, the rich information stored in DNA, and its retrieval capability has established the paradigm of "DNA computers and DNA-based computations". The seminal suggestion of Adleman using DNA for parallel computation [48] was followed by concepts of enhanced computation complexity. This topic and its practical utility are critically reviewed in Chapter 14. Photonically triggered biomolecules offer additional possibilities for optical information storage and processing. The future applications of light-triggered biomolecules for computing, reversible sensor design, targeted therapeutics, or light signal amplifiers are addressed in Chapter 11.

While organizing the book, we tried to combine well-established principles of bioelectronics that are ripe for practical and commercial applications, with new scientific disciplines that are at an embryonic phase of development, and their practical utility stands for future evaluation. Although we made efforts to balance and present the different facets of bioelectronics, we are sure that some topics are missing from this essay, and we certainly apologize for any unbalanced presentation. The progress in the field turned bioelectronics into an interdisciplinary area that combines research efforts of biologists, chemists, physicists, material scientists and electronic engineers. In view of the impressive advances in the field and its promising perspectives, we anticipate a bright future for bioelectronics. We would like to thank all contributors for their efforts to bring this book into reality, and we believe that it will provide a comprehensive background to researchers and scholars active in the field.

References

1. I. Willner, *Science*, **2002**, *298*, 2407–2408.
2. C.A. Nicolini (Ed.), *Biophysics of Electron Transfer and Molecular Bioelectronics*, Plenum Press, New York, **1998**.
3. K.-H. Hoffmann (Ed.), *Coupling of Biological and Electronic Systems*, Springer-Verlag, Berlin, **2002**.
4. L. Habermüller, M. Mosbach, W. Schuhmann, *Fresenius' J. Anal. Chem.*, **2000**, *366*, 560–568.
5. A. Heller, *Acc. Chem. Res.*, **1990**, *23*, 128–134.
6. F.A. Armstrong, G.S. Wilson, *Electrochim. Acta*, **2000**, *45*, 2623–2645.
7. A. Heller, *Phys. Chem. Chem. Phys.*, **2004**, *6*, 209–216.
8. E. Katz, A.N. Shipway, I. Willner, in *Handbook of Fuel Cells – Fundamentals, Technology, Applications* (Eds.: W. Vielstich, H. Gasteiger, A. Lamm), Vol. 1, Part 4, Wiley, Chichester, **2003**, Chapter 21, pp. 355–381.
9. P. Fromherz, *Chem Phys Chem*, **2002**, *3*, 276–284.
10. C.M. Niemeyer, *Angew. Chem., Int. Ed.*, **2001**, *40*, 4128–4158.
11. E. Katz, I. Willner, J. Wang, *Electroanalysis*, **2004**, *16*, 19–44.
12. E. Katz, I. Willner, *ChemPhysChem*, **2004**, *5*, 1194–1104.
13. I. Willner, E. Katz, *Angew. Chem., Int. Ed.*, **2000**, *39*, 1180–1218.
14. R.A. Marcus, N. Sutin, *Biochim. Biophys. Acta*, **1985**, *811*, 265–322.
15. M. Bixon, J. Jortner, *Adv. Chem. Phys.*, **1999**, *106*, 35–202.
16. H.B. Gray, J.R. Winkler, *Q. Rev. Biophys.*, **2003**, *36*, 341–372.
17. H.B. Gray, J.R. Winkler, *Annu. Rev. Biochem.*, **1996**, *65*, 537–561.
18. S.O. Kelley, J.K. Barton, *Science*, **1999**, *283*, 375–381.
19. M. Ratner, *Nature*, **1999**, *397*, 480–481.
20. H.-W. Fink, C. Schönenberger, *Nature*, **1999**, *398*, 407–410.
21. A.Y. Kasumov, M. Kociak, S. Gueron, B. Reulet, V.T. Volkov, D.V. Klinov, H. Bouchiat, *Science*, **2001**, *291*, 280–282.
22. P. Tran, B. Alavi, G. Gruner, *Phys. Rev. Lett.*, **2000**, *85*, 1564–1567.
23. P.J. de Pablo, F. Moreno-Herrero, J. Colchero, J. Gómez-Herrero, P. Herrero, A.M. Baró, P. Ordejón, J.M. Soler, E. Artacho, *Phys. Rev. Lett.*, **2000**, *85*, 4992–4995.
24. J. Jortner, M. Bixon, A.A. Voityuk, N. Rosch, *J. Phys. Chem. A*, **2002**, *106*, 7599–7606.
25. G.B. Schuster, U. Landman, *Top. Curr. Chem.*, **2004**, *236*, 139–161.
26. A. Riklin, E. Katz, I. Willner, A. Stoker, A.F. Bückmann, *Nature*, **1995**, *376*, 672–675.
27. I. Willner, V. Heleg-Shabtai, R. Blonder, E. Katz, G. Tao, A.F. Bückmann, A. Heller, *J. Am. Chem. Soc.*, **1996**, *118*, 10321–10322.
28. F. Patolsky, E. Katz, I. Willner, *Angew. Chem., Int. Ed.*, **2002**, *41*, 3398–3402.
29. A. Heller, *Annu. Rev. Biomed. Eng.*, **1999**, *1*, 153–175.
30. O.S. Willemsen, M.M.E. Snel, A. Cambi, J. Greve, B.G. De Grooth, C.G. Figdor, *Biophys. J.*, **2000**, *79*, 3267–3281.
31. C. Albrecht, K. Blank, M. Lalic-Multhaler, S. Hirler, T. Mai, I. Gilbert, S. Schiffmann, T. Bayer, H. Clausen-Schaumann, H.E. Gaub, *Science*, **2003**, *301*, 367–370.
32. D.S. Ginger, H. Zhang, C.A. Mirkin, *Angew. Chem., Int. Ed.*, **2004**, *43*, 30–45.
33. J. Hyun, J. Kim, S.L. Craig, A. Chilkoti, *J. Am. Chem. Soc.*, **2004**, *126*, 4770–4771.
34. R.F. Khairutdinov, *Colloid J.*, **1997**, *59*, 535–548.
35. P. Mulvaney, *Langmuir*, **1996**, *12*, 788–800.
36. L.N. Lewis, *Chem. Rev.*, **1993**, *93*, 2693–2730.
37. E. Katz, A.N. Shipway, I. Willner, in *Nanoparticles – From Theory to Applications* (Eds.: G. Schmid), Wiley-VCH, Weinheim, **2003**, Chapter 6, pp. 368–421.
38. Y. Xiao, F. Patolsky, E. Katz, J.F. Hainfeld, I. Willner, *Science*, **2003**, *299*, 1877–1881.
39. S.J. Park, T.A. Taton, C.A. Mirkin, *Science*, **2002**, *295*, 1503–1506.

40 J. WANG, D. XU, R. POLSKY, *J. Am. Chem. Soc.*, **2002**, *124*, 4208–4209.

41 J. WANG, G. LIU, A. MERKOÇI, *J. Am. Chem. Soc.*, **2003**, *125*, 3214–3215.

42 V. PARDO-YISSAR, E. KATZ, J. WASSERMAN, I. WILLNER, *J. Am. Chem. Soc.*, **2003**, *125*, 622–623.

43 J. RICHTER, *Physica E*, **2003**, *16*, 157–173.

44 K. KEREN, M. KRUEGER, R. GILAD, G. BEN-YOSEPH, U. SIVAN, E. BRAUN, *Science*, **2002**, *297*, 72–75.

45 K. KEREN, R.S. BERMAN, E. BUCHSTAB, U. SIVAN, E. BRAUN, *Science*, **2003**, *302*, 1380–1382.

46 M. RECHES, E. GAZIT, *Science*, **2003**, *300*, 625–627.

47 R. SEGEV, M. BENVENISTE, Y. SHAPIRA, E. BEN-JACOB, *Phys. Rev. Lett.*, **2003**, *90*, 168101-1–168101-4.

48 R.S. BRAICH, N.C. HELYAPOV, C.J. JOHNSON, P.W.K. ROTHEMUND, L.A. ADLEMAN, *Science*, **2002**, *296*, 499–502.

2
Electron Transfer Through Proteins

Jay R. Winkler, Harry B. Gray, Tatiana R. Prytkova, Igor V. Kurnikov, and David N. Beratan

2.1
Electronic Energy Landscapes

The electron-localizing groups in protein electron transfer pathways fall in an energy window of about two electron volts [1]. Simple electron transferases include flavodoxins, blue-copper proteins, iron-sulfur proteins, and cytochromes [2]. While tyrosines and tryptophan radicals may serve as ET (electron transfer) intermediates, the redox window is not sufficiently oxidizing or reducing to react with the peptide backbone and the other side chains. As such, proteins are – to a first approximation – wide band-gap semiconductors with "dopants" (redox active species) that provide electron localizing sites.

How, then, is it possible for electron transfer in proteins to proceed between cofactors on timescales as short as picoseconds? The answer lies in the dual wave-particle nature of the electron. Despite the insulating characteristics of the protein, the wave character of the electron allows it to "mix" between the cofactors *via* the protein. This mixing facilitates electrons-tunneling across the protein between localization sites, even though the transient population of the protein by the electron is highly disfavored on energetic grounds. The electronic propagation assisted by the protein is known as *super-exchange* [2, 3]. The large energy mismatch between electronic orbital energies of the protein and cofactors causes the probability of finding the tunneling electron in the protein to be extremely small but nonzero. The decay of this probability is approximately exponential with distance from the localizing sites. Simple single exponential decay is expected to arise for tunneling in one dimension through flat and high or wide barriers [3]. The picture is somewhat more complex in three-dimensions, but the same physical description of rapidly decaying probability as a function of distance remains.

2.2
Theory of Electron Tunneling

The golden rule provides a theoretical foundation to describe the rates of (nonadiabatic) tunneling transitions [2, 3]. The transition rate is proportional to an electronic coupling matrix element (H_{DA}) squared and a nuclear (Franck–Condon) factor [2]. Hopfield introduced a qualitative description of the protein-mediated superexchange reaction based on the energetics of the protein electronic states [4]. Assuming that the donor and acceptor are localized over N_a and N_b atoms, respectively, he estimated the tunneling barrier to be 2 eV. With an interaction of 1 eV between two orbitals in bonded contact, the coupling element is:

$$H_{DA} = \frac{2.7\,\text{eV}}{\sqrt{N_a N_b}} \exp[-0.72 R] \tag{1}$$

The distance-decay exponent is $\sqrt{2 m_e E_{\text{barrier}}}/\hbar$ with $E_{\text{barrier}} = 2.0$ eV (a number that is less than the maximum barrier estimate of 3 eV, which is one-half of the bonding–antibonding optical gap in the protein). R is the edge-to-edge distance between donor and acceptor cofactors (as shown in Figure 2.1), and the 2.7-eV prefactor reduces to an interaction energy of 1 eV when $R \sim 1.5$ Å. The fact that the 2-eV value is *much* less than the ionization potential of the cofactors ($\sim$5 eV) is critically important in the physical theory of protein-mediated tunneling. This 3-eV difference indicates that the through-bond tunneling barrier height is about *one-half* of the through-space barrier. This large difference causes the through-bond and through-space tunneling decay *exponents* to differ by factors of $2.5^{1/2} \sim 1.6$. This is a difference in decay exponents, and the influence on long-distance electron transfer rates is enormous.

It is useful to note the difference between the distance decay of H_{DA} and decay of H_{DA}^2. The squared matrix element appears in the nonadiabatic ET reaction the rate. As such, the decay exponent discussed above (0.72 Å^{-1}) is doubled (1.44 Å^{-1}) when the coupling element is squared. At shorter distances, the electron transfer reorganization energy is expected to be distance dependent [5]. This can give the overall rate a stronger distance dependence (in the "normal" Marcus regime) than would be predicted from the decay of the squared electronic factor alone [2, 5].

Equation (1) displays the proper short- and long-range behavior of the coupling within a structureless barrier model: at long distances the decay is exponential, at short distances it matches typical atomic interaction energies. Moreover, the square root normalization factors in Eq. (1) assume that the coupling is dominated by the interaction between the two nearest atoms in the cofactors. The square barrier model of electron tunneling through proteins is a "coarse-grained" description; it neglects variations of tunneling barriers on the length scale of chemical bonds.

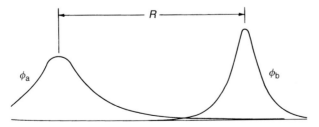

Fig. 2.1 Sketch indicating localized wave functions as they decay from donor and acceptor localized sites [4].

The protein medium between donors and acceptors contains bonded and nonbonded interactions. It is structured and inhomogeneous. As described above, the tunneling decay exponent associated with through-space propagation is about 1.6 times the decay exponent for tunneling through space. How can we describe protein electron transfer that takes into account these two length scales without performing large-scale electronic structure calculations?

2.3 Tunneling Pathways

The decay of H_{DA} follows from the decay of donor/acceptor localized states. In periodic potentials, localized states drop approximately by a constant factor, ε, per repeating unit in the periodic potential. If there is just empty space, then the decay across a distance a is $\varepsilon_S = \exp[-\beta a]$ and over a distance R, the decay is ε_S^N where the donor–acceptor distance is $R = Na$, that is, the overall decay across that distance R is $\exp[-\beta_S R_S]$. If the medium contains a bonded chain with translational symmetry and repeating units of length ℓ_B, the decay per bond is $\varepsilon_B = \exp[-\beta_B \ell_B]$. Across N bonds, the decay is $\varepsilon_B^N = \exp[-\beta_B R_B]$. The idea of writing a decay factor per repeating unit was first used for chemical bridges in the 1960s by McConnell to describe the delocalization of spins [6]. A perturbative expression for the electronic decay parameter is $\varepsilon_B = V/\Delta E$, where V is the interaction between neighboring units of the bridge and ΔE is the energy gap between each bridge element and the donor/acceptor energy. The McConnell model is valid when $\varepsilon_B \ll 1$, while the translational symmetry arguments made above are valid for larger values of the decay factor. Indeed, electron transfer studies of rigid saturated hydrocarbons suggest $\varepsilon_B \approx 0.6$ [7].

The pathway model for tunneling in aperiodic structures identifies tunneling steps as either through-bond or through-space. Motivated by the repeating structure of the peptide backbone, it treats through-bond propagation as if it occurs in a periodic potential and assigns a single through-bond decay factor (Figure 2.2). To the extent that through-space connections are weak perturbative links between the nearly periodic potential tunneling through bond,

$$H_{DA} = A \prod_i \varepsilon_i^B \prod_j \varepsilon_j^S \qquad (2)$$

Parameterization of the model is simple, ε_B is taken from model compound studies (0.6) [9], and the through-space decay is drawn from ionization-potential data. Initial estimates of the through-space decay exponent were $\beta_s \approx 1.7$ Å^{-1}. We have revised this exponent to $\beta_s \approx 1.1$ Å^{-1} to reflect a cofactor binding energy of ~5 eV (a value in the mid-range of physiological cofactor binding energies).

The through-bond decay factor of 0.6, applied to a typical extended chain, produces an average decay exponent (in H_{DA}) of $\beta_B \approx 0.5$ Å^{-1}. *The key parameter of the pathway model is the ratio of the through-bond and through-space exponents.* If the ratio β_s/β_B

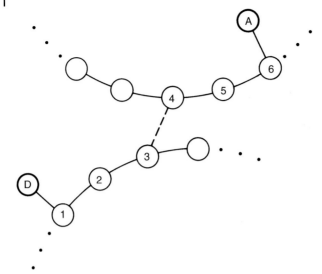

Fig. 2.2 The links in a covalent tunneling pathway (solid lines) are treated as periodic and are associated with a universal through-bond decay factor. A noncovalent jump, shown with a dashed line, links the covalent legs [8]. While the through-space link may be weak, pure covalent routes from donor to acceptor can be prohibitively long or may not exist at all.

is very large, through-bond propagation dominates; if the ratio is close to one, there is no preference for bond-mediated transport, and proteins are "isotropic". Our analysis suggests that $\beta_s/\beta_B \approx 2.2$. Since the ratio is larger than one, through-bond propagation is highly favored. For example, coupling 3.5 Å through space is 10 times weaker than propagating the same distance along a bonded network, and the impact on the rate is 100-fold. On the basis of the binding energy arguments above, one might expect unsaturated bridging orbitals to be superior tunneling mediators. In proteins, the pi-symmetry effect is expected to be offset by a symmetry penalty for σ–π mixing. Moreover, since a π-system may not be aligned directly along the donor–acceptor axis, and π-orbitals of side chains generally constitute a small fraction of the tunneling bridge, the influence on the tunneling rate is not expected to be particularly large. The framework is different in unsaturated polymers and in DNA [10].

With the penalty factors defined for bonded and nonbonded contacts, pathway estimates of the coupling, H_{DA}, identify the chain of interactions that maximizes the product in Eq. (2). Pathway searches are usually carried out using atomic positions assigned from X-ray data and a graph search algorithm [11]. The coupling prefactor, A, is cofactor dependent. As an estimate, A can be drawn from the Hopfield factor in Eq. (1) (2.7 eV/$\sqrt{N_a N_b}$). More specific quantum chemical approaches to computing H_{DA} are discussed below. Pathway-coupling estimates often examine relative values and avoid the prefactor issue.

2.4
Coupling-limited ET Rates and Tests of the Pathway Model

We have introduced low-resolution (square barrier) and higher-resolution (pathway) theories of electron tunneling in proteins (illustrated in Figure 2.3). The lowest-resolution model defines one decay parameter β (Eq. 1). Using this parameter, the donor–acceptor distance R defines the coupling. Higher-resolution (pathway) estimates are based on the through-bond and through-space composition of the medium between donor and acceptor. At this resolution, protein-specific "pathway effects" influence the donor–acceptor interaction, in addition to the overall distance. Pathway effects arise from protein connectivity (covalent, hydrogen bonded and van der Waals); pathway couplings strongly reflect secondary and tertiary structure [9c, 12]. Fully quantum calculations of couplings include multiple-pathway contributions to the coupling and support the basic secondary motif effects found using pathway analysis [13]. Indeed, the interferences among coupling pathways (and their fluctuations over time) need to be captured in quantum calculations to make quantitative estimates of the rate. Fully quantum calculations of coupling are challenging because the interferences are "delicate" – they fluctuate strongly with geometry. Nonetheless, the required computational methods are becoming accessible (*vide infra*) [13, 14].

Investigations of tailored proteins, especially ruthenium-modified proteins, with donor and acceptor groups at fixed and known geometries [15–20], allowed detailed tests of the electron tunneling theories for proteins. These experiments continue to challenge theory to develop fully quantum descriptions of electron tunneling in proteins and to make quantitative rate predictions [13, 16, 21, 22]. Key features of these modified protein systems are that the redox groups weakly perturb the structure (as demonstrated with X-ray crystallography) and that the appended redox species are not substitution labile on oxidation/reduction. Ruthenium complexes are particularly well suited since the accessible redox potentials are within the range relevant to biology. In Ru-modified proteins electron transfer rates can be obtained readily from flash-quench kinetics experiments [16].

The tunneling pathway model makes testable predictions for modified proteins. First, there are qualitative differences predicted in the tunneling decay exponent for traversal through "well connected" versus "weakly connected" protein links (see Figures 2.4 and 2.5). For example, Figure 2.5a shows a compact representation of the dominant pathways from heme to surface groups in cytochrome c.

Strong pathway effects are also observed in ruthenated high-potential iron–sulfur proteins (HiPIP) from *Chromatium vinosum*. Ru was attached to four surface sites (His18, 42, 50, 81) that differ in distance from the FeS cluster by no more than 2 Å (see Figure 2.5b). A 2 order of magnitude range in electron-transfer rates was explained by differences in the structure of the intervening protein medium in the four species [24]. A quantitative description of the relative electron transfer rates requires a multipathway quantum calculation [24].

In addition to pathway-specific effects, protein structural motif effects are a key prediction of the analysis. For example, average distance-decay parameters for tunneling rates through highly helical proteins versus sheet proteins are predicted to be 1.4 Å^{-1}

2 Electron Transfer Through Proteins

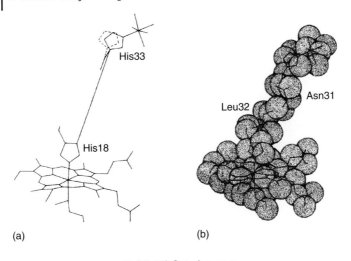

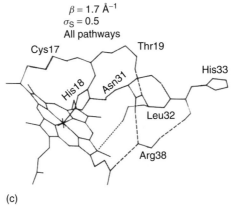

Fig. 2.3 Conceptual models for electron tunneling (here, in cytochrome c) treat the intervening medium in an average manner (a) [15b]. Explicit models (b) introduce some amount of tunneling "medium" near the line of sight joining Fe and Ru [15b]. Strong tunneling pathways (c) link the heme to His33 by a combination of bonded (solid lines), hydrogen bonded (dashed), and through-space (dotted) links [12a].

and 1.0 Å^{-1} respectively [9c]. Pathways are more direct in the β-strand motifs, and the hydrogen bond cross-linking in the barrels provides strong pathways in many directions. Pathways through an α-helix are somewhat less direct than in a β-strand, and the linkage between helical segments may be very weak indeed (less highly cross-linked than in β-sheet or barrel structures). These generic effects are shown in Figure 2.6 for idealized structural elements.

The pathway model rate predictions do not simply drop exponentially as a function of distance because the through-bond and through-space tunneling penalties are

2.4 Coupling-limited ET Rates and Tests of the Pathway Model

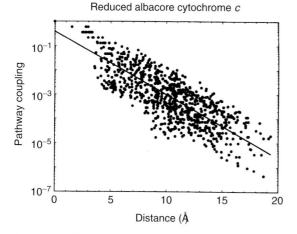

Fig. 2.4 Coupling strength versus distance in cytochrome *c*. The order of magnitude scatter around the single exponential line arises from donor–acceptor pairs at a fixed distance that have coupling pathways of different strengths because of tunneling pathway effects [23]. We refer to those points above the best-fit exponential decay line as *hot spots*. They are "hot" as a consequence of strong coupling pathways [9c].

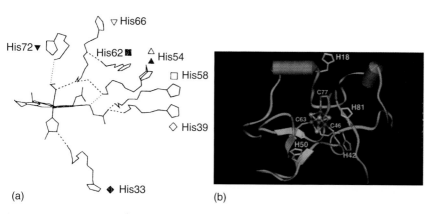

Fig. 2.5 (a) Strongest tunneling pathways from the heme of cytochrome *c* to surface Ru-modification sites [17, 18]. The His72 site, while closer to the heme than the His39 site by 4 Å (measuring distances edge-to-edge), has a threefold slower rate. Structureless barrier models (Eq. 1) would suggest a rate in the His72 derivative *faster* by 300-fold. Measuring the distance between metal atoms (appropriate for ground state electron transfer) produces an even larger disparity between experimental and theoretical rates if the structureless barrier model is used.
(b) Four surface histidine modification sites in HiPIP. The distances from the FeS cluster edge to the histidine sites differ by less than 2 Å, yet the electron transfer rates vary by 2 orders of magnitude.

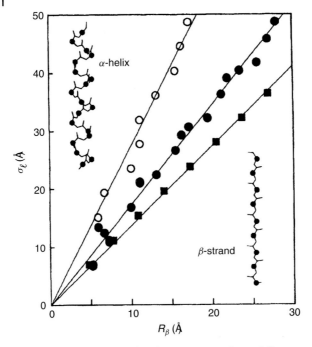

Fig. 2.6 Tunneling pathway length (σ_ℓ) versus end-to-end distance for an α-helix neglecting hydrogen bond links (open circles), α-helix including hydrogen bonds (dots), and β-strand (squares) based on tunneling pathway analysis [18]. Tunneling pathway length (in Å) is defined by $\sigma_\ell = 1.4 \times \ln(\Pi\varepsilon)/\ln(0.6)$ [12b].

substantially different, and the medium between donor and acceptor varies with protein structure. Dutton and coworkers recently suggested that packing density provides a predictor of protein effects on electron tunneling rates [25]. The packing density model draws lines between donor and acceptor cofactors and measures the fraction of these lines that are enclosed in the van der Waals spheres of the protein atoms. The coupling decay is then constructed by weighting the through-bond and through-space decay exponents based on the assessed bonded and nonbonded content of these pathways. This method of assessing pathway effects was shown to be essentially identical with the pathway model as shown in Figure 2.7 [26], except for cases where (1) strong paths stray just outside the region sampled by the packing analysis, (2) strong paths exist in otherwise low packing density regions, or (3) water (often neglected in the pathway model) provides strong coupling pathways.

In most cases, the two methods make the same predictions – within the precision of the parameter choice [26]. More accurate predictions of coupling require inclusion of multiple coupling pathways and geometry fluctuations associated with pathway dynamics. Inclusion of these effects is now becoming accessible through the quantum chemical calculation of electronic couplings over a range of snapshot geometries sampled along molecular dynamics trajectories [13c].

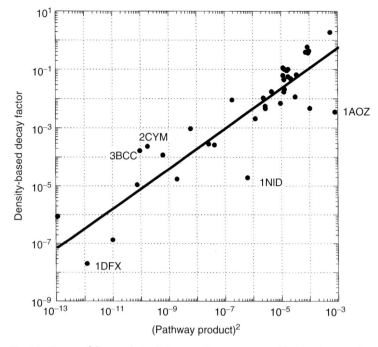

Fig. 2.7 Survey of the correlation between pathway versus packing density–based coupling decay factors for proteins. The results are identical (within the intrinsic uncertainties of the model parameters) except for a few special cases that can be understood in the context of specific medium effects [26].

The wide range of experimental data available for coupling-limited tunneling rates in proteins can be put into a broader context by examining tunneling timetables that include rates through water [27] and vacuum (theoretical values) (Figure 2.8) [27].

2.5
Multiple Tunneling Pathway Models

The pathway models succeeded at: (1) linking average distance-decay exponent parameters with helix and sheet content and (2) predicting anomalously strongly or weakly coupled derivatives. Some intriguing chemical questions are not addressed by the pathway model: What is the influence of multiple paths and pathway interferences on the protein-mediated coupling? Do certain motifs favor constructive or destructive interference among paths? What dynamical effects arise in fluctuating proteins? What is the impact of cofactor structure on donor–acceptor coupling? Are there chemical effects specific to the 20 amino acids that make some of them stronger or weaker mediators of tunneling? A yet "finer-grained" theory than the pathways approach is needed to address these questions.

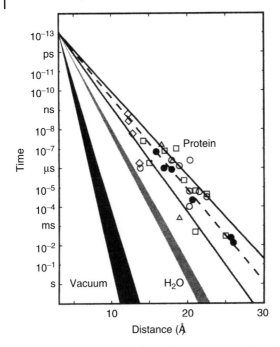

Fig. 2.8 Timetable (coupling-limited times versus Ru-Cu or Ru-Fe distances) for electron tunneling through vacuum, water and Ru-modified proteins: azurin (●); cytochrome c (○); myoglobin (△); cytochrome b_{562} (□); high-potential iron-sulfur protein (◊). The scatter of the protein data reflects the diversity of pathways associated with the specific structure of the intervening polypeptide medium [27].

Figure 2.9 shows the essential aspect of multipathway coupling. This figure indicates families of pathways between the cytochrome c heme and two surface modification sites. Each of the many paths propagate amplitude of a given magnitude and sign. The donor/acceptor groups "integrate" all of this amplitude. Multipathway models can analyze the interferences among these paths and their sensitivity to structural fluctuations.

Simple tight-binding models – in the spirit of the pathway method – have been applied to electron transfer proteins to explore multiple-pathway interferences [13a, 19, 22]. Regan and Onuchic argued that nontrivial interference effects arise when multiple-pathway bundles, or pathway "tubes" [22], mediate interactions between donor and acceptor. These and other simple tight-binding Hamiltonians have been used to explore multiple-pathway effects. Some of these methods retain elements of the pathway model that allow simplicity of analysis, while others produce results that are more challenging to analyze.

Self-consistent field calculations show that destructive interference among pathways is "delicate" [28]. In proteins, interferences appear and disappear with structural fluctuations. Not only do the couplings change with geometry but the tunneling energy

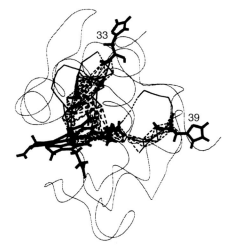

Fig. 2.9 Tunneling pathway families in cytochrome c [13a]. These families are added together with proper signs in fully quantum calculations of the donor–acceptor interaction. Moreover, the couplings are averaged over structural fluctuations of the protein.

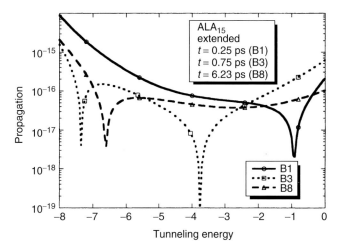

Fig. 2.10 Tunneling propagation fluctuations in an extended alanine chain at three different times. Note that the dips (zeros of the propagation) shift in energy as the peptide structure fluctuates [28].

dependence of the coupling also is strongly geometry-dependent. Qualitatively similar pathway interferences have been found in calculations on Ru-modified azurins. These flickering interferences are shown in Figure 2.10 for a model peptide.

The implications of these flickering interferences are clear: small changes in protein geometry can generate large changes in couplings. For example, for thermally accessible conformations, the donor–acceptor coupling may be zero. Indeed, nodes in the protein-mediated electronic propagation are reported in Ref. [14]. If H_{DA} values fluctuate rapidly compared to the ET rate, the mean-square coupling value will appear in the golden rule rate expression. Since the mean-square coupling will be dominated by the conformers with larger coupling values, it is essential that the squared couplings be averages over accessible geometries. Indeed, Balabin and Onuchic recently defined a coherence parameter to identify structures with a propensity for destructive interference among coupling pathways [14a]. In structures of this kind, fluctuations in coupling values may be essential to induce electron transfer.

The snapshots of Figure 2.10 indicate the need to average over conformations. In redox cofactors with metal containing donors and acceptors, a second kind of averaging is also needed. We have recently found that fluctuations of the Ru-histidine complex cause the interchange in energy ordering of the metal-centered orbitals. This interchange is rapid on the ET timescale. Indeed, "double averaging" – averaging over geometry and over ligand-field orbital energy crossings – is essential to reach quantitative agreement between computed and experimental rates [13c]. Figure 2.11 shows computed and experimental ET rates in Ru-modified azurins [13c] derived from this approach.

The tunneling decay parameter in the periodic bridge (McConnell) model [6] depends on the tunneling barrier: $\beta = (1/a)\ln(\Delta E/V)$ for repeating units of length a. If ΔE is large, the dependence of β on ΔE is weak. This far off–resonance regime is probably the one that exists in proteins. As of now, there is no clear evidence that the average distance decay in proteins can be tuned by changing the absolute

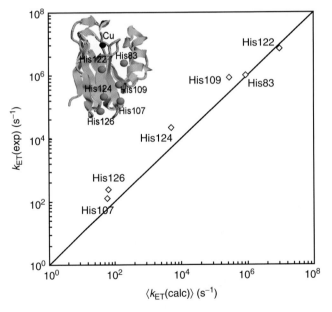

Fig. 2.11 Preliminary correlation of theoretical and experimental electron transfer rates in six Ru-modified azurins based on "double averaging" over ligand-field states and molecular geometries [13c] using an *ab initio* Hartree–Fock (protein fragment) approach. Backbone positions of histidine ligands in six azurins are shown in the upper left corner.

positions of the donor and acceptor redox potentials. The β-parameter is most likely to be tunable in the "near-resonance" regime. This near-resonance regime has been observed recently in DNA electron transfer, because the donor and acceptor species are often modified bases [10]. This effect is distinct from the transition from single to multistep tunneling [16d].

2.6
Interprotein Electron Transfer: Docking and Tunneling

Rates of interprotein electron transfer can be limited by protein diffusion, electron transfer, or gated events [2b, 29], as indicated in Figure 2.12. In the simple docking electron transfer–limited regime, the rates are given by the product of the binding constant and the tunneling rate. Recent studies of myoglobin-cytochrome b_5 electron transfer have suggested a "dynamical docking" paradigm [30, 31] for some of these rate processes. In this regime, protein–protein docking favors a range of docked structures, a minority of which are electron-transfer active. The fact that few docked structures are electron-transfer capable leads to a decoupling between the binding constant and the bimolecular electron transfer rate, as indicated in Figures 2.12 and 2.13.

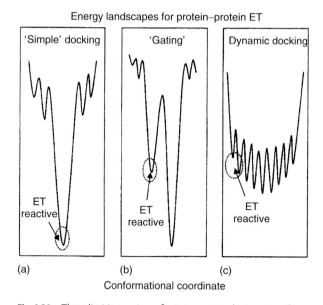

Fig. 2.12 Three limiting regimes for interprotein electron transfer: (a) simple docking in which the most stable conformation is ET active, (b) gating where a slow conformational change is required for electron transfer, and (c) dynamic docking where a minority of rapidly exchanging populations accounts for the electron transfer reaction [30].

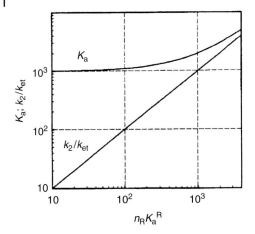

Fig. 2.13 Plot of binding constant (K_a) and bimolecular rate (k_2) versus the binding affinity of *reactive* conformers assuming the affinity for nonreactive conformers is held constant [30]. As long as reactive complex species are in the vast *minority*, increasing their population (through protein charge mutation) increases the bimolecular rate without impacting the binding constant. This is the signature of the dynamic docking regime.

Building a quantitative theoretical description of interprotein electron transfer is challenging because both docking energetics and electron tunneling must be described. Searching conformational space is costly, and performing electronic structure calculations in each of the geometries is prohibitive. Thus, it is of interest that interprotein electron transfer rates have been analyzed recently using pathway-coupling biased geometry sampling [31]. This approach considers both the pathway-coupling strength and the electrostatic energy of binding in a pathway biased Monte Carlo sampling scheme, as shown in Figure 2.14.

Electron tunneling between proteins also has been investigated in crystals containing photoactivatable donors and acceptors at specific lattice sites. In crystal lattices of tuna cyt c, chains of cyt c molecules form helices with a 24.1 Å separation between neighboring metal centers [32]. All other metal–metal distances in the lattice are greater than 30 Å. By doping Zn-cyt c into this lattice, Tezcan and coworkers found a rate constant of $4(1) \times 10^2$ s^{-1} for ET from the triplet-excited Zn-porphyrin to a neighboring Fe(III)-cyt c; the rate of charge recombination was about 5 times faster ($2.0(5) \times 10^3$ s^{-1}) [32]. The Zn-Fe separation in doped tuna cyt c crystals is similar to that in Zn-Fe-Hb hybrids (24.7 Å, T-state), although the tetrameric heme protein has many more contacts between subunits and a greater atom density at the interface. Interestingly, *Zn-porphyrin → Fe(III) and Fe(II) → Zn-porphyrin$^{+\bullet}$ ET rates in Hb hybrids [33] and Zn-doped tuna cyt c crystals are quite similar and fall well within the range that has been established in studies of Ru-modified proteins [16]. The protein crystal ET data demonstrate that small interaction zones of low density can be quite effective in mediating protein–protein redox reactions.

2.7
Some New Directions in Electron Transfer Theory and Experiment

The tools that were developed to investigate electron tunneling in proteins have many potential applications. For example, electron transfer is being used both as a fast trigger

2.7 Some New Directions in Electron Transfer Theory and Experiment

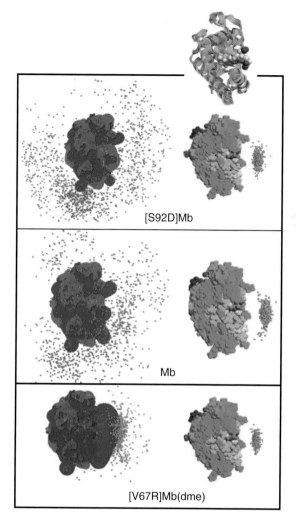

Fig. 2.14 Monte Carlo docking profiles for cytochrome b_5-myoglobin electron transfer (left, without pathway bias; right, with pathway bias) for native, mutant and chemically modified proteins [31]. Biasing the sampling toward geometries with a substantial donor–acceptor tunneling interaction allows a computationally more intensive treatment of the protein–protein interactions and electronic coupling than would be accessible otherwise. At left are rendered electrostatic maps of the myoglobin surface and at right are coupling surface maps for myoglobin. The gray dots show the docking sites of cytochrome b_5 without pathway bias (left) and with pathway bias (right) for the various derivatives.

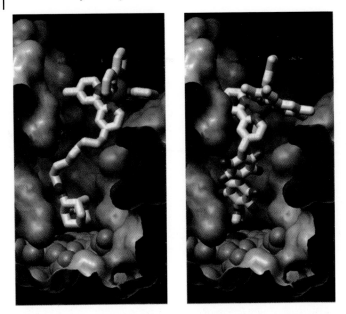

Fig. 2.15 Ru-wires bind the active-site channel of cytochrome P450$_{cam}$ [35].

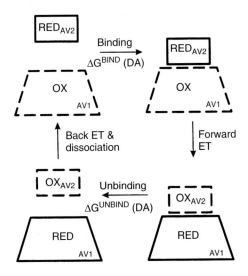

Fig. 2.16 Protein–protein association, triggered by ATP binding, is believed to desolvate the [4Fe4S] cluster of the iron–protein (Av2) in nitrogenase, thus enhancing the electron transfer driving force and the forward electron transfer rate [36].

and detector for protein folding [34]. Electron transfer pathways are being "wired" to deliver electrons at prescribed rates to catalytic centers [35]. Ru-wires allow the ET pathway between the Ru-diimine and the heme to be altered while retaining the surrounding protein matrix. Photoinduced ET through a perfluorobiphenyl bridge in a Ru-wire : P450$_{cam}$ conjugate, shown in Figure 2.15, is 1500 times faster than through an alkyl chain, in accordance with a through-bond model of Ru:heme electronic coupling [35c].

A reason for such great interest in protein electron transfer is our desire to establish molecular-level descriptions of free energy flow in biology. We have recently explored ATP-coupled electron transfer in nitrogenase. On the basis of simple electrostatic arguments (Figure 2.16), we believe that the role of ATP binding and protein–protein docking in nitrogenase is to enhance the ET driving force and to accelerate the rate through a Marcus [2, 5] free-energy effect. Analysis of this scheme shows how ATP energy is fed into an electron transfer pathway.

2.8
Concluding Remarks

A range of useful models, from simple to complex, is available to describe protein-mediated electron tunneling. Structured-medium models capture the difference between through-bond and through-space tunneling. At the pathway level, these models do not adequately describe the cofactor electronic structure, multiple-pathway interferences, or dynamical effects. Because of subtle pathway interferences and near degenerate ligand-field states, more advanced electronic structure methods must include geometry sampling as a part of the calculations, and methods that accomplish this task appear promising. These tools for the synthesis and design of electron transfer macromolecules are enabling a wide range of applications in energy transduction, protein folding and biocatalysis.

Acknowledgments

We thank the National Institutes of Health (GM-48043 and DK-19038) for support of this research. We thank José Onuchic and Brian Hoffman for their collaboration on some of these projects.

References

1. W.A. CRAMER, D.B. KNAFF, *Energy Transduction in Biological Membranes*, Springer-Verlag, New York, **1990**.
2. (a) I. BERTINI, H.B. GRAY, S.J. LIPPARD, J.S. VALENTINE, *Bioinorganic Chemistry*, University Science Books, Mill Valley, **1994**. (b) D.S. BENDALL, *Protein Electron Transfer*, BIOS Scientific Publishers, Oxford, **1996**. (c) V. BALZANI Ed., *Electron Transfer in Chemistry*, Vols. 1–5, Wiley, Weinheim, Germany. (d) A.M. KUZNETSOV, J. ULSTRUP, *Electron Transfer in Chemistry and Biology: An Introduction to the Theory*, John Wiley & Sons, New York, **1998**.
3. E. MERTZBACHER, *Quantum Mechanics*, 3rd ed., Wiley, New York, **1998**.
4. J.J. HOPFIELD, *Proc. Natl. Acad. Sci. U.S.A.*, **1974**, *71*, 3640–3644.
5. R.A. MARCUS, N. SUTIN, *Biochim. Biophys. Acta*, **1985**, *811*, 265–322.
6. H.M. MCCONNELL, *J. Chem. Phys.*, **1961**, *35*, 508–515.

7 G. Closs, L.T. Calcaterra, N.J. Green, K.W. Penfield, J.R. Miller, *J. Phys. Chem.*, **1986**, *90*, 3673–3683.

8 D.N. Beratan, J.N. Onuchic, *Photosynth. Res.*, **1989**, *22*, 173–186.

9 (a) D.N. Beratan, J.N. Onuchic, J.J. Hopfield, *J. Chem. Phys.*, **1987**, *86*, 4488–4498. (b) J.N. Onuchic, D.N. Beratan, *J. Chem. Phys.*, **1990**, *92*, 722–733. (c) D.N. Beratan, J.N. Betts, J.N. Onuchic, *Science*, **1991**, *252*, 1285–1288.

10 G. Schuster Ed., *Long-Range Charge Transfer in DNA*, Topics in Current Chemistry, Volumes 236–237, Springer-Verlag, Berlin, Germany, **2004**.

11 J.N. Betts, D.N. Beratan, J.N. Onuchic, *J. Am. Chem. Soc.*, **1992**, *114*, 4043–4046.

12 (a) D.N. Beratan, J.N. Onuchic, J.N. Betts, B.E. Bowler, H.B. Gray, *J. Am. Chem. Soc.*, **1990**, *112*, 7915–7921. (b) J.N. Onuchic, D.N. Beratan, J.R. Winkler, H.B. Gray, *Annu. Rev. Biophys. Biomol. Struct.*, **1992**, *21*, 349–377.

13 (a) J.J. Regan, S.M. Risser, D.N. Beratan, J.N. Onuchic, *J. Phys. Chem.*, **1993**, *97*, 13083–13088. (b) I.V. Kurnikov, D.N. Beratan, *J. Chem. Phys.*, **1996**, *105*, 9561–9573. (c) T. Prytokva, I.V. Kurnikov, D.N. Beratan, **2004**, in press.

14 (a) I.A. Balabin, J.N. Onuchic, *Science*, **2000**, *290*, 114–117. (b) T. Kawatsu, T. Kakitani, T. Yamato, *J. Phys. Chem. B*, **2002**, *106*, 11356–11366.

15 (a) J.R. Winkler, D.G. Nocera, K.M. Yocom, E. Bordignon, H.B. Gray, *J. Am. Chem. Soc.*, **1982**, *104*, 5798–5800. (b) S.L. Mayo, W.R. Ellis, R.J. Crutchley, H.B. Gray, *Science*, **1986**, *233*, 948–952.

16 (a) M.J. Bjerrum, D.R. Casimiro, I.-J. Chang, A.J. Di Bilio, H.B. Gray, M.G. Hill, R. Langen, G.A. Mines, L.K. Skov, J.R. Winkler, D.S. Wuttke, *J. Bioenerg. Biomembr.*, **1995**, *27*, 295–302. (b) H.B. Gray, J.R. Winkler, *Annu. Rev. Biochem.*, **1996**, *65*, 537–561. (c) J.R. Winkler, A.J. Di Bilio, N.A. Farrow, J.H. Richards, H.B. Gray, *Pure Appl. Chem.*, **1999**, *71*, 1753–1764. (d) H.B. Gray, J.R. Winkler, *Q. Rev. Biophys.*, **2003**, *36*, 341–372.

17 D.S. Wuttke, M.J. Bjerrum, J.R. Winkler, H.B. Gray, *Science*, **1992**, *256*, 1007–1009.

18 R. Langen, I.-J. Chang, J.P. Germanas, J.H. Richards, J.R. Winkler, H.B. Gray, *Science*, **1995**, *268*, 1733–1735.

19 J.J. Regan, A.J. Di Bilio, R. Langen, L.K. Skov, J.R. Winkler, H.B. Gray, J.N. Onuchic, *Chem. Biol.*, **1995**, *2*, 489–496.

20 B.R. Crane, A.J. Di Bilio, J.R. Winkler, H.B. Gray, *J. Am. Chem. Soc.*, **2001**, *123*, 11623–11631.

21 S.S. Skourtis, D.N. Beratan, *Adv. Chem. Phys.*, **1999**, *106*, 377–452.

22 J. Regan, J.N. Onuchic, *Adv. Chem. Phys.*, **1999**, *107*, 497–553.

23 D.N. Beratan, J.N. Betts, J.N. Onuchic, *J. Phys. Chem.*, **1992**, *96*, 2852–2855.

24 E. Babini, I. Bertini, M. Borsari, F. Capozzi, C. Luchinat, X.-Y. Zhang, G.L.C. Moura, I.V. Kurnikov, D.N. Beratan, A. Ponce, A.J. Di Bilio, J.R. Winkler, H.B. Gray, *J. Am. Chem. Soc.*, **2000**, *122*, 4532–4533.

25 C.C. Page, C.C. Moser, X.X. Chen, P.L. Dutton, *Nature*, **1999**, *402*, 47–52.

26 M. Jones, I.V. Kurnikov, D.N. Beratan, *J. Phys. Chem. A*, **2002**, *106*, 2002–2006.

27 A. Ponce, H.B. Gray, J.R. Winkler, *J. Am. Chem. Soc.*, **2000**, *122*, 8187–8191.

28 J. Wolfgang, S.M. Risser, S. Priyadarshy, D.N. Beratan, *J. Phys. Chem. B*, **1997**, *101*, 2987–2991.

29 V.L. Davidson, *Acc. Chem. Res.*, **2000**, *33*, 87–93.

30 Z.-X. Liang, J.M. Nocek, K. Huang, R.T. Hayes, I.V. Kurnikov, D.N. Beratan, B.M. Hoffman, *J. Am. Chem. Soc.*, **2002**, *124*, 6849–6859.

31 Z.-X. Liang, I.V. Kurnikov, J.M. Nocek, A.G. Mauk, D.N. Beratan, B.M. Hoffman, *J. Am. Chem. Soc.*, **2004**, *26*, 2785–2798.

32 F.A. Tezcan, B.R. Crane, J.R. Winkler, H.B. Gray, *Proc. Natl. Acad. Sci. U.S.A.*, **2001**, *98*, 5002–5006.

33 J.L. McGourty, N.V. Blough, B.M. Hoffman, *J. Am. Chem. Soc.*, **1983**, *105*, 4470–4472.

34 (a) T. Pascher, J.P. Chesick, J.R. Winkler, H.B. Gray, *Science*, **1996**, *271*, 1558–1560. (b) J.R. Telford, P. Wittung-Stafshede, H.B. Gray, J.R. Winkler, *Acc. Chem. Res.*, **1998**, *31*, 755–763. (c) J.C. Lee, H.B. Gray, J.R. Winkler, *Proc. Natl. Acad. Sci. U.S.A.*, **2001**, *98*, 7760–7764. (d) J.C. Lee,

I.-J. Chang, H.B. Gray, J.R. Winkler, *J. Mol. Biol.*, **2002**, *320*, 159–164. (e) I.-J. Chang, J.C. Lee, J.R. Winkler, H.B. Gray, *Proc. Natl. Acad. Sci. U.S.A.*, **2003**, *100*, 3838–3840.

35 (a) I.J. Dmochowski, B.R. Crane, J.J. Wilker, J.R. Winkler, H.B. Gray, *Proc. Natl. Acad. Sci. U.S.A.*, **1999**, *96*, 12987–12990. (b) A.R. Dunn, I.J. Dmochowski, A.M. Bilwes, H.B. Gray, B.R. Crane, *Proc. Natl. Acad. Sci. U.S.A.*, **2001**, *98*, 12420–12425. (c) A.R. Dunn, I.J. Dmochowski, J.R. Winkler, H.B. Gray, *J. Am. Chem. Soc.*, **2003**, *125*, 12450–12456.

36 I.V. Kurnikov, A.K. Charnley, D.N. Beratan, *J. Phys. Chem. B*, **2001**, *105*, 5359–5367.

3
Reconstituted Redox Enzymes on Electrodes: From Fundamental Understanding of Electron Transfer at Functionalized Electrode Interfaces to Biosensor and Biofuel Cell Applications

Bilha Willner and Itamar Willner

3.1
Introduction

The electrical contacting of redox proteins with electrodes is a key issue in bioelectronics. Redox proteins usually lack direct electrical communication with electrodes, and thus the direct electrochemical activation of the proteins is prohibited. The lack of electrical contact between the redox centers of proteins and electrodes can be explained by the electron-transfer (ET) Marcus theory [1]. The ET rate constant between a donor and acceptor pair is given by Eq. 1, where d and d_o are the distance separating the electron and donor, and the van der Waals distance, respectively, β is the electron-coupling constant and $\Delta G°$ and λ are the free energy change and the reorganization energy accompanying the electron-transfer process, respectively.

$$k_{ET} \propto \exp[-\beta(d - d_o)] \cdot \exp[-(\Delta G° + \lambda)^2/(4RT\lambda)] \qquad (1)$$

Since the electrode and the protein redox–center may be considered as a donor–acceptor pair, the distance, or spatial, separation of the enzyme redox–center from the electrode by means of the protein shell prohibits the direct electrical communication between the redox site and the electrode. Substantial research efforts were directed during the last two decades to overcome the insulating protein barrier surrounding the redox centers and to develop means to electrically contact redox proteins and electrodes [2]. Diffusional molecular electron–transfer mediators such as ferricyanides [3], quinones [4], bipyridinium salts [5] or transition metal complexes [6], were widely applied to contact redox proteins with the electrodes by a diffusional path. In these systems, the redox active–electron mediator is reduced (or oxidized) at the electrode surface, and the reduced (or oxidized) species penetrate the protein backbone by a diffusional mechanism leading to short distances in respect to the enzyme-active center. The short distances between the electron mediator and the enzyme redox site permits the reduction (or oxidation) of the biocatalyst redox centers. A different approach employed molecular modifiers on electrodes (promoters) that interact with the proteins and align them on the surface in a configuration that facilitates electrical contact [7]. Using this method, hemoproteins such as cytochrome c [7] or myoglobin [8] were activated toward ET communication with conductive supports. A different method to establish ET between redox enzymes and electrodes involved the chemical modification of the protein with electron-relay units

Bioelectronics. Edited by Itamar Willner and Eugenii Katz
Copyright © 2005 WILEY-VCH Verlag GmbH & Co. KGaA, Weinheim
ISBN: 3-527-30690-0

that shorten the ET distances [9]. The tethered relay units activate the electrical contacting of the redox centers and the electrodes by an intraprotein ET hopping process. This method was successfully applied to develop integrated, electrically contacted, layered enzyme-electrodes that included tethered relay units [10]. The loading of the protein by the relay groups is, however, an important parameter that controls the ET effectiveness. Increased loading of the relay units enhances the ET hopping probability by shortening the intraprotein ET distances. On the other hand, the introduction of the foreign relay units adversely affects the protein structure, and its catalytic activity usually decreases as the loading of the tethered redox units increases. Hence, an appropriate balance of the extent of loading has to be optimized for the systems. A further method of electrically contacting redox enzymes with electrode surfaces involves the immobilization of the biocatalysts in conductive polymer matrices [11, 12] or redox-active polymers and, specifically, hydrogels [13, 14]. The flexibility of the polymer chains, and the fact that the polymer may contain functionalities such as charged units or hydrogen bond–forming groups in addition to the redox functions, facilitates their interaction with the protein and the penetration into it to form short ET distances relative to the enzyme redox center. Furthermore, often the redox mediating groups are tethered to the polymer by long-chain alkyl bridges. The flexibility of the tethered redox groups permits their incorporation in the protein to close positions to the biocatalytic redox centers that enable effective mediated electron transfer between the enzyme and the macroscopic redox-active polymer matrix. Indeed, it was found that the ET communication between the enzyme redox sites, and the mediating redox-active polymer is controlled by the length of the tethering chains that link the molecular relay units to the polymer [15]. It was found that as the tethering bridge is longer, the mediated electron transfer is enhanced.

Although the different methods discussed so far are successful in establishing ET communication between the redox centers and electrodes, they suffer from intrinsic limitations. The bioelectrocatalytic features of the resulting enzyme-electrodes represent the collective properties of numerous configurations of enzyme molecules of variable degrees of loading with electron mediator groups that reach a variety of orientations in respect to the conductive support. These difficulties limit the ET communication of the biocatalysts and the conductive supports, as compared to the ET efficiency between the enzyme redox sites and their natural ET substrates or cofactors. For example, the most efficient electrically contacted ferrocene-tethered glucose oxidase revealed an ET exchange rate that corresponded to 2 s^{-1} between the redox center and the electrode [16], while the ET turnover rate between the glucose oxidase redox center and its native oxygen electron acceptor [17] is ca 650 s^{-1}. The effectiveness of ET communication between the enzyme redox centers and the electrodes has important implications on enzyme-electrodes performances and their practical utility in bioelectronic devices. The efficiency of ET communication determines the current output of the electrodes, and thus controls their sensitivities, their specificities and their sensing processes. Also, the electrical communication efficiency between redox enzymes and electrode surfaces controls the current and extractable voltage of enzyme-electrodes. These are two invaluable parameters that regulate the output power of biofuel cells [18]. Thus, to improve the electrical communication between the redox centers of enzymes and electrodes, the alignment of the redox sites on the electrode must be accomplished, and the positioning of the electron mediator(s) between the redox center of the protein and the electrode needs to be optimized. Such organized nanoarchitecturing of the enzyme on the conductive support will lead to equivalent protein configurations on the electrode supports with optimized electrical communication properties. Besides the fundamental importance of such ordered redox proteins on electrodes in understanding the parameters controlling ET at enzyme/electrode interfaces, the development of such protein nanoarchitectures will find important applications in biosensor and biofuel cell technologies.

The present article addresses recent advances in the assembly of structurally aligned redox protein layers on electrodes by means of surface reconstitution and surface cross-linking of structurally oriented enzyme/cofactor complexes on electrodes. The ET properties of the nanostructured interfaces are discussed, and the possible application of the systems in tailoring bioelectronic devices such as biosensors or biofuel cell elements is addressed. A further aspect that will be considered is the control of ET at redox protein/electrode interfaces by means of external stimuli such as electrical or magnetic signals.

The reconstitution process is a well-practiced methodology to examine the structure–function relationship of proteins. The reconstitution method, Figure 3.1, (exemplified for myoglobin) involves the exclusion of the native active center from the protein, for example, an ion or a cofactor, to yield the respective apoprotein (or hollow-protein). The implanting (or reconstitution) of a structurally related cofactor or ion generates the semi-synthetic reconstituted protein that might exhibit new tailored functions that are not present in the native protein. The reconstitution process has been extensively used in studying mechanistic aspects of photoinduced ET in proteins [19–21]. Hemoproteins such as cytochrome c [22] or myoglobin [23] were reconstituted with photoactive Zn(II)-protoporphyrin IX or different metal analog systems that enabled the systematic elucidation of the effects of free energy and distance dependencies on intraprotein ET. The reconstitution of hemoproteins with synthetically modified heme units [24] or with metal protoporphyrin analogs [25] enabled the synthesis of proteins of new designed photochemical, binding and catalytic functions. For example, Figure 3.1 depicts several structures of the functional reconstituted proteins. The reconstitution of apomyoglobin with the Ru(II)-*tris*-bipyridine-heme dyad, (**1**), yields [26] a reconstituted photoactive protein that photocatalyzes the reduction of O_2. The reconstitution of apomyoglobin with the quinone-Zn(II)-protoporphyrin IX, (**2**) [27], or the bipyridinium-Zn(II)-protoporphyrin IX, (**3**) [28], generated photoactive reconstituted proteins that revealed protein-regulated ET properties. For example [28], the Zn(II)-protoporphyrin IX-*bis*-bipyridinium molecule, (**3**), represents a photosensitizer-electron acceptor dyad. The flexibility of the molecular structure results in, upon the photoexcitation of the chromophore, the efficient quenching of the singlet excited state of the photosensitizer. This effective quenching of the singlet state prohibits the formation of the triplet excited state. The reconstitution of (**3**) into apomyoglobin results in rigidification of the chromophore electron-acceptor units. The spatial separation of the components eliminates the singlet-state quenching, and the triplet photosensitizer is formed. The latter long-lived triplet is quenched, Eq. 2,

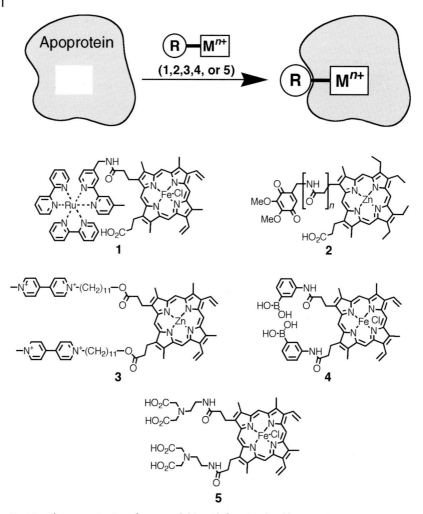

Fig. 3.1 The reconstitution of apomyoglobin with functionalized heme units.

by the bipyridinium units, $k_q = 1.55 \times 10^6$ s^{-1}. The resulting photogenerated redox species, recombine, Eq. 3, with a rate constant that corresponds to $k_b = 1.3 \times 10^6$ s^{-1}. Introduction of the secondary acceptor, Ru(NH$_3$)$_6$$^{3+}$, to the system, Figure 3.2A, allows the ET from the bipyridinium radical cation unit to the secondary acceptor, Eq. 4. This structurally separates the reduced secondary acceptor from the protein-embedded oxidized site leading to the very slow recombination of the redox products, Eq. 5, $k'_b = (6.5 \pm 0.5) \times 10^7$ M$^{-1}\cdot$ s^{-1}. The ET transfer cascade that occurs in the artificial myoglobin reconstituted with (**3**), mimics the functions of the photosynthetic reaction center. The stepwise ET leads to the spatial separation of the redox products and to their stabilization against back ET.

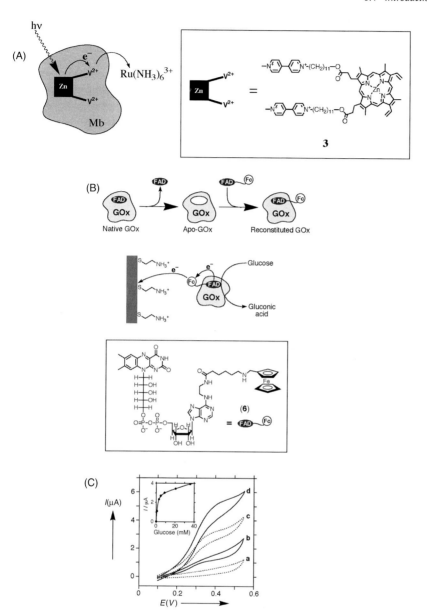

Fig. 3.2 (A) Photoinduced vectorial electron transfer in a Zn(II)-protoporphyrin IX-*bis*-bipyridinium, (**3**), reconstituted myoglobin. (B) The reconstitution of apo-glucose oxidase with ferrocene- functionalized FAD, (**6**), and the bioelectrocatalytic functions of the reconstituted enzyme. (C) Cyclic voltammograms corresponding to the bioelectrocatalayzed oxidation of glucose by the (**6**)-reconstituted glucose oxidase in the presence of different concentrations of glucose: (**a**) 0 mM, (**b**) 1 mM, (**c**) 3 mM, (**d**) 20.5 mM. Inset: Calibration curve corresponding to the electrocatalytic currents at different glucose concentrations of the system. Data recorded in 0.1 M phosphate buffer, pH = 7.3, 35 °C, scan-rate 2 mV s^{-1}.

$$\text{Mb}-^{\text{T}}\text{Zn(II)}-\text{P}-\text{V}^{2+} \xrightarrow{k_q = 1.55 \times 10^6 \text{ s}^{-1}} \text{Mb}-\text{Zn(II)}-\text{P}^{\bullet+}-\text{V}^{\bullet+} \quad (2)$$

$$\text{Mb}-\text{Zn(II)}-\text{P}^{\bullet+}-\text{V}^{\bullet+} \xrightarrow{k_b = 1.3 \times 10^6 \text{ s}^{-1}} \text{Mb}-\text{Zn(II)}-\text{P}-\text{V}^{2+} \quad (3)$$

$$\text{Mb}-\text{Zn(II)}-\text{P}^{\bullet+}-\text{V}^{\bullet+} + \text{Ru(NH}_3)_6^{3+} \longrightarrow \text{Mb}-\text{Zn(II)}-\text{P}^{\bullet+}-\text{V}^{2+} + \text{Ru(NH}_3)_6^{2+} \quad (4)$$

$$\text{Mb}-\text{Zn(II)}-\text{P}^{\bullet+}-\text{V}^{2+} + \text{Ru(NH}_3)_6^{2+} \xrightarrow{k'_b = (6.5 \pm 0.5) \times 10^7 M^{-1} \cdot s^{-1}}$$

$$\text{Mb}-\text{Zn(II)}-\text{P}-\text{V}^{2+} + \text{Ru(NH}_3)_6^{3+} \quad (5)$$

The reconstitution of apomyoglobin with the heme unit tethered to the phenyl boronic acid ligand, (4), or the amino-diacetic acid ligand, (5), generated proteins with sugar binding sites [29] and metal-ion [30] binding functions, respectively. Similarly, the reconstitution of apomyoglobin with Co(III)-protoporphyrin IX and the covalent attachment of eosin chromophores to the protein, generated a photoactive hydrogenation biocatalyst that stimulated the light-induced hydrogenation of acetylenes [24].

The reconstitution of apo-flavoenzymes with semi-synthetic flavin adenine dinucleotide cofactors tethered to a ferrocene electron mediator group, (6), led to the electrical contacting of the redox proteins and to the electrochemical activation of their biocatalytic functions, Figure 3.2B [31].

The reconstitution of apo-glucose oxidase, apo-GOx, with the ferrocene-amino-FAD (FAD or flavin adenine dinucleotide) dyad, (6), led to the formation of an electrically contacted enzyme [31]. At a cystamine monolayer-functionalized Au-electrode, Figure 3.2B, the reconstituted enzyme was electrochemically activated toward the oxidation of glucose. As the concentration of glucose increased, the bioelectrocatalyzed oxidation of glucose was enhanced, and this was reflected by the formation of higher electrocatalytic anodic currents. The bioelectrocatalytic oxidation of glucose was attributed to the mediated electron transfer from the FAD cofactor to the electrode. Electrochemical oxidation of the ferrocene unit to the ferrocenylium cation is followed by the oxidation of the FAD cofactor that activates the biocatalyzed oxidation of glucose, Figure 3.2B. The intermediary functions of the ferrocene units in the bioelectrocatalyzed oxidation of glucose is supported by the fact that the electrocatalytic anodic currents originating from the bioelectrocatalyzed oxidation of glucose appear at the oxidation potential of the ferrocene units. Similarly, the reconstitution of apo-amino acid oxidase with (6) yielded a bioelectrocatalytically active enzyme that stimulated the electrochemical oxidation of L-alanine.

The rapid progress in nanotechnology paves the way for developing nanobiotechnology or nanobioelectronics as new scientific disciplines that combine biomaterials with the unique functions of nano-objects such as nanoparticles, nano-rods, nanowires or nanotubes [32, 33]. Nanoparticles (metallic or semiconductor) exhibit unique size-controlled optical [34], electronic [35] and catalytic [36] features as a result of the quantum size effect on the material properties. Similarly, ingenious methods were reported for the preparation of metallic or semiconductor nano-rods [37–40]. The height aspect ratio

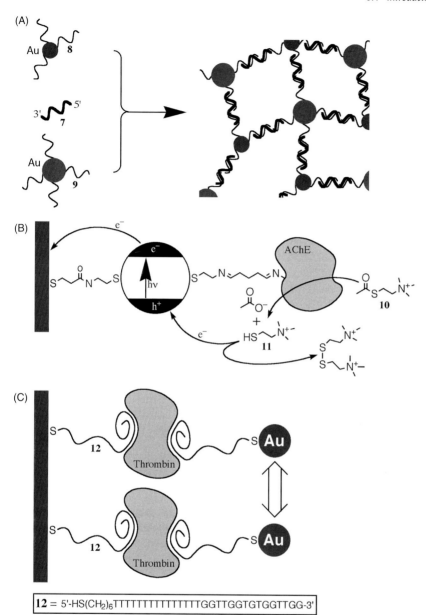

Fig. 3.3 Functional biomolecule-nanoparticle architectures: (A) aggregation of nucleic acid–functionalized Au-nanoparticles by means of a complementary DNA. (B) Photocurrent generation by an acetylcholine esterase-functionalized CdS-nanoparticle film on an electrode in the presence of thioacetylcholine (**10**). (C) Aggregated aptamer-thrombin Au-nanoparticle structure on a surface for the optical detection of thrombin.

of these nano-objects leads to polarized optical and electronic properties along the axes of the rods [41]. Among the different nanomaterials, carbon nanotubes (CNTs) attract substantial research effort. The discovery that the electronic properties of the CNTs are controlled by the folding symmetry of the respective graphite sheets leads to fascinating new electronic materials for bioelectronic applications. While the symmetric folding of the graphite sheets leads to carbon nanotubes of ballistic conductivity, the asymmetric folding of the graphite layers yield chiral CNTs of semiconductor properties [42]. Biomaterials such as proteins (enzymes, antigen/antibodies) or nucleic acids exhibit structural dimensions comparable with the nanoparticles (or nano-rods), and thus the integration of the biomaterials with the nanoparticles (or nano-rods) may lead to hybrid systems of new photonic or electronic properties. Figure 3.3 exemplifies the photonic and electronic properties of several nanoparticle/biomaterial hybrid systems. The hybridization of the nucleic acid (**7**) with two kinds of Au-nanoparticles modified with the nucleic acids (**8**) and (**9**) complementary to the 3′ and 5′ ends of the DNA, (**7**), leads to hybridization between the components and the formation of a three-dimensional Au-nanoparticle aggregate, Figure 3.3A. The three-dimensional aggregation of the Au-nanoparticles leads to an interparticle coupled–plasmon exciton that shifts the individual Au-nanoparticle absorbance (red color) to the interparticle coupled plasmon absorbance (blue color). These optical changes of the Au-nanoparticles resulting from the hybridization process, were extensively applied as a colorimetric method for DNA detection [43, 44]. Figure 3.3B depicts the assembly of a CdS-nanoparticle/acetylcholine esterase hybrid system on an Au-electrode [45]. Photoexcitation of the semiconductor nanoparticle yields an electron-hole pair that rapidly recombines to the ground state. The biocatalytic hydrolysis of thioacetylcholine (**10**) to thiocholine (**11**) generates an electron donor for the valence-band holes. The scavenging of the valence-band holes by (**11**) stabilizes the conduction-band electrons, and their injection to the electrode led to the formation of a photocurrent. Inhibition of the acetylcholine esterase by specific inhibitors blocked the catalytic activity of the enzyme and the formation of (**11**), and this retarded the intensity of the resulting photocurrent. The latter system was suggested as a sensor for biological warfare nerve gases that act as acetylcholine esterase inhibitors. Figure 3.3C shows the photonic detection of thrombin by an aptamer-functionalized Au-nanoparticle hybrid system [46]. The aptamer, (**12**), specifically recognizes thrombin, and the interaction of the (**12**)-modified Au-nanoparticles with the thrombin dimer leads to the formation of the thrombin/nucleic acid/Au-nanoparticle aggregate. The resulting Au-nanoparticle aggregate exhibits the characteristic interparticle coupled plasmon absorbance that enables the colorimetric detection of aptamer/protein interactions.

Nanotechnology also provides unique tools for the structural characterization of nanostructures. Scanning probe microscopes such as scanning tunneling microscopy (STM), scanning force microscopy (AFM), near-field scanning optical microscopy (NSOM) and other scanning probe techniques, as well as other microscopy imaging methods (such as TEM or SEM), provide unique tools to image the nanoparticle/biomaterial hybrid systems and to examine their functional operation at the single molecule level. Other surface techniques, such as X-ray photoelectron spectroscopy (XPS), surface plasmon resonance (SPR), ellipsometry, microgravimetric quartz crystal–microbalance (QCM) analyses, provide invaluable methods to characterize the structural features of the nanoparticle (or nano-rod)/biomaterial hybrid systems and to elucidate the structure-function relationships of the assemblies at the single molecule level.

3.2
Electrodes Functionalized with Reconstituted Redox Proteins

Two general strategies to assemble the reconstituted enzymes on electrodes were developed, Figure 3.4. By one method a relay-cofactor dyad is assembled on the electrode, and the respective apoprotein is reconstituted on the surface to yield an aligned protein that is linked to the conductive surface by the relay component. The second method involves the synthesis of the relay-cofactor unit and the reconstitution of the apoprotein in solution. The specific immobilization of the enzyme on the electrode by the relay unit provides the structurally organized enzyme-electrodes. While the first method is technically easier, the second methodology that involves tedious synthetic and separation steps permits the fundamental structural characterization of the reconstituted protein. In the two configurations, the redox enzymes are anticipated to be electrically contacted with the electrode by means of the relay, a conductive molecular or polymer wire, a conductive nanoparticle, or even conductive carbon or doped silicon nanotubes.

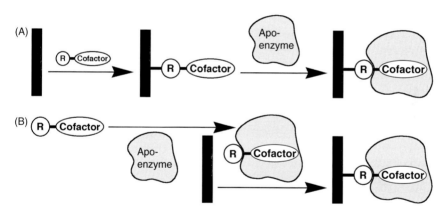

Fig. 3.4 Methods for the assembly of reconstituted relay-cofactor enzyme assemblies (A) by the surface reconstitution of the apoenzyme; (B) by the attachment of the pre-reconstituted enzyme on the electrode.

3.2.1
Reconstituted Flavoenzyme-Electrodes Using Molecular or Polymer Relay Systems

Integrated electrically contacted enzyme-electrodes were prepared by the surface reconstitution of different apoenzymes on electrode surfaces [47]. The pyrroloquinoline

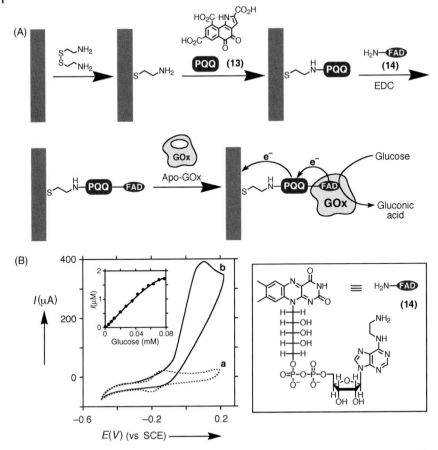

Fig. 3.5 (A) Surface reconstitution of apo-glucose oxidase on a PQQ-FAD monolayer associated with a Au-electrode. (B) Cyclic voltammograms corresponding to (a) the reconstituted glucose oxidase monolayer in the absence of glucose, and (b) in the presence of glucose, 80 mM. Data recorded in 10 mM phosphate buffer, pH = 7.0, at 35 °C, scan-rate 5 mV s^{-1}. *Inset*: Calibration curve corresponding to the chronoamperometric currents at E = 0.2 V at different glucose concentrations.

quinone, PQQ, (**13**), was covalently linked to a cystamine monolayer associated with a Au-electrode, and N^6-(2-aminoethyl-FAD), (**14**), was covalently attached to the PQQ units, Figure 3.5A. The integrated enzyme-electrode was then prepared by the reconstitution of apo-GOx on the FAD units. The surface coverage of the PQQ-FAD units was estimated to be 5.5×10^{-10} mole cm^{-2}, whereas the surface coverage of the reconstituted enzyme was determined to be 1.5×10^{-12} mole cm^{-2}. Figure 3.5B shows the cyclic voltammogram corresponding to the integrated enzyme-electrode in the presence of glucose of 80 mM. The electrocatalytic anodic current observed at $E = -0.125$ V versus SCE (the redox potential of the PQQ units) indicates that the reconstituted enzyme is electrically contacted with the electrode. Since the reconstitution of apo-GOx onto an FAD-monolayer linked to the Au-electrode did not yield an electrically contacted enzyme-electrode, and since the electrocatalytic anodic current observed for the integrated electrode was observed at the redox potential of the PQQ units, the electrical communication between the enzyme redox center and the electrode and its bioelectrocatalytic activation was attributed to the PQQ-mediated electron transfer in the system, Figure 3.5A. The FAD centers of reconstituted GOx are reduced by glucose, and the reduced cofactors are then oxidized by the PQQ units to form $PQQH_2$. Electrochemical oxidation of $PQQH_2$ that is not shielded by the protein shell regenerates the ET mediator, and this enables the cyclic biocatalyzed oxidation of glucose as long as the $PQQH_2$ units are electrochemically oxidized. Figure 3.5B, inset, shows the calibration curve that corresponds to the anodic currents observed at variable concentrations of glucose. The current increases as the glucose concentration is elevated, and it levels off at a concentration of glucose that corresponds to 80 mM. The saturated anodic current corresponds to the maximum efficiency of the bioelectrocatalytic process at the electrode interface. Knowing the value of the saturation current, and the surface coverage of the enzyme, the ET turnover rate between the FAD center and the electrode was calculated to be $900 \pm 150\ s^{-1}$ at 35 °C. This is an unprecedetedly high turnover rate that equals, and eventually exceeds, the ET turnover rate between GOx and its native electron acceptor (O_2) [17]. This effective electrical contacting of the redox enzyme and the electrode has important consequences on the bioelectrocatalytic sensoric features of the electrode. The efficient ET turnover rate leads to an electrode that is unaffected by oxygen or common glucose-sensing interferants such as ascorbic acid or uric acid.

The organization of the reconstituted integrated glucose-sensing electrode has required the use of the scarce semi-synthetic FAD cofactor (14). A modified procedure depicted in Figure 3.6A used the native FAD cofactor for reconstituting the enzyme to resolve this limitation [48]. The PQQ electron mediator (13) was coupled to the cystamine monolayer-functionalized electrode, and 3-aminophenyl boronic acid (15) was covalently coupled to the PQQ units. A boronate complex was generated between the boronic acid ligands and the ribose units of FAD, and the resulting complex was used to reconstitute apo-GOx on the FAD sites. This procedure yielded the integrated, electrically contacted, enzyme-electrode. The kinetics of the reconstitution process on the cofactor-functionalized monolayer electrode was characterized by electrochemical means [48], Figure 3.6. Since the reconstitution of the apoprotein on the electrodes yields a hydrophobic insulating protein layer, the interfacial electron-transfer resistance is expected to increase upon the surface reconstitution process. Figure 3.6B shows the Faradaic impedance spectra of the enzyme-electrodes at time intervals of reconstitution. The interfacial electron-transfer resistance increases as reconstitution proceeds, and it reaches a saturation value after ca 4 h of reconstitution (see inset, Figure 3.6B). A similar conclusion was derived by following the bioelectrocatalytic functions of the enzyme-electrode at time intervals of reconstitution in the presence of a fixed concentration of glucose, 8×10^{-2} M, Figure 3.6C. As reconstitution proceeds, the surface coverage of the bioelectrocatalyst increases, and the resulting anodic current is enhanced. The anodic current reaches a saturation value after ca 4 h of reconstitution that corresponds to the time interval required for generating the optimal surface coverage of the protein at the stated conditions. The saturated surface coverage of

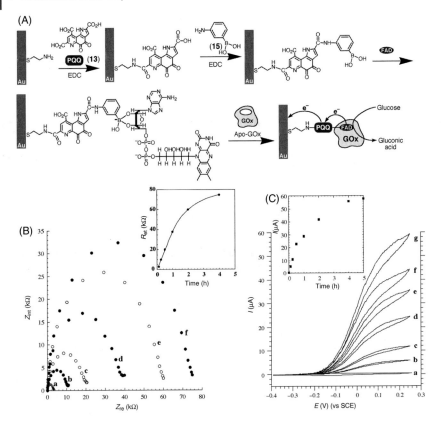

Fig. 3.6 (A) Assembly of reconstituted glucose oxidase on a PQQ-FAD monolayer linked to a Au-electrode. (B) Faradaic impedance spectra of the modified electrode at time intervals of reconstitution. (a) 0.1 h, (b) 0.25 h, (c) 0.5 h, (d) 1 h, (e) 2 h, (f) 4 h. *Inset*: Interfacial electron-transfer resistance of the modified electrode at time intervals of reconstitution. (C) Cyclic voltammograms corresponding to the bioelectrocatalyzed oxidation of glucose, 80 mM, by the enzyme-functionalized electrode at time intervals of reconstitution: (a) 0 h, (b) 0.1 h, (c) 0.25 h, (d) 0.5 h, (e) 1 h, (f) 2 h, (g) 4 h. *Inset*: Electrocatalytic currents transduced by the enzyme-modified electrode at time intervals of reconstitution.

the enzyme was elucidated by electrochemical means and microgravimetric quartz crystal–microbalance measurements [48] to be ca 2×10^{-12} mol cm^{-2}. Realizing that the footprint of GOx is 58 nm^2, the experimental surface coverage is translated into the value of 70% of an ordered densely packed monolayer. The electrocatalytic anodic currents of the electrically contacted enzyme-electrode were analyzed at variable concentrations of glucose, Figure 3.7. The catalytic current levels off to a saturation value at glucose concentrations higher than 6×10^{-2} M, conditions at which the maximal bioelectrocatalytic performance of the reconstituted enzyme-electrode is observed. Knowing the saturation value of the catalytic anodic current, I_{cat}^{sat}, the surface coverage of GOx, Γ_{GOx}, and the electrode area, A, the ET turnover rate from the active center to

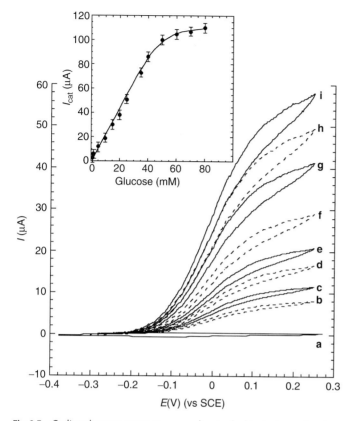

Fig. 3.7 Cyclic voltammograms corresponding to the bioelectrocatalyzed oxidation of variable concentrations of glucose by the reconstituted glucose oxidase- functionalized electrode shown in Figure 3.6A. Glucose concentrations correspond to (a) 0 mM, (b) 5 mM, (c) 10 mM, (d) 15 mM, (e) 20 mM, (f) 25 mM, (g) 35 mM, (h) 40 mM (i) 50 mM. *Inset*: Calibration curve corresponding to the transduced electrocatalytic currents at different concentrations of glucose.

the electrode TR_{max}, was estimated to be ca 700 s^{-1}, using Eq. 6. This ET turnover rate is similar to the ET rate between GOx and O_2.

$$TR_{max} = I_{cat}^{sat}/(Fn\Gamma_{GOx}A) \qquad (6)$$

A different method to electrically wire a redox enzyme by a molecular electroactive relay unit involved the use of molecular rotaxane architectures [49]. Rotaxanes are supramolecular architectures that include an interlocked molecular ring on a molecular wire that is stoppered at its two ends by bulky molecular components [50, 51]. The rotaxanes are usually prepared by the formation of an affinity complex between the ring and a molecular site associated with the wire followed by a chemical reaction

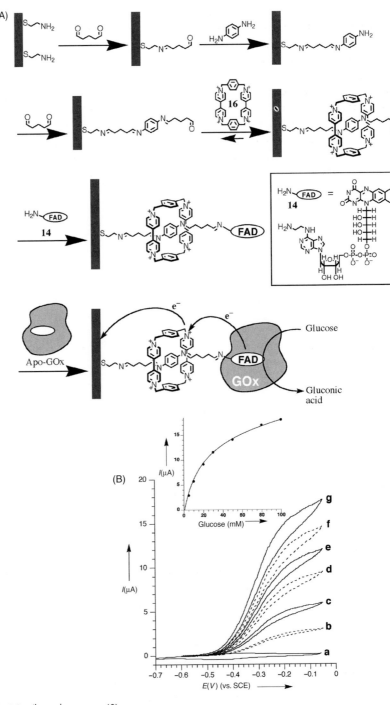

Fig. 3.8 (legend see page 49)

at the wire ends that stopper the ring on the wire, and this leads to a noncovalent stable supramolecular structure. Ingenious rotaxane structures were developed in solutions [52–54] and on surfaces [55], and the signal-controlled translocations of the ring-component to distinct specific positions on the molecular wire were demonstrated using photonic [56], electrochemical [57] or chemical [58] stimuli. An electrically contacted enzyme-stoppered rotaxane that includes a threaded electron relay in the rotaxane structure was synthesized [49], Figure 3.8A. A monolayer consisting of chains that include the *bis*-iminophenylene π-donor unit was synthesized on the Au-surface by a sequence of condensation reactions that included 1,4-phthaldialdehyde and 1,4-diaminobenzene. The monolayer-functionalized electrode was then interacted with *bis*-(paraquat)phenylene (**16**) to form the supramolecular π-donor-acceptor complex, and the resulting supramolecular assembly was *in situ* stoppered with the amino-FAD cofactor unit, (**14**). The surface reconstitution of apo-GOx on the FAD sites yielded the surface-reconstituted, aligned, electrically contacted enzyme-electrode. Microgravimetric quartz – crystal–microbalance measurements indicated a surface coverage of the enzyme that corresponds to 2×10^{-12} mol cm^{-2} on the electrode surface. Figure 3.8B shows the cyclic voltammograms corresponding to the bioelectrocatalyzed oxidation of different concentrations of glucose. The bioelectrocatalyzed oxidation of glucose is observed at -0.4 V versus SCE, the lowest potential ever observed for glucose oxidation. The potential at which glucose oxidation proceeds in this system is only ca 100 mV more positive than the redox potential of FAD associated with the enzyme. Figure 3.8B inset, shows the derived calibration curve that shows the amperometric responses of the nano-engineered enzyme-electrode at variable concentrations of glucose. The turnover rate of electron transfer in the system was estimated to be ca 450 s^{-1}. The very low potential for the oxidation of glucose in the system has enormous significance for the development of biofuel cells of high power output (see Section 3.4). The effective electrical contacting of the redox enzyme in the rotaxane structure is attributed to the dynamic freedom of the electron relay in the system. The bioelectrocatalyzed oxidation of glucose results in the ET from the FAD cofactor to the threaded bipyridinium relay in the rotaxane configuration, Figure 3.8A. The reduction of the relay perturbs the donor–acceptor complex, and the reduced relay freely moves on the molecular wire leading to its oxidation at the electrode surface, and the reorganization of the oxidized relay at the π-donor site. This system reveals a rigid electrically contacted enzyme configuration with the advantages of a dynamically free mediated ET transfer by the electroactive rotaxane shuttle.

A different approach to electrically contact reconstituted flavoenzymes on electrodes involved the use of the conductive redox-polymer polyaniline as a redox mediator for the electrical contacting of the enzyme with the electrode [59]. Aniline was

---

Fig. 3.8 (A) Assembly of a rotaxane-based reconstituted glucose oxidase electrode. (B) Cyclic voltammograms corresponding to the bioelectrocatalyzed oxidation of different concentrations of glucose by the GOx-reconstituted electrode in the rotaxane structure: (a) 0 mM, (b) 5 mM, (c) 10 mM, (d) 20 mM, (e) 30 mM, (f) 50 mM, (g) 80 mM. *Inset*: Calibration curve derived from the cyclic voltammograms at -0.1 V (vs SCE).

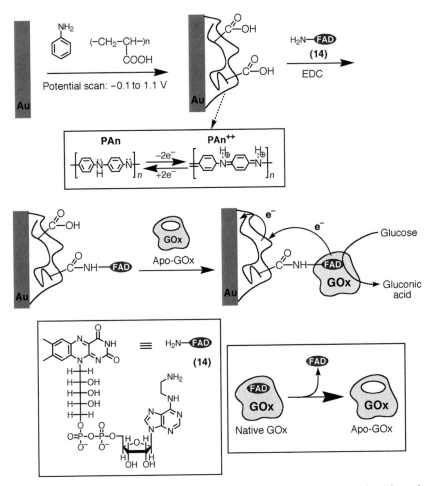

Fig. 3.9 Assembly of an electrically contacted polyaniline-reconstituted glucose oxidase electrode.

electropolymerized on Au-electrodes in the presence of polyacrylic acid. Polymerization yielded a polyaniline film with entangled polyacrylic acid chains. In contrast to a bare polyaniline film that reveals redox activities only in acidic aqueous solutions, the polyaniline/polyacrylic acid composite film was reported [60] to be electrochemically active in neutral aqueous solutions, and thus the polymer film could be coupled to bioelectrocatalyzed transformations. The enzyme-electrode was constructed as depicted in Figure 3.9. The amino-FAD semi-synthetic cofactor (**14**) was covalently linked to the polyacrylic acid chains entangled in the polymer film, and apo-glucose oxidase was reconstituted onto the FAD units to generate the integrated enzyme-electrode. In the specific construction assembled on the electrode, the polyaniline/polyacrylic acid film exhibited a thickness of ca 90 nm and the surface coverage of the FAD units and of the reconstituted GOx components was estimated

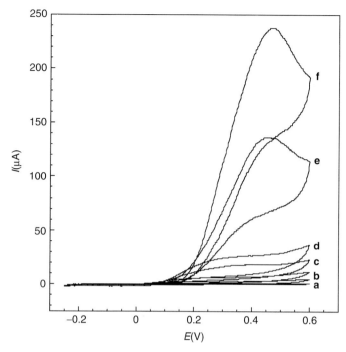

Fig. 3.10 Cyclic voltammograms corresponding to the bioelectrocatalyzed oxidation of variable concentrations of glucose by the integrated, electrically contacted polyaniline-reconstituted glucose oxidase electrode. Glucose concentrations correspond to (a) 0 mM, (b) 5 mM, (c) 10 mM, (d) 20 mM, (e) 35 mM, (f) 50 Mm.

to be 2×10^{-11} mol cm^{-2} and 3×10^{-12} mol cm^{-2} respectively. The resulting enzyme-electrode revealed bioelectrocatalytic activities toward the oxidation of glucose, Figure 3.10. Knowing the enzyme content in the film and using the saturated current density observed in the system, $i = 0.3$ mA cm^{-2}, the maximum ET turnover rate for the biocatalyst was estimated to be $TR_{max} \approx 1000$ s^{-1}. In a control experiment, the lysine residues of native GOx, with its naturally implanted FAD cofactor, were directly coupled to the polyacrylic acid chains of the polyaniline/polyacrylic acid film. The resulting electrode revealed very poor bioelectrocatalytic activities, and the maximum current output was ca 100-fold lower. These results highlight the importance of alignment of the biocatalyst in respect to the redox-active polymer film. The electrostatic attraction of the polyacrylic acid chains to the oxidized polyaniline simultaneously attracts the aligned tethered enzyme components, thus enabling the mediated ET from the active site through the oxidized polyaniline to the electrode, Figure 3.9. This mediated ET electrically contacts the biocatalyst toward the biocatalyzed oxidation of glucose.

The system was further characterized by *in situ* electrochemical/SPR measurements [59], and SPR spectroscopy was introduced as a means to follow bioelectrocatalytic transformations. Surface plasmon resonance spectroscopy is a useful method for the

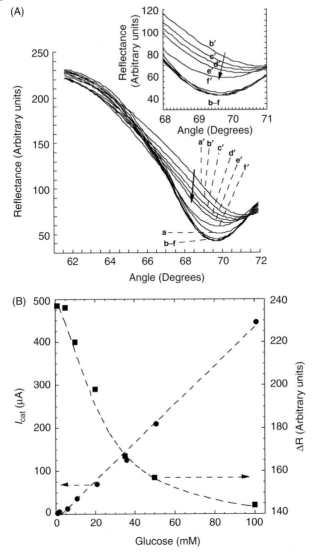

Fig. 3.11 (A) Surface plasmon resonance (SPR) spectra corresponding to (a) the reduced polyaniline film with covalently linked reconstituted glucose oxidase in the absence of glucose (applied potential $E = -0.3$ V vs SCE), and (b) to (f) the reduced polyaniline-reconstituted GOx film upon addition of 5 mM, 10 mM, 20 mM, 35 mM, 50 mM of glucose, respectively. The (a′) spectrum corresponds to the oxidized polyaniline-reconstituted GOx film ($E = 0.6$ V vs SCE) in the absence of glucose whereas spectra (b′) to (f′) correspond to the oxidized polyaniline-reconstituted GOx film in the presence of 5, 10, 20, 35 and 50 mM, respectively. (B) calibration curves corresponding to (a) transduced anodic currents in the presence of variable concentrations of glucose. (b) the changes in the reflectance intensities at the minimum reflectivity angles, ΔR, of the oxidized polyaniline-reconstituted GOx-modified Au-surface in the presence of different concentrations of glucose.

characterization of the refractive index and thickness of interfaces associated with Au surfaces (or Ag surfaces) [61, 62]. In fact, SPR spectroscopy is a common practice in biosensing, and numerous examples that follow the formation of biorecognition complexes on SPR active surfaces have been reported [63, 64]. The redox transformations of polyaniline films on a Au-surface were examined in detail [65]. The minimum reflectivity angle of the oxidized polyaniline was found to be shifted to higher values. Thus, under conditions in which the applied potential oxidizes the polyaniline, PAn, to its oxidized state PAn^{2+}, and provided that the mediated biocatalyzed oxidation of glucose is fast, a steady state equilibrium that is controlled by the glucose concentration, between the oxidized polyaniline PAn^{2+} and reduced polyaniline, PAn, will be generated in the film. As the concentration of glucose is elevated, the concentration of PAn will increase, and the ratio PAn^{2+}/PAn will decrease. As a result, as the concentration of glucose is elevated, the SPR spectra of the polymer film would shift to lower reflectivity angles that resemble the spectrum of PAn, even though the potential applied on the electrode is thermodynamically adequate to oxidize the polymer film. Figure 3.11A shows the SPR of the reduced polymer film, curve (**a**), and oxidized (PAn^{2+}) film in the absence of glucose. While the effect of addition of glucose on the SPR spectra of the reduced PAn film is minute, curves (**b**)–(**f**), addition of glucose to the film subjected to the oxidative potential, resulted in a shift to lower values in the minimum reflectivity angles as the concentration of glucose increases. This is consistent with the enrichment of the polymer film with the reduced state of PAn. Figure 3.11B shows the derived calibration curves that correspond to the amperometric detection of glucose and to the SPR transduction of glucose sensing by the integrated PAn/reconstituted GOx electrode.

3.2.2
Electrical Contacting of Flavoenzymes by Reconstitution on Carbon Nanotubes and Conducting Polymer Wires

Carbon nanotubes (CNTs) in single-wall or multiwall configurations provide new graphite-like materials or cylindrical nano-dimensions. CNTs exhibit unique electron transport properties that are controlled by the topological structure of the wrapped graphite sheet [66]. While "armchair" or "zigzag" edge joining of the graphite sheets yield CNTs of metallic conductivity the asymmetric, chiral folding of the graphite sheet generates CNTs of semiconductive properties. The integration of biomaterials such as proteins [67] or DNA [68] with CNTs attracts substantial research efforts, directed to the design of biomolecule-CNTs hybrid systems of new electronic and structural properties. Several redox proteins [69] such as cytochrome c or myoglobin were electrically contacted with electrode surfaces using CNTs, but in these systems random immobilization of the proteins on the CNTs support was employed. The undecapeptide microperoxidase, MP-11, that consists of the cytochrome c active-site microenvironment, was covalently attached to the ends of standing carbon nanotubes linked to a Au-surface, and the CNT-mediated biocatalyzed reduction of H$_2$O$_2$ was demonstrated by the system [70]. Also, semiconductive CNTs functionalized at the sidewalls with glucose oxidase were

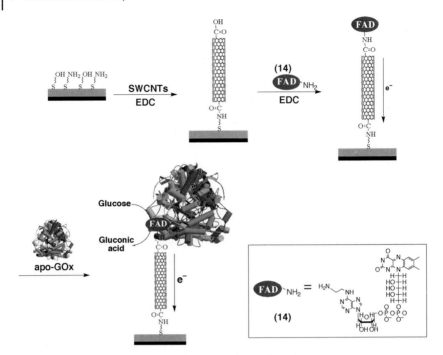

Fig. 3.12 Assembly of the CNT electrically contacted GOx electrode.

used as building elements of a field-effect transistor (FET) [71]. It was found that the CNT-gate conductance was controlled by the biocatalytic oxidation of glucose.

The reconstitution of a flavoenzyme on the ends of standing CNTs associated with electrodes led to the electrical contacting of the enzyme with the bulk electrode surface [72]. Figure 3.12 depicts the methodology to assemble the enzyme-electrode. The oxidative treatment of single wall carbon nanotubes (SWCNTs) led to the chemical digestion of the SWCNT to CNT of variable lengths terminated at their ends with carboxylic acid functionalities [73]. The shortened and functionalized carbon nanotubes could then be chromatographically separated to a narrow size distribution of CNTs. The functionalized CNTs were covalently tethered to a cystamine-monolayer-modified electrode, and the amino-FAD cofactor (14) was then covalently associated to the carboxylic acid residues linked to the ends of the CNTs. Apo-GOx was then reconstituted on the cofactor-modified CNTs to yield the integrated enzyme-electrode. Figure 3.13 depicts the microscopic characterization of the electrode and of the CNT-enzyme hybrid system. Figures 3.13A and B show the AFM images of the standing CNTs on the Au-electrode obtained upon coupling the CNTs for different time intervals. Clearly, an increased surface coverage of the CNTs is observed as the modification time interval is prolonged. A surface with a low CNT coverage was used for the reconstitution process to allow maximum space for the formation of a densely packed reconstituted GOx monolayer. Figure 3.13C shows the image of the enzyme-reconstituted surface. Bumps

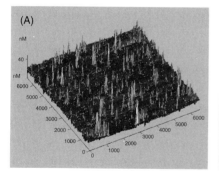
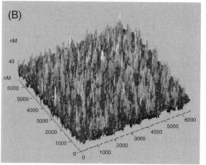
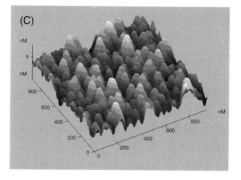

Fig. 3.13 (A) and (B) AFM images of standing CNTs on the Au-surface upon chemical coupling for 30 and 90 min respectively. (C) AFM image of GOx reconstituted on the CNTs shown in image (A) (length of CNTs ca 50 nm).

of heights consistent with the dimensions of GOx are observed on the surface. The surface coverage of GOx on the electrode was determined by microgravimetric quartz crystal–microbalance measurements, as well as by the analysis of the AFM images to be ca 1×10^{-12} mol cm^{-2}. The reconstituted GOx on the CNTs revealed direct electrical contact with the electrode support. Figure 3.14 shows the cyclic voltammograms observed upon the bioelectrocatalyzed oxidation of variable concentrations of glucose by the enzyme reconstituted on 25 nm long CNT. Figure 3.14, (inset), depicts the derived calibration curve. From the saturation current, $i_{max} = 60$ µA and knowing the surface coverage of the enzyme, the turnover ET rate was estimated to be ca 4100 s^{-1} in the system.

Although the results imply that the CNTs act as conductive nano-needles that transport the electrons from the FAD cofactor to the electrode, it was found that the electrical contacting efficiency was controlled by the length of the connecting CNTs. Figure 3.15A shows the current responses of enzyme-electrodes consisting of GOx monolayers reconstituted on CNTs of different lengths in the presence of variable concentrations of glucose. The electrical contacting efficiency decreases as the CNT length increases, and a linear relationship between the ET turnover rate and the reciprocal of the CNT length is observed, Figure 3.15B. A plausible explanation for

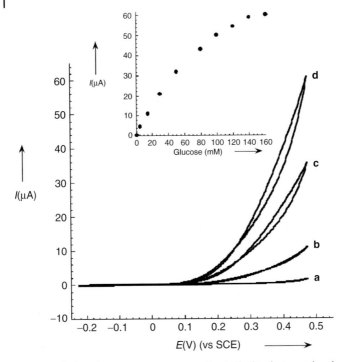

Fig. 3.14 Cyclic voltammograms corresponding to the bioelectrocatalyzed oxidation of different concentrations of glucose by the GOx-reconstituted CNT electrode: (**a**) 0 mM, (**b**) 20 mM, (**c**) 60 mM, (**d**) 160 mM.
Inset: derived calibration curve corresponding to the amperometric responses of the reconstituted electrode at 0.45 V vs. SCE in the presence of different concentrations of glucose.

these results was based on theoretical predictions on ET along CNTs [74]. The chemical oxidative digestion of the CNTs yields defect sites on the sidewalls of the CNTs. The encounter of electrons transported along the CNTs with defect sites results in their backscattering to alternative conductive conjugated paths, leading to slower transport rate. The probability of defects sites increases with the CNTs length and the theory predicts [74] that the ET rate along the CNT will inversely relate to the CNT length.

The electrical contacting of redox enzymes is not limited to CNTs, and other conductive (or redox-active) nanowires mediate ET between the enzyme redox center and electrodes. Polyaniline wrapped over double-stranded DNA was found to exhibit redox functions in neutral aqueous solutions [75]. This property was used to synthesize polyaniline wires for the electrical contacting of glucose oxidase [76]. A copolymer wire of polyaniline-anilineboronic acid was prepared on a λ-phage DNA template. The flavin adenine dinucleotide native cofactor was then coupled to the boronic acid ligand and the FAD-functionalized copolymer was assembled on a cysteic acid monolayer-modified Au-electrode by a layer-by-layer deposition process, Figure 3.16A. Apo-GOx was then reconstituted on the FAD cofactor units to form an electrically contacted

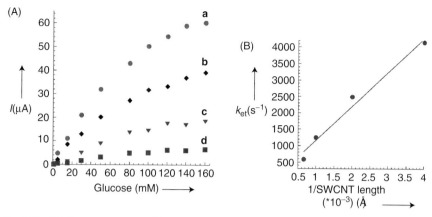

Fig. 3.15 (A) Current responses of the GOx-reconstituted CNT electrodes at variable glucose concentrations using CNTs of variable lengths as connectors: (**a**) ca 25 nm, (**b**) ca 50 nm, (**c**) ca 100 nm, (**d**) ca 150 nm (B) Dependence of the electron transfer–turnover rate of the reconstituted GOx on the reciprocal of the CNT length (L) used as a connector.

configuration. Figure 3.16B shows the cyclic voltammograms corresponding to the bioelectrocatalyzed oxidation of different concentrations of glucose, and the resulting derived calibration curve. The electrocatalytic anodic currents demonstrate that the polyaniline wire mediates the transport of electrons from the enzyme redox center to the electrode surface. Interestingly, the binding of the native glycoprotein to the boronic acid ligand did not lead to an electrically contacted enzyme configuration, implying that the alignment of the enzyme on the polymer wire by the reconstitution process is mandatory to yield the electrical communication between the redox enzyme and the electrode.

The available advanced imaging techniques enable us to probe the structural features of the nanotube (or nanowire)/reconstituted enzyme hybrid systems. For example, Figure 3.17A shows the AFM image of CNTs functionalized at their ends with the FAD cofactor units, and reconstituted with apo-Gox [72]. The individual enzyme molecules at the end of the CNTs are clearly visible. Figure 3.17B shows the stained TEM image of the analogous CNT/GOx hybrid systems. Thus, the possibilities of imaging and characterizing these hybrid systems at the single-molecule level paves the way for integrating the systems in the form of nanoscale single-molecule devices.

3.2.3
Electrical Contacting of Flavoenzymes by Means of Metallic Nanoparticles

Significant progress in tailoring electrically contacted enzyme-electrodes was accomplished by the application of Au-nanoparticles (diameter 1.2 nm) as electron mediator units for the enzyme glucose oxidase [77]. The Au-nanoparticle/reconstituted GOx electrodes were prepared by two alternative methods, Figure 3.18, and yield integrated

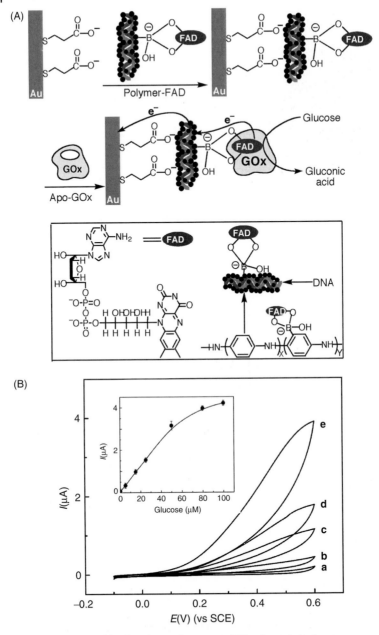

Fig. 3.16 (A) Assembly of an electrically contacted GOx electrode by the reconstitution of apo-GOx on FAD-modified polyaniline wires generated on DNA templates. (B) Cyclic voltammograms corresponding to the bioelectrocatalyzed oxidation of different concentrations of glucose by the GOx-reconstituted electrode: (a) 0 mM, (b) 5 mM, (c) 15 mM, (d) 25 mM, (e) 50 mM. *Inset*: Derived calibration curve.

3.2 Electrodes Functionalized with Reconstituted Redox Proteins | 59

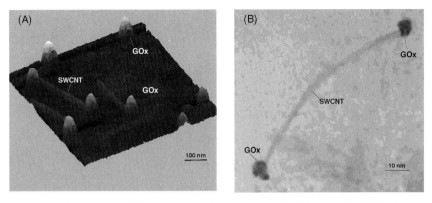

Fig. 3.17 (A) AFM image of CNTs reconstituted at their ends with GOx units. (B) HRTEM image of a CNT modified at its end with GOx unit (Enzyme was stained with uranyl acetate.).

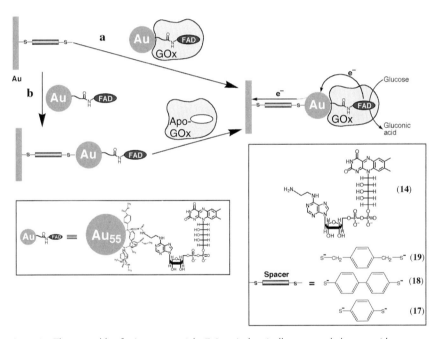

Fig. 3.18 The assembly of a Au-nanoparticle (1.2 nm) electrically contacted glucose oxidase electrode by (**a**) the primary reconstitution of apo-GOx on the FAD-functionalized Au-nanoparticle, and the immobilization of the enzyme-nanoparticle hybrid on an electrode surface; (**b**) the primary immobilization of the FAD-modified Au-nanoparticle on the electrode surface and the surface reconstitution of apo-GOx on the functionalized electrode.

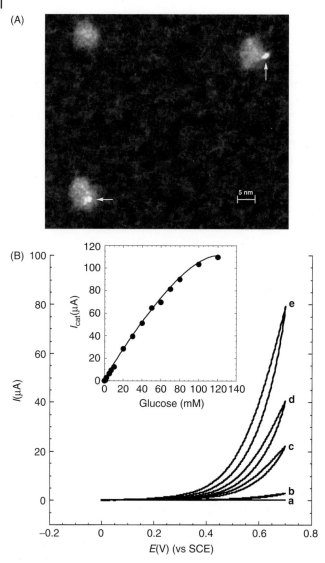

Fig. 3.19 (A) Scanning transmission electron microscopy image of the Au-nanoparticle (1.2 nm)-reconstituted GOx hybrid. (Arrow indicates Au-nanoparticle). (B) Cyclic voltammograms corresponding to the bioelectrocatalyzed oxidation of variable concentrations of glucose by the electrically contacted Au nanoparticle–reconstituted GOx-modified electrode. Glucose concentrations correspond to (**a**) 0 mM, (**b**) 1 mM, (**c**) 2 mM, (**d**) 5 mM, (**e**) 10 mM. *Inset*: Calibration curve corresponding to the electrocatalytic currents at different glucose concentrations.

enzyme-electrodes of similar composition and functions. By one method, route **a**, the amino-FAD semi-synthetic cofactor was attached to the maleimide-functionalized Au_{55}-nanoparticle (diameter 1.2 nm). Apo-GOx was then reconstituted on the particle and the resulting nanoparticle-enzyme hybrid was linked to a dithiol monolayer associated with the Au-electrode. The scanning transmission electron micrograph (STEM) of the nanoparticle-enzyme conjugate is shown in Figure 3.19A that visualizes nicely the composition of the nanoparticle-enzyme hybrid. The second method to organize the enzyme-electrode, route **b**, has involved the attachment of the FAD-functionalized Au-nanoparticles on the dithiol monolayer, and the surface reconstitution of apo-GOx onto the FAD units. The surface coverage of the reconstituted enzyme was estimated to be 1×10^{-12} mol·cm^{-1}. In contrast to reconstituted GOx units on flat Au-electrodes that lack direct electrical contact with electrodes, the Au-nanoparticles serve as electron mediating units that transport the charge between the active center and the bulk electrode surface. Figure 3.19B shows the cyclic voltammograms that correspond to the bioelectrocatalyzed oxidation of glucose by the Au-nanoparticle-reconstituted GOx linked to the bulk Au-electrode by the benzene dithiol bridge (**17**). The inset in Figure 3.19B corresponds to the calibration curve that corresponds to the resulting electrocatalytic anodic current at variable concentrations of glucose. From the observed saturation current, and knowing the surface coverage of the biocatalyst, the maximum ET turnover rate is calculated to be $TR_{max} = 5000$ s^{-1}. This value is ca sixfold higher than the electron-transfer turnover rate of the enzyme with its native O_2 electron acceptor. This unprecedented effective electrical contacting of the enzyme's redox center with the electrode has important consequences on the functional properties of the electrode. The electrode was found to be insensitive to common glucose-sensing interferants such as ascorbic acid or uric acid, and it was unaffected by the presence of oxygen. The mechanistic analysis of the electron-transfer process in the Au-nanoparticle/reconstituted enzyme system revealed that the ET transport through the molecular bridge linking the Au-nanoparticle to the electrode is the rate-limiting process. It was found that benzene dithiol, (**17**), was the most efficient electron transporting bridge, whereas the biphenyl dithiol, (**18**), and the p-xylene dithiol, (**19**), were less efficient bridging elements. This was attributed to the different electron tunneling properties of the different bridges, while the conjugated planar π-system in (**18**) and the SP^3 methylene units in (**19**) increase the tunneling barriers, resulting in slower ET turnover rates. The results suggest that the use of conductive molecular wires such as oligophenyl acetylenes to bridge the functional nanoparticles with the surface could further enhance the electrical contacting efficiency between the reconstituted enzyme and the electrode. Also, the use of other electrical contacting units such as carbon nanotubes for the reconstitution of the enzyme on electrodes could be an attractive method to generate integrated electrically contacted enzyme-electrodes. Recent theoretical studies [78] have supported the conductivity pattern of the molecular wires (**17**), (**18**) and (**19**).

The biopumping of electrons into Au-nanoparticles functionalized with reconstituted glucose oxidase has been examined by SPR and electrochemical methods [79]. The charging of the Au-nanoparticles with electrons is known to significantly alter the localized surface plasmon (LSP) [80] and thus to influence the coupling between the localized surface plasmon and the surface plasmon wave of a bulk Au-surface [81, 82].

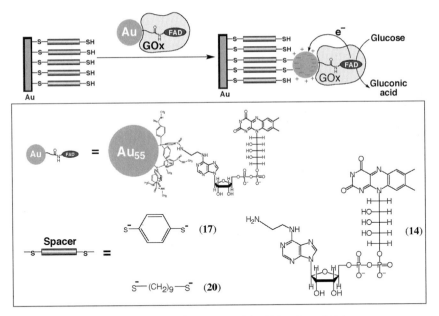

Fig. 3.20 Assembly of GOx-reconstituted Au-nanoparticles (1.2 nm) on dithiol tunneling barriers associated with Au-electrode supports.

The enzyme glucose oxidase, GOx, was reconstituted on Au-nanoparticles (1.2 nm) functionalized with the flavin adenine dinucleotide that were linked to a bulk Au-surface with the long-chain 1,9-nonanedithiol (**20**) or 1,4-benzene dithiol, (**17**), Figure 3.20. The monolayer film separating the Au-nanoparticles from the bulk surface provides a tunneling barrier for electron transfer from the nanoparticle to the bulk surface. In the presence of glucose, the bioelectrocatalyzed oxidation of the substrate leads to the steady state charging of the Au-nanoparticles. The steady state charging of the particles resulted in changes in the dielectric properties, and consequently the refractive index features of the interface. These interfacial perturbations led to shifts in the surface plasmon–resonance reflectivity angles of the bulk Au-surface, Figure 3.21A. The shifts in the minimum reflectivity angles in the SPR spectra were found to be controlled by the concentration of glucose. Upon exclusion of the medium effects on the SPR spectra, the angle shifts originating from added glucose could be derived, and a calibration curve that correlates the shifts in the minimum reflectivity angle, as a result of charging the Au-nanoparticles, and the glucose concentrations, was extracted, Figure 3.21B. It was found that the charging process was enhanced as the dielectric tunneling barrier separating the Au-nanoparticle and the bulk electrode was higher. Interestingly, no charging of the Au-nanoparticles, and no shifts in the minimum reflectivity angles of the SPR spectra were observed for Au-nanoparticles linked to the bulk electrode with (**20**) as a linker, and randomly functionalized with GOx, upon the addition of glucose. Thus, the electrical contacting of the FAD redox-sites of the enzyme with the Au-nanoparticles by the reconstitution process was found to have a key role in

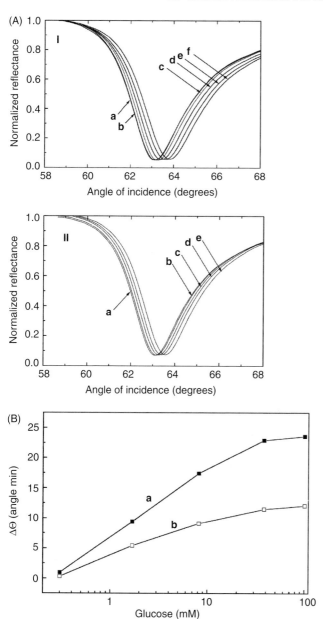

Fig. 3.21 (A) SPR spectra corresponding to the GOx-reconstituted Au-nanoparticle interfaces assembled on the tunneling barriers: (I) (**20**) and (II) (**17**), in the presence of different concentrations of glucose: (**a**) 0 mM, (**b**) 0.3 mM, (**c**) 1.6 mM, (**d**) 8 mM, (**e**) 40 mM, (**f**) 100 mM. (B) Calibration curves corresponding to the changes in the minimum reflectivity angles in the SPR spectra at different glucose concentrations in the presence of (**a**) (**20**) as a linker, (**b**) (**17**) as a linker.

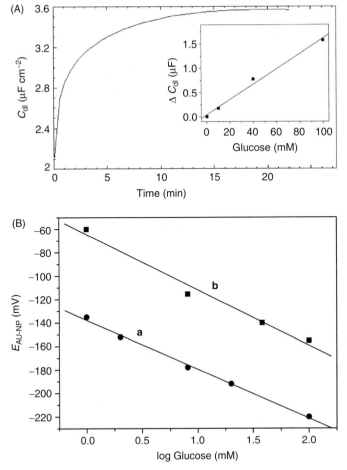

Fig. 3.22 (A) Time-dependent capacitance changes of the GOx-reconstituted Au nanoparticle–surface assembled on a Au-electrode using (**20**) as a linker upon the addition of glucose, 100 mM. *Inset*: Calibration curve corresponding to the capacitance changes of the modified electrode in the presence of different concentrations of glucose. (B) Potentials generated on the GOx-reconstituted Au-nanoparticle electrodes using (a) (**20**) as a linker, (b) (**17**) as a linker, in the presence of different concentrations of glucose.

the charging of the nanoparticles. The biocatalytic charging of the Au-nanoparticles was further characterized by following the capacitance of the GOx/Au-nanoparticle interface, and by monitoring the potential generated on the bulk electrode. It was found that the charging of the Au-nanoparticles increased the capacitance of the interface by ca 80%, Figure 3.22A, and that the capacitance changes related to the glucose concentration in the system, Figure 3.22A, inset, and to the extent of charging of the nanoparticles. Similarly, the potentials generated on the bulk electrode as a result of charging of the nanoparticles was found to be controlled by the concentration of glucose, Figure 3.22B. By the simultaneous analysis of the charging of the Au-nanoparticles by the external potential biasing of the bulk Au-electrode, the degree of charging of the Au-nanoparticles was quantitatively assayed. It was found that at a concentration of glucose that corresponded to 1×10^{-2} M, the steady state charging of the Au-nanoparticles lead to an average number of 10 and 6 electrons per particle using (**20**) and (**17**) as tunneling barriers, respectively.

3.2.4
Integrated Electrically Contacted Electrodes Composed of Reconstituted Quinoproteins

Redox proteins that include quinone cofactor units play important roles in biological ET processes [83, 84]. Some of the quinoproteins include the quinone cofactor in a noncovalently linked configuration, such as the PQQ dependent enzymes, whereas other quinoproteins include the quinone cofactor covalently linked to the protein, for example, topaquinone (2,4,6-trihydroxyphenyl-alanine quinone, TPQ) dependent enzymes. A number of quinoproteins include in addition to the quinone cofactor an ET cofactor unit in another protein subunit. These cofactors may be metal ions or a cytochrome-type heme cofactor such as D-fructose dehydrogenase that is a heme containing PQQ-dependent enzyme [85].

The reconstitution of apo-glucose dehydrogenase on a PQQ-monolayer-functionalized Au-electrode led to an active configuration of the enzyme on the electrode support, but it lacked direct electrical contact with the electrode surface [86]. An integrated electrically contacted glucose dehydrogenase electrode was organized [87] by the reconstitution of the apoenzyme on PQQ units linked to a polyaniline/polyacrylic acid redox-active polymer composite associated with an electrode, Figure 3.23. Electropolymerization of aniline in the presence of polyacrylic acid led to the formation of a polyaniline/polyacrylic acid composite film on the electrode that exhibited redox activity at neutral pH values. The covalent coupling of 1,4-diaminobutane to the carboxylic acid residues, followed by the attachment of the PQQ cofactor units and the reconstitution of apo-glucose dehydrogenase on the PQQ sites led to an electrically contacted enzyme-electrode for the bioelectrocatalyzed oxidation of glucose. The surface coverage of the PQQ units and the enzyme components on a 90-nm thick polymer film was estimated to be 1×10^{-11} mol cm^{-2} and 2×10^{-12} mol cm^{-2}, respectively. The bioelectrocatalyzed oxidation of glucose was followed electrochemically and by *in situ* electrochemical surface plasmon resonance measurements [87]. Figure 3.24 exemplifies the cyclic voltammograms that correspond to the bioelectrocatalyzed oxidation of glucose by the integrated electrode. The resulting calibration curve is depicted in Figure 3.24, inset. It was assumed that the redox-active polyaniline chains penetrate the protein matrix and reach sufficiently close distances in respect to the reconstituted PQQ units that facilitate the ET contact. Control experiments reveal that the direct covalent attachment of the glucose dehydrogenase to the polymer film does not yield an electrically contacted enzyme, implying that the alignment of the biocatalyst on the polymer via the

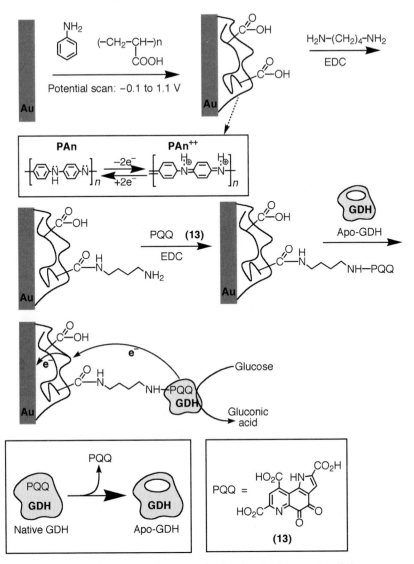

Fig. 3.23 Assembly of the electrically contacted polyaniline/PQQ-reconstituted glucose dehydrogenase electrode.

reconstitution process is important to establish electrical communication between the PQQ redox center and the electrode.

An interesting approach to electrically contact quinoproteins and electrodes involved the use of molecular wires to align the enzyme on the electrode and to conduct electrons between the quinone center and the electrode [88]. Amine oxidase catalyzes the oxidation of amines to aldehydes using copper ions and the topaquinone, TPQ,

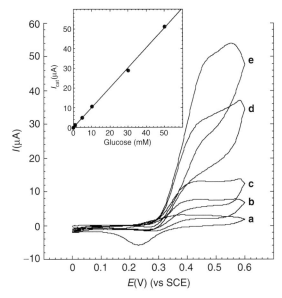

Fig. 3.24 Cyclic voltammograms corresponding to the bioelectro-catalyzed oxidation of different glucose concentrations by the PQQ-reconstituted glucose dehydrogenase polyaniline–functionalized electrode. Glucose concentrations: (**a**) 0 mM, (**b**) 5 mM, (**c**) 10 mM, (**d**) 20 mM, (**e**) 50 mM. *Inset*: Calibration curve corresponding to the electrocatalytic currents as a function of glucose concentration.

as redox centers. The fact that the enzyme is inhibited by diethylaniline that binds to the redox-site pocket suggests that the diethylaniline-terminated oligophenylacetylene thiol (**21**) may act as a wire for the alignment of the enzyme on the electrode and the electrical contacting of its redox center by charge transport through the wire. Indeed, the assembly of the thiol-functionalized molecular wire on a Au-electrode followed by the association of the protein onto the inhibiting diethylaniline units, led to an electrically contacted protein where the TPQ moiety embedded in the protein revealed quasi-reversible electrochemical properties.

(**21**)

3.2.5
Reconstituted Electrically Contacted Hemoproteins

Hemoproteins are a broad class of redox proteins that act as cofactors, for example, cytochrome *c*, or as biocatalysts, for example, peroxidases. Direct ET between peroxidases such as horseradish peroxidase [89], lactoperoxidase [90], or chloroperoxidase [91] and electrode surfaces, mainly carbonaceous materials, were extensively studied. The mechanistic aspects related with the immobilized peroxidases on electrode surfaces and their application as active ingredients in developing biosensor devices were reviewed in detail [92, 93]. The direct electrical contact of peroxidases with electrodes was attributed to the location of the heme site at the exterior of the protein that yields close contact with the electrode surface even though the biocatalyst is randomly deposited on the electrode. For example, it was reported [94, 95] that nonoriented randomly deposited horseradish peroxidase on a graphite electrode resulted in 40 to 50% of the adsorbed biocatalyst in an electrically contacted configuration. For other hemoproteins such as cytochrome c, it was found that the surface modification of the electrodes with promoter units [96, 97] such as pyridine units induced the binding of the hemoproteins in an orientation that facilitated direct electron transfer. The promoter sites bind the protein to the electrode surface by an equilibrium process, and the electron transfer occurring at the electrode surface results in the dissociation of the protein to the bulk solution. For example, an association constant of $K_a = 8.5 \times 10^3$ M^{-1} was reported for the interaction of cytochrome c with thiol pyridine sites linked to a Au-surface [98]. Alternatively, the site-specific covalent attachment of hemoproteins such as cytochrome c resulted in the orientation of the protein on the electrode surfaces and direct ET communication [99]. A series of cytochrome c mutants that include inserted cysteine units in different protein positions were covalently tethered to electrode supports. The interfacial electron transfer between the heme site and the electrode was found to be controlled by the distance separating the cysteine anchoring site and the hemoprotein redox center.

The reconstitution process was adopted for enhancing ET of peroxidase and for establishing direct electrical communication of horseradish peroxidase with electrodes. Apo-horseradish peroxidase (apo-HRP) was reconstituted [100] with a dyad consisting of ferrocene tethered to hemin, (22), Figure 3.25. The resulting reconstituted HRP revealed

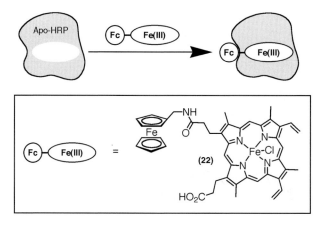

Fig. 3.25 The reconstitution of apo-horseradish peroxidase, apo-HRP, with the ferrocene-tethered hemin (**22**).

enhanced ET features and the oxidation of 2,2′-azinobis(3-ethyl-benzothiazoline-6-sulfonic acid), ABTS, by H_2O_2 was ca 2.5-fold faster in the presence of the (**22**)-reconstituted HRP as compared to the native HRP.

The surface reconstitution of apo-HRP on a hemin-functionalized monolayer was reported to yield a surface-oriented HRP, exhibiting direct ET with the electrode [101]. A mercaptobutanoic acid monolayer was assembled on a Au-electrode, followed by the covalent coupling of 1,12-diaminododecane to the base monolayer. Hemin was then covalently linked to the surface, and apo-HRP was reconstituted on the hemin sites, Figure 3.26. The resulting surface-reconstituted HRP revealed direct ET with the electrode, and the bioelectrocatalyzed reduction of H_2O_2 was activated by the reconstituted HRP.

3.2.6
Reconstituted *de novo* Hemoproteins on Electrodes

Extensive research efforts were directed in recent years toward the synthesis of *de novo* proteins [102, 103]. Specifically, the reconstitution of *de novo* proteins with metal porphyrin attracted interest as a biomimetic approach to tailor synthetic hemoproteins [104, 105].

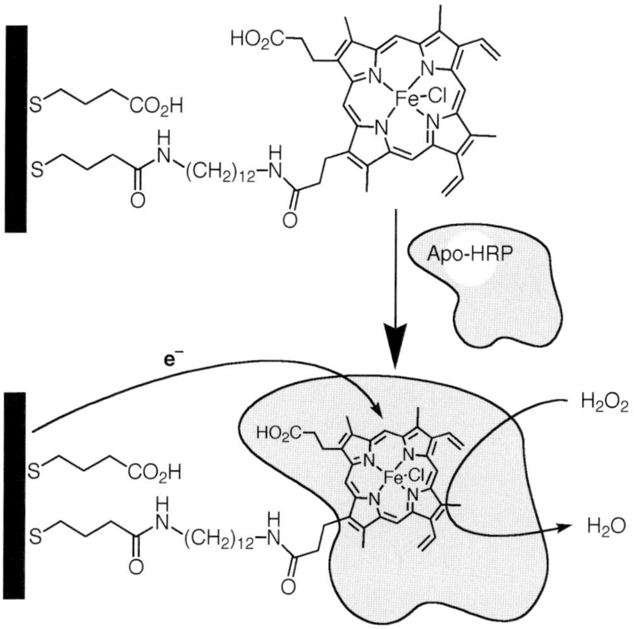

Fig. 3.26 The surface-reconstitution of apo-horseradish peroxidase on a hemin monolayer assembled on a Au-electrode.

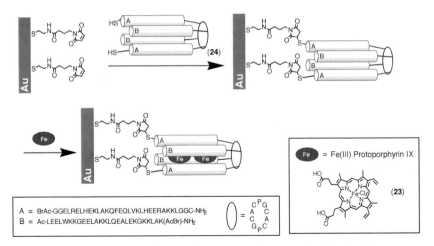

Fig. 3.27 The reconstitution of Fe(III)-protoporphyrin IX, **(23)**, in the four-helix bundle *de novo* protein, **(24)**, bound to a functionalized Au-electrode.

The reconstitution of a tailored *de novo* protein with Fe(III)-protoporphyrin IX units, **(23)**, on an electrode surface generates a hemoprotein-type assembly that mimics ET functions of cytochrome *c* [106, 107]. A four-helix bundle *de novo* protein, **(24)**, consisting of 128 amino acids and a mass of 14 728 was synthesized [108]. It included two pairs of identical helices A and B. The B helices are terminated with Gly-Gly-Cys units for the assembly of the structures on surfaces, whereas the helices A each include two histidines acting as ligand. The position of the histidine ligands on the opposite helices A permits, in principle, the axial ligation of two Fe(III)-heme sites into the protein structure. The four helix–bundle *de novo* protein was assembled as a monolayer on a Au-electrode as depicted in Figure 3.27. Interaction of the functionalized-monolayer electrode with Fe(III)-protoporphyrin IX, **(23)**, resulted in the reconstitution of the protein with the two heme units. The surface coverage of the reconstituted *de novo* hemoprotein was found to be 2.5×10^{-11} mol cm^{-2}. Assuming that the protein footprint area is ca 25 Å^2, the saturated surface coverage of the *de novo* protein corresponds to 3.5×10^{-11} mol cm^{-2}, and thus the experimental value corresponds to ca 70% of a saturated densely packed monolayer. Detailed electrochemical characterization of the heme-reconstituted *de novo* protein indicated that the two heme units integrated in the protein exhibited slightly different redox potentials and different kinetics of reconstitution. The heme site close to the electrode surface exhibited a more negative redox potential that corresponded to −0.43 V versus SCE, and it was reconstituted into the protein slowly, whereas the heme center that was reconstituted in the remote position was incorporated into the protein by a faster process and revealed a slightly more positive redox potential that corresponded to −0.36 V versus SCE. The ordering of the redox potentials of the two heme sites resulted in a vectorial directional electron transfer from the electrode to the first heme site that acted as a relay unit for ET to the second heme center [107].

3.2 Electrodes Functionalized with Reconstituted Redox Proteins | 71

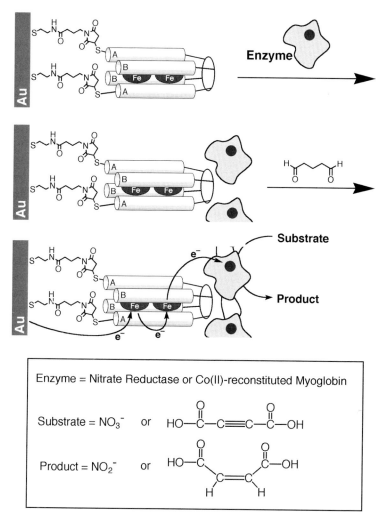

Fig. 3.28 The assembly and bioelectrocatalytic functions of integrated, electrically contacted, electrode consisting of the Fe(III)-protoporphyrin IX-reconstituted *de novo* protein and surface cross-linked redox proteins.

The heme-reconstituted *de novo* protein associated with the electrode was found to act as an effective ET mediator to other hemoproteins. Also, it acted as an artificial heme cofactor that substituted cytochrome *c* in activating cytochrome *c*-dependent enzymes. The cytochrome *c*-dependent enzyme nitrate reductase, NR (E.C. 1.9.6.1 from *Escherichia coli*) was found to form an affinity complex with the surface-reconstituted hemo-*de novo* protein. The cross-linking of the affinity complex generated on the electrode surface with glutaric dialdehyde led to the formation of an integrated bioelectrocatalytic electrode, Figure 3.28. The reduction of the *de novo* hemoprotein

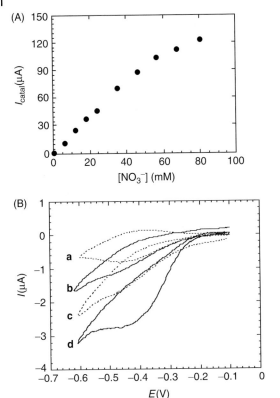

Fig. 3.29 (A) Calibration curve corresponding to the bioelectrocatalytic currents generated by the Fe(III)-protoporphyrin–reconstituted *de novo* protein/nitrate reductase electrode in the presence of different concentrations of nitrate (NO_3^-). (B) Cyclic voltammograms corresponding to the bioelectrocatalyzed hydrogenation of different concentrations of acetylene dicarboxylic acid in the presence of Fe(III)-protoporphyrin IX-reconstituted *de novo* protein/Co(II)-reconstituted electrode. The concentration of acetylene dicarboxylic acid corresponds to (a) 0 M, (b) 5×10^{-6} M, (c) 1×10^{-5} M, (d) 2×10^{-5} M.

activated an ET cascade to NR that induced the bioelectrocatalyzed reduction of nitrate (NO_3^-) to nitrite (NO_2^-). Figure 3.29A shows the derived calibration curve that corresponds to the amperometric responses of the enzyme-electrode at different concentrations of nitrate. This system indicates that the heme-reconstituted *de novo* protein acts as a synthetic cofactor analog that mimics the functions of cytochrome *c* in mediating ET to the biocatalyst. Interestingly, this concept was further broadened by the biomimetic assembly of reconstituted bioelectrocatalytic protein structures that stimulate electrocatalytic transformations that do not proceed in nature. Apomyoglobin was reconstituted with Co(II)-protoporphyin. While the Co(II)-reconstituted myoglobin lacked direct electrical communication with the electrode, it was found that the heme-reconstituted *de novo* hemoprotein activated the Co(II)-reconstituted myoglobin toward the bioelectrocatalytic hydrogenation of acetylene dicarboxylic acid to malic acid. It was found [107] that cross-linking of the affinity complex generated between Co(II)-reconstituted myoglobin and the hemo-*de novo* protein monolayer associated with the electrode yielded an integrated bioelectrocatalytic electrode for the hydrogenation of acetylene dicarboxylic acid. Figure 3.29B shows the cyclic voltammograms generated by the integrated electrode at different concentrations of the acetylenic substrate. Mechanistic analysis of the reaction has revealed that the hemo-*de novo* protein mediates ET to the Co(II)-reconstituted myoglobin. The resulting Co(I)-hydride myoglobin acted as the reactive species in the hydrogenation of acetylene dicarboxylic acid.

3.3
Electrical Contacting of Redox Proteins by Cross-linking of Cofactor-Enzyme Affinity Complexes on Surfaces

The previous sections have addressed the electrical contacting of redox proteins by means of the reconstitution process. The concept was based on the exclusion of the firmly associated cofactor from the protein, and the reconstitution of the resulting apoprotein on cofactor units tethered to the electrode surface. The reconstitution process regenerated a tight cofactor–protein complex on the surfaces, and this facilitated the mediated ET by alignment of the protein on the surface. ET cofactors, such as nicotinamide dinucleotide, NAD^+, nicotinamide dinucleotide phosphate, $NADP^+$, or cytochrome c, act, however, as diffusional ET mediators that form labile equilibrium complexes with the respective redox enzymes. The orientation of the enzyme's redox center in respect to the cofactor in the complex results in mediated ET and the biocatalytic activation of the enzyme. A reconstitution-related method for the electrical contacting of redox enzymes dependent on diffusional cofactors was developed, as schematiclly depicted in Figure 3.30. The cofactor linked to the electrode surface is either directly electrically communicated to the conductive support or is used for communicating the ET between the cofactor and the electrode. The enzyme generates the labile complex with the cofactor interface, and in order to generate an integrated rigid bioelectrocatalytic interface, the complexed proteins are cross-linked with a molecular cross-linker. Since the primary cofactor–enzyme complex exhibits the appropriate orientation for ET, the resulting array is anticipated to reveal bioelectrocatalytic functions. The lateral cross-linking of the enzyme rigidifies the electrically contacted biocatalytic assembly and yields a nondetachable biocatalytic interface on the electrode.

3.3.1
Integrated $NAD(P)^+$-Dependent Enzyme-Electrodes

This method was applied to assemble integrated electrically contacted $NAD(P)^+$-dependent enzyme-electrodes. The direct electrochemical reduction of $NAD(P)^+$ cofactors or the electrochemical oxidation of NAD(P)H cofactors are kinetically unfavored [109]. Different diffusional redox mediators such as quinones, phenazine, phenoxazine, ferrocene or Os-complexes were employed as electrocatalysts for the oxidation of NAD(P)H cofactors [110, 111]. An effective electrocatalyst for the oxidation of the NAD(P)H is PQQ (13), and its immobilization on electrode surfaces led to efficient electrocatalytic interfaces (particularly in the presence of Ca^{2+} ions) for the oxidation of the NAD(P)H cofactors [112, 113]. This observation led to the organization

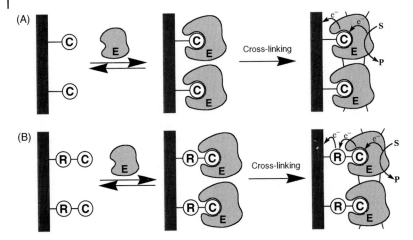

Fig. 3.30 (A) Electrical contacting of redox enzymes by cross-linking cofactor/enzyme affinity complexes on electrode supports. (B) Electrical contacting of redox enzymes by cross-linking cofactor/enzyme affinity complexes on electrode supports using a relay as an ET mediator.

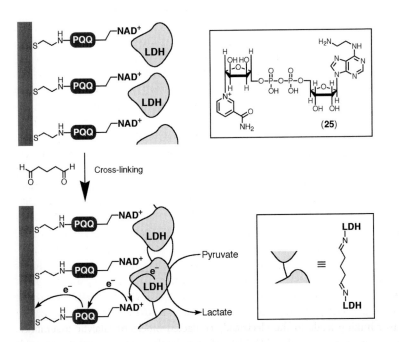

Fig. 3.31 Assembly of an integrated lactate dehydrogenase, LDH, electrode for the bioelectrocatalyzed oxidation of lactate by the surface cross-linking of an affinity complex formed between LDH and a PQQ-NAD$^+$ monolayer assembled on an Au-electrode.

of integrated electrically contacted enzyme-electrodes as depicted in Figure 3.31 for the organization of a lactate dehydrogenase electrode [114]. The PQQ, (13), cofactor was covalently linked to a cystamine monolayer associated with a Au-electrode. The modified cofactor N^6-(2-aminoethyl)-NAD$^+$, (25), was then linked to the PQQ monolayer. The NAD$^+$-dependent enzyme lactate dehydrogenase, LDH, was further interacted with the modified electrode to yield the NAD$^+$-LDH affinity complex on the electrode support. Cross-linking of the labile surface-associated complex with glutaric dialdehyde led to the formation of rigid and firmly integrated electrically contacted LDH electrodes [113]. The integrated electrode revealed bioelectrocatalytic activities toward the oxidation of lactate to pyruvic acid, and lactate could be amperometrically analyzed by the electrode in the concentration range of 1×10^{-3} M – 1×10^{-2} M. A similar approach was also applied to assemble an electrically contacted NAD$^+$/alcohol dehydrogenase electrode for the electrocatalyzed oxidation of ethanol. The biocatalyzed reduction of the NAD$^+$ cofactors by the lactate or ethanol substrates in the systems is followed by the electrocatalyzed oxidation of NADH by PQQ.

A related method for the preparation of integrated NAD(P)$^+$-dependent enzyme-electrodes employed [48] 3-aminophenyl boronic acid (15) as coupling reagent of the native NAD(P)$^+$ cofactors to the PQQ monolayer, Figure 3.32. This is exemplified by the preparation of the electrically contacted NADP$^+$-dependent malate dehydrogenase, MalD, electrode, Figure 3.32A, and of the integrated, electrically contacted NAD$^+$-dependent lactate dehydrogenase, LDH, electrode, Figure 3.32B. Aminophenyl boronic acid (15) was covalently linked to the PQQ monolayer associated with the electrode. The ribose units of NADP$^+$ or NAD$^+$ that include vicinal hydroxyl functions were bound to the boronic acid ligands to form the respective boronate complexes. The MalD and LDH proteins were then interacted with the PQQ-NADP$^+$ and the PQQ-NAD$^+$ interfaces to form the respective labile complexes. Cross-linking of the respective affinity complexes associated with the electrode with glutaric dialdehyde generated nondetachable electrically contacted enzyme-electrodes. The detailed surface characterization of the systems revealed that the surface coverage of the PQQ and NADP$^+$ units corresponded to 1.4×10^{-10} mol cm^{-2} and 1.8×10^{-10} mol cm^{-2}, respectively, and the surface coverage of MalD was 1.0×10^{-12} mol cm^{-2}, that corresponded to 70% of a random densely packed enzyme monolayer. Similarly, for the integrated LDH electrode, the surface coverage of the PQQ and NAD$^+$ units was estimated to be 1.9×10^{-10} mol cm^{-2} and 2.0×10^{-10} mol cm^{-2}, respectively, and the surface coverage of LDH was 7.0×10^{-12} mol cm^{-2}. Figure 3.33A shows the electrocatalytic anodic currents observed in the presence of the MalD integrated electrode at variable concentrations of malate, and the respective calibration curve (inset). Figure 3.33B depicts the cyclic voltammograms corresponding to the oxidation of variable concentrations of lactate by the integrated LDH-electrode, and the respective calibration curve (inset). These systems reveal that the alignment of the NAD(P)$^+$–dependent enzymes on the NAD(P)$^+$ cofactors and their lateral cross-linking leads to the electrically contacted bioelectrocatalytic assemblies. The biocatalytic reduction of the NAD(P)$^+$ cofactors by the respective substrates yields the reduced NAD(P)H cofactors. The electrocatalytic oxidation of NAD(P)H by PQQ activates the continuous cyclic bioelectrocatalytic functions of the enzymes. Impedance measurements performed on the electrically contacted MalD electrode enabled the

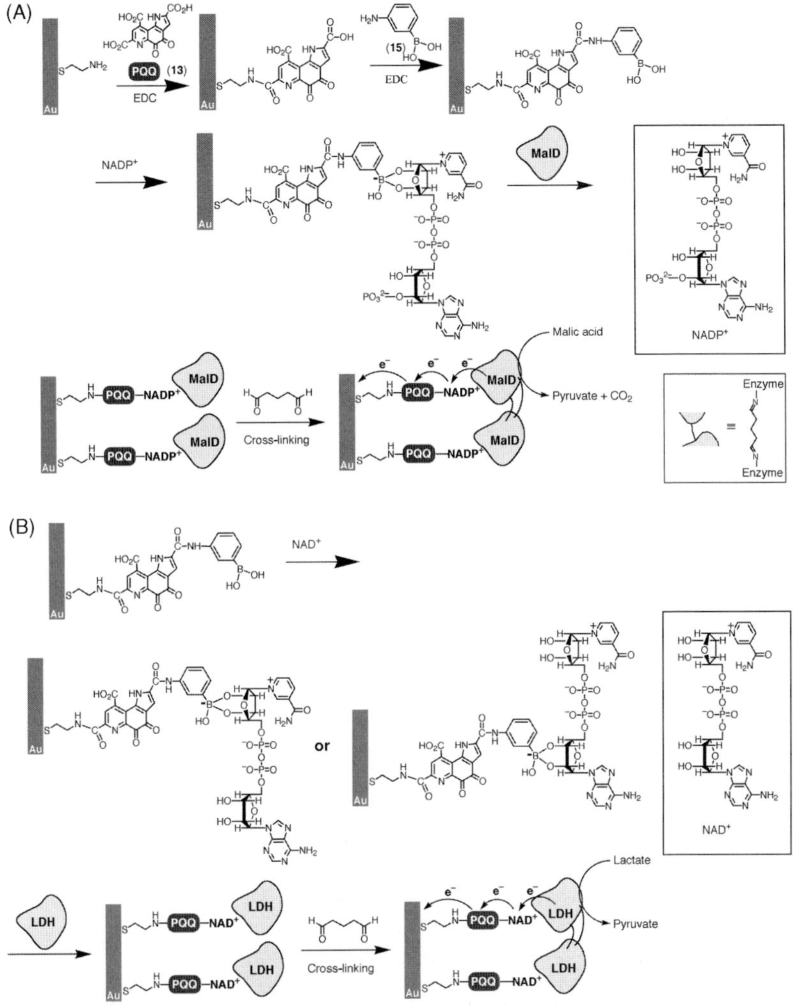

Fig. 3.32 (A) Assembly of an integrated malate dehydrogenase, MalD, electrode for the bioelectrocatalyzed oxidation of malate by the surface cross-linking of an affinity complex formed between MalD and a boronate-linked PQQ-NADP$^+$ monolayer. (B) Assembly of an integrated lactate dehydrogenase, LDH, electrode for the bioelectrocatalyzed oxidation of lactate by the surface cross-linking of an affinity complex formed between LDH and different structures of a boronate-linked PQQ-NAD$^+$ monolayer.

extraction of the heterogeneous electron transfer rate–constant $k_{et} \approx 3.8 \times 10^{-6}$ cm s^{-1}. Knowing the surface coverage of MalD, the overall rate constant reflecting the reaction of malate with the enzyme-electrode was estimated to be $k_{overall} = 8.5 \times 10^5$ M^{-1} s^{-1}.

The alignment and electrical contacting of NAD$^+$-dependent enzymes on electrodes was also accomplished by the generation of the NAD$^+$-enzyme complex and its

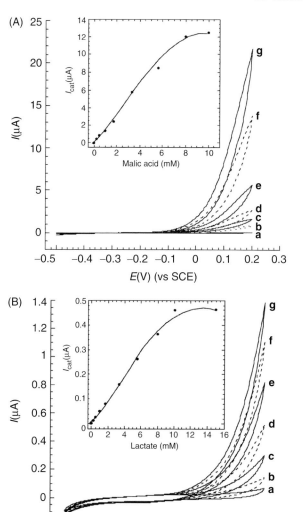

Fig. 3.33 (A) Cyclic voltammograms corresponding to the bioelectrocatalyzed oxidation of different concentrations of malate by the electrode configuration shown in Figure 3.32(A). Malate concentrations are (**a**) 0 mM, (**b**) 0.25 mM, (**c**) 0.5 mM, (**d**) 1 mM, (**e**) 1.7 mM, (**f**) 3.3 mM. (**g**) 5.6 mM. *Inset*: Calibration curve corresponding to the electrocatalytic anodic currents as a function of malate concentration.

(B) Cyclic voltammograms corresponding to the bioelectrocatalyzed oxidation of different concentrations of lactate by the electrode configuration shown in Figure 32(B). Lactate concentrations are (**a**) 0 mM, (**b**) 0.5 mM, (**c**) 1.7 mM, (**d**) 3.3 mM, (**e**) 5.6 mM, (**f**) 8.0 mM, (**g**) 10 mM. *Inset*: Calibration curve corresponding to the electrocatalytic anodic currents as a function of lactate concentration.

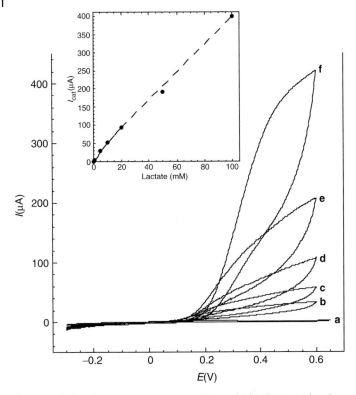

Fig. 3.34 Cyclic voltammograms corresponding to the bioelectrocatalyzed oxidation of different concentrations of lactate by the electrode configuration shown in Figure 23. Lactate concentrations are (a) 0 mM, (b) 5 mM, (c) 10 mM, (d) 20 mM, (e) 50 mM, (f) 100 mM. *Inset*: Calibration curve corresponding to the electrocatalytic currents as a function of lactate concentration.

cross-linking on a conductive, redox-active, polymer that electrically contacts the cofactor-enzyme assembly with the electrode [59]. A polyaniline/polyacrylic acid composite that exhibits redox activities in neutral pH media, was assembled on an Au-electrode, and the N^6-aminoethyl-NAD^+ cofactor, (25), was covalently linked to the acrylic acid residues. The labile complex generated between lactate dehydrogenase, LDH and NAD^+ was cross-linked on the polymer-modified surface. Figure 3.34 shows the cyclic voltammograms of the electrode in the presence of different concentrations of lactate, and the respective calibration curve is depicted in the Figure. From the highest value of the observed current density, 0.27 mA cm^{-2}, and knowing the enzyme content on the electrode, 4×10^{-2} mol cm^{-2}, the electron-transfer turnover rate between the enzyme's redox site and the electrode was estimated to be ca 350 s^{-1}.

The electrical contacting of NAD^+-cofactor dependent enzymes with the macroscopic environment by means of structural alignment of the enzyme on relay-NAD^+ units

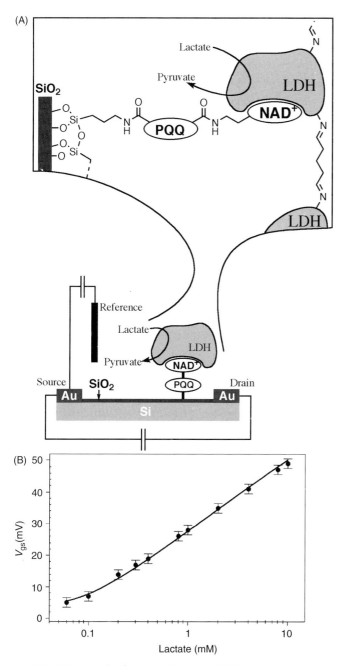

Fig. 3.35 (A) Assembly of a cross-linked relay (PQQ)-cofactor/enzyme (NAD$^+$/LDH) affinity complex on the gate interface of an ISFET device. (B) Gate to source potentials of the enzyme-functionalized ISFET device at variable concentrations of lactate.

was used to develop an enzyme-based field-effect transistor (ENFET) for the analysis of lactate [115]. The field-effect transistor consisted of a thin insulating layer supported on a Si-semiconductor that separates the Au source and drain electrodes, Figure 3.35A. The SiO_2 layer acts as a gate, and its potential, V_{gs}, controls the current flow from the source-to-drain upon the application of an appropriate potential, V_{sd}, between the source and drain electrodes. The SiO_2 gate interface was functionalized with the NAD^+-dependent lactate dehydrogenase by the surface-alignment of the biocatalyst on a relay (PQQ)-NAD^+ (cofactor) interface. This was accomplished by the primary functionalization of the gate with a thin film of 3-aminopropylsiloxane, and the covalent linkage of PQQ (**13**) to the amine residues. Subsequently, the N^6-aminoethyl-NAD^+ cofactor (**25**), was covalently tethered to the PQQ units. The affinity complex between the surface-confined NAD^+ and lactate dehydrogenase was then cross-linked with glutaric dialdehyde to yield an integrated electrically contacted enzyme-based field-effect transistor device. The sequence of reactions that occurs on the gate surface is summarized in Eqs. 7 to 9. In the primary step, Eq. 7, the lactate dehydrogenase, LDH, catalyzed reduction of NAD^+ by lactate occurs. The subsequent oxidation of the gate-confined NADH by the PQQ relay yields the reduced quinone, $PQQH_2$, Eq. 8, that is reoxidized by oxygen to yield PQQ and H_2O_2, Eq. 9. This sequence of reactions generates a steady state ratio of $PQQ/PQQH_2$ on the gate interface and this controls the potential of the gate interface. As the concentration of lactate increases, the $PQQ/PQQH_2$ ratio is anticipated to decrease. The function of the ENFET device is depicted in Figure 3.35B, where the changes in the gate-to-source potential required to retain a constant current flow to the device upon analyzing different concentration of glucose are presented.

$$CH_3\overset{\overset{\displaystyle OH}{|}}{CH}CO_2H + NAD^+ \xrightarrow{LDH} CH_3\overset{\overset{\displaystyle O}{\|}}{C}CO_2H + NADH + H^+ \tag{7}$$

$$PQQ + NADH + H^+ \xrightarrow{Ca^{2+}} PQQH_2 + NAD^+ \tag{8}$$

$$PQQH_2 + O_2 \longrightarrow PQQ + H_2O_2 \tag{9}$$

3.3.2
Integrated Electrically Contacted Hemoprotein Electrodes

Cytochrome *c*, Cyt *c*, represents a further diffusional cofactor that activates many ET processes with proteins. While cytochrome *c* lacks direct electrical contact with electrodes, the modification of electrodes with promoter molecules such as pyridine proved as an effective means to establish electrical communication between Cyt *c* and the electrode [7]. It was suggested that the interaction of Cyt *c* with the promoter units via a diffusional equilibrium process aligns the hemoprotein on the electrode surface in a configuration that facilitates ET between the heme center and the electrode. A different approach to establish direct ET between Cyt *c* and the electrode included the use of the wild type Cyt *c* (from *Saccharomyces cerivisiae*), that includes a single cysteine residue, or Cyt c mutants with specific cysteine residues introduced by genetic engineering, for the directional immobilization of the hemoprotein on Au-electrodes [57]. The electrical contacting of Cyt c by the covalent linkage of the hemoprotein to the electrode enables the organization of integrated, electrically contacted Cyt c-enzyme-electrodes [116]. For example, a Cyt c-cytochrome oxidase, COx, integrated electrode for the bioelectrocatalyzed reduction of O_2 to water was developed [116], Figure 3.36A. The wild type Cyt c was covalently tethered to a maleimide-functionalized monolayer

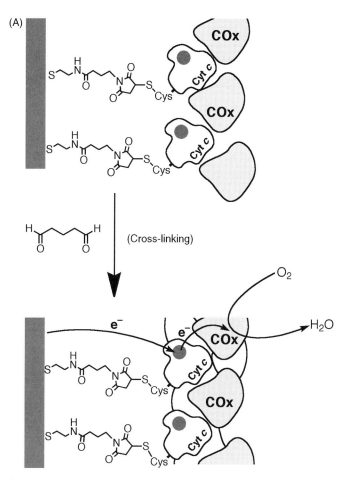

Fig. 3.36 (A) The assembly of an integrated electrically contacted electrode for the reduction of O_2 to water generated by the cross-linking of a cytochrome c/cytochrome oxidase (Cyt c/COx) monolayer associated with the electrode. (B) The organization of a function electrode for the bioelectrocatalyzed hydrogenation of acetylene dicarboxylic acid, (**26**), to maleic acid, (**27**), by the surface cross-linking of an affinity complex formed between Co(II)-reconstituted myoglobin and an MP-11 monolayer associated with an electrode.

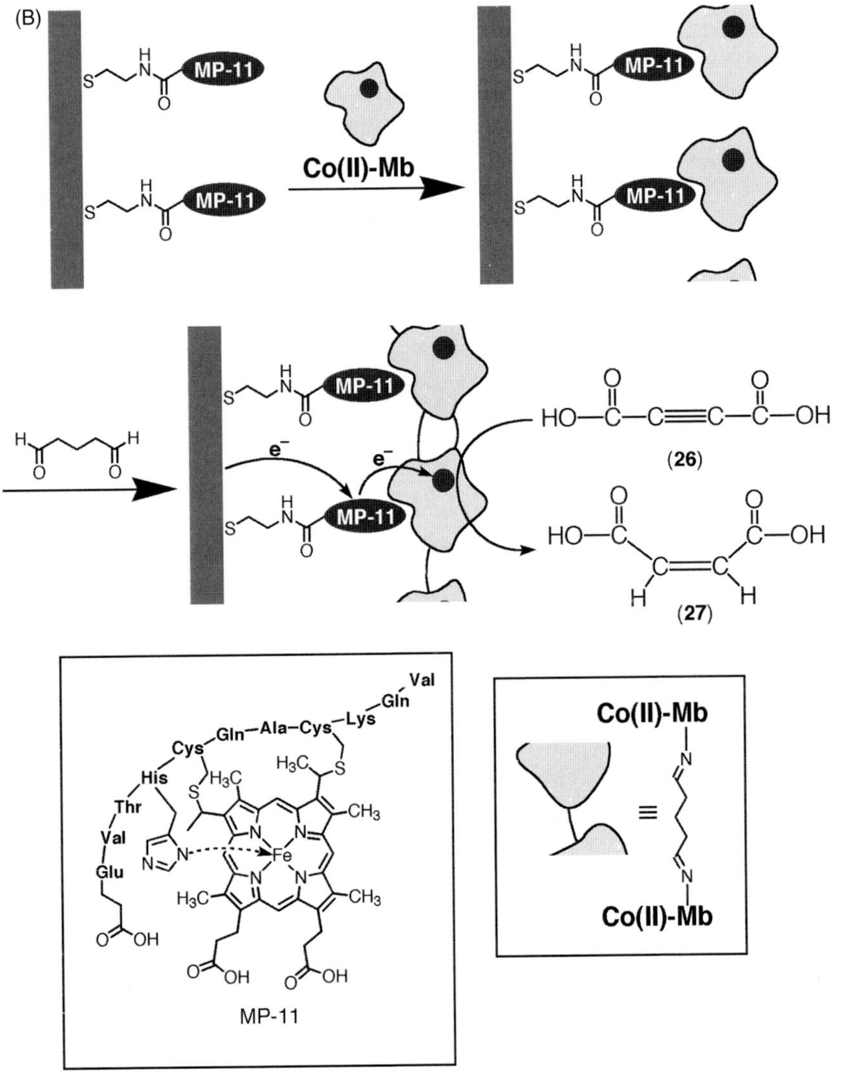

Fig. 3.36 (Continued)

assembled on a Au-electrode. The protein–protein complex generated between Cyt c associated with the surface and COx was cross-linked with glutaric dialdehyde to yield the Cyt c-COx electrically contacted electrode that acted as an effective surface for the bioelectrocatalyzed reduction of O_2 to H_2O.

Microperoxidase-11, MP-11, is a heme-undecapeptide that is prepared by the digestion of cytochrome c and it includes the active surrounding of cytochrome c [117]. MP-11 was immobilized on electrode surfaces and its electrochemistry was characterized [118]. The MP-11-modified electrodes were reported [119] to act as effective electron mediator interfaces for the reduction of cytochrome c, hemoglobin, myoglobin and nitrate reductase (cytochrome c-dependent). The MP-11-mediated activation of nitrate reductase, NR, was employed to assemble an integrated MP-11/NR electrode for the bioelectrocatalyzed reduction of nitrate (NO_3^-) to nitrite (NO_2^-). An affinity complex between an MP-11-functionalized electrode and NR ($K_a = 3.7 \times 10^3$ M^{-1}) was cross-linked with glutaric dialdehyde to yield the electrically contacted electrode for the bioelectrocatalyzed reduction of nitrate to nitrite. In this system, the reduced MP-11 mediates ET to NR and activates the enzyme toward the reduction of NO_3^-.

A related study employed [120] an MP-11 monolayer-functionalized electrode for the electrochemical activation of Co(II)-protoporphyrin IX toward the hydrogenation of acetylene dicarboxylic acid, (**26**), in an integrated bioelectrocatalytic electrode configuration, Figure 3.36B. A MP-11 monolayer was linked to an Au-electrode. It was found [118] that the association of MP-11 to the electrode surface occurs by two binding modes that include the covalent linking of the terminal carboxylic acid functionality of the peptide, or the binding of carboxylic residues that substitute the heme site. Co(II)-protoporphyrin IX-reconstituted myoglobin, Co(II)-Mb, was then interacted with the MP-11 interface, and the resulting complex was cross-linked with glutaric dialdehyde. The modified electrode revealed bioelectrocatalytic activities toward the hydrogenation of acetylene dicarboxylic acid (**26**) to maleic acid (**27**). Mechanistic studies revealed that the reduced MP-11 mediated the reduction of Co(II)-Mb to a Co(I)-hydride-Mb species that was the active intermediate in the hydrogenation process. The electrocatalyzed hydrogenation proceeded with a current yield that corresponded to ca 80%.

3.4
Reconstituted Enzyme-Electrodes for Biofuel Cell Design

The previous sections outlined the importance of electrically contacted enzymes for the designing of amperometric or potentiometric biosensor systems, and for the tailoring of bioelectrocatalytic interfaces for synthetic applications. A great promise for electrically contacted enzyme-electrodes is, however, their use as functional components of biofuel cell devices.

The development of biofuel cells that catalyze the conversion of chemical energy stored in abundant organic raw materials, or biomass, to electrical energy is a continuing challenge in bioelectronics [18].

Numerous waste materials such as lactic acid, or biomass products such as ethanol, are essentially fuel products. While their spontaneous burning in the presence of O_2 is thermodynamically feasible, the process is prohibited by kinetic barriers. The use of appropriate catalysts, and specifically biocatalysts immobilized on electrode supports, may, however, overcome the kinetic barriers, and lead to the conversion of the chemical energy stored in the organic material into electrical energy. Figure 3.37 schematically depicts the configuration of a biofuel cell device. The anode consists of an electrically contacted enzyme that stimulates the biocatalyzed oxidation of the substrate, resulting in the transfer of electrons from the active site through the mediator, R_1, to the electrode.

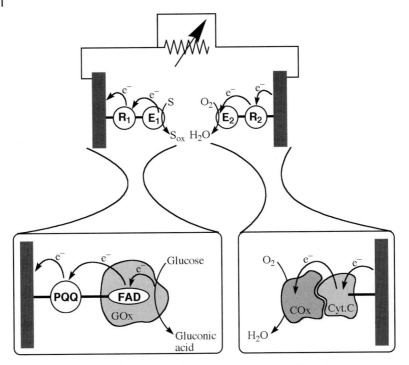

Fig. 3.37 Schematic configuration of a biofuel cell element that utilizes glucose as a fuel and O_2 as the oxidizer.

The cathode in the system consists of a biocatalyst that reduces O_2 to water by an enzyme that is electrically contacted to the electrode through the electron mediator R_2 (other oxidants such as H_2O_2 may be applied similarly in the biofuel cell). The linkage of the anode with the cathode through an external circuit results in the formation of an electrical current, or the development of a potential across an external resistor placed in the external circuit. The electrical power output of the biofuel cell is given by Eq. 10, where V_{oc} is the open-circuit potential and I_{sc} is the short circuit current generated between the electrodes. The maximum potential that is extractable from the biofuel cell corresponds to the difference between the thermodynamic redox potential corresponding to the oxidation of the fuel substrate and the reduction of the oxidizer at the biocatalytic anode and cathode, respectively.

$$P = V_{oc} \cdot I_{sc} \tag{10}$$

Nonetheless, since the electron mediators, R_1 and R_2, are used as electron shuttles that electrically contact the redox center and the respective electrodes, the difference in the reduction potentials of the intermediate electron mediators control the extractable potential from the biofuel cell. Thus, for efficient power output electron mediators operating at potentials close to the active sites of the enzymes associated with the anode and cathode, respectively, should be selected. The current output of the biofuel cell is controlled by the loading degree of the respective electrodes with the biocatalysts, and most importantly, by the interfacial electron transfer between the biocatalysts and the respective electrodes. The slowest interfacial electron-transfer rate at either the anode or the cathode dominates the extractable current from the device. The increase of the content of the electrically contacted enzymes on the electrodes may be accomplished by physical means such as increasing the electrode area by roughening, or by the deposition of thin enzyme films on the electrode surface. On the other hand, enhancing the interfacial electron-transfer kinetics at the electrode surfaces requires optimal electrical contacting of the enzyme redox centers and the electrodes. This improved electrical communication between the enzyme redox centers and the electrodes may then be achieved by the nanoscale alignment of the biocatalyst on the electrode by surface reconstitution.

The use of body fluids such as blood as a source of fuel substrates, for example, glucose, is particularly attractive. The use of glucose oxidizing electrodes and O_2 reducing electrodes as an implanted biofuel cell device may utilize the body fluid as a fuel reservoir for generating electricity that could activate prosthetic units, pace makers or insulin pumps. The advances in electrically contacting the enzyme glucose oxidase (GOx) with electrodes by means of the surface reconstitution method [47, 48], and the development of an integrated electrically contacted cytochrome c/cytochrome oxidase (COx) electrode for the reduction of O_2 to water [116], led to the assembly of a noncompartmentalized glucose-O_2 biofuel cell [116].

Figure 3.37 depicts the schematic configuration of the biofuel cell that includes the reconstituted glucose oxidase (GOx) on the PQQ-FAD monolayer electrode as anode and the cross-linked Cyt c/COx affinity complex on the Au-electrode as cathode. Glucose is used as the fuel substrate in the system, whereas O_2 acts as the oxidizer. The biocatalyzed oxidation of glucose to gluconic acid injects electrons into the electrode (anode), whereas electrons from the cathode are transferred to O_2 via the Cyt c/COx biocatalyst. The current flow through the external circuit, or the formation of a potential difference across the external resistor, provides the route for the conversion of the chemical energy released upon the biocatalytic oxidation of glucose into electrical energy. Figure 3.38 shows the current-voltage behavior of the biofuel cell at different external loads. The power of the cell at different external loads is shown in Figure 3.38, inset. The maximum power output is 4 µW at an external load of 0.9 kΩ. The ideal voltage–current relationship for an electrochemical generator is rectangular. The experimental voltage–current curve deviates from the rectangular shape and the fill factor, $f = P_{max} I_{sc}^{-1} V_{oc}^{-1}$, corresponds to ca 40%. The power output of the cell is low, but the fundamental understanding of the features of the system may stimulate routes to improve the system. In the configuration of the biofuel cell presented in Figure 3.37, the anode and cathode consisted of enzymes in monolayer assemblies. Increasing the enzyme content by roughening of the electrode surface and/or the reconstitution of the biocatalyst in a thin film configuration rather than a by monolayer, may enhance the power output. Indeed, in several studies the content of electrically contacted enzymes associated with electrodes was increased by the incorporation of the biocatalysts in redox-active polymer hydrogels [121, 122]. Also, improved biocatalysts such as laccase [123]

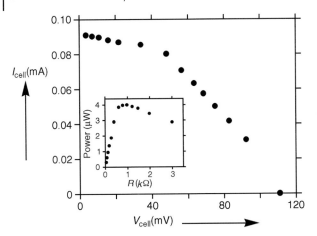

Fig. 3.38 *I–V* curves at different external resistances of the biofuel cell configuration shown in Figure 3.37. Glucose concentration corresponds to 1 mM. *Inset*: Extracted electrical power from the biofuel cell at different external resistances.

or bilirubin oxidase [124] for the reduction of O_2 to water at electrode surfaces were reported. For the system described in Figure 3.37, the ET turnover rate at the anode is ca 650 s^{-1}, while the ET turnover rate at the Cyt c/COx cathode is only ca 20 s^{-1}. Thus, the electrocatalytic reduction of O_2 is the rate-limiting step in the current generation process. The substitution of the biomaterial on the cathode, that is, by laccase, and the organization of the reconstituted enzymes on roughened electrodes, may thus increase the power output from the cell.

Recently, biofuel cell systems were suggested as self-powered biosensor devices [125]. The fact that the voltage output of the cell is controlled by concentrations of the fuel suggests that the open-circuit voltage of the cell may be used as a signal for the sensing of the fuel product. Indeed, the biofuel cell consisting of the PQQ/reconstituted GOx as anode, and the Cyt c/COx as cathode (Figure 3.37) was used as a self-powered glucose-sensing system. Figure 3.39A shows the open-circuit voltages of the cell in the presence of variable concentrations of glucose. Similarly, a biofuel cell configuration consisting of an integrated, electrically contacted, lactate dehydrogenase, LDH, electrode acting as the anode for the oxidation of lactate, and of the Cyt c/COx cathode, was employed for the self-powered detection of lactate [125]. Figure 3.39B shows the open-circuit voltages of the biofuel cell at variable concentrations of lactate. The derived calibration curves corresponding to the sensing of glucose and lactate by the two biofuel cell structures are depicted in Figure 3.39, inset. Thus, the two electrodes that are the active components of the biofuel cells may be implanted into body fluids for the continuous monitoring of glucose or lactate levels. Such methods of glucose and lactate sensing may eliminate the perturbation of the read-out signal by interferants since the potentials at the electrodes are essentially determined by the biocatalytic processes.

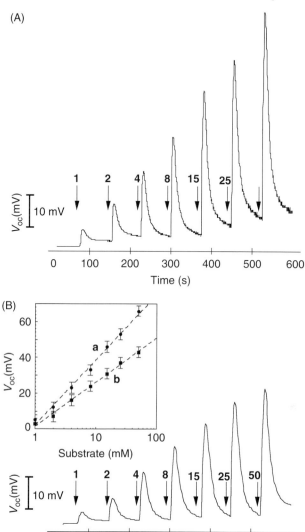

Fig. 3.39 (A) Open-circuit voltages of a glucose-based biofuel cell consisting of the anode and cathode shown in Figure 3.37 in the presence of 1 to 50 mM concentrations of glucose. (B) Open-circuit voltages of a lactate-based biofuel cell consisting of the anode and cathode configurations shown in Figures 3.31 and 3.37 (right side) in the presence of 1 to 50 mM of lactate. *Inset*: Calibration curves corresponding to the analysis of: (a) glucose and (b) lactate by the biofuel cell–based systems.

The electrochemical switching of bioelectrocatalytic transformations at electrode surfaces was recently developed by the organization of the electrically contacted enzymes in polymers of electroswitchable conductivity. These systems were used for designing electronically activated biosensors and electroswitchable biosensor devices [126]. The electrochemically induced generation of conductive Cu^0-clusters or their dissolution, provides the mechanism for the electrical contacting of the redox enzymes with the electrodes, Figure 3.40. A polyacrylic acid/Cu^{2+} film was coated by a protecting polyethyleneimine (PEI) layer, and the PQQ (**13**) and the amino-FAD cofactor (**14**) were covalently linked to the PEI layer. Apo-glucose oxidase, apo-GOx, was then reconstituted onto the FAD cofactor sites, Figure 3.40A. Similarly, Cyt c was covalently linked to the PEI coating and the affinity complex of the Cyt c/cytochrome oxidase (COx) complex was cross-linked on the modified polymer film, Figure 3.40B. Electrochemical reduction ($E = -0.5$ V) of the Cu^{2+} ions in the polymer films associated with the electrode assemblies generated Cu^0 metallic clusters that provided the electrical contact between reconstituted GOx or the cross-linked Cyt c/COx complex and the respective electrodes.

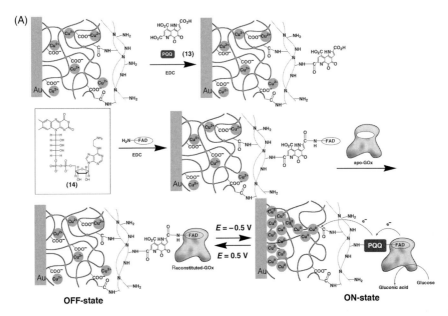

Fig. 3.40 (A) The assembly of an electroswitchable, electrically contacted, glucose oxidase electrode by the reconstitution of apo-glucose oxidase, apo-GOx, on PQQ-FAD units linked to a polyacrylic acid film that includes Cu^{2+}-ion attached to the film. The conductivity of the film and the electrical contacting of the enzyme are accomplished by applying a potential of -0.5 V vs SCE and the generation of Cu-clusters in the film. (B) The assembly of an electroswitchable, electrically contacted electrode for the bioelectrocatalyzed reduction of O_2 consisting of a cross-linked affinity complex of cytochrome c/cytochrome oxidase (Cyt c/COx) on a polyacrylic acid-polyethyleneimine film that includes incorporated Cu^{2+}-ions. The conductivity of the film and the electrical contacting of the redox proteins are accomplished by applying a potential of -0.5 V vs SCE and the generation of Cu-clusters in the film.

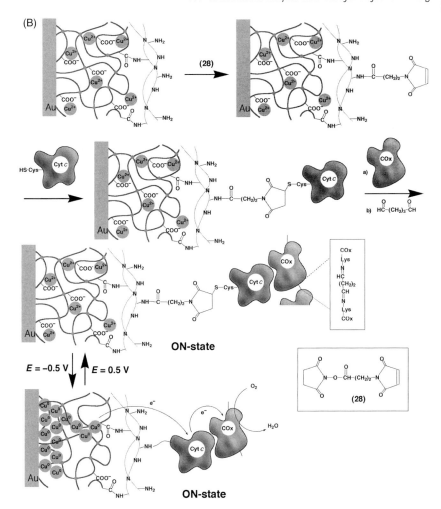

Fig. 3.40 (Continued)

Upon the oxidation of the Cu^0 clusters ($E = 0.5$ V) to the Cu^{2+} ions complexed to the polyacrylic acid film, the electrical contact between the redox proteins was blocked. Figure 3.41A shows the scanning electron micrograph of the conductive Cu^0-nanowire assembly generated in the polymer film upon the reduction of the Cu^{2+}-ions. A three-dimensional network of interconnected Cu-nanowires of a thickness that corresponds to ca 280 nm is generated in the polymer. This leads to the three-dimensional conductivity of the film. Figure 3.41B shows the "ON" and "OFF" short circuit current developed in the biofuel cell in the presence of an 80-mM glucose solution as a fuel and upon the reduction of the film to the Cu^0-polyacrylic acid state and the oxidation to the Cu^{2+}-polyacrylic acid configuration respectively. The switched-on state of the biofuel cell can be applied not only for the generation of electrical power but also as a potentiometric

3 Reconstituted Redox Enzymes on Electrodes

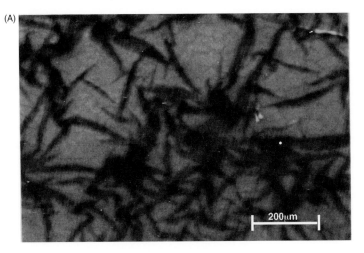

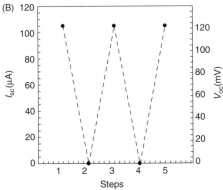

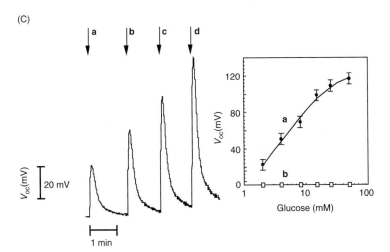

Fig. 3.41 (legend see page 91)

biosensor for glucose. Figure 3.41C shows the open-circuit voltage of the biofuel cell in the presence of different glucose concentrations and the respective calibration curve for analyzing glucose. Such electroswitchable biofuel cells may find important future applications as implantable electrical power generator devices that use a body fluid (blood) as the fuel source, that is, the consumption of the fuel of the body fluid (e.g. glucose in blood) will occur only upon the electrical activation of the biofuel cell. One may argue that the current flow could be switched off by the application of a high external resistor. Nonetheless, this latter method would result in the development of voltage on the anode and cathode. This may activate nonspecific electrochemical processes of other ingredients in the body fluid, a process that is eliminated by the switchable electrical triggering of the tailored electrodes of the biofuel cell.

3.5
Conclusions and Perspectives

The present article has described the method of surface reconstitution of apoenzymes on cofactor-modified electrodes, and the surface cross-linking of cofactor-enzyme affinity complexes on electrode supports as means to structurally align and electrically contact the biocatalysts with the electrodes. The method represents a new approach to nano-engineer the electrode surfaces with the enzymes for tailoring the bioelectrocatalytic functionalities of the biocatalysts. The effectiveness of the electrical communication between the redox enzymes and the electrodes was found to be far higher than the electrical contacting methods of the biocatalysts by any other chemical methodology. Besides the fundamental importance of the reconstitution process in the basic understanding of the structural factors that control electron transfer between the biocatalyst and the electrode, the method has important practical implications toward the future development of biosensor systems and biofuel cell devices. The high electron-transfer exchange rates between enzymes and electrodes yields high current outputs from the modified electrodes, and this is anticipated to enable the assembly of sensitive biosensor

Fig. 3.41 (A) SEM image of the electrogenerated Cu^0-nanowires in the PAA film. (B) Reversible "ON" and "OFF" switching of the short circuit current, I_{sc} and the open-circuit voltage, V_{oc}, generated by the biofuel cell consisting of the anode and cathode shown in Figure 3.32A and B respectively. The cell output is switched-on at steps 1, 3 and 5 by the application of a potential of −0.5 V vs SCE on both electrodes and the generation of Cu-clusters in the respective polymer films. At points 2 and 4, the biofuel cell is switched off by the application of a potential of +0.5 V vs SCE on both electrodes, which transforms the Cu-clusters into Cu^{2+}-ions. The biofuel cell operated in an air-saturated solution that included 80 mM glucose. (C) Open-circuit voltages (V_{oc}) at variable concentrations of glucose consisting of the Cu-clusters activated anode and cathode shown in Figure 3.40A and B, respectively, upon the injection of different glucose concentrations: (a) 2 mM, (b) 3 mM, (c) 8 mM, (d) 40 mM. *Inset*: Calibration curve corresponding to the open-circuit voltage of the electrochemically activated biofuel cell at variable glucose concentrations. (The system is always saturated with air)

devices, to organize miniaturized biosensors, and to construct efficient biofuel cell elements.

Recent advances in nanotechnology provide new opportunities for the integration of redox proteins and electrode supports. Metal and semiconductor nanoparticles or metallic or other conductive or semiconductive nanowires exhibit dimensions comparable to redox enzymes. Thus, the integration of redox proteins with these nano elements may combine the unique electronic and charge transport properties of nanomaterials with the superb catalytic properties of enzymes to yield hybrid materials of new bioelectrocatalytic functions. Indeed, the recent reports on the electrical contacting of redox enzymes by means of nanoparticles [77], carbon nanotubes [72] or polymeric wires [76] spark imaginative challenging possibilities for the future development of nanostructured enzyme systems. The recent demonstration of the electrical contacting of the redox enzyme glucose oxidase by its reconstitution on polyaniline wires associated with electrodes [76], suggests that other conductive polymers such as polythiophenes may be used as electrical contacting wires. Also, the semiconductive properties of conductive polymers may be coupled with redox enzymes to yield new biomaterial-based transistor systems or photovoltaic devices. Also, the recent report [127] on the fabrication of chiral polyaniline wires suggests that the reconstitution of apo-flavoenzymes on FAD-functionalized chiral polyaniline wires may lead to chiroselective electrical contacting between redox proteins and electrode surfaces. Of particular interest would be the analysis of the electrical contacting phenomenon at the single molecule level. Extensive recent research efforts are directed to the analysis of the bioelectrocatalytic functions of single redox proteins [128] and the development of biomaterial-based transistors [71]. Although these studies are at their infancy, the use of the reconstitution process to align redox proteins on surfaces might be an important step to analyze electron transport properties through proteins at the molecular level. Similarly, the bioelectrocatalytic charging of Au-nanoparticles by surface-reconstituted redox enzymes [79] suggests that systems that "bio-pump" electrons into a single Au-nanoparticle and reveal quantized single-electron charging events and coulomb blockades will be assembled in the near future.

Another important broadening of the reconstitution process is in the area of *de novo* synthesized proteins. The present study has demonstrated the feasibility of organizing bioelectronic systems based on reconstituted *de novo* proteins (see Section 3.2.4). These examples spark the future possibilities in the field. By the reconstitution of new electroactive synthetic cofactors into predesigned *de novo* proteins, new manmade bioelectrocatalysts may be envisaged. In fact, the reconstitution processes might be extended to other biomimetic systems such as pretailored nucleic acids to yield biocatalytic DNAzymes systems. Indeed, the reconstitution of Fe(III)-protoporphyrin IX into a G-quadruplex nucleic acid structure has been reported to act as a biocatalytic assembly that mimics peroxidase function [129]. The reconstitution of nucleic acids (aptamers) with other molecular components such as FAD or NAD(P)$^+$ cofactors, and the integration of these systems with electrode surfaces, are anticipated to lead to new man-made bioelectrocatalytic systems.

References

1. R.A. Marcus, N. Sutin, *Biochim. Biophys. Acta*, **1985**, *811*, 265–322.
2. [a] I. Willner, E. Katz, *Angew. Chem. Int. Ed.*, **2000**, *39*, 1180–1218; [b] A. Heller, *J. Phys. Chem.*, **1992**, *96*, 3579–3587; [c] A. Heller, *Acc. Chem. Res.*, **1990**, *23*, 128–134.
3. P.N. Bartlett, P. Tebbutt, R.G. Whitaker, *Prog. React. Kinetics*, **1991**, *16*, 55–155.
4. [a] D.L. Williams, A.P. Doig, Jr., A. Korosi, *Anal. Chem.*, **1970**, *42*, 118–121; [b] P. Janada, J. Weber, *J. Electroanal. Chem.*, **1991**, *300*, 119–124.
5. [a] M.R. Tarasevich, *Bioelectrochem. Bioenerg.*, **1979**, *6*, 587–597; [b] H.A.O. Hill, I.J. Higgins, *Philos. Trans. R. Soc. London, Ser. A*, **1981**, *302*, 267–273.
6. T. Matsue, N. Kasai, M. Narumi, M. Nishizawa, H. Yamada, H. Uchida, *J. Electroanal. Chem.*, **1991**, *300*, 111–118.
7. [a] F.A. Armstrong, H.A.O. Hill, N.J. Walton, *Acc. Chem. Res.*, **1988**, *21*, 407–413; [b] I. Taniguchi, K. Toyosawa, H. Yamaguchi, K. Yasukouchi, *J. Chem. Soc., Chem. Commun.*, **1982**, 1032–1033.
8. [a] J. Ye, R.P. Baldwin, *Anal. Chem.*, **1988**, *60*, 2263–2268; [b] A.-E.F. Nassar, W.S. Willis, J.F. Rusling, *Anal. Chem.*, **1995**, *67*, 2386–2392.
9. [a] W. Schuhmann, T.J. Ohara, H.-L. Schmidt, A. Heller, *J. Am. Chem. Soc.*, **1991**, *113*, 1394–1397; [b] Y. Degani, A. Heller, *J. Am. Chem. Soc.*, **1988**, *110*, 2615–2620; [c] W. Schuhmann, *Biosens. Bioelectron.*, **1995**, *10*, 181–192; [d] I. Willner, N. Lapidot, A. Riklin, R. Kasher, E. Zahavy, E. Katz, *J. Am. Chem. Soc.*, **1994**, *116*, 1428–1441.
10. I. Willner, A. Riklin, B. Shoham, D. Rivenson, E. Katz, *Adv. Mater.*, **1993**, *5*, 912–915.
11. [a] P.D. Hale, T. Inagaki, H.I. Karan, Y. Okamoto, T.A. Skotheim, *J. Am. Chem. Soc.*, **1989**, *111*, 3482–2484; [b] T. Kaku, H.L. Karan, Y. Okamoto, *Anal. Chem.*, **1994**, *66*, 1231–1235.
12. I. Willner, E. Katz, N. Lapidot, *Bioelectrochem. Bioenerg.*, **1992**, *29*, 29–45.
13. R. Maidan, A. Heller, *Anal. Chem.*, **1992**, *64*, 2889–2896.
14. B.A. Gregg, A. Heller, *J. Phys. Chem.*, **1991**, *95*, 5970–5975.
15. I. Willner, R. Kasher, E. Zahavy, N. Lapidot, *J. Am. Chem. Soc.*, **1992**, *114*, 10 963–10 965.
16. A. Badia, R. Carlini, A. Fernandez, F. Battaglini, S.R. Mikkelsen, A.M. English, *J. Am. Chem. Soc.*, **1993**, *115*, 7053–7060.
17. H.G. Eisenwiener, G.V. Schultz, *Naturwissenschaften*, **1969**, *56*, 563–564.
18. E. Katz, A.N. Shipway, I. Willner, in *Handbook of Fuel Cell Technology* (Eds.: W. Vielstich, S. Lamm, H.A. Gasteiger), Vol. 1, Part 4, John Wiley & Sons, Chichester, **2003**, 355–381.
19. [a] J.R. Winkler, A.J. Di Bilio, N.A. Farrow, J.H. Richards, H.B. Gray, *Pure Appl. Chem.*, **1999**, *71*, 1753–1764; [b] H.B. Gray, J.R. Winkler, *Annu. Rev. Biochem.*, **1996**, *65*, 537–561.
20. L.S. Fox, M. Kozik, J.R. Winkler, H.B. Gray, *Science*, **1990**, *247*, 1069–1071.
21. J.R. Winkler, H.B. Gray, *Chem. Rev.*, **1992**, *92*, 369–379.
22. G. McLendon, *Acc. Chem. Res.*, **1988**, *21*, 160–167.
23. I. Hamachi, S. Shinkai, *Eur. J. Org. Chem.*, **1999**, 539–549.
24. [a] I. Willner, E. Zahavy, V. Heleg-Shabtai, *J. Am. Chem. Soc.*, **1995**, *117*, 542–543; [b] E. Zahavy, I. Willner, *J. Am. Chem. Soc.*, **1996**, *118*, 12 499–12 514.
25. D.R. Casimirio, L.L. Wong, J.L. Colon, T.E. Zewert, J.H. Richards, I.-J. Chang, J.R. Winkler, H.B. Gray, *J. Am. Chem. Soc.*, **1993**, *115*, 1485–1489.
26. I. Hamachi, S. Tanaka, S. Shinkai, *J. Am. Chem. Soc.*, **1993**, *115*, 10 458–10 459.
27. T. Hayashi, Y. Hisaeda, *Acc. Chem. Res.*, **2002**, *35*, 35–43.
28. V. Heleg-Shabtai, T. Gabriel, I. Willner, *J. Am. Chem. Soc.*, **1999**, *121*, 3220–3221.
29. [a] I. Hamachi, Y. Tajiri, S. Shinkai, *J. Am. Chem. Soc.*, **1994**, *116*, 7437–7438; [b] I. Hamachi, T. Nagase, Y. Tajiri,

S. Shinkai, *J. Chem. Soc., Chem. Commun.* **1996**, 2205–2206; [c] I. Hamachi, Y. Tajiri, T. Nagase, S. Shinkai, *Chem. Eur. J.*, **1997**, *3*, 1025–1031.

30 I. Hamachi, T. Matsugi, K. Wakigawa, S. Shinkai, *Inorg. Chem.* **1998**, *37*, 1592–1597.

31 A. Riklin, E. Katz, I. Willner, A. Stocker, A.F. Bückmann, *Nature*, **1995**, *376*, 672–675.

32 [a] E. Katz, I. Willner, J. Wang, *Electroanal.*, **2004**, *16*, 19–44; [b] C.M. Niemeyer, *Angew. Chem. Int. Ed.*, **2001**, *40*, 4128–4158.

33 [a] E. Katz, I. Willner, *ChemPhysChem*, **2004**, *5*, 1084–1104; [b] E. Katz, I. Willner, in *Nanotechnology – Concepts, Applications and Perspectives* (Eds.: C.M. Niemeyer, C.A. Mirkin), Wiley-VCH, Weinheim, Germany, **2004**, Chapter 14, 200–226.

34 [a] P. Mulvaney, *Langmuir*, **1996**, *12*, 788–800; [b] A.P. Alivisatos, *J. Phys. Chem.*, **1996**, *100*, 13 226–13 239; [c] L.E. Brus, *Appl. Phys. A*, **1991**, *53*, 465–474.

35 H. Grabert, M.H. Devoret, (Eds.), *Single Charge Tunneling and Coulomb Blockade Phenomena in Nanostructures*, NATO ASI Ser. B, Plenum Press, New York, **1992**.

36 [a] V. Kesavan, P.S. Sivanand, S. Chandrasekaran, Y. Koltypin, A. Gedanken, *Angew. Chem. Int. Ed.*, **1999**, *38*, 3521–3523; [b] R. Schlögl, S.B.A. Hamid, *Angew. Chem. Int. Ed.*, **2004**, *43*, 1628–1637.

37 C.J. Johnson, E. Dujardin, S.A. Davis, C.J. Murphy, S. Mann, *J. Mater. Chem.*, **2002**, *12*, 1765–1770.

38 [a] C.J. Murphy, N.R. Jana, *Adv. Mater.*, **2002**, *14*, 80–82; [b] H.A. Becerril, R.M. Stoltenberg, C.F. Monson, A.T. Woolley, *J. Mater. Chem.*, **2004**, *14*, 611–616.

39 C.C. Chen, C.Y. Chao, Z.H. Lang, *Chem. Mater.*, **2000**, *12*, 1516–1518.

40 [a] B.D. Busbee, S.O. Obare, C.J. Murphy, *Adv. Mater.*, **2003**, *15*, 414–416; [b] T. Mokari, E. Rothenberg, I. Popov, R. Costi, U. Banin, *Science*, **2004**, *304*, 1787–1790.

41 [a] M.H. Huang, S. Mao, H. Feick, H.Q. Yan, Y.Y. Wu, H. Kind, E. Webber, R. Russo, P.D. Yang, *Science*, **2001**, *292*, 1897–1899; [b] J.T. Hu, L.S. Li, W.D. Yang, L. Manna, L.W. Wang, A.P. Alivisatos, *Science*, **2001**, *292*, 2060–2063; [c] N.R. Jana, L. Gearheart, C.J. Murphy, *J. Phys. Chem. B*, **2001**, *105*, 4065–4067.

42 [a] S.J. Tans, M.H. Devoret, H. Dai, A. Thess, R.E. Smalley, L.J. Geerligs, C. Dekker, *Nature*, **1997**, *386*, 474–477; [b] A. Bezryadin, A.R.M. Verschueren, S.J. Tans, C. Dekker, *Phys. Rev. Lett.*, **1998**, *80*, 4036–4039.

43 [a] C.A. Mirkin, R.L. Letsinger, R.C. Mucic, J.J. Storhoff, *Nature*, **1996**, *382*, 607–609; [b] L.M. Demers, C.A. Mirkin, R.C. Mucic, R.A. Reynolds, III, R.L. Letsinger, R. Elghanian, G. Viswanadham, *Anal. Chem.* **2000**, *72*, 5535–5541.

44 G.P. Mitchell, C.A. Mirkin, R.L. Letsinger, *J. Am. Chem. Soc.*, **1999**, *121*, 8122–8123.

45 V. Pardo-Yissar, E. Katz, J. Wasserman, I. Willner, *J. Am. Chem. Soc.*, **2003**, *125*, 622–623.

46 V. Pavlov, Y. Xiao, B. Shlyahovsky, I. Willner, *J. Am. Chem. Soc.*, **2004**, in press.

47 I. Willner, V. Heleg-Shabtai, R. Blonder, E. Katz, G. Tao, A.F. Bückmann, A. Heller, *J. Am. Chem. Soc.*, **1996**, *118*, 10 321–10 322.

48 M. Zayats, E. Katz, I. Willner, *J. Am. Chem. Soc.*, **2002**, *124*, 14 724–14 735.

49 E. Katz, L. Sheeney-Haj-Ichia, I. Willner, *Angew. Chem. Int. Ed.*, **2004**, *43*, 3292–3300.

50 [a] D. Philp, J.F. Stoddart, *Angew. Chem. Int. Ed.*, **1996**, *35*, 1155–1196; [b] D.B. Amabilino, M. Asakawa, P.R. Ashton, R. Ballardini, V. Balzani, M. Belohradsky, A. Credi, M. Higuchi, F.M. Raymo, T. Shimizu, J.F. Stoddart, M. Venturi, K. Yase, *New J. Chem.*, **1998**, *22*, 959–972.

51 M.C.T. Fyfe, J.F. Stoddart, *Acc. Chem. Res.*, **1997**, *30*, 393–401.

52 S. Menzer, A.J.P. White, D.J. Williams, M. Belohradsky, C. Hamers, F.M. Raymo, A.N. Shipway, J.F. Stoddart, *Macromolecules*, **1998**, *31*, 295–307.

53 S.J. Rowan, J.F. Stoddart, *Polym. Adv. Technol.*, **2002**, *13*, 777–787.

54 A.S. Lane, D.A. Leigh, A. Murphy, *J. Am. Chem. Soc.*, **1997**, *119*, 11 092–11 093.
55 I. Willner, V. Pardo-Yissar, E. Katz, K.T. Ranjit, *J. Electroanal. Chem.*, **2001**, *497*, 172–177.
56 [a] V. Balzani, M. Gómez-López, J.F. Stoddart, *Acc. Chem. Res.*, **1998**, *31*, 405–414; [b] M. Asakawa, P.R. Ashton, V. Balzani, C.L. Brown, A. Credi, O.A. Matthews, S.P. Newton, F.M. Raymo, A.N. Shipway, N. Spencer, A. Quick, J.F. Stoddart, A.J.P. White, D.J. Williams, *Chem. Eur. J.*, **1999**, *5*, 860–875.
57 [a] C.P. Collier, E.W. Wong, M. Belohradsky, F.M. Raymo, J.F. Stoddart, P.J. Kuekes, R.S. Williams, J.R. Heath, *Science*, **1999**, *285*, 391–394; [b] Y. Luo, C.P. Collier, J.O. Jeppesen, K.A. Nielsen, E. Delonno, G. Ho, J. Perkins, H.R. Tseng, T. Yamamoto, J.F. Stoddart, J.R. Heath, *Chem. Phys. Chem.*, **2002**, *3*, 519–525.
58 R.A. Bissel, E. Cordova, A.E. Kaifer, J.F. Stoddart, *Nature*, **1994**, *369*, 133–137.
59 O.A. Raitman, E. Katz, A.F. Bückmann, I. Willner, *J. Am. Chem. Soc.*, **2002**, *124*, 6487–6496.
60 P.N. Bartlett, E. Simon, *Phys. Chem. Chem. Phys.*, **2002**, *2*, 2599–2606.
61 W. Knoll, *Annu. Rev. Phys. Chem.*, **1998**, *49*, 569–638.
62 A.G. Frutos, R.M. Corn, *Anal. Chem.*, **1998**, *A70*, 449A–455A.
63 [a] S. Sasaki, R. Nagata, B. Hock, I. Karube, *Anal. Chim. Acta*, **1998**, *368*, 71–76; [b] J.M. McDonnell, *Curr. Opin. Chem. Biol.*, **2001**, *5*, 572–577.
64 D.M. Disley, D.C. Cullen, H.Y. You, L.R. Lower, *Biosens. Bioelectron.*, **1998**, *13*, 1213–1225.
65 V. Chegel, O. Raitman, E. Katz, R. Gabai, I. Willner, *Chem. Commun.* **2001**, 883–884.
66 J.E. Fischer, H. Dai, A. Thess, R. Lee, N.M. Hanjani, D.L. Dehaas, R.E. Smalley, *Phys. Rev. B*, **1997**, *55*, R4921–R4924.
67 [a] W. Huang, S. Taylor, K. Fu, Y. Lin, D. Zhang, T.W. Hanks, A.M. Rao, Y.-P. Sun, *Nano Lett.*, **2002**, *2*, 311–314; [b] K. Jiang, L.S. Schadler, R.W. Siegel, X. Zhang, H. Zhang, M. Terrones, *J. Mater. Chem.*, **2004**, *14*, 37–39.
68 [a] C.V. Nguyen, L. Delzeit, A.M. Cassel, J. Li, J. Han, M. Meyyappan, *Nano Lett.*, **2002**, *2*, 1079–1081; [b] S.E. Baker, W. Cai, T.L. Lasseter, K.P. Weidkamp, R.J. Hamers, *Nano Lett.*, **2002**, *2*, 1413–1417; [c] M. Hazani, R. Naaman, F. Hennrich, M.M. Kappes, *Nano Lett.*, **2003**, *3*, 153–155.
69 [a] F.L. Cheng, S. Du, B.K. Lin *Chinese J. Chem.*, **2003**, *21*, 436–441; [b] C. Cai, J. Chen, *Anal. Biochem.*, **2004**, *325*, 285–292; [c] L. Zhang, G.-C. Zhao, X.-W. Wei, Z.-S. Yang, *Chem. Lett.*, **2004**, *33*, 86–87.
70 J.J. Gooding, R. Wibowo, J. Liu, W. Yang, D. Losic, S. Orbons, F.J. Mearns, J.G. Shapter, D.B. Hibbert, *J. Am. Chem. Soc.*, **2003**, *125*, 9006–9007.
71 K. Besteman, J.O. Lee, F.G.M. Wiertz, H.A. Heering, C. Dekker, *Nano Lett.*, **2003**, *3*, 727–730.
72 F. Patolsky, Y. Weizmann, I. Willner, *Angew. Chem. Int. Ed.*, **2004**, *43*, 2113–2117.
73 P. Diao, Z. Liu, B. Wu, X. Nan, J. Zhang, Z. Wie, *Chem. Phys. Chem.*, **2002**, *3*, 898–901.
74 V. Mujica, A. Nitzan, S. Datta, M.A. Ratner, C.P. Kubiak, *J. Phys. Chem. B*, **2003**, *107*, 91–95.
75 Y. Xiao, A.B. Kharitonov, F. Patolsky, Y. Weizmann, I. Willner, *Chem. Commun.* **2003**, 1540–1541.
76 L. Shi, Y. Xiao, I. Willner, *Electrochem. Commun.*, **2004**, *6*, 1057–1060.
77 Y. Xiao, F. Patolsky, E. Katz, J.F. Hainfeld, I. Willner, *Science*, **2003**, *299*, 1877–1881.
78 F. Remacle, R.D. Levine, *Chem. Phys. Lett.*, **2004**, *383*, 537–543.
79 O. Lioubashevski, V.I. Chegel, F. Patolsky, E. Katz, I. Willner, *J. Am. Chem. Soc.*, **2004**, *126*, 7133–7143.
80 [a] A. Henglein, J. Lilie, *J. Am. Chem. Soc.*, **1981**, *103*, 1059–1066; [b] T. Ung, M. Giersig, D. Dunstan, P. Mulvaney, *Langmuir*, **1997**, *13*, 1773–1782.
81 [a] A.C. Templeton, J.J. Pietron, R.W. Murray, P. Mulvaney, *J. Phys. Chem. B*, **2000**, *104*, 564–570; [b] P. Mulvaney, *Lnagmuir*, **1996**, *12*, 788–800.

82 [a] D.A. Schultz, Curr. Opin. Biotechnol., 2003, 14, 13–22; [b] P. Englebienne, A.V. Hoonacker, M. Verhas, Analyst, 2001, 126, 1645–1651.
83 V.L. Davidson, L.H. Jones, Anal. Chim. Acta, 1991, 249, 235–240.
84 V.L. Davidson, Principles and Applications of Quinoproteins, Dekker, New York, 1993.
85 T. Ikeda, Bull. Electrochem., 1992, 8, 145–159.
86 E. Katz, D.D. Schlereth, H.-L. Schmidt, A.A.J. Olsthoorn, J. Electroanal. Chem., 1994, 368, 165–171.
87 O.A. Raitman, F. Patolsky, E. Katz, I. Willner, Chem. Commun., 2002, 1936–1937.
88 C.R. Hess, G.A. Juda, D.M. Dooley, R.N. Amii, M.G. Hill, J.R. Winkler, H.B. Gray, J. Am. Chem. Soc., 2003, 125, 7156–7157.
89 A.I. Yaropolov, V. Malovic, S.D. Varfolomeev, I.V. Berezin, Dokl. Akad. Nauk SSSR, 1979, 249, 13 999–11 401.
90 E. Csöregi, G. Jönsson, L. Gorton, J. Biotechnol., 1993, 30, 315–317.
91 T. Ruzgas, L. Gorton, J. Emnéus, E. Csöregi, G. Marko-Varga, Anal. Proc., 1995, 32, 207–208.
92 A.L. Ghindilis, P. Atanasov, E. Wilkins, Electroanalysis, 1997, 9, 661–675.
93 T. Ruzgas, E. Csöregi, J. Emnéus, L. Gorton, G. Marko-Vargas, Anal. Chim. Acta, 1996, 330, 123–138.
94 T. Ruzgas, L. Gorton, J. Emneús, G. Marko-Varga, J. Electroanal. Chem., 1995, 391, 41–49.
95 A. Lindgren, F.-D. Munteau, I.G. Gazaryan, T. Ruzgas, L. Gorton, J. Electroanal. Chem., 1998, 458, 113–120.
96 F.A. Armstrong, H.A.O. Hill, N.J. Walton, Q. Rev. Biophys., 1986, 18, 261–322.
97 P.M. Allon, H.A.O. Hill, N.J. Walton, J. Electroanal. Chem., 1984, 178, 69–86.
98 M. Lion-Dagan, I. Ben-Dov, I. Willner, Colloids Surfaces B: Biointerfaces, 1997, 8, 251–260.
99 V. Pardo-Yissar, E. Katz, I. Willner, A.B. Kotlyar, C. Sanders, H. Lill, Faraday Discussions, 2000, 116, 119–134.
100 A.D. Ryabov, V.N. Goral, L. Gorton, E. Csöregi, Chem. Eur. J, 1995, 5, 961–967.
101 H. Zimmermann, A. Lindgren, W. Schuhmann, L. Gorton, Chem. Eur. J, 2000, 6, 592–599.
102 [a] W.F. DeGrado, Z.R. Wassermann, J.D. Lear, Science, 1989, 243, 622–628; [b] R.B. Hill, W.F. DeGrado, J. Am. Chem. Soc., 1998, 120, 1138–1145.
103 F. Rabanal, W.F. DeGrado, P.L. Dutton, J. Am. Chem. Soc., 1996, 118, 473–474.
104 T. Arai, K. Kobata, H. Mihara, T. Fujimoto, N. Nishino, Bull. Chem. Soc. Jpn., 1995, 68, 1989–1998.
105 B.R. Gibney, S.E. Mulholland, F. Rabanal, P.L. Dutton, Proc. Natl. Acad. Sci. U.S.A, 1996, 93, 15 041–15 046.
106 E. Katz, V. Heleg-Shabtai, I. Willner, H.K. Rau, W. Haehnel, Angew. Chem. Int. Ed. Engl., 1998, 37, 3253–3256.
107 I. Willner, V. Heleg-Shabtai, E. Katz, H.K. Rau, W. Haehnel, J. Am. Chem. Soc., 1999, 121, 6455–6468.
108 H.K. Rau, W. Haehnel, J. Am. Chem. Soc., 1998, 120, 468–476.
109 [a] J. Moiroux, P.J. Elving, J. Am. Chem. Soc., 1980, 102, 6533–6538; [b] H.-L. Schmidt, W. Schuhmann, Biosens. Bioelectron., 1996, 11, 127–135.
110 [a] J. Kulys, G. Gleixner, W. Schuhmann, H.-L. Schmidt, Electroanalysis, 1993, 5, 201–207. [b] D.D. Schlereth, E. Katz, H.-L. Schmidt, Electroanalysis, 1995, 7, 46–54.
111 M. Ohtani, S. Kuwabata, H. Yoneyama, J. Electroanal. Chem., 1997, 422, 45–54.
112 E. Katz, T. Lötzbeyer, D.D. Schlereth, W. Schuhmann, H.-L. Schmidt, J. Electroanal. Chem., 1994, 373, 189–200.
113 I. Willner, A. Riklin, Anal. Chem., 1994, 66, 1535–1539.
114 A. Bardea, E. Katz, A.F. Bückmann, I. Willner, J. Am. Chem. Soc., 1997, 119, 9114–9119.
115 M. Zayats, A.B. Kharitonov, E. Katz, A.F. Bückmann, I. Willner, Biosens. Bioelectron., 2000, 15, 671–680.
116 E. Katz, I. Willner, A.B. Kotlyar, J. Electroanal. Chem., 1999, 479, 64–68.
117 P.A. Adams, in Peroxidases in Chemistry and Biology (Eds.: J. Everse, K.E.

EVERSE), Vol. 2, CRC Press, Boston, **1991**, Chapter 7, 171–200.
118 E. KATZ, I. WILLNER, *Langmuir*, **1997**, *13*, 3364–3373.
119 A. NARVAEZ, E. DOMINGUEZ, I. KATAKIS, E. KATZ, K.T. RANJIT, I. BEN-DOV, I. WILLNER, *J. Electroanal. Chem.*, **1997**, *430*, 227–233.
120 V. HELEG-SHABTAI, E. KATZ, S. LEVI, I. WILLNER, *J. Chem. Soc., Perkin Trans.*, **1997**, *2*, 2645–3652.
121 [a] A. HELLER, *Acc. Chem. Res.*, **1990**, *23*, 128–134; [b] A. HELLER, *J. Phys. Chem.*, **1992**, *96*, 3579–3587.
122 R. RAJAGOPALAN, A. AOKI, A. HELLER, *J. Phys. Chem.*, **1996**, *100*, 3719–3727.
123 S.C. BARTON, H.H. KIM, G. BINYAMIN, Y. ZHANG, A. HELLER, *J. Am. Chem. Soc.*, **2001**, *123*, 5802–5803.
124 [a] N. MANO, H.H. KIM, A. HELLER *J. Phys. Chem. B*, **2002**, *106*, 8842–8848.
[b] H.H. KIM, N. MANO, X.C. ZHANG, A. HELLER, *J. Electrochem. Soc.*, **2003**, *150*, A209–A213.
125 E. KATZ, A.F. BÜCKMANN, I. WILLNER, *J. Am. Chem. Soc.*, **2001**, *123*, 10752–10753.
126 E. KATZ, I. WILLNER, *J. Am. Chem. Soc.*, **2003**, *125*, 6803–6813.
127 A.C. TEMPELTON, W.P. WUELFING, R.W. MURRAY, *Acc. Chem. Res.*, **2000**, *33*, 27–36.
128 G. MARUCCIO, R. CINGOLANI, R. RONALDI, *J. Mater. Chem.*, **2004**, *14*, 542–554.
129 [a] P. TRAVASCIO, Y. LI, A.J. BENNET, D.Y. WANG, D. SEN, *Chem. Biol.*, **1999**, *6*, 779–787; [b] P. TRAVASCIO, P.K. WITTING, A.G. MAUK, D. SEN, *J. Am. Chem. Soc.*, **2001**, *123*, 1337–1348; [c] V. PAVLOV, Y. XIAO, R. GILL,, A. DISHON M. KOTLER, I. WILLNER, *Anal. Chem.*, **2004**, *767*, 2152–2156.

4
Application of Electrically Contacted Enzymes for Biosensors

Frieder W. Scheller, Fred Lisdat, and Ulla Wollenberger

4.1
Introduction

Electron transfer plays a central role in biological energy conversion, for example, in photosynthesis or respiratory chain, and metabolism, for example, in glycolysis or citric acid cycle, and also in the regulation of gene expression where so-called redox switches are involved. The charge transfer is either accomplished by low-molecular redox mediators, such as ubiquinone and plastoquinone, and also by oxygen/superoxide and the NAD^+/NADH couple, or by the direct interaction of the redox centers of proteins. Both types of biochemical charge transfer reactions have been adopted to *electrically* couple proteins to redox electrodes.

1. Low-molecular redox mediators have been applied for the estimation of the redox potential of biomolecules for a long time. More recently, mediators have served as a key component in many amperometric biosensors.
2. Since the time of Heyrovski and Berezin [1, 2], it has been an attractive goal to achieve "mediatorless" communication between electrodes and the prosthetic groups of redox proteins and enzymes. The "artificial" reactions may improve the basic knowledge of the mechanism of biological redox reactions at interfaces, for example, at biological membranes. Furthermore, the combination of biological recognition processes is most promising for designing sensors and bioelectronic functionalities [3–6].

4.2
Biosensors – Precursors of Bioelectronics

Traditionally, enzymes are used as analytical reagents to measure substrate molecules by catalyzing the turnover of these species to detectable products [7]. In addition, compounds modifying the rate of the enzyme reaction, such as activators, prosthetic groups, inhibitors, and enzymes themselves, are accessible to the measurement. Owing to their excellent chemical specificity, enzymes allow the determination of minute amounts in complex media and thus avoid the need of highly sophisticated

Bioelectronics. Edited by Itamar Willner and Eugenii Katz
Copyright © 2005 WILEY-VCH Verlag GmbH & Co. KGaA, Weinheim
ISBN: 3-527-30690-0

instrumentation. Furthermore, when enzymes are employed as labels in binding assays using antibodies, binding proteins, lectins, and so on, the inherent chemical amplification properties of the enzyme's catalytic activity can be exploited to realize extremely sensitive assay methods [8].

As early as 1956, the principle of the litmus paper used for pH measurement was adapted to simplify the enzymatic determination of glucose [9]. By impregnating filter paper with the glucose-converting enzymes, the "enzyme test strip" was invented. It can be regarded as the predecessor of optical biosensors, and, at the same time, initiated the development of the so-called *dry chemistry*. Nowadays, highly sophisticated enzyme and immuno test strips are commercially available for the determination of about 15 low-molecular metabolites and drugs as well as the activities of 10 enzymes.

In parallel, analytical enzyme reactors have been developed [10] where the progress of the reaction is indicated in the reactor effluent calorimetrically or electrochemically: In *packed bed reactors*, the enzyme-catalyzed reaction is carried out in a column of 100 μL to 10 mL volume filled with tiny particles bearing the immobilized enzyme. In contrast, in *open tubular reactors*, the enzyme is attached in a monolayer to the inner walls. Such reactors permit a higher measuring frequency.

Electrochemical sensors are well-established tools in the determination of gases, ion activities, and oxidizable and reducible organic substances down to the submicromolar concentration range. The analysis of many other important substances by electrochemical sensors requires *coupling with an enzymatic reaction*, which involves an electroactive species [11] (Table 4.1).

Tab. 4.1 Coupling of enzymes with electrochemical sensors

Biocomponents	Indicated species	Electrode type
Oxidoreductases: Dehydrogenases Oxidases Peroxidases Electron transferases	Cosubstrates: NAD(P)H O_2/H_2O_2 Mediators	Amperometric Electrodes
	Products: Phenols/Quinones Redox dyes	
	Prosthetic Groups: Heme PQQ FAD Cu^{2+}	
Hydrolases: Proteases Esterases Glucosidases	Products: H^+ HCO_3^- NH_4^+ I^- F^-	Potentiometric Electrodes

In the early 1960s, Clark introduced the concept of the *enzyme electrode* [12]. He placed an enzyme solution (an oxidase) immediately in front of an oxygen electrode and prevented loss of the enzyme by diffusion into the background solution by covering the reaction chamber with a semipermeable membrane. The same enzyme preparation could then be used for several samples. The next stage was reached in 1967 by Updike and Hicks [13] who entrapped the enzyme in a gel, thus increasing the working stability and simplifying the sensor preparation. In 1975, Yellow Springs Instruments Co. (USA) commercialized a glucose analyzer, which was based on Clark's invention.

The first step in the biosensing process is the specific complex formation of the immobilized biological recognition element with the analyte. The biological part of a biosensor is often submitted to a conformational change in context with the binding of its partner. In nature, this effect may immediately be used for transduction and amplification, for example, in the ion channels of nerve tissue. The effects of interaction between the analyte molecule and the biological system are quantified by the transducer and electronic part of the biosensor. As transducers, chemical sensors, that is, potentiometric, amperometric, and impedimetric electrodes, optical detectors using indicator dyes, as well as physical sensors, such as piezoelectric crystals, thermistors, and plain optical sensors, have been combined with appropriate biocomponents [14] (Table 4.2). In analogy to affinity chromatography, in so-called *binding* or *affinity sensors*, dyes, lectins, antibodies, or hormone receptors are used in matrix-bound form for the molecular recognition of enzymes, glycoproteins, antigens, and hormones. The

Tab. 4.2 Combinations of biocomponents with transducers

Recognition elements	*Transducers*	*Signal*	*Configuration*
Biological Enzymes Antibodies Receptors Nucleic acids Cells	*Electrochemical* Amperometric Potentiometric Conductometric Capacitive ISFET		
	Optical Photometers Interferometers Refractometers Fluorimeters Luminometers	*Threshold Values* Concentration Biological Effect	*Sensor* Microanalysis System (μTAS) Array on Chip
Biomimetic Aptamers Ribozymes Synzymes Imprints	*Mechanical Thermodynamic* Thermistors Viscosimeters Oscillating quartz crystals	Substance "Pattern"	
	Magnetic Toroids		

complex formation changes the magnitude of physicochemical parameters, such as layer thickness, refractive index, light absorption, or electrical charge, which may then be indicated by means of optical sensors, potentiometric electrodes, or field effect transistors. After the measurement, the initial state is regenerated by splitting of the complex. On the other hand, the molecular recognition by enzymes, which can also be applied in the form of organelles, microorganisms, and tissue slices, is accompanied by the chemical conversion of the analyte to the respective products. Therefore, this type of sensor is termed *catalytic* or *metabolism sensor*. It usually returns to the initial state when the analyte conversion is completed. Under appropriate conditions, catalytic biosensors are capable of determining cosubstrates, effectors, and enzyme activities via substrate determination. Amperometric and potentiometric electrodes, fiber optics, and thermistors are the preferred transducers here, the former being by far the most important.

In 2001, IUPAC [15] proposed to define a biosensor as follows:

An electrochemical biosensor is a self-contained integrated device, which is capable of providing specific quantitative and semiquantitative analytical information using a biological recognition element (biochemical receptor), which is in direct spatial contact with a transduction element.

A biosensor should be clearly distinguished from a bioanalytical system, which requires additional processing steps, such as reagent addition.

A device that is both disposable after one measurement, that is, single use, and unable to monitor the analyte concentration continuously or after rapid and reproducible regeneration should be designated a single-use biosensor.

Owing to the rapid technological progress, essential parts of it appear ambiguous today [16]. The borderline between recognition tools that are biological in nature and synthetic (organic) receptor molecules is no longer well definable; these two classes of recognition elements are merging. This is especially true of synthetic binder molecules, for example, aptamers based on (synthetic) oligonucleotides, imprinted polymers, and also for ionophores, which mimic the function of ion channels. Furthermore, in the light of the ability of miniaturized bioreactors to interact with transducers (without additional sample processing elements) on a single chip, the term 'direct spatial contact' of the elements is amenable to different interpretations.

4.3
Via Miniaturization to Sensor Arrays – The Biochip

An enzyme electrode (Figure 4.1) is a dense package of a bioanalyzer consisting of dialyzer, enzyme reactor, and detector. A typical example would be a glucose or lactate electrode comprising the appropriate oxidase entrapped in a small reaction chamber that contains an oxygen or hydrogen peroxide detecting electrode.

The choice of the *indicator electrode* is largely determined by the species involved in the sensing reaction. Oxygen and H_2O_2, which are the cosubstrate and product of oxidases,

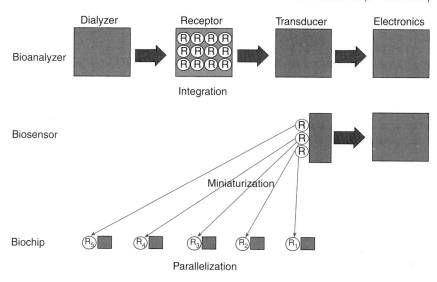

Fig. 4.1 The route from bioanalyzer via biosensor to the biochip.

as well as NAD(P)H, the cosubstrates of about 300 pyridine nucleotide-dependent dehydrogenases, can be determined amperometrically. Hydrolases are mostly coupled to ion-selective electrodes.

For the repeated use of enzymes in analytical devices, numerous techniques for fixing them to carrier materials including membranes and electrode surfaces have been developed [10, 11].

The techniques for immobilizing enzymes comprise physical and chemical methods as well as combinations of both. The main physical methods are adsorption to water-insoluble carriers or surfaces and entrapment in water-insoluble polymeric gels. Chemical immobilization is effected by covalent coupling to derivatized carriers or by intermolecular cross-linking of the biomolecules.

Immobilization brings about a number of further advantages for their application in analytical chemistry:

- In many cases the enzyme is stabilized.
- The immobilized enzyme may be easily separated from the sample.
- The stable and largely constant enzyme activity renders the enzyme an integral part of the analytical instrument.

Structured deposition of enzymes on the surface of microelectrodes or ion-selective field effect transistors (ISFETs) is a major problem of mass production. Layers of polymers, for example, polypyrrole or polyaniline, are deposited on conducting areas by electropolymerization. Biomolecules are either entrapped in the polymer matrix while the layer is being formed or coupled to the layer via specific chemical reactions. The structuring of a uniform layer containing a biocomponent is possible when passive regions are photodeactivated. Alternatively, enzymes are deposited only at the sensitive region by using an enzyme solution in a photoresist. Using microspotting techniques, the active areas of arrays are covered with different receptor molecules or cells.

The immediate adsorptive or covalent binding of the biocatalyst at the electrode surface or the inclusion within the electrode body permits the elimination of semipermeable membranes. This direct binding of the biocatalyst to the transducer is the basis for further miniaturization and parallelization. Independently addressable microelectrodes, or ISFETs, are the transducer units of these electrochemical biochips, which allow a *parallel* determination of a broad spectrum of analytes [16].

4.4
The Route to Electrically Contacted Enzymes in Biosensors

The electronic coupling between redox enzymes and electrodes for the construction of enzyme electrodes has been based on the electroactivity of the enzyme cosubstrate or product (*first generation*) or through the use of redox mediators (*second generation*) [17] most typically illustrated by numerous biosensors using glucose oxidase (Figure 4.2). The drawbacks with first-generation biosensors put the focus on the use of mediators, small redox-active molecules that could diffuse in and react with the active site of the enzyme and diffuse out and react with the electrode surface, thus shuttling the electrons between the enzyme and the electrode. Diffusional electron mediators, such as ferrocene derivatives, ferricyanide, conducting organic salts (particularly tetrathiafulvalene-tetracyanoquinodimethane, TTF/TCNQ, phenothiazine

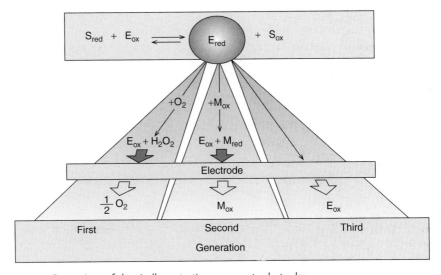

Fig. 4.2 Generations of electrically contacting enzymes to electrodes.

and phenoxazine compounds, or quinone compounds have thus been widely used to electrically contact glucose oxidase (GOD) [18]. The use of mediators made it possible to decrease the applied potential for hydrogen peroxide–producing oxidases, thus decreasing the influence from bias signals caused by electrochemically easily oxidizable interfering compounds present in real samples. The use of mediators also opened up the possibility for the use of various dehydrogenases, peroxidases, and even whole cells. However, the redox mediators are general redox catalysts facilitating not only the electron transfer between electrode and enzyme but also various interfering reactions. Further progress in the development of second-generation biosensors was achieved with the use of flexible polymers onto which mediating functionalities were covalently bound [19, 20].

Chemical conversion of a single-center redox enzyme, for example, by introducing a mediator molecule (ferrocene [21], pyrrolo quinoline quinone (PQQ)) [5] located half way between the prosthetic group and the molecule surface produced electroactive hybrids. The procedure was demonstrated to result in almost oxygen-independent electrodes, probably due to the fact that an artificially introduced redox center is shielded by the protein as well as the rapid electron exchange with the electrode. Obviously, the rate of electron transfer via the 'wired' route is high compared to the rate of reaction of the enzyme with endogenous oxygen. Most mediator-based sensors involving oxidases do not meet this requirement and exhibit parasitic effects of oxygen on their response. The approach requires, however, a high-skilled manipulation for unfolding-modifying-refolding of the enzyme.

Efficient direct electron transfer – the basic principle of *third generation* enzyme electrodes – has been reported for several electron transport proteins but only for a restricted number of redox enzymes [17, 22]. Proteins and enzymes transferring electrons to other proteins typically function in ordered structures such as mitochondria, and the redox-active centers are generally accessible to the outer surface of the protein and therefore able to communicate with electrodes (Table 4.3).

It has been demonstrated that promoters, such as aminoglycosides, are effective for accelerating the electron transfer of cytochrome c peroxidase, flavohemoproteins, and methylamine dehydrogenase at graphite electrodes and modified gold electrodes. In the absence of promoters, peroxidase, laccase, cytochrome c peroxidase, methylamine dehydrogenase, and ferredoxin-NADP oxidoreductase produce catalytic currents by the bioelectrocatalytic oxidation or by reduction of their substrates at carbon electrodes. Flavohemoproteins and quinohemoproteins from bacterial cytoplasmic membranes immobilized by adsorption produce catalytic currents in the presence of the substrates based on direct electron transfer [23]. These enzymes contain a subunit of a c-type cytochrome and another subunit of flavine adenine dinucleotide (FAD), or PQQ. The heme group accepts electrons from the FAD or PQQ by intramolecular electron transfer and donates them to the respiratory chain *in vivo*. Approximate orientation of the enzymes in the electrodes allows bioelectrocatalytic oxidation of the substrates based on the direct electron transfer between the electrode and the heme group of the adsorbed enzyme.

Tab. 4.3 Redox enzymes for which direct mediator-free reactions with electrodes have been shown

Enzyme	Prosthetic group	Substrates
Ascorbate oxidase	Cu	O_2
Laccases	Cu	O_2
Theophylline oxidase	Cu	Cytochrome c
Superoxide dismutase	Cu-Zn	O_2^-
	Fe	
	Mn	
Diaphorase	FMN	NADH
Pentachlorophenol hydroxylase	FAD	O_2, pentachlor phenol
Putidaredoxin reductase	FAD	Putidaredoxin
Methylamine dehydrogenase	TTQ	Methylamine
Phospholipidhydroperoxide Glutathione peroxidase	Selenocysteine	Glutathione, H_2O_2
Catalase	Heme	H_2O_2/O_2
Cytochrome P450	Heme b	O_2, aminopyrine, benzphetamine
Methane monooxygenase	Binuclear heme	Acetonitrile, methane
Peroxidases	Heme	H_2O_2
Chloroperoxidase	Heme	H_2O_2
Cytochrome c peroxidase	Heme b	H_2O_2
Fungal peroxidase	Heme	H_2O_2
Horseradish peroxidase	Heme	RHO_2
Lignin peroxidase	Heme	H_2O_2
Manganese peroxidase	Heme	H_2O_2
Peanut peroxidase	Heme	H_2O_2
Soybean peroxidase	Heme	H_2O_2
Sweet potato peroxidase	Heme	H_2O_2
Tobacco peroxidase	Heme	H_2O_2
Pentachlorophenol hydroxylase	FAD	O_2, PCP
Putidaredoxin reductase	FAD	Putidaredoxin
Multi Center Enzymes		
Amine oxidase	Cu, topa quinone	Amines
Cytochrome c oxidase	Cu_A, Cu_B, heme a3,	O_2, cytochrome c
Cellobiose dehydrogenase	FAD, heme c	Cellobiose, lactose
p-Cresolmethylhydrolase	FAD, heme c	p-Cresol
L-lactate dehydrogenase (flavocytochrome b_2)	FMN, heme b_2	Lactate
Flavocytochrome c_{522}	FAD, heme	Sulfide
Flavocytochrome c3	FAD, heme c	Fumarate
Fumarate reductase	FAD, Fe-S FAD, heme c	Fumarate/succinate
D-gluconate dehydrogenase	FAD, heme c, Fe-S	D-gluconate
Alcohol dehydrogenase	PQQ, heme c	Ethanol
Aldose dehydrogenase	PQQ, heme	Aldose
D-fructose dehydrogenase	PQQ, heme c	D-fructose
Amine oxidase	Cu, topa quinone	Amines
Cytochrome c oxidase	Cu_A, Cu_B, heme a3,	O_2, cyt c
Nitrite reductase	Cu, multi heme	NO_2^-
DMSO-reductase	Mo-pterin, Fe-S	DMSO

Tab. 4.3 (Continued)

Enzyme	Prosthetic group	Substrates
Sulfite oxidase	Mo-pterin heme b5	SO_3^{2-}
Sulfite dehydrogenase	Mo-pterin, heme c	SO_3^{2-}
Hydrogenase	Fe-S Ni, (Se), Fe-S	H_2, H^+, NAD

4.5
Routine Applications of Enzyme Electrodes

Routine application of enzyme electrodes is presently restricted to the determination of low-molecular metabolites.

Redox enzymes catalyzing the reactions between small molecules are characterized by the catalytic center buried well within the protein. Communication between this center and an electrode serving as a source or sink for electrons is frequently poor or nonexistent. Heterogeneous electron transfer rates for this type of enzymes are vanishingly low so that the enzyme redox activity must be coupled to the electrode using cosubstrates or mediators (first and second generation).

The most relevant fields of practical application of enzyme electrodes are medical diagnostics, followed by process control, food analysis, and environmental monitoring [24]. The first commercial enzyme electrode-based analyzer (Yellowsprings 1975) was developed to meet the high demand for glucose determination in the blood of diabetic patients. Since 1975, analyzers for about 12 different analytes have been commercialized. As compared to conventional enzymatic analysis, the main advantages of such analyzers are the extremely low enzyme demand (a few milliunits per sample), their simplicity of operation, the high sample throughput rate, and high analytical quality.

The selective determination of blood glucose is of extraordinary importance for the screening and treatment of diabetes. In the medical field, a trend toward handheld devices based on disposable glucose enzyme electrodes for home control of diabetes and for on-site monitoring of surgery and exercise is evident.

Personal blood glucose meters are based on disposable (screen-printed) enzyme electrode test strips. Such single-use disposable strips are mass-produced by the thick-film (screen-printing) microfabrication technology. Each strip contains the printed working and reference electrodes, with the working one coated with the necessary reagents (i.e., enzyme, mediator, stabilizer, linking agent).

The first product was a pen-style device (the Exatech), launched by Medisense Inc. in 1987, that relied on the use of a ferrocene-derivative mediator [25]. Various commercial strips and pocket-sized test meters, for self-monitoring of blood glucose – based on the use of ferricyanide or ferrocene mediators – have since been introduced (Table 4.4). The Johnson & Johnson Company, which is currently the market leader in blood glucose self-testing, launched its first biosensor product – the Fast Take™ system, in April 1998. The biggest suppliers to this market all now sell biosensor-based products.

Tab. 4.4 Glucose sensors for patient self-control

Electron acceptor	Enzyme	Company	Product
O_2	GOD	BST	Glucometer 3000
Ferricyanide	GOD	Bayer	Glucometer Elite
			Glucometer Elite XL
	(PQQ)GDH	Roche	ACCUcheck Advantage
	(NAD)GDH	Abbott	SOFT-Tact
		LifeScan	One Touch Ultra
Ferrocene adsorbed	GOD	Abbott	Precision QID
			Exac Tech
Os Redox polymer	(PQQ)GDH	TheraSense	FreeStyle

The handheld devices for self-control of blood glucose are based on the combination of the enzymes glucose oxidase and PQQ-dependent or NAD^+-dependent glucose dehydrogenase (GDH) with different redox mediators. Whenever GOD is still dominating, both GDHs present the advantage of abolishing parasitic oxygen reactions.

For glucose measurement by the first device of the electrochemical meters – the Medisense glucose pen – the sample is put directly on the surfaces of the enzyme electrode. The glucose-converting reaction requires the dissolution of the oxidized ferrocene mediator within the electrode-near layer in order to reoxidize the GOD, which is adsorbed at the carbon electrode. The electrode current is generated by the anodic oxidation of the enzymatically reduced mediator.

However, the majority of commercial glucose meters perform the sample application by filling a microreaction chamber using capillary forces. The reagents are dissolved from the walls of the chamber by the inflowing sample liquid, and the formation of the reduced mediator is quantified by the amperometric electrode.

The capillary-based meters do not fulfill the IUPAC definition in the strict sense because the enzyme is not in direct contact with the electrochemical transducer but is trapped in a confined space.

These arrangements prevent misfunctions of the device by inappropriate sample application such as insufficient volume or mechanical stress of the sensor and allow the decrease of the sample volume from almost 20 µL to a minimum of 300 nL. Furthermore, the inconvenience of taking 10 to 20 µL of blood by punching the finger is avoided because the sample can be taken almost painlessly with the recently introduced Free Style (Thera Sense) device. The integration of the sample transport into the device by a vacuum pump is realized in the Soft-Tact (Abbott). It represents a further breakthrough because the handling and the reliability of the whole measuring is improved.

Even less painful is the iontophoretic detection with the Gluco Watch™ automatic glucose biographer from Cygnus. This device controls blood glucose over a range of 2.2 to 22.2 mmol L^{-1} (40–400 mg dL^{-1}) for up to 12 h using a single-point calibration. The automatic, frequent, and noninvasive measurements provide more information about glucose levels than the current standard of care.

Tighter glycemic control through *continuous in vivo* monitoring is desired for triggering proper alarm in cases of hypo- and hyperglycemia and for making valid therapeutic decisions. Most of the recent attention has been given to the development of subcutaneously implantable needle-type electrodes. Such devices are designed to operate for a few days, for example, the recently introduced CGMS unit of Minimed. Inc. (Sylmar), which offers 72 h of subcutaneous monitoring with measurement of tissue glucose every 5 min.

For a continuous *in vivo* monitoring of glucose, long-term stability, selectivity and sensitivity in the mMolar range is needed. *In vivo* devices are mediatorless because of potential leaching and toxicity of the mediator. Extremely fast-responding glucose sensors with submicrometer tip diameter have been constructed by using platinum-deposited, flame-etched carbon fiber electrodes additionally coated with glucose oxidase [26].

4.6
Research Applications of Directly Contacted Proteins

Proteins and enzymes transferring electrons to other proteins typically have redox-active centers accessible to the outer surface on the protein and are therefore able to communicate with electrodes (Table 4.3).

Electrochemical sensors using direct electron transfer between proteins and electrodes are not yet commercially available but they are useful research tools. They mostly apply electron-transferring proteins, for example, c-type cytochromes, to couple a solution redox reaction with the amperometric electrode. Prominent examples are the sensors for the signaling molecules superoxide and nitric oxide (see Chapter 6.1).

Several substrate-converting oxidoreductases that have a redox protein as coreactant are also able to communicate directly with electrodes (Table 4.3). An efficient transfer of electrons from the redox electrode substitutes the electron-supplying protein of mixed functional oxidases, for example, of the cytochrome P-450 family (see Chapter 6.2). This concept has a high potential for several bioelectronic devices, for example, sensors, biofuel cells, and electroenzyme reactors, respectively.

4.6.1
Protein Electrodes for the Detection of Oxygen-derived Radicals

The need for detection elements for reactive species originates from the progress in biomedical research demonstrating the involvement of oxygen-derived radicals such as superoxide or nitric oxide in a number of biochemical key processes such as vasodilation, carcinogenesis, inflammatory processes, immunodeficiencies, and so on [27–30]. Under physiological conditions there is a balance between the generation and decomposition of these species. If this balance is disturbed, significantly higher concentrations occur. This may lead to degenerative processes and even to diseases. Thus, the locally produced and time-dependent concentration increase has to be detected by appropriate means. The analytical signals obtained can serve as medical indicators for pathological situations and can help develop strategies to prevent the rise in radical concentration and thus tissue damage.

The main advantage of the sensorial approach for radical analysis using protein electrodes is the combination of a selective protein-radical interaction with the spatial and time resolution of the measurement, which allows the online observation of concentration changes. Different principles have been followed and reviewed [31]. Besides the construction of conventional enzyme electrodes using superoxide dismutase (SOD) as the biocatalyst (for superoxide sensing [32, 33]) or the use of electrodes, which were modified with transition metal complexes (for NO sensing [34–36]), the interaction of both radicals with heme proteins can be exploited. The interaction can result in a simple coordination at the heme site or redox reactions, depending on the nature of the reaction partners. For analytical purpose, this radical-induced modification of heme proteins can be used since it changes the electrochemical redox conversion of the immobilized protein.

For superoxide analysis, mainly cytochrome c was used since the radical is easily oxidized by the protein [37]. This is also the basis for the classical optical test [38].

For nitric oxide, another small heme protein – myoglobin – can be used to facilitate the reduction of the signal molecule [39, 40]. A special group of heme proteins are cytochromes c', which have been found in phototropic, sulfur-oxidizing, and denitrifying bacteria. These proteins have been successfully applied for optical NO-sensors [41], but they can also be immobilized on modified gold electrodes, and an efficient electrical communication with the redox center can be established as a basis for an electrochemical NO sensor [42].

4.6.1.1 Superoxide Sensor Based on a Cytochrome c Electrode

Cytochrome c can be efficiently immobilized at promoter-modified electrodes, allowing quasi-reversible redox transformations of the protein. For sensor construction, the long-chain thiol mercaptoundecanoic acid (MUA) can be applied as promoter [43, 44]. It forms a rather dense layer by self-assembly and thus provides an efficient barrier for potentially interfering substances. The promoter layer is negatively charged at neutral pH and thus able to attract the positive heme environment of cytochrome c. So the protein can be adsorbed at low ionic strength and further covalently linked to MUA by activation with N-(3-Dimethylaminopropyl)-N'-ethylcarbodiimide (EDC) [45, 46].

Electron transfer of immobilized proteins to the electrode is influenced by several factors. Besides the promoter layer thickness, it was found that protein orientation is quite important. This can be influenced by the use of mixed thiols (with different head groups) [47, 48]. A mixed promoter layer of hydroxy- and carboxy-terminated thiols resulted in a significantly enhanced electron transfer for the covalently fixed protein (35 s^{-1} compared to 4 s^{-1} for only carboxy-terminated thiol layers) and still effectively rejected interfering substances from the electrode surface. The mixed thiol approach proved to be valuable not only for an optimum orientation of the protein with respect to an efficient electron transfer but also for an enhanced reaction rate with superoxide radicals compared to a promoter layer using only carboxy-groups (C11 – hydroxy-, carboxy-terminated: $k_{mixed} = 3.1 \times 10^4$ M^{-1}s^{-1} compared to C11 – carboxy-terminated: $k = 1.4 \times 10^4$ M^{-1}s^{-1}). Furthermore, the amount of electroactive protein can be drastically increased after the covalent linkage of cytochrome c to the promoter layer. This directly enhances the sensitivity of the protein electrode.

When superoxide is generated in solution, it can reduce the immobilized cytochrome c, which subsequently can be oxidized at an appropriate electrode potential [49]. Thus, a signal chain was constructed, transferring electrons from the radical via cytochrome c to the electrode. This results in an enhanced oxidation current that can be used as the sensor signal. Figure 4.3 shows a cyclic voltammogram for illustration.

Calibration of the cytochrome c electrode is not trivial since the radical is a short-lived species. Investigations of the biocatalytic generation of superoxide by the xanthine oxidase catalyzed conversion of hypoxanthine to uric acid showed that a steady state situation can be achieved in solution [53]. The continuous generation of the radical is counterbalanced by the spontaneous dismutation. This situation continues until the substrate of the generation reaction (hypoxanthine) will be consumed. The steady state radical concentration in solution corresponds to a steady state oxidation current detected at the electrode. A kinetic analysis showed that this sensor current can be linearly correlated to the square root

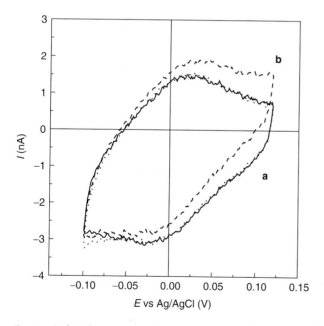

Fig. 4.3 Cyclic voltammogram of cytochrome c immobilized on a modified gold electrode (mercaptoundecanoic acid/mercaptoundecanol 1:3) in phosphate buffer pH 7.5 with 160 μmol hypoxanthine and catalase (2 U mL^{-1}) (**a**) ----- before and (**b**) ——— after addition of xanthine oxidase (XOD, 160 mU mL^{-1}), scan rate 10 mV s^{-1}.

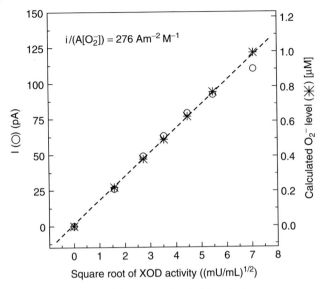

Fig. 4.4 Dependence of the current signal of the cytochrome c sensor on the XOD activity used for the generation of superoxide radicals. According to [42] the steady state superoxide concentration was calculated. (0.1 M Na- phosphate buffer pH 7.5, 100 μM hypoxanthine, Au/MUA/MU/cytochrome c needle electrode, $E = +130$ mV (vs Ag/AgCl).

of the xanthine oxidase (XOD) activity used for the radical generation [48]. This allows a calibration of the sensor as shown in Figure 4.4. Alternatively, also the generation rate of superoxide radicals can be used for the calibration of the sensor [51].

The analysis also showed that the cytochrome c electrode is sensitive to superoxide in the nanomolar concentration range. Because of the use of surface immobilization techniques, the response rate is rather fast. The electrode responded to concentration changes in solution within seconds.

Selectivity of the radical interaction with the protein electrode is a complex matter, which is not simply determined by the protein properties. This can be exemplified by the hydrogen peroxide effect on the sensor signal. At electrode potentials <100 mV versus Ag/AgCl the pseudo-peroxidative activity of cytochrome c resulted in a reduction current that subsequently interfered with the superoxide signal in the presence of H_2O_2. This is particularly relevant since both species are coexisting. At higher potentials, however, (~190 mV versus Ag/AgCl), H_2O_2 oxidation can be observed. So there is only a small potential window in which limited interference can be guaranteed. For practical measurements, specificity of the signal can be easily tested by addition of superoxide dismutase – the most effective scavenger of superoxide radicals producing hydrogen peroxide and oxygen. Well-prepared sensors showed an instantaneous decline of the sensor current to the baseline.

With the background of a covalently immobilized cytochrome c amount in the range of a monolayer, a fast electron transfer of the protein to the electrode, a high reaction rate with superoxide radicals and limited interference *in vivo* radical measurement become feasible.

Oxygen radicals presumably play a key role in the pathobiochemical mechanism of reperfusion injury after ischemic conditions. The online detection of superoxide concentrations during reperfusion after a variable time of ischemia is of special interest. Measurements were performed by placing the cytochrome c sensor into the gastrocnemius muscle tissue. Ischemia was induced by clamping the vena and arteria femoralis. Current response of the sensor was recorded continuously as an equivalent for superoxide concentration. Ischemia times varied from 5 to 120 min. The experiments showed that with the modified biosensor, reliable and stable signals for superoxide production could be obtained using this animal model. The time between removal of the microclips (to allow recirculation of the blood) and the first superoxide signal obtained can be seen as the reperfusion time of the tissue. It takes between 30 and 90 s for the blood flow to reach the muscle and to reoxygenate the cells, so that they are able to generate O_2^- from the XOD formed during ischemia. The minimum ischemia time to record superoxide production in the postischemic period was 10 min. By inducing longer periods of ischemia, an increase in superoxide concentration reached its highest levels at 2 h. This is illustrated in Figure 4.5. Furthermore, the total time of superoxide production after reperfusion depended on the total time of ischemia, which was found to vary from minutes to more than one hour. These results clearly indicate that radical

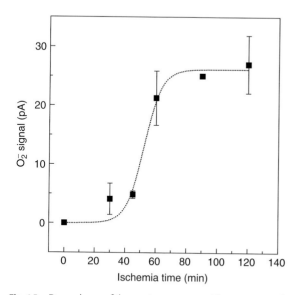

Fig. 4.5 Dependence of the maximum superoxide sensor signal after reperfusion on the time period of ischemia (*in vivo* measurements within the legs of Whistar rats, cytochrome c-gold needle electrode).

production is not a short-time process. Further investigations will show the effect of systematic application of substances that can scavenge superoxide radicals to prevent tissue damage during reperfusion injury. Thus, the cytochrome c–based sensor is a valuable analytical tool in radical research.

Besides the *in vivo* detection of superoxide radicals, the cytochrome c electrode was found valuable for the study of the interaction of NO with the radical [52], the detection of small amounts of superoxide dismutase [53, 54], and the quantification of the different efficiency of substances to scavenge the superoxide radical [55].

For the latter field of application, it is important that the cytochrome c–based sensor can be applied in organic solvent containing mixtures [56]. The electrochemical behavior of immobilized cytochrome c has been investigated with increasing content of organic solvent until no protein redox conversion could be observed. The highest possible content of organic solvent still sustaining protein electrochemistry, which can be termed as threshold content, was found to depend on the solvent used: 70% for butanediol, 50% for DMSO and methanol, and 30% for THF. In mixtures of butanediol, methanol, or DMSO, voltammetric peak currents remained stable. THF, however, caused a rather rapid deterioration of the voltammetric signal.

The formal potential of the immobilized protein was found to decrease linearly with increasing content of organic solvent (see Figure 4.6). However, the slope of these dependencies was different for each solvent. In addition to the thermodynamic

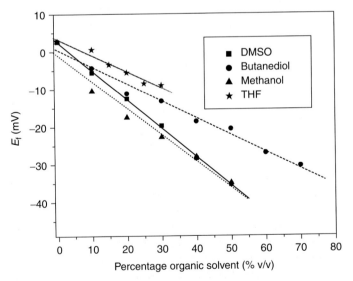

Fig. 4.6 Change of the formal potential of cytochrome c immobilized on SAM-modified gold (mercaptoundecanoic acid mercaptoundecanol 1 : 3) with increasing amount of organic solvent in solution (determined by cyclic voltammetry vs Ag/AgCl/1MKCl) 100 mV s^{-1}; 0.1 M phosphate buffer; DMSO, butanthiol + 0.1 M tetra butylammonium chloride; methanol + 0.1 M NaCl; THF + 0.1 M NaClO$_4$).

properties of cytochrome c, the kinetics of the electrochemical conversion was influenced by the organic solvent as well. The heterogeneous electron transfer rate constant k_s decreased first rather sharply with increasing content of organic solvent. Further increase in solvent proportion slowed this decrease down. Despite this general tendency, the decline in k_s differed in the individual organic solvents. However, since the ratio of anodic and cathodic peak currents also kept close to 1, even with high organic solvent content, the redox conversion of cytochrome c maintained its quasi-reversible character in the mixed media. Together with the small change in the driving force for redox reactions (shift of E_f), this allows the application of a cytochrome c electrode for interaction studies of superoxide and potential scavengers even if they are of more hydrophobic nature. DMSO turned out to be an excellent solvent for sparsely soluble polyphenols or creamlike products.

Furthermore, it can be stated that arguments have been collected that the nature of the organic solvent – in addition to the content – influences the electrochemical conversion of cytochrome c fixed as a monolayer on gold.

4.6.1.2 Nitric Oxide Sensor Based on Cytochrome c' Electrodes

Nitric oxide can coordinate at the iron center of cytochrome c'–type proteins [57–60]. In contrast to mitochondrial cytochrome c, the iron of these proteins is mainly high-spin and pentacoordinated. It has an empty coordination site that is buried within the protein limiting the access to small molecules. Most of the cytochrome c' are dimers, with a subgroup showing the tendency to dissociate upon ligand binding. Cytochrome c' from *Chromatium vinosum* and *Rhodocyclus gelatinosus* are typical examples.

Direct cytochrome c' electrochemistry can be observed at mercaptosuccinic acid (MSA) modified gold electrodes [42, 61]. Mercaptoundecanoic acid and also shorter thiols such as mercaptopropionic acid were not effective as promoter molecules. These electrodes were also capable of adsorbing cytochrome c'. In contrast to cytochrome c, however, the adsorption was found to be dominated by nonionic interactions since adsorption could be observed at higher ionic strength (100 mM) and was facilitated at low pH (pH 5–6) where both the promoter and the protein become less charged. This conclusion was also supported by the slow rate of adsorption, which required several hours.

The immobilized cytochrome c' is stable if transferred to a protein-free buffer. A stable redox behavior at the functionalized electrodes was observed. The formal potential was -132 mV versus Ag/AgCl, and the heterogeneous electron transfer rate constant was determined to be $30 \, s^{-1}$ (*Rh. gelatinosus*). In comparison to cytochrome c with promoters of comparable chain length, the electron transfer is less effective but still fast and quasi-reversible. The shape of voltammetric peaks indicated a one-electron step for the redox transformation. The amount of protein charge was calculated from the cyclic voltammogram to be $4.5 \, \mu C \, cm^{-2}$. A study of the immobilized protein amount using surface plasmon resonance ($16 \, pmol \, cm^{-2}$) allowed the conclusion that both parts of the dimer molecule are electroactive.

When the cytochrome c'–functionalized electrode was exposed to a nitric oxide solution, an increase in the reduction current was observed. This is indicative of a signal chain from nitric oxide in solution to cytochrome c' by coordination and subsequent oxidation followed by an electrocatalytic reduction of the biomolecule. This is an opposite situation compared to the cytochrome c–superoxide system. The functionalized electrode has a biocatalytic activity for the reduction of the signal molecule. This can also be used in an amperometric mode. Consequently, the electrode potential was fixed at negative values (−220 mV vs Ag/AgCl) and the response, which is here a reduction current, was recorded. The electrode reaction was found to be rather fast, resulting in response times of a few seconds. The sensor signal depended linearly on the NO concentration in the nanomolar concentration range. This is illustrated in Figure 4.7.

The different schemes of interaction with both cytochrome electrodes can be discussed on the basis of the different redox potentials of the proteins. Whereas immobilized cytochrome c showed a quasi-reversible redox behavior at about −15 mV versus Ag/AgCl for cytochrome c' (*Rh. gelatinosus*), this was found at −132 mV versus Ag/AgCl. Thus, cytochrome c' is more easily oxidized compared to cytochrome c. Superoxide is only a moderate reducing species and therefore can only reduce the stronger oxidant – cytochrome c. In contrast, nitric oxide can be activated for reduction by interaction with cytochrome c' since the protein is a better reductant than cytochrome c.

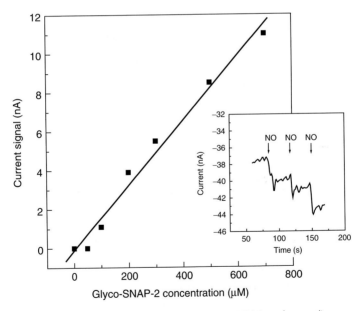

Fig. 4.7 Response at the cytochrome c' sensor (Au/MSA/cytochrome c') to additions of the NO-donor glyco-SNAP ($E = -220$ mV vs Ag/AgCl). The highest glyco-SNAP concentration corresponds to a NO concentration of about 600 nM. The insert shows the response kinetics for repeated additions of the NO-donor compound.

4.6.2
Cytochrome P 450 – An Enzyme Family Capable of Direct Electrical Communication

4.6.2.1 Mechanism and Direct Electrochemistry

Cytochromes P450 (P450) form a large family of heme enzymes that catalyze a diversity of transformations, among which are epoxidation, hydroxylation, and heteroatom oxidation. The enzymes are involved in the metabolism of many drugs and harmful compounds found in the human's environment and are responsible for bioactivation. Furthermore, their appearance indicates the presence of a respective class of compounds as P450 enzymes are induced on their exposure [62–64]. Since the first three-dimensional structures of the bacterial P450cam were elucidated by Poulos [65], several other structures have been resolved, including P450 2B4 [66] (see also [67]).

The active center is the iron-protoporphyrin IX with an axial thiolate of a cystein residue as fifth ligand to iron. Resting P450 is in the 6-coordinate low-spin ferric form.

The overall reaction of substrate hydroxylation is insertion of one atom of an oxygen molecule into a substrate RH, the second atom of oxygen being reduced to water and consumption of two reducing equivalents under formation of ROH (Eq.1).

$$RH + O_2 + 2e^- + 2H^+ \longrightarrow ROH + H_2O \tag{1}$$

The electrons are delivered by flavoproteins or ferredoxin-like proteins and NAD(P)H in a complex electron transfer system. The most generally accepted mechanism for substrate hydroxylation by P450 includes the following steps although some details remain still unsolved [68]: (1) Substrate binding to the 6-coordinate low-spin ferric enzyme excludes water from the active site, causing a change to the 5-coordinate high-spin state. The accompanied positive shift of the redox potential by about 130 mV makes the first reduction step thermodynamically favorable. (2) The transfer of one electron from a redox partner reduces the ferric iron to the ferrous enzyme. (3) This can now bind molecular oxygen forming a ferrous-dioxy (FeII-O_2) complex. (4) The second electron is transferred along with a proton, gaining an iron-hydroperoxo (FeIII-OOH) intermediate. (5) The O–O bond is cleaved to release a water molecule and a highly active iron–oxo ferryl intermediate. (6) This intermediate abstracts one hydrogen atom from the substrate to yield a one-electron reduced ferryl species (FeIV-OH) and a substrate radical, followed by enzyme-product complex formation and (7) release of product ROH to regenerate the initial low-spin state.

P450s are highly relevant to the bioanalytical area. The main problem, however, is the complexity of the monooxygenase systems, which require flavoproteins or ferredoxin-like proteins, and NAD(P)H as electron-supplying components. Furthermore, the low redox potential along with the need for oxygen makes an unfavorable working potential necessary at which oxygen reduction also takes place.

For a potential application of cytochromes P450 in bioreactors or biosensors, an interesting alternative is to substitute the biological electron donors by artificial ones like electrochemical [69, 70] or photochemical systems [71, 72]. Mediated spectroelectrochemistry based on the use of antimony-doped tin oxide electrodes has been used to determine the influence of mutations on the redox potentials [73].

Although some electrochemical aspects of P450 were reported more than 20 years ago [74, 75] the *direct, nonmediated* electrochemistry of P450 is rather difficult to obtain. On unmodified electrodes, enzymes tend to denature and to passivate the electrode. However, P450s are naturally are involved in electron transport pathways of protein redox partners, which require specific docking sites. Therefore, electrical contact to P450-enzymes should be possible at suitable surface modifications of electrodes.

In recent years, reversible one-electron transfer could be achieved with P450s assembled in or at biomembrane-like films, inorganic and organic polyion layers, and surfactant modifiers.

For example, the enzyme has been immobilized on electrodes modified with sodium montmorillonite [76] and additional detergent [77], synthetic membrane film (didodecyldimethylammonium bromide) [78, 79] and dimyristoyl-L-α-phosphatidylcholine film [80] and in thin protein-polyanion film on mercaptopropane sulfonate-coated Au electrodes [81]. In most cases, carbon electrode material has been used, but in a few cases gold with self-assembled modifier layers and tin oxide electrodes were reported. However, if gold electrodes were used with monolayers of short thiols, which were most successful for other heme proteins such as cytochrome c and peroxidase, only a very slow reduction of P450 could be obtained [82], and they could therefore not be used for sensor preparation. Direct electron transfer has also been observed using multilayer modified Au and pyrolytic graphite electrodes based on the use of alternating P450cam/polycation layers [83].

Reversible direct one-electron transfer has been reported for P450cam [76, 78–82, 84] and various mutants [85] at edge-plane graphite by using P450cam in solution. Furthermore, direct electron transfer has been shown with different P450 types such as P450 1A1 [86], P450 1A2 [87, 88], P450 3A4 [89], P450 4A1 [90], P450 2B4 [87, 91], P450 BM3 [92], P450cin [93] and P450scc [87].

The potential for the nonmediated first electron-redox transition in a number of P450s and at different electrodes is in the range of −450 mV to −100 mV versus Ag/AgCl.

At montmorillonite modified glassy carbon electrodes [76] reversible and very fast heterogeneous redox reaction of substrate-free cytochrome P450cam with a formal potential of −361 mV (vs Ag/AgCl) was obtained (Figure 4.8). The heterogeneous electron transfer rate constants reached values as high as 152 s^{-1} for P450cam [76] very similar to rates between 27–84 s^{-1} reported for the transfer of the first electron from putidaredoxin to P450cam [94, 95]. This similarity suggests that the negatively charged clay [96] obviously mimics the electrostatics of the natural redox partner putidaredoxin and may hold the P450 in a productive orientation. In this orientation, the active site of the adsorbed P450cam is still accessible for small iron ligands like CO and dioxygen and also the larger metyrapone indicated by a positive shift of the formal potential. The apparent surface coverage of the electroactive P450cam was calculated to be about 3.54×10^{-12} mol cm^{-2}, which was about 35% of the total immobilized P450 amount. The observed linearity of the peak current with the scan rate is characteristic of a surface process.

The formal potential of substrate-free P450cam is approximately 160 mV more positive than the solution value, but close to the value of the camphor-bound species in solution ($E^{\circ\prime} = -407$ mV vs Ag/AgCl 1 M KCl [69]. Thus, the interaction with the

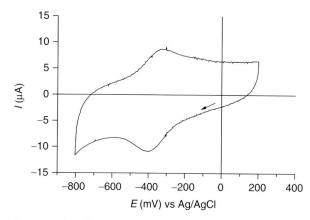

Fig. 4.8 Cyclic voltammogram of cytochrome P450cam adsorbed on montmorillonite modified glassy carbon electrode in 33 mM potassium phosphate buffer, pH 7.

matrix may force displacement of solvent in the local environment of the heme or conformational changes. Changes of the secondary structures were, however, not identified with IR-spectroscopy [76].

In a spectroelectrochemical experiment at modified gold for substrate-bound P450cam, an $E^{\circ\prime}$ of -373 mV was found (Figure 4.9). The surface interaction and direct electrochemical transformation does not effect the enzyme structure as was proved with parallel spectroscopy. The spectral changes, both upon direct electrochemical reduction and upon ligand binding, clearly indicated the native state of P450cam during reversible reduction and oxidation [82].

The liver microsomal phenobarbital induced P450 2B4 has also been incorporated in montmorillonite on glassy carbon electrodes [77]. In contrast to P450cam, this enzyme has a flavoenzyme as redox partner and does not need an iron–sulfur helping protein for delivery of electrons. Using cyclic voltammetry at low scan rates, a reduction peak is observed at around -430 mV (vs Ag/AgCl), which disappears at higher rates. The electron transfer reaction is obviously very slow. However, this process is enhanced in the presence of a nonionic detergent such as Tween 80. P450 2B4 is a membrane-bound enzyme and detergent is needed to monomerize P450 2B4 [97] as was confirmed also by AFM-studies. From the cyclic voltammograms the amount of electroactive protein of 40.5 pmol cm^{-2} was calculated. Cyclic voltammetry also demonstrates a reversible one-electron surface redox reaction with a formal potential of about -295 mV versus Ag/AgCl 1 M KCl and a heterogeneous electron transfer rate constant of 80 s^{-1}. The positive shift of the formal potential compared to the oligomeric P450 2B4 and the enzyme in the native system indicates conformational changes, which could, however, not be identified in solutions containing clay.

In many cases, the formal potential of P450 determined by the heterogeneous redox reaction is more positive than the redox potential in solution. Furthermore, the values

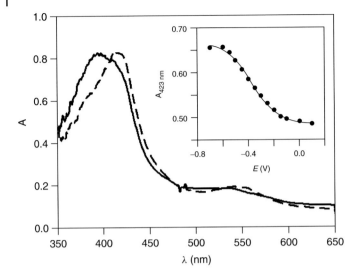

Fig. 4.9 Spectroelectrochemical picture of P450cam (12.5 µM) in 100 mM K-phosphate buffer, 50 mM KCl, 1 mM (1R)-camphor. Solid line: oxidized P450cam at +100 mV. Dashed line: reduced P450cam at −700 mV. Inset: Absorbance at 423 nm at stepwise reduction of the applied potential E. The cuvette is a cylindrical 4,4′-dithiodipyridin modified Au electrode. Light path length: 1 cm. A Ag/AgCl-reference electrode and Pt-counterelectrode was used [82].

$E^{o\prime}$ in films containing P450cam showing a distribution ranging from −231 to −350 mV versus Ag/AgCl 1 M KCl [80]. A positive shift of the redox potential may be indicative for low to high spin-state conversion that has been ascribed to strong interaction of P450 with surfaces [98]. The positive shifts of the redox potential are generally observed when water is excluded from the heme pocket as in the case of camphor binding [99] and therefore we suggested that the adsorption process leads to a dehydration of the P450 structure.

The studies of direct heterogeneous electron transfer have been carried out in most cases by cyclic and square wave voltammetry. In these studies, the first of the two electrons required for the catalytic reaction has been transferred despite the authors not seeing the shift of the reduction potential upon substrate addition as has been reported [100] and is known for the reaction in solution. In all cases, catalytic oxygen reduction is observed but only in very few cases could catalytic substrate conversion be achieved.

4.6.2.2 Catalytic P450 Sensors

An early development describes the use of liver microsomes for aniline determination in both the monooxygenase and the peroxide dependent reaction.

P450cam predominantly catalyzes the regio- and stereo-specific hydroxylation of (1R)-camphor to exclusively 5-exo-hydroxycamphor. Compounds other than camphor, such as compounds of environmental and industrial interest, have also been identified as substrates for P450cam.

Electrochemically driven epoxidation of styrene was observed at polyion-multilayers containing P450cam [81] or P450 1A2 [88]. The mechanism involves a peroxide-activated reaction step. Epoxidation is initiated by a single-electron reduction of the heme. In an electrocatalytic oxygen reduction, peroxide is formed, which activates the enzyme for styrene epoxidation. The role of peroxide has been proved by the lack of styrene oxide formation in the presence of catalase [88, 101]. Acceleration of styrene epoxidation and dehalogenation of hexachloroethane, carbon tetrachloride, and other polyhalomethanes was successful using a genetic engineering approach [73, 102, 103].

Human P450 3A4 (quinine 3-monooxygenase) is electrocatalytically active at a polycationic film-loaded gold electrode prepared with polydimethyldiallylammonium chloride adsorbed to mercaptopropane sulfonate–activated gold [89]. This immobilization causes a drastic anodic potential shift to about +98 mV (vs NHE) indicating conformational changes. For drug-sensing dealkylation of verapamil, midazolam, progesterone, and quinidin was followed. Addition of mMolar amounts of verapamil or midazolam to oxygenated solution increased the reduction current. In the case of the human P450 3A4, peroxide had only a minor effect. Furthermore, product analysis after electrolysis at 500 mV under aerobic conditions confirmed the demethylation and dealkylation of verapamil at a rate of about 4.23 and 5.05 min^{-1}.

Attempts are being made to contact P450-enzymes to electrodes by introducing electroactive bridges covalently coupled to the protein [70, 85]. Such redox relay has been introduced at specifically selected sites generated by protein engineering or randomly. In the case of the P450 2B4, the native redox partner is a flavoenzyme. Therefore, riboflavin was proposed to be a possible electron donor for P450 2B4. Riboflavin was covalently attached to the enzyme and immobilized on rhodium-graphite screen-printed electrodes [70]. At a constant potential of −500 mV (vs Ag/AgCl) substrates such as aminopyrine, aniline, and 7-pentoxyresorufin were converted.

We succeeded in developing biosensors based on mediator-free P450 2B4 catalysis by immobilizing monomerized P450 2B4 in montmorillonite [77]. When aminopyrine was added to air saturated buffer solution, there was an increase in the reduction current (Figure 4.10). Figure 4.11 presents a dependence of the chronoamperometric response upon aminopyrine addition with a detection limit of 1 mM of aminopyrine and 1.2 mM of benzphetamine. The reaction was inhibited by metyrapone. This indicates that P450 2B4 possesses catalytic activity in the presence of substrate. Further evidence was delivered by product analysis. After 1 h of controlled potential electrolysis at −500 mV versus Ag/AgCl, formaldehyde was measured. The apparent catalytic rate related to the amount of electroactive protein is $k'_{cat} = 1.54$ min^{-1}, which is comparable to the value $k_{cat} = 3.5$ min^{-1} of the microsomal system [87].

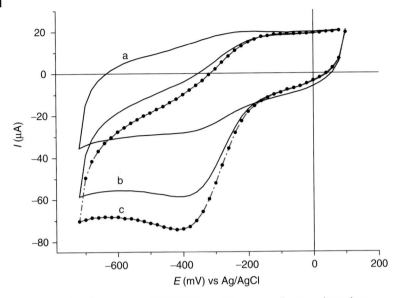

Fig. 4.10 Cyclic voltammetry at P450 2B4/Tween80/montmorillonite – electrode in 100 mM phosphate buffer, 50 mM KCl, pH 7.4, (a) argon saturated, (b) air saturated, and (c) air saturated after introduction of 8 mM aminopyrine scan rate 10 V s^{-1}.

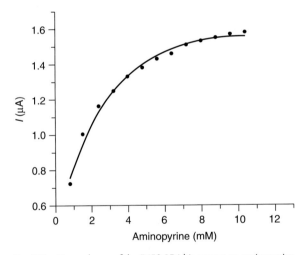

Fig. 4.11 Dependence of the P450 2B4 biosensor on aminopyrine and benzphetamine. Measurements were made in air saturated stirred 100 mM phosphate buffer solution, 50 mM KCl, pH 7.4, at a constant potential −500 mV versus Ag/AgCl.

4.7
Conclusions

Electrically contacted enzymes play the leading role in personal blood glucose monitoring, which at present represents a world market of almost five billion US dollars. The principal advantage of electrically contacting enzymes is their high sensitivity combined with the absence of electrically interfering sample constituents. These features also have a high potential for electrochemical affinity probes, for example, immunosensors or nucleic acid hybridization. In fact, laccase and peroxidase have been used for signal generation via the direct electron transfer in binding assays. In the future, two directions might be envisaged: (1) single-molecule bio-electrochemistry and (2) bioelectronic switching and storage functionalities.

References

1 S.D. Varfolomeev, I.V. Berezin, *J. Mol. Catal.*, **1978**, *4*, 387–399.
2 M.R. Tarasevich, V.A. Bogdanovskaya, *Topics in Bioelectrochemistry and Bioenergetics* (Ed.: G. Milazzo), Vol. 5, John Wiley & Sons, New York, **1983**, 225–260.
3 V.J. Razumas, J.J. Jasaitis, J.J. Kulys, *Bioelectrochem. Bioenerg.*, **1984**, *12*, 297–322.
4 I. Willner, E. Katz, B. Willner, *Electroanalysis*, **1997**, *9*, 965–977.
5 I. Willner, E. Katz, B. Willner, R. Blonder, V. Heleg-Shabtai, A.F. Bückmann, *Biosens. Bioelectron.*, **1997**, *12*, 337–356.
6 F.A. Armstrong, G.S. Wilson, *Electrochim. Acta*, **2000**, *45*, 2623–2645.
7 H.J. Bergmeyer, M. Grassl, *Methods of Enzymatic Analysis*, VCH Verlagsgesellschaft, Weinheim, **1986**.
8 T.G.M. Schalkhammer, *Analytical Biotechnology, Methods and Tools in Biosciences and Medicine*, Birkhäuser Verlag, Basel, **2002**.
9 O. Sonntag, *Trockenchemie*, Georg Thieme Verlag, Stuttgart, **1988**.
10 P.W. Carr, L.D. Bowers, *Immobilized Enzymes in Analytical and Clinical Chemistry, Chemical Analysis, Fundamentals and Applications* (Eds.: P.J. Elving, J.D. Winefordner, I.M. Kolthoff), Vol. 56, John Wiley & Sons, New York, **1982**.
11 F.W. Scheller, U. Wollenberger, *Encyclopedia of Electrochemistry* (Eds.: A.J. Bard, M. Stratmann), Vol. 9 *Bioelectrochemistry* (Ed.: G.S. Wilson), Wiley-VCH, Weinheim, **2002**.
12 L.C. Clark, C. Lyons, *Ann. N.Y. Acad. Sci.*, **1962**, *102*, 29–36.
13 S.J. Updike, G.P. Hicks, *Nature*, **1967**, *214*, 986–988.
14 F.W. Scheller, F. Schubert, *Biosensor, Techniques and Instrumentation in Analytical Chemistry*, Vol. 11, Elsevier, Amsterdam, **1992**.
15 D.R. Thévenot, K. Toth, R.A. Durst, G.S. Wilson, *Biosens. Bioelectron.*, **2001**, *16*, 121–131.
16 F.W. Scheller, U. Wollenberger, A. Warsinke, F. Lisdat, *Curr. Opin. Biotech.*, **2001**, *12*, 35–40.
17 L. Gorton, A. Lindgren, T. Larsson, F.D. Munteanu, T. Ruzgas, I. Gazaryan, *Anal. Chim. Acta*, **1999**, *400*, 91–108.
18 J. Wang, *Electroanalysis*, **2001**, *13*, 983–988.
19 I. Katakis, A. Heller, *Frontiers in Biosensors I, Fundamental Aspects* (Eds.: F.W. Scheller, F. Schubert, J. Fedrowitz), Birkhäuser Verlag, Basel, **1997**, 229–242.
20 W. Schuhmann, *Rev. Mol. Biotech.*, **2002**, *82*, 425–442.
21 Y. Degani, A. Heller, *J. Phys. Chem.*, **1987**, *91*, 1285–1291.

22 A.L. Ghindilis, P. Atanasov, E. Wilkins, *Electroanalysis*, **1997**, *9*, 661–674.

23 T. Ikeda, *Frontiers in Biosensorics, I, Fundamental Aspects* (Eds.: F.W. Scheller, F. Schubert, J. Fedrowitz), Birkhäuser Verlag, Basel, **1997**.

24 D. Pfeiffer, *Frontiers in Biosensorics, I Fundamental Aspects; II Practical Applications* (Eds.: F.W. Scheller, F. Schubert, J. Fedrowitz), Birkhäuser Verlag, Basel, **1997**.

25 A. Cass, G. Davis, G. Francis, H.A. Hill, W. Aston, I.J. Higgins, E. Plotkin, L. Scott, A.P. Turner, *Anal. Chem.*, **1984**, *56*, 667–673.

26 V. Poitout, D. Moatti-Sirat, G. Reach, Y. Zhang, G.S. Wilson, F. Lemonnier, J.-C. Klein, *Diabetologia*, **1993**, *36*, 658–665.

27 M. Feelisch, J.S. Stamler, Eds., *Methods in Nitric Oxide Research*, John Wiley & Sons, Chichester, **1996**.

28 J.G. Scandalios, Ed., *Oxidative Stress and the Molecular Biology of Antioxidant Defenses*, Cold Spring Harbor Laboratory Press, **1997**.

29 J.S. Beckman, W.H. Koppenol, *Am. J. Physiol.*, **1996**, *271*, C1424–C1437.

30 D.A. Wink, J.B. Mitchell, *Free Radical Biol. Med.*, **1998**, *25*, 434–456.

31 F. Lisdat, F.W. Scheller, *Anal. Lett.*, **2000**, *33*, 1–16.

32 M.I. Song, F.F. Bier, F.W. Scheller, *Bioelectrochem. Bioenerg.*, **1995**, *38*, 419–426.

33 V. Lvovich, A. Scheeline, *Anal. Chem.*, **1997**, *69*, 454–462.

34 T. Malinski, Z. Taha, *Nature*, **1992**, *358*, 676–678.

35 F. Bedioui, S. Trevin, J. Devnynck, F. Lantoine, A. Brunet, M.A. Devnynck, *Biosens. Bioelectron.*, **1997**, *12*, 205–212.

36 O. Raveh, N. Peleg, A. Bettleheim, I. Silberman, J. Rishpon, *Bioelectrochem. Bioenerg.*, **1997**, *43*, 19–23.

37 C.J. McNeil, K.A. Smith, *Free Radical Res. Commun.*, **1989**, *7*(2), 89–94.

38 J.M. McCord, I. Fridovich, *J. Biol. Chem.*, **1969**, *244*, 6049–6055.

39 M. Bayachou, R. Lin, W. Cho, P.J. Farmer, *J. Am. Chem. Soc.*, **1998**, *120* 9888–9895.

40 S. Kröning, F.W. Scheller, U. Wollenberger, F. Lisdat, *Electroanalysis*, **2004**, *16*, 253–259.

41 S.L.R. Barker, R. Kopelman, T.E. Meyer, M.A. Cusanovich, *Anal. Chem.*, **1998**, *70*, 971–976.

42 B. Ge, T. Meyer, M. Schoening, U. Wollenberger, F. Lisdat, *Electrochem. Comm.*, **2000**, *2*, 557–561.

43 S. Song, R.A. Clark, E.F. Bowden, M.J. Tarlov, *J. Phys. Chem*, **1993**, *97*, 6564–6572.

44 Z.Q. Feng, S. Imabayashi, T. Kakiuchi, K. Niki, *J. Electroanal. Chem.*, **1995**, *394*, 149–154.

45 M. Collinson, E.F. Bowden, *Langmuir*, **1992**, *8*, 1247–1250.

46 W. Jin, U. Wollenberger, E. Kaergel, W.-H. Schunck, F.W. Scheller, *J. Electroanal. Chem.*, **1997**, *433*, 135–139.

47 A.E. Kasmi, J.M. Wallace, E.F. Bowden, S.M. Binet, R.J. Linderman, *J. Am. Chem. Soc.*, **1998**, *120*, 225–226.

48 F. Lisdat, B. Ge, M.E. Meyerhoff, F.W. Scheller, *Probe Microsc.*, **2001**, *2*, 113–120.

49 J.M. Cooper, K.R. Greenough, C.J. McNeil, *J. Electroanal. Chem.*, **1993**, *347*, 267–275.

50 F. Lisdat, B. Ge, E. Ehrentreich-Förster, R. Reszka, F.W. Scheller, *Anal. Chem.*, **1999**, *71*, 1359–1365.

51 K. Tammeveski, T.T. Tenno, A.A. Mashirin, E.W. Hillhouse, P. Manning, C.J. McNeil, *Free Radical Biol. Med.*, **1998**, *25*, 973–978.

52 B. Ge, F. Lisdat, *Anal. Chim. Acta*, **2002**, *454*, 53–64.

53 R. Büttemeyer, A. Philipp, B. Ge, F.W. Scheller, F. Lisdat, *Microsurgery*, **2002**, *22*, 108–113.

54 F. Lisdat, B. Ge, R. Reszka, E. Kozniewska, *Fresenius' J. Anal. Chem.*, **1999** *365*, 494–498.

55 S. Ignatov, D. Shishniashvili, B. Ge, F.W. Scheller, F. Lisdat, *Biosens. Bioelectron.*, **2002**, *17*, 191–199.

56 M. Beissenhirtz, F.W. Scheller, F. Lisdat, *Electroanalysis*, **2003**, *15*, 1425–1435.

57 T. Yoshimura, H. Iwasaki, S. Shidara, S. Suzuki, A. Nakahara, T. Matsubara, *J. Biochem.*, **1988**, *103*, 1016–1023.

58 T. Yoshimura, S. Fuji, H. Kamada, K. Yamaguchi, S. Suzuki, S. Shidara, S. Takakuwa, *Biochim. Biophys. Acta,* **1996**, *1292,* 39–46.

59 Zh. Ren, T. Meyer, D.E. McRee, *J. Mol. Biol.,* **1993**, *234,* 433–445.

60 S. Benini, W.R. Rypniewski, K.S. Wilson, S. Ciurli, *Acta Crystallogr.,* **1998**, *D54,* 284–287.

61 T. Erabi, Sh. Ozawa, Sh. Hayase, M. Wada, *Chem. Lett.,* **1992**, *11,* 2115–2118.

62 T.L. Poulos, *Curr. Opin. Struct. Biol.,* **1995**, *5,* 767–774.

63 P.R. Ortiz de Montellano, Ed., *Cytochrome P450: Structure, Mechanism, and Biochemistry,* Plenum Press, New York, **1995**.

64 D.F.V Lewis, *Cytochromes P450 – Structure, Function and Mechanism,* Taylor & Francis, London, **1996**.

65 T.L. Poulos, B.C. Finzel, I.C. Gunsalus, G.C. Wagner, J. Kraut, *J. Biol. Chem.,* **1985**, *260,* 16122–16130.

66 E.E. Scott, Y.A. He, M.R. Wester, M.A. White, C.C. Chin, J.R. Halpert, E.F. Johnson, C.D. Stout, *Proc. Natl. Acad. Sci. U.S.A.,* **2003**, *100,* 13196–13201.

67 D. Werck-Reichhart, R. Feyereisen, *Genome Biol.,* **2000**, *1,* 3003.1–3003.9.

68 K. Auclair, Z. Hu, D.M. Little, P.R. Ortiz de Montellano, J.T. Groves, *J. Am. Chem. Soc.,* **2002**, *124,* 6020–6027.

69 R.W. Estabrook, K.M. Faulkner, M.S. Shet, C.W. Fisher, *Adv. Enzymol.,* **1996**, *272,* 44–51.

70 V.V. Shumyantseva, T.V. Bulko, T.T. Bachmann, U. Bilitewski, R.D. Schmid, A.I. Archakov, *Arch. Biochem. Biophys.,* **2000**, *376,* 43–48.

71 M.J. Hintz, J.A. Peterson, *J. Biol. Chem.,* **1980**, *255,* 7317–7325.

72 J. Contzen, C. Jung, *Biochemistry* **1999**, *38,* 16253–16260.

73 V. Reipa, M.P. Mayhew, M.J. Holden, V.L. Vilker, *Chem. Commun.,* **2002**, 318–319.

74 F.W. Scheller, R. Renneberg, G. Strnad, K. Pommerening, P. Mohr, *Bioelectrochem. Bioenerg.,* **1977**, *4,* 500–507.

75 G. Dryhurst, K.M. Kadish, F.W. Scheller, R. Renneberg, *Biological Electrochemistry,* Vol. 1, Academic Press, New York, **1982**, 398–521.

76 C. Lei, U. Wollenberger, C. Jung, F.W. Scheller, *Biochem. Biophys. Res. Commun.,* **2000**, *268,* 740–744.

77 V.V. Shumyantseva, Y.D. Ivanov, N. Bistolas, F.W. Scheller, A.I. Archakov, U. Wollenberger, **2004**, *76,* in press.

78 E.I. Iwuoha, S. Joseph, Z. Zhang, M.R. Smyth, U. Fuhr, P.R. Ortiz de Montellano, *J. Pharm. Biomed. Anal.,* **1998**, *17,* 1101–1110.

79 E.I. Iwuoha, M.R. Smyth, *Biosens. Bioelectron.,* **2003**, *18,* 237–244.

80 Z. Zhang, A.E.F. Nassar, Z. Lu, J.B. Schenkman, J.F. Rusling, *J. Chem. Soc. Faraday Trans.,* **1997**, *93,* 1769–1774.

81 X. Zu, Z. Lu, Z. Zhang, J.B. Schenkman, J.F. Rusling, *Langmuir,* **1999**, *15,* 7372–7377.

82 N. Bistolas, A. Christenson, T. Ruzgas, C. Jung, F.W. Scheller, U. Wollenberger, *Biochem. Biophys. Res. Commun.,* **2004**, *314,* 810–816.

83 Y.M. Lvov, Z. Lu, J.B. Schenkman, X. Zu, J.F. Rusling, *J. Am. Chem. Soc.,* **1998**, *120,* 4073–4080.

84 J. Kazlauskaite, A.C.G. Westlake, L.L. Wong, H.A.O. Hill, *Chem. Commun.,* **1996**, 2189–2190.

85 K.K. Lo, L.L. Wong, H.A.O. Hill, *FEBS Lett.,* **1999**, *451,* 342–346.

86 M. Hara, Y. Yasuda, H. Toyotama, H. Ohkawa, T. Nozawa, J. Miyake, *Biosens. Bioelectron.,* **2002**, *17,* 173–179.

87 V.V. Shumyantseva, T.V. Bulko, S.A. Usanov, R.D. Schmid, C. Nicolini, A.I. Archakov, *J. Inorg. Biochem.,* **2001**, *87,* 185–190.

88 C. Estavillo, Z. Lu, I. Jansson, J.B. Schenkman, J.F. Rusling, *Biophys. Chem.,* **2003**, *104,* 291–296.

89 S. Joseph, J.F. Rusling, Y.M. Lvov, T. Friedberg, U. Fuhr, *Biochem. Pharmacol.,* **2003**, *65,* 1817–1836.

90 K.M. Faulkner, M.S. Shet, C.W. Fisher, R.W. Estabrook, *Proc. Natl. Acad. Sci. U.S.A.,* **1995**, *92,* 7705–7709.

91 V.V. Shumyantseva, Y.D. Ivanov, N. Bistolas, F.W. Scheller, A.I. Archakov, U. Wollenberger, **2004**, submitted.

92 B.D. Flemming, Y. Tian, S.G. Bell, L.L. Wong, V. Urlacher, H.A.O. Hill, *Eur. J. Biochem.*, **2003**, *270*, 4082–4088.

93 K.-F. Aguey-Zinsou, P.V. Bernhardt, J.J. De Voss, K.E. Slessor, *Chem. Commun.*, **2003**, *3*, 418–419.

94 M.T. Fisher, S.G. Sligar, *J. Am. Chem. Soc.*, **1985**, *107*, 5018–5019.

95 C. Mouro, A. Bondon, C. Jung, G. Hui Bon Hoa, J.D. De Certaines, R.G.S. Spencer, G. Simonneaux, *FEBS Lett.*, **1999**, *455*, 302–306.

96 D. Ege, P.K. Ghosh, J.R. White, J.-F. Equey, A.J. Bard, *J. Am. Chem. Soc.*, **1985**, *107*, 5644–5652.

97 O.I. Kiselyova, I.V. Yaminsky, Y.D. Ivanov, I.P. Kanaeva, V.Y. Kuznetsov, A.I. Archakov, *Arch. Biochem. Biophys.*, **1999**, *371*, 1–7.

98 K. Niki, *Encyclopedia of Electrochemistry* (Eds.: A.J. Bard, M. Stratmann), Vol. 9 *Bioelectrochemistry* (Ed.: G.S. Wilson), Wiley-VCH, Weinheim, **2002**.

99 T.L. Poulos, B.C. Finzel, A.J. Howard, *Biochemistry*, **1986**, *25*, 5314–5322.

100 B. Munge, C. Estavillo, J.B. Schenkman, J.F. Rusling, *Chem. Biochem.*, **2003**, *4*, 101–108.

101 M.E. Walsh, P. Kyritsis, N.A.J. Eady, H.A.O. Hill, L.L. Wong, *Eur. J. Biochem.*, **2000**, *267*, 5815–5820.

102 M.P. Mayhew, V. Reipa, M.J. Holden, V.L. Vilker, *Biotechnol. Prog.*, **2000**, *16*, 610–616.

103 I. Fridovich, *J. Biol. Chem.*, **1970**, *245*, 4053–4058.

5
Electrochemical DNA Sensors

Emil Palecek and Miroslav Fojta

5.1
Introduction

In 1955, E. Chargaff and J. N. Davidson summarized the current knowledge on nucleic acids (NAs) in two volumes of their "NUCLEIC ACIDS. Chemistry and Biology". It was declared in this book that among the NA bases only adenine (A) is reducible at mercury electrodes (at strongly acidic pHs) while other bases are inactive [1]. In 1957, H. Berg [2] used polarographic methods to study proteins in DNA and RNA samples and he concluded, in agreement with the above data, that in difference to proteins, DNA and RNA are polarographically inactive species. At the same time, experiments by one of us (Emil Palecek) were showing something very different. Cytosine (C) and A were reducible at neutral or weakly acid pHs and guanine (G) produced a characteristic anodic signal, if the so-called oscillographic polarography at controlled a.c. (constant a.c. chronopotentiometry according to the present terminology) was used [3–6]. These phenomena were then studied for almost a decade and formed a solid background for studies of NAs by other electrochemical methods and different electrodes (reviewed in [7, 8]). Polarographic research of DNA with mercury electrodes brought an early evidence of DNA premelting and polymorphy of the DNA double helix many years before the DNA X-ray crystal structures became available (reviewed in [9]).

At present we know that NA bases undergo reduction and oxidation processes at electrodes, and studies of long chain DNA and RNA molecules showed that the electrochemical signals of these molecules can be significantly influenced by their ordered higher structures (reviewed in [10–15]). Electrochemical methods can be thus applied to detect minor damages to the DNA double helix and to study structural transitions in nucleic acids. Various kinds of NA interactions can be studied by electrochemical methods. Binding of electroactive substances to DNA can be manifested by changes in their signals. Both electroactive and electroinactive compounds may affect the NA signals, particularly if their binding results in changes of the DNA structure. In addition to the natural electroactivity of NAs, electroactive markers can be introduced into them. Particularly end-labeling of DNA with various redox-active compounds appears important in the development of the DNA hybridization sensors [13, 16, 17]. The first electroactive labels, based on osmium complexes, were introduced in DNA in the beginning of the 1980s [18–22]. They are convenient for easy DNA end-labeling as

Bioelectronics. Edited by Itamar Willner and Eugenii Katz
Copyright © 2005 WILEY-VCH Verlag GmbH & Co. KGaA, Weinheim
ISBN: 3-527-30690-0

well as for labeling of all pyrimidines in target DNA (Sections 5.2.3.2 and 5.3.4). To our knowledge, the first DNA-modified electrode was used in the middle of the 1980s [23] and few years later, instead of linear DNA, supercoiled (sc) DNA was immobilized on the electrode surface [24–26] to create a new tool for sensing DNA damage (Section 5.4.4.1).

5.1.1
Indicator Electrodes

NAs and particularly DNA have been analyzed primarily with liquid mercury and solid carbon electrodes [27–33] (reviewed in [7, 11, 14, 16, 20, 34]), including the recently applied boron-doped diamond electrodes [35]. Some work has also been done with other electrodes such as gold, copper, silver and platinum. In recent times, gold, carbon, indium tin oxide (ITO), solid amalgam electrodes (SAE) and mercury film electrodes (MFE) are gaining importance in connection with the development of DNA sensors (see Sections 5.3 and 5.4). Mercury electrodes and some SAE operate usually between 0 and -2 V against the saturated calomel electrode at neutral and weakly alkaline pHs. Compared to mercury-containing electrodes, the potential windows of most of the solid electrodes are shifted by approximately 1 V to more positive values. The latter electrodes are thus better suited for studying NA oxidation, while mercury-containing electrodes (both liquid and solid) are better suited for investigating reduction of NAs and their binders. The atomically smooth and highly reproducible surfaces of liquid mercury are particularly suitable for a.c. impedance measurements, including the impedance spectroscopy, which can provide important information about DNA and RNA interfacial properties ([36] and references therein).

5.1.2
Electrochemical Methods

Starting in the middle of the 1960s, in addition to the oscillographic polarography with controlled a.c., various electrochemical methods have been applied for the analysis of NAs, including differential pulse polarography (DPP) [37] and differential pulse voltammetry (DPV), square-wave voltammetry (SWV) [38], cyclic voltammetry (CV) [39] a.c. voltammetry (a.c.V) [40–42] and constant current chronopotentiometry (CP) [31]. It has been known for decades that G and A residues in DNA and RNA molecules are oxidizable at carbon electrodes; their voltammetric peaks were, however, poorly developed, providing insufficient sensitivity in the DNA analysis. Recent application of sophisticated baseline correction techniques, mainly in connection with CP and SWV [31, 43, 44], produced well-developed oxidation peaks of DNA and RNA at carbon electrodes (Figure 5.1C,D). Elimination voltammetry was used in DNA analysis with mercury [45, 46] and solid electrodes [47, 48]. AC impedance methods were widely applied in the analysis of NAs [36, 49–56]) and their components [57]. In the first part of this article, we wish to briefly summarize present knowledge on DNA natural

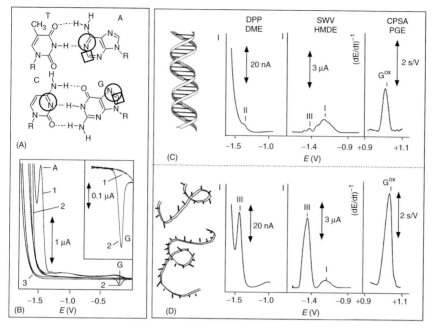

Fig. 5.1 Reduction and oxidation of single-stranded and double-stranded nucleic acids at electrodes. (A) Schematic representation of Watson–Crick base pairs and electroactive groups (Sections 5.2.1.1 and 5.2.2.1). Primary reduction and oxidation sites at mercury (circles) and carbon electrodes (rectangles) are shown. (B) Cyclic voltammograms of biosynthetic polyribonucleotides (100 µg mL^{-1}) measured at the hanging mercury drop electrode (HMDE). (1), polyadenylic acid; (2), polyguanylic acid; (3), background electrolyte; *inset*, detail of peak G. (C and D) Redox signals (C), obtained with 100 µg^{-1} mL of native (double-stranded) or (D), with 50 µg mL^{-1} of denatured ssDNA by differential pulse polarography (DPP) (Section 5.1.2) at the static dropping mercury electrode (DME) (Section 5.1.1), by adsorptive stripping square-wave voltammetry (SWV) at a hanging mercury drop electrode (HMDE), and by constant current chronopotentiometric stripping analysis (CPSA) at a pyrolytic graphite electrode (PGE). Adapted with permission from [16]. Copyright 2001 American Chemical Society.

electroactivity and DNA labeling (Section 5.2) and then give a survey on sensors for DNA hybridization (Section 5.3) and DNA damage (Section 5.4) in the following Sections.

5.2
Natural Electroactivity and Labeling of Nucleic Acids

5.2.1
Electroactivity of Nucleic Acid Components

Among NA components, purine and pyrimidine bases are electroactive. At most types of electrodes deoxyribose and ribose in nucleosides and nucleotides as well as phosphate groups in the sugar-phosphate backbone are electroinactive, while bases are able to undergo electroreduction and/or electrooxidation (reviewed in [10, 13]). Adsorption/desorption properties of NA components were studied by different techniques, including a.c. impedance [50, 58–60].

5.2.1.1 Reduction and Oxidation of Bases

Primary reduction sites of A, C and G (reviewed in [10, 13]) as well as oxidation sites of A and G (reviewed in [13]) at mercury and carbon electrodes are shown in (Figure 5.1A). Signals caused by reduction of A and C at mercury electrodes are strongly influenced by the DNA secondary structure. They are thus well suited for studies of changes in DNA structure in solution (Section 5.2.2.1) and at the electrode surface (Section 5.2.2.3). Carbon electrodes appear less sensitive to changes in DNA structure [11, 14]. Small differences in the intensities of the double-stranded (ds) and single-stranded (ss) DNA oxidation were explained by the difference in flexibility of these two DNA forms rather than by the direct influence of the DNA secondary structure. For a long time, DNA was supposed to be electroinactive at a gold electrode. Recently, Ferapontova and Domingues [61] reported oxidation of G at polycrystalline Au electrodes.

5.2.1.2 Sparingly Soluble Compounds with the Electrode Mercury

It was shown that all NA bases and a number of purine and pyrimidine derivatives produced sparingly soluble compounds with mercury and could be determined by the cathodic stripping voltammetry (CSV) at nanomolar and subnanomolar concentrations ([62–65]; reviewed in [13]).

Groups responsible for the formation of the mercury compounds may involve exocyclic and ring nitrogens in dependence of the nature of the base and experimental conditions ([65] and references therein). For example, at alkaline pH the mercury binding site of adenine is the 6-aminogroup. Pyrimidine nucleosides are inactive because of substitution of the pyrimidine ring by a sugar residue at N-1, involved in the mercury binding in bases. In difference to other pyrimidine nucleosides, pseudouridine (in which the sugar moiety is bound to C-5 while N-1 is free) produces electrochemical responses typical of mercury compounds [64]. Purine nucleosides (the sugar residues bound to N-9) behave similarly to their parent bases.

5.2.1.3 Analysis of Nucleic Acid Components

A and G can be analyzed at carbon electrodes and A and C at mercury electrodes by DPV, DPP and elimination voltammetry at micromolar concentrations [66]. Determination of bases by CSV (see above) gives substantially better sensitivities (see Section 5.2.1.2). At nanomolar concentrations, 100-fold excess of various substances, including proteins and NAs, had little influence on the CSV peak height of the NA bases. Mixtures of bases with NAs can be analyzed. Sensitivity of the determination of purine and pyrimidine bases can be increased by 1 to 3 orders of magnitude if the analysis is performed in alkaline media in the presence of copper ions or at copper SAE (CuSAE, [67] and references therein).

5.2.2
Analysis of Unlabeled Nucleic Acids

C and A residues in ssDNA and RNA are reduced at mercury electrodes, yielding reduction signals at about −1.4 V at acid and neutral pH. In cyclic modes, G produces an anodic signal close to −0.3 V, for which oxidation of the G reduction product (formed at the highly negative potentials) is responsible (Figure 5.1B) (reviewed in [13]). G and A residues are oxidized at carbon electrode at about 1.0 and 1.2 V, respectively.

5.2.2.1 Electrochemical Signals Respond to DNA Structure

In Watson–Crick base pairing, the primary reduction sites of A and C are hidden in the interior of the DNA double helix, forming a part of the hydrogen bonding system (Figure 5.1A). The DNA double-helical structure strongly influences the reduction signals of A and C; the reduction DPP peak of denatured (ss) calf thymus DNA can be almost 100 times higher than that of the parental duplex (native) calf thymus DNA (reviewed in [10–13]). Similarly, the DNA adsorption, at mercury electrodes, reflected by a.c. impedance (nonfaradaic) signals, is highly sensitive to changes in DNA structure, as accessibility of bases for their interaction with the electrode surface plays an important role in DNA adsorption (reviewed in [13]). High sensitivities of electrochemical signals to changes in the DNA structure were utilized for detection of local conformational changes and DNA helix-coil transitions (reviewed in [9, 10, 13]). Neither the reduction site of G at mercury electrodes nor the oxidation sites of A and G at carbon electrodes are involved in Watson–Crick hydrogen bonding (Figure 5.1A). The sensitivity of these signals to DNA structural changes is rather limited. For example, differences in the peak heights of ss and dsDNAs are much smaller as compared to those of the DNA cathodic at mercury electrodes (Figure 5.1C,D), and small changes in the DNA structure such as the single-strand breaks (ssb) are not detectable at carbon electrodes (Section 5.4).

Mercury electrodes reflect single-strand interruptions (see Section 5.6.4) in linear and circular DNA molecules (reviewed in [10, 13]), as well as superhelix density-dependent structural transitions in DNA [68]. These electrodes are able to discriminate between RNA and DNA, making possible the most sensitive label-free determination of traces of RNA in a large excess of DNA [69–71]. Anodic signal of G at mercury electrodes and oxidation signals of A and G at carbon electrodes showed lower sensitivity to changes in NA structure (reviewed in [13]). The behavior of ss and ds RNAs resembled that of ss and dsDNAs (reviewed in [10, 13]) but under certain conditions potentials of voltammetric peaks of RNAs differed from those of DNA [69–71]. RNAs were studied to a lesser extent than DNA.

5.2.2.2 DNA-modified Electrodes

Since its beginning (by the end the 1950s), electrochemical analysis of DNA was performed with the electrode immersed into the DNA solution [10, 11]. This arrangement was acceptable when the experiments were performed with chromosomal

DNA and biosynthetic polynucleotides, which were easy to prepare in relatively large quantities. When the interest of scientists turned to well-defined DNA molecules with known nucleotide sequences, such as the viral and plasmid DNAs, DNA fragments and synthetic oligodeoxyribonucleotides (ODN), whose preparation was expensive and/or laborious, working with 1 to 10 mL of the analyte became unreasonable.

In 1986, we proposed a simple method of preparation of DNA-modified electrodes [23, 24, 72]. In this method (Figure 5.2), the bare electrode was immersed into a small drop of a DNA solution (3–10 μl) for a short time (usually 60 s), the electrode was then washed and the electrochemical measurement was performed in a usual electrolytic cell with the DNA-modified electrode dipped in the blank background electrolyte. Both ss and dsDNAs were irreversibly adsorbed either at mercury or at some carbon electrodes such as pyrolytic graphite electrode (PGE), carbon paste electrode (CPE) or screen-printed electrode (SPE), resisting washing and other treatments [34, 72]. This process did not work with gold electrodes where DNA was not so strongly adsorbed. At present, DNA-modified electrodes prevail in electrochemical analysis of NAs. Various kinds of covalent immobilization of NAs at electrodes were used. For example, DNA end-labeled with – SH groups can be conveniently bound to gold and mercury electrodes (Section 5.6.3.3.1). With the mercury electrodes, such immobilization is much faster than with the gold ones [73]. Positively charged carbon electrodes with electrostatically bound probe DNA can be used in DNA hybridization sensors. On the contrary, strong

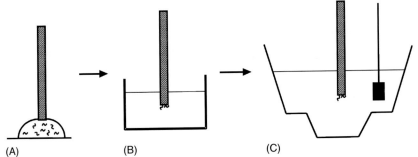

Fig. 5.2 Scheme of simple preparation of DNA-modified electrodes (Section 5.2.2.2). A bare electrode (A) is immersed in a small (usually 3–10 μL) drop of a DNA solution for a short time (e.g. 60 s). (B) After removing from the DNA solution, the DNA-modified electrode is washed and (C) immersed into a blank background electrolyte in an electrolytic cell to perform the electrochemical measurements. This procedure is also called *Adsorptive Transfer Stripping* (AdTS) [11, 24, 34] reviewed in [13]. For such a simple preparation of DNA-modified electrodes, mercury, solid amalgam and carbon, but not gold, electrodes can be used. A number of carbon electrodes were tested; GCE appeared least suitable for this purpose while pyrolytic graphite, carbon paste and some screen-printed electrodes yielded excellent results. The DNA immobilization can be performed at open current circuit but with carbon electrodes application of positive potential may be beneficial. With microelectrodes, much smaller DNA volumes can be used. The above procedure works well also with RNA, oligonucleotides, proteins, peptides and other biomacromolecules. In addition to the DNA, determination with DNA-modified electrodes, interactions with NA complementary strands, specific proteins, drugs, DNA-damaging agents and so on, can be studied.

hydrophobic interactions of bases of the probe DNA with the mercury surface may prevent an efficient DNA hybridization at the mercury electrode.

5.2.2.3 DNA Structure at the Electrode Surface

It was shown that dsDNA could be adsorbed at the mercury electrode in a wide range of potentials without any detectable structural change (reviewed in [8, 9]). However in 1977, relatively slow changes in electrochemical signals of dsDNA were observed in a narrow potential range ([20, 74] and references therein). At neutral and weakly alkaline pHs, prolonged contact of dsDNA with the mercury surface charged to potentials around -1.2 V resulted in changes in the DNA electrochemical responses (reviewed in [10, 13]). These changes were explained by opening of the DNA double helix at the electrode surface. They were relatively slow (tens to hundreds of s for the duplex opening of a chromosomal DNA) and were detectable by various techniques. To avoid these changes, methods working with small voltage excursions during the (DME) drop lifetime should be applied. No such changes were observed in covalently closed circular (ccc) DNAs [25, 26, 68] regardless of the applied technique. Direct current field–induced denaturation of short double-stranded DNA molecules was recently detected by surface plasmon resonance at gold electrodes [75].

5.2.2.4 Electrocatalysis

The term "electrocatalysis" refers to catalysis of an electrochemical reaction.

The effect of an electrocatalyst is to increase the rate of the electrochemical reaction, which is usually expressed by the electrolytic current [76]. By means of electrocatalysis, quantitation of the catalyst itself or of some substrate compounds (that are not amenable to direct electrochemical determination) is possible. Electrocatalysis can proceed at bare and modified electrodes. Engineering of catalytic layers on electrode surfaces resulted in a great variety of chemical and biochemical sensors (for DNA biosensors involving electrocatalysis, see also Sections 5.3.3 and 5.3.4). A characteristic feature of bare-electrode processes is the occurrence of both the substrate and the catalyst in the bulk of solution. Three types of electrocatalytic processes at bare electrodes can be distinguished: (a) processes with a complex-forming regenerative step (catalytic reduction of some metal ions), (b) redox-type catalytic processes (mediated electron transfer) and (c) processes with proton transfer as a regenerative step (catalytic hydrogen evolution). Catalytic hydrogen evolution has been utilized in the analysis of peptides and proteins [10, 43, 77] while in natural (unmodified) NAs no analytically useful signals of catalytic hydrogen evolution have been observed. On the other hand, DNA modified with osmium tetroxide complexes produced well-developed, analytically useful signals (see Sections 5.2.3.2 and 5.3.4.4). Recently, a new electrocatalytic scheme based on interaction of azine dyes (such as phenothiazines, phenoxazines and phenazines) with ds and ssDNA has been described [78]. These dyes can act as efficient catalysts of NADH oxidation at carbon electrodes. Complexes of dsDNA with organic dyes gave higher electrocatalytic NADH oxidation signal than ssDNA-organic dyes. In NA electrochemistry, the mediated electron transfer has been so far the most widely applied type of electrocatalysis (see Sections 5.2.3.1 and 5.3.3.9).

Mediators in DNA Electrochemistry

Thorp and coworkers [79–91] pioneered the use of mediators in electrochemistry of nucleic acids. Among the NA bases, G is known to be most easily oxidizable by different oxidants such as hydroxyl radicals, singlet oxygen, transition-metal complexes and alkylating agents. Electrochemical oxidation of guanine and adenine residues in DNA and RNA at carbon electrodes was demonstrated more than 25 years ago [92] (reviewed in [13]). Oxidizability of these bases has been applied for various purposes, including the detection of the DNA hybridization [13]. To increase the sensitivity of the guanine-based electrochemical analysis, Thorp and coworkers (reviewed in [17, 90]) investigated the application of transition-metal complexes for mediating oxidation of guanine in DNA at ITO electrodes. They found that ruthenium trisbipyridyl, $[Ru(bipy)_3]^{2+}$ was an effective oxidation catalyst of guanine. Addition of DNA to a solution of $[Ru(bipy)_3]^{2+}$ resulted in a catalytic enhancement of the $[Ru(bipy)_3]^{2+}$ oxidation peak (Figure 5.3). The reaction proceeded according to the following two-step mechanism (Eqns. (1–2)):

$$[Ru(bipy)_3]^{2+} \longrightarrow [Ru(bipy)_3]^{3+} + e^- \qquad (1)$$

$$[Ru(bipy)_3]^{3+} + DNA \longrightarrow DNA_{ox} + [Ru(bipy)_3]^{2+} \qquad (2)$$

where DNA_{ox} is a G-containing DNA molecule oxidized by $[Ru(bipy)_3]^{3+}$ or by Ru(III) generated at the electrode. Ru(II) mediator is regenerated making the reaction catalytic.

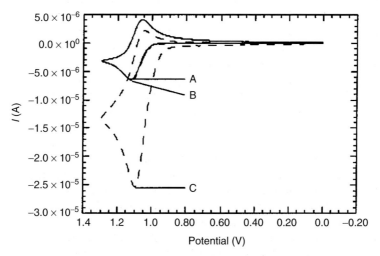

Fig. 5.3 Cyclic voltammograms (25 mV s^{-1}) of 200 mM $[Ru(bpy)_3]^{2+}$ at a (A, solid) bare ITO working electrode, (B, dotted) 1,12-dodecanedicarboxylic acid (DDCA)-modified ITO working electrode, and (C, dashed) DDCA-modified electrode following attachment of calf thymus DNA. Reference electrode, Ag/AgCl; auxiliary electrode, Pt wire; supporting electrolyte, 50 mM phosphate buffer, pH 7 (Sections 5.2.2.4 and 5.3.3.9). Preparation of the DDCA and DDCA/DNA electrodes is described in the original paper. Voltammograms in (A) and (B) are not distinguishable. (Reprinted with permission from M.E. Napier and H.H. Thorp, Langmuir **13**, p. 6343 [81]. Copyright 1997 American Chemical Society).

Guanine residues form radical cations that quickly deprotonate back to guanine, which can undergo further reactions. The rate constant for oxidation of G residues in calf thymus DNA solution by Ru(III) was determined as 9.0×10^3 M^{-1} s^{-1} [80]. Reactivity of G increased in the following way: dsDNA < mismatched dsDNA < ssDNA. Guanines in perfectly matched duplex DNA were 200 times less reactive than those in ssDNA. Adsorption of ssDNA via its phosphate groups and the oxide surface of ITO resulted in submonolayer coverages [86, 88]. In variance to its behavior at carbon electrodes [34] ssDNA did not show any significant oxidation signal but acted as a substrate for the electrocatalytic oxidation by the mediator $[Ru(bipy)_3]^{2+}$. For DNA irreversibly adsorbed to the ITO electrode, a sensitivity of 44 amol mm^{-2} or $\sim 3 \times 10^9$ molecules was reported for the 1497 bp polymerase chain reaction product [86]. Most efficient electron transfer between the mediator and ITO can be obtained at polycrystalline ITO films perhaps because of the higher density of defects.

5.2.2.5 Conducting Polymers

In contrast to model studies (using ODNs as target DNAs) in experiments with real DNA samples, nonspecific interactions at the electrode surfaces (involving DNA impurities, long ssDNA chains, etc.) can obscure the hybridization signals. Efficient interfacing between the DNA system and the electrode surface is thus necessary. Thiols (see Section 5.3.3) and conducting polymers have been used for this purpose.

Individual chains of organic conjugated polymers may possess a high intrinsic conductivity (reviewed in [93, 94]). Conducting polymers can be electropolymerized as thin films onto an electrode (e.g., polythiophenes and polypyrroles) or synthesized chemically (e.g., polyphenylenes). Experimentally measured conductivities of conjugated polymers are between 1 and 10^4 S cm^{-1}, that is several orders of magnitude lower than those of metals but sufficiently high to consider these macromolecules (several micrometers in length) as molecular wires [93]. Deposited as thin films at electrodes, the reversible redox processes in these polymers can be potential-controlled. Cyclic voltammograms can provide electrochemical signatures of the studied polymers. Such signatures are very sensitive to the nearby environment; any modification of the groups along the chain can be electrically transduced to the electrode and manifested by modified polymer electrochemical signature. Different functionalities can be inserted in the polymer known to show selectivity for solution species such as NAs, antibodies and enzymes. Specific chemical recognition (resembling affinity chromatography) can be thus built through the proper functioning of the conducting polymers.

Conducting polymers such as copolymer functionalized with an osmium complex [95–100], polypyrroles, polyazines and polyanilines, [93, 101–103] or polythiophenes [104] may be used for modulation of the DNA interactions at surfaces and for generating signals resulting from such interactions and also for blocking and interfacing the transducer (reviewed in [105–107]). Conducting polypyrrole functionalized with bulky oligonucleotides remains electroactive in aqueous media; on interacting with a complementary, but not with noncomplementary, ODN changes in the voltammetric current were observed [103]. Doping of the NA probes within electropolymerized polypyrrole films and monitoring the current changes provoked by the hybridization event were suggested as a promising label-free biosensing strategy [108]. Adsorption of NAs on polypyrrole-coated glassy carbon electrode (GCE) was used for amperometric ODN detection in flowing streams [101]. Quite recently, Thompson and coworkers [109] reported on a new approach to electrochemical detection of DNA hybridization based on electrostatic modulation of ion-exchange kinetics of polypyrrole film. They used surface-immobilized unlabeled ssODN as a probe to interact with the unlabeled complementary target DNA. The probe was linked to the conducting polymer by bidentate complex between Mg^{2+} and alkylphosphonate group on the polymer and the DNA phosphate group. The activated electrode was immersed in the target DNA solution for a short time, and the CV was recorded. Marked difference between voltammograms obtained with complementary and noncomplementary DNAs was observed. The results indicated that the addition of negative charge to the surface of the electrode in the form of the complementary target DNA further hinders the chloride ion exchange, which is manifested on the resulting cyclic voltammetry. Advances in polymer-based electrodes in DNA research were recently reviewed [17]. See Section 5.3.3 for further details.

5.2.3
Electroactive Labels of Nucleic Acids

It has been shown that natural NAs are electroactive but their oxidation and reduction are electrochemically irreversible, occurring at highly positive or highly negative potentials. To obtain electrochemical signals at potentials closer to the potential of zero charge and to increase the sensitivity of the analysis, electroactive markers were introduced into DNA. Most of these markers either undergo reversible electrode reaction at less extreme potentials or produce catalytic signals of high-electron yield. The number of the known electroactive markers is quite large: in this chapter, only some examples will be given.

5.2.3.1 Noncovalently Bound DNA Labels

A large number of inorganic ions and small organic molecules interact reversibly with DNA. These interactions can proceed in three basic modes: (a) nonspecific electrostatic binding involving cations that interact along the exterior of the DNA double helix, primarily via electrostatic interactions with negatively charged phosphate groups. This binding mode is not restricted to dsDNA; (b) groove binding, which involves direct contacts of the bound molecule with the edges of base pairs in the minor or major grooves of the DNA double helix. Depending on the nature of the interacting species, electrostatic and/or van der Waals contacts as well as hydrogen bonding may take part in this mode of binding; (c) intercalation of planar, mostly positively charged aromatic ring systems between DNA base pairs, primarily involving base stacking interactions. Small molecules on binding to duplex DNA change their electrochemical responses because of slower mass transport of the resulting DNA complex and/or decreased accessibility of the small molecule redox site for the electrode process (see Section 5.4.6). In some cases, changes in DNA electrochemical responses resulted from small molecule binding to

DNA [110] (reviewed in [111]). A large portion of the substances capable of interacting with DNA and especially the intercalators and groove binders are electroactive.

To our knowledge, the first electrochemical studies of the DNA interactions with small molecules were published by Berg and coworkers [112–114]. In the early 1970s, they pioneered the research of noncovalent interactions of electroactive intercalators and other compounds with DNA by means of electrochemical methods.

Inorganic Cations

Nature and concentration of inorganic cations may strongly affect adsorption of DNA and RNA at electrodes because of their shielding of the negative charges of the NA sugar-phosphate backbone as well as because of their effect on the electrode double layer. Interaction of DNA with inorganic cations was recently reviewed [13]. Here we wish to briefly summarize the DNA interactions with intercalators and groove binders, which can serve as redox indicators in the DNA hybridization sensors.

Groove Binders and Intercalators

The dye Hoechst 33258 represents a typical compound binding to the minor groove of dsDNA. Hashimoto and coworkers [115] showed that the oxidation signal of this dye at a gold electrode is increased as a result of the dye interaction with dsDNA. They used this principle in the detection of the DNA hybridization (see Section 5.3.3).

Daunomycin can serve as an example of an intercalator whose interaction with dsDNA is well understood [116]. Wang and coworkers [110] obtained complex results when they studied the interaction of daunomycin with DNA in solution and on the carbon electrode with immobilized DNA (see Section 5.4.6). Daunomycin was successfully applied as a redox indicator of DNA duplex formation in DNA sensors [29, 117]. Using reversible electrochemistry of daunomycin site specifically coupled to an ODN, Kelley and coworkers [118] investigated long-range electron transfer through dsODNs anchored at a gold electrode via end thiol groups. They showed that perfectly matched DNA double helix effectively conducted electrons from daunomycin bound to the ODN at a distant site relative to the electrode surface. Presence of single-base mismatches in the DNA duplexes disturbed the electron transfer through base stack (see Section 5.3.3.9). Such effects were observed with DNA intercalators (daunomycin, methylene blue or a metallointercalator) but not with groove or electrostatic binders [118–120].

Electrochemical behavior of DNA complexes with bis-intercalators (e.g., a bis-9-acridinyl derivative with a viologen linker [121] and a naturally occurring antibiotics echinomycin (ECHI) [122, 123]) and threading intercalators (e.g., a ferrocenyl-modified naphthalene diimide) [124, 125] was studied. With the above threading intercalator, a sensor capable of discrimination between ss, duplex and hairpin DNA at the gold electrode surface, and a sensor for DNA hybridization [121, 124–127] were described. A quinoxaline antibiotic ECHI was electrochemically active and yielded several cyclic voltammetry (CV) signals down to submicromolar concentrations [123]. Interaction of ECHI with dsDNA attached to hanging mercury drop electrode (HMDE) was studied by

CV [123] and impedance measurement [122]. Specific DNA and ECHI signals were obtained in agreement with the strong binding of ECHI to dsDNA by bis-intercalation ([122, 123] and references therein). Under the same conditions interaction of ECHI with ssDNA resulted in high DNA but very small or no ECHI signals, suggesting only very weak binding of ECHI to ssDNA at the electrode surface. ECHI appears to be a good candidate for a redox indicator in electrochemical DNA hybridization sensors (see Section 5.3).

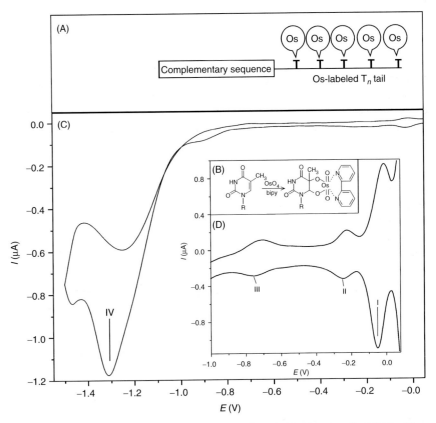

Fig. 5.4 (A) Osmium tetroxide complexes with nitrogen ligands (Os,L) such as Os,2,2'-bipyridine (Os,bipy) react preferentially with thymine residues (B), forming stable adducts in ssDNA [130]. Os,bipy can be used for different DNA labeling (Sections 5.2.3.2, 5.3.4.2), including (A) end-labeling of reporter probes with multiple electroactive labels. Electrochemical signals of Os,bipy DNA labels can be measured with various methods. (C,D) Cyclic voltammograms of Os,bipy-labeled ssDNA. In (C), peak IV is shown, which is due to the catalytic hydrogen evolution; it was obtained with mercury and solid amalgam electrodes. Redox couples I–III in (D) were measured with carbon, mercury and solid amalgam electrodes. Cyclic voltammograms in (C and D) of calf thymus ssDNA (20 μg mL^{-1}, modified with 2 mM Os,bipy at neutral pH, 25 °C) obtained by AdTS (Figure 5.2). Initial potential 0.125 V, switching potential −1.5 V, accumulation from a 5-μL drop of DNA solution for 60 s. 0.3 M ammonium formate with 0.05 M sodium phosphate (pH 7.0) was used as a background electrolyte. (C) Scan rate 2.56 Vs^{-1}, (D) 0.08 Vs^{-1} (L. Havran and E. Palecek, unpublished).

5.2.3.2 Covalently Bound Labels

The first electroactive labels covalently bound to DNA were proposed already in the beginning of the 1980s [19–22, 128]. On the basis of osmium tetroxide complexes with nitrogen ligands (Os,L), these labels were used for a simple and fast DNA modification under conditions close to physiological [19–22, 128–130]. Os,L-modified DNA produced three redox couples between 0 and at about −0.6 V, and a signal due to the catalytic hydrogen evolution (Figure 5.4). The redox couples were observed both with mercury and carbon electrodes [131–133], while the catalytic signal at about −1.2 V was found only with mercury [19–22, 128–130] and SAE [134]. It was shown [21, 129] that osmium tetroxide, 2,2'-bipyridine (Os,bipy) is covalently bound to pyrimidine bases (preferentially to thymine) in ssDNA. Later Os,bipy and other Os,L complexes have been widely applied as probes of the DNA structure *in vitro* and *in vivo* in connection with nonelectrochemical methods, including DNA sequencing techniques and immunoassays (reviewed in [130, 135–137]). Most of the DNA-Os,L adducts can be determined electrochemically at or below ppb concentrations. Os,bipy-modified target DNA (tDNA) and end-labeled reporter probes (RP) can be conveniently used in DNA sensors (Section 5.3.4.4).

In recent years, electroactively labeled oligonucleotides were prepared by covalent linkage of a ferrocenyl group to the amino hexyl-terminated ODN [138, 139]. Using HPLC equipped with an electrochemical detector DNA and RNA were determined at femtomole level [138]. Uridine-conjugated ferrocene ODNs were synthesized, and ferrocene-labeled signaling probes with different redox potentials were prepared for reliable detection of point mutations [140, 141]. Other compounds, such as daunomycin [118], viologen and thionine [142] were used as electroactive labels of ODNs. These labels required solid-state organic chemistry, and their use was limited to ODN modification. On the other hand the Os,L labels can be used for an easy labeling of ODNs and plasmid and chromosomal DNAs in a laboratory not equipped with any organic chemistry facilities [130, 131] (Section 5.3.4.4). End-labeling of DNA with biotin and thiol groups has been widely applied (see Section 5.3.3. for details). Electroactive labels were introduced in NAs not only by chemical methods but also by means of enzymes such as DNA polymerases capable to introduce synthetic ferrocene tethered to dUTP [143].

Other Nucleic Acid Modifications

Potentialities of electrochemistry in the analysis of DNA-drug adducts are mentioned in Section 5.4.

Peptide Nucleic Acid

Peptide (or polyamide) nucleic acid (PNA) is a useful tool in DNA biotechnologies and one of the candidates for diagnostic and therapeutic application in medicine of this century [144–146]. In PNA, the entire sugar-phosphate backbone is replaced by N-(2-amino-ethyl) glycine units. In variance to RNA and DNA (with negatively charged backbones), the backbone of PNA is electrically neutral. PNA yields similar electrochemical responses as DNA and RNA (i.e., A, C and G are reduced at mercury electrodes, G reduction product is oxidized back to G, producing an anodic CV peak [43] and A and G are oxidizable at carbon electrodes [32]). Compared to ssDNA, peak potentials of ssPNA at HMDE are shifted to more negative values. Different adsorption behavior of PNA and DNA as detected by a.c. impedance [36] at mercury and by CP measurements at carbon electrodes [32] are related to differences in backbones of these two NAs. At higher surface concentrations, PNA molecules are associated at the mercury surface; prolonged exposure of PNA to highly negative potentials does not induce PNA desorption under conditions when almost all DNA molecules are removed from the surface. Adsorption of DNA increases with salt concentration; in contrast, adsorption of PNA decreased under the same conditions [36]. In DNA hybridization sensors, PNA has proved to be an excellent probe especially for the detection of point mutations [30] (see Section 5.3.3.8). PNA-functionalized Au-coated tips were used to probe the DNA hybridization using chemical force microscopy [147].

5.2.4
Signal Amplification

In addition to electrocatalysis (Section 5.2.2.4) and biocatalysis (Section 5.3.3.7), other approaches have been used to amplify the NAs electrochemical signals. Willner and coworkers [51–56, 143, 147–164] have developed highly sophisticated amplification schemes on the basis of various principles, utilizing biotin-labeled or NA-labeled liposomes, NA-labeled CdS or gold nanoparticles for dendritic-type analysis of DNA and magnetic particles. They used microgravimetric quartz-crystal-microbalance analysis, photoelectrochemical and electrochemical measurements to transduce the DNA recognition events on surfaces. Those approaches using electrochemical detection will be mentioned in Sections 5.3.3.5, 5.3.3.7 and 5.3.4.

5.3
Sensors for DNA and RNA Hybridization

For several decades, electrochemistry of NAs was a purely academic field, not considered to be very exciting both by electrochemists and biochemists. In recent years, the situation has dramatically changed in connection with the great progress in genomics and particularly with the success of the Human Genome Project. Microarrays or "gene chips" featuring the dense arrays of ODNs have been successfully applied to highly parallel analyses of single nucleotide polymorphisms and transcriptional profiling studies (reviewed in [165]). The fluorescence detection of the DNA hybridization used in these microarrays requires expensive and precise instrumentation including a laser source and charge coupled device camera. Thus, the microarrays have been limited to research laboratories.

The recent changes in the interest of scientists in NA electrochemistry is related not only to the high sensitivity of electrochemical methods for small changes in the DNA

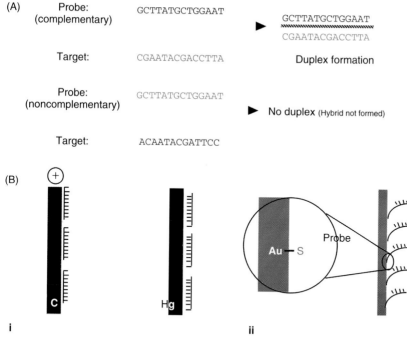

Fig. 5.5 Principles of DNA hybridization and its electrochemical detection by single-surface techniques (Section 5.3.3). (A) Scheme of the formation of duplex DNA (DNA hybridization) from two complementary ss oligonucleotides (ODNs). Duplex DNA is not formed when target DNA sequence is noncomplementary. In an electrochemical hybridization detector, one of the two strands is used as a probe, which is immobilized at the electrode surface. (B) Scheme of immobilization of the probe at the electrode surface (Sections 5.3.3.1 and 5.3.3.2) (i) at positively charged carbon electrodes, DNA is adsorbed electrostatically via the negatively charged sugar-phosphate backbone. At the mercury electrodes, hydrophobic bases are strongly bound to the hydrophobic electrode surface. (ii) Covalent binding of thiol-end-labeled ODN at gold electrodes securing oriented ODN immobilization and self-assembled monolayer (SAM) formation (see Figure 5.6). (C) Detection of the DNA hybridization: (i) based on the formation of the hybrid duplex DNA using either (a) a redox indicator that binds preferentially to the hybrid duplex DNA and produces an oxidation (or reduction) signal (Sections 5.2.3.1 and 5.3.3.4) or (b) direct measurements of changes in DNA properties (such as conductivity) due to the duplex formation. Alternatively, a reporter (signaling) probe (Section 5.3.3.6) can be used as shown in Figure 5.9A (Section 5.3.3.7); (ii–iii) based on the determination of the target DNA at the electrode surface; (ii) a suitable enzyme such as horse radish peroxidase (HRP) is coupled to the target DNA. Upon hybridization, an electrocatalytic current due to hydrogen peroxide reduction is measured. (iii) After DNA hybridization, the DNA probe serves as a primer, which is extended at the electrode surface by DNA polymerase incorporating biotinylated nucleotides in the newly synthesized strand. HRP is then attached to the DNA using streptavidin-biotin chemistry followed by measurements as in ii; (iv) DNA probe not containing guanine is used, and presence of guanine in target DNA is detected after the hybridization using guanine oxidation at carbon electrodes (Figure 5.1C,D) or catalytic signals on indium tin oxide electrodes (Figure 5.3, Section 5.2.2.4 and 5.3.3.9). Adapted from [16] with permission.

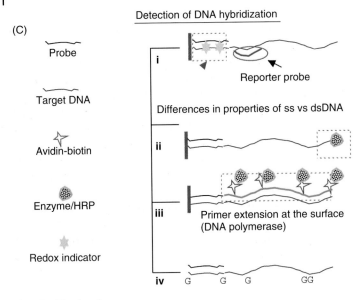

Fig. 5.5 (Continued)

structure mentioned in previous paragraphs (Section 5.2.2.1). It is primarily due to expectations that electrochemical DNA detection can complement optical detection in DNA hybridization sensors. In contrast to optical devices, an electrochemical device requires only a simple electric circuitry connected to an electrode system. Such a device would be well suited for decentralized detection of DNA damage (see Section 5.4) and DNA hybridization. Rapid and inexpensive molecular diagnostics of various diseases in the physician's office may thus become a reality in a relatively short time. Such development would resemble the changes of glucose sensing in diabetes monitoring from optical to electrochemical methods. Recent progress in the development of electrochemical sensors for DNA hybridization and damage was summarized in numerous review articles [12, 13, 16, 17, 29, 82, 90, 108, 111, 166–172].

5.3.1
DNA Hybridization

Original methods of DNA sequencing are laborious and expensive. New technologies are therefore sought and among them sequencing by DNA hybridization appears most promising. The ability of denatured ssDNA to reform its double-helical structure was discovered by Julius Marmur and Paul Doty more than 40 years ago [173] and termed *DNA renaturation or hybridization* (Figure 5.5A). This ability of DNA represents one of the most important principles of the present molecular biology, which is incorporated in many DNA biotechnologies.

5.3.2
Electrochemical Detection in DNA Sensors

A biosensor consists of a selective biological recognition element associated with a transducer, which translates the recognition event into a physically measurable value [174]. In an electrochemical DNA hybridization sensor, usually a short ss ODN (DNA probe, about 15–30 nucleotides long) is immobilized on an electrode (transducer) (Figure 5.5B) or bound to some other surface suitable for the DNA hybridization to create a DNA recognition element. This element is then immersed into a solution of target DNA whose nucleotide sequence is to be tested. When the target DNA contains a sequence, which exactly matches that of the probe DNA (based on the complementarity principle stating that C pairs with G and T with A), a hybrid duplex DNA is formed (Figure 5.5A) at the surface of the electrode. If no complementarity between the probe and target DNAs exists, no duplex is formed. The formation of the hybrid duplex DNA has to be translated into an electrical signal, which is utilized in the DNA analysis (Figure 5.5B). In the following sections, we shall show that two different strategies can be used in the electrochemical DNA sensors. In the most commonly used strategy, DNA hybridization and DNA detection is performed at the same surface. This surface must necessarily be an electrode. Techniques based on this principle can be called *single-surface techniques* (SST).

5.3.3
Single-surface Techniques

Further we shall focus on the two most important steps in the process of detection of the complementary DNA nucleotide sequence: (a) the formation of the DNA recognition layer and DNA hybrid formation and (b) the transformation of the hybridization event into an electrical signal. In both steps, a critical role is played by the material of the electrode and by the method of DNA immobilization at the surface.

5.3.3.1 Immobilization of DNA at Electrode Surfaces

Both covalent and noncovalent binding of probe DNA to the electrode surface has been used. Adsorption forces were utilized for binding DNA to carbon electrodes [29, 90, 167, 175, 176]. Electrostatic binding of DNA to the positively charged carbon electrode was sufficiently strong and bases were accessible for the hybridization with target DNA. Covalent binding to carbon electrodes utilized carbodiimide or silane chemistries [177]. Avidin (or streptavidin)-biotin binding was used to attach DNA to various surfaces (reviewed in [17]). Only recently this technique was used to immobilize biotinylated probe DNA to carbon electrodes [178].

Generally, covalent binding of the DNA probe via one (or both) of its ends is advantageous because it can be well controlled and the nonspecific molecules weakly bound to the surface can be easily removed. Covalent binding of DNA probes to gold and ITO was used (reviewed in [17, 90, 170]). ITO electrodes can be modified via phosphonate self-assembled monolayers (SAMs) [88] and silane overlayers [89]. Carbon and gold surfaces are well suited as electrode materials in DNA sensors; with the latter electrode, covalent attachment of DNA is necessary. Various aspects of DNA immobilization on different surfaces and particularly at gold electrodes were recently summarized in an excellent review by Tarlov and Steel [17]. Here, only a brief summary on DNA immobilization at gold electrodes will be given.

Among a number of approaches the well-known thiol-gold interaction prevails as a method of the DNA probe immobilization [17]. Thiol or disulfide end-labeled DNA probe is directly attached to gold via its – SH group and an SAM is formed resembling self-assembly of alkanethiols [179–183]. After DNA self-assembly (taking several hours), a second SAM is created using alkanethiol molecules that limits nonspecific adsorption of other compounds including the noncomplementary DNA or RNA (see [184] for a recent review on SAMs). Moreover, the alkanethiol SAM prevents interactions between the probe and gold surface, leaving probes in largely upright position accessible for DNA hybridization. Alkanethiol monolayers can easily be assembled not only on gold but also at mercury surfaces [185–189].

At coverages of 5×10^{12} cm^{-2} nearly 100% hybridization efficiency (Figure 5.6) was reported [181, 190]. Thermal stability of thiol-attached probes was limited to 75 °C [17]. Two component films containing a thiol-derived ssDNA probe and a diluent thiol, mercaptohexanol, were prepared on the gold surface [191]. The thickness and dielectric constant of the film were determined by two-color surface plasmon resonance and the DNA amount tethered to the surface was quantified. Measurements of the kinetics of hybridization and thermally induced dehybridization indicated a high efficiency of the hybridization process. SAM containing a viologen group was formed on a gold electrode via Au–S bonds [192]. Binding of dsDNA to this layer resulted in a positive shift of the redox potential of the viologen signals, indicating hydrophobic interactions.

5.3.3.2 Efficiency of Hybridization

There are a number of factors that influence the hybridization efficiency. One of them is the DNA probe length. The probe has to be sufficiently long to secure the binding specificity in the presence of a large number of nonspecific sequences. On the other hand, too long probe may be less favorable for the DNA hybridization. It was shown by Steel and coworkers [193] that the random-coil nature of DNA probes directly influences the probe-packing density. Surface coverage of thiol-labeled DNA probes on gold was determined as a function of probe length (Figure 5.6). It was observed that the surface-packing density increased with the probe length up to 24 bases and decreased in probes containing 30 or 40 bases. This result was explained by highly extended configuration of the probes whose length did not exceed 24 bases, while in longer probes, a more random-coil configuration was assumed (Figure 5.6).

It was found that the hybridization yields could be increased up to 2 orders of magnitude by introducing the spacer (linker) between the oligonucleotide probe and the surface. One of the most important are the properties of the probe linker; they include

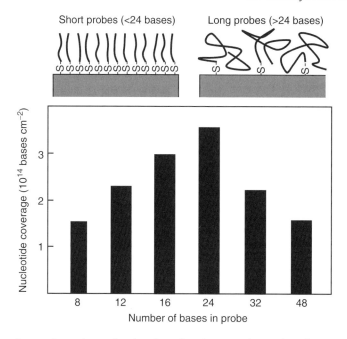

Fig. 5.6 Dependence of nucleotide surface density on the number of nucleotides in a given thiolated DNA probe adsorbed on gold surfaces (Sections 5.3.3.1 and 5.3.3.2). Surface densities were determined by measuring of radioactively (^{32}P) labeled probes and multiplying the probe density by the number of nucleotides in the probe. The probe was non-self-complementary 4-base repetitive unit ACTG. Schemes above the graph represent two approximate structural regimes of the probes. Shorter probes (<25 nucleotides) assume a primarily extended configuration, while the longer ones (>25 nucleotides) adopt a more random-coil-like configuration. Data taken from [17].

its length, charge and hydrophobicity [194–199]. Among them the spacer length appears the most important in affecting the hybridization [196]. At polypropylene surfaces, an optimal spacer length of 40 atoms was found beyond which the hybridization yield decreased. On the other hand, nearly 100% hybridization efficiencies were achieved with probes linked to gold by a six-methylene linker [181, 190]. The reason for the above discrepancy is not clear; the oligonucleotide coupling chemistry and different substrates may play a role. Experimentally determined values for optimal probe coverage are in the range of 10^{12} to 10^{13} probes cm^{-2} while packing density of the n-alkanethiol SAMs on gold is about 5×10^{14} cm^{-2}, that is almost 2 orders of magnitude higher than that of DNA probes [200]. Melting temperatures of the DNA duplexes covalently bound via organosilane chemistry to fused silica optical fibers depended on the surface density of immobilized DNA with the highest surface density (about 5×10^{12} probes cm^{-2}), showing the lowest stability.

5.3.3.3 Detection of DNA Hybridization at the Electrode Surface

In early studies, redox indicators interacting preferentially with dsDNA such as simple intercalators (e.g. metal intercalators or daunomycin) and minor-groove binders (e.g. Hoechst 33258) were applied to differentiate between dsDNA and ssDNA at electrodes (Figure 5.5Ci) (reviewed in [177]). These indicators interacted, however, not only with dsDNA but also with ssDNA thus substantially decreasing the selectivity of the analysis. Higher selectivity of redox indicators for duplex DNA was therefore sought. Threading intercalators and bis-intercalators appeared to be an interesting alternative for simple intercalators (see Section 5.2.3.1).

At present there are a large number of methods of electrochemical detection of the DNA hybridization at the electrode surface. Some of them are shown in Figure 5.5C. In addition to redox indicators, natural electroactivity of the NA bases alone [167, 175, 176] or in combination with the mediator electrochemistry (Sections 5.2.2.4 and 5.3.3.9) were applied for this purpose. Simple chemical labeling of target [131, 201] or reporter probe DNAs [138, 202] (Section 5.3.3.5), as well as label-free [67] and reagent-free [203] techniques are currently used. Amplification of the electrochemical signals by applying catalytic methods [90, 133] (Section 5.2.2.4 and 5.3.3.7) or other ways of signal amplification are gaining importance. Here only some of the electrochemical detection methods, currently applied in the DNA hybridization sensors, will be mentioned in greater detail.

5.3.3.4 Threading Intercalators and Bis-Intercalators

Potentialities of the threading intercalators and bis-intercalators showing higher specificity for dsDNA have not been yet fully tested (Section 5.2.3.1). Threading intercalators have usually high DNA-binding constants, indicating that once their bulky side substituents slide between the base pairs, DNA assumes a conformation that gives a very favorable free energy of complex formation. In bis-intercalators, two intercalating rings are covalently linked to connecting chains of variable length. Electrochemical studies of the synthetic threading intercalator ferrocenyl naphthalene diimide [127] and the naturally occurring bis-intercalator ECHI (antibiotics and antitumor agent) were undertaken [122, 123]. Both compounds bound to dsDNA more tightly than the usual simple intercalators and showed low affinity for ssDNA. By means of the ferrocenyl naphthalene diimide yeast choline transport gene has recently been detected [204].

In principle, target DNA can be end-labeled covalently with a redox indicator producing a signal when the target DNA is hybridized. This is rather inconvenient in SST because in a long target DNA (tDNA) molecule, the label may not always get oxidized (or reduced) at an electrode as many of the redox labels may be too far from the electrode surface. Moreover, covalent end-labeling of long target (chromosomal) DNA molecules may be expensive and/or difficult.

5.3.3.5 Reporter and Signaling Probes

Compared to redox indicators noncovalently bound to duplex DNA or covalently attached to tDNA, covalent binding of a redox indicator to an ODN DNA complementary to tDNA (reporter or signaling probe, Figure 5.5Ci) appears more promising. Ferrocene-labeled reporter probes were used by Ihara and coworkers [139] who immobilized their probe DNA on a gold electrode via five successive phosphorothioate units on the 5′-terminus. A disadvantage of the reporter probes in single-surface techniques is that they have to be complementary only to sequences on target DNA located close to the electrode surface. Nevertheless, signaling probes are used in the first commercialized electrochemical DNA hybridizations sensors offered by Motorola (Section 5.3.3.9). In double-surface (DS) techniques (Section 5.3.4), reporter probes can be applied even with a greater efficiency.

5.3.3.6 Labeling of DNA at the Electrode Surface

Various kinds of DNA prelabeling were employed to facilitate and amplify electrochemical detection of tDNA. They include functionalization of DNA with nanoparticles (reviewed in [172]) or liposomes [149, 154], NA-enzyme conjugates (Sections 5.3.3.7) and biocatalytic conjugates that associate to DNA and stimulate the precipitation of an insoluble product on the electrode [52, 53]. Recently a series of biocatalyzed transformations of surface-attached NAs, including ligation, replication and restriction cleavage, was reported by Willner's group [153]. It was shown that DNA and RNA can be synthesized by DNA polymerase or reverse transcriptase at a gold electrode [155]. Using biotin-labeled dCTP these biotin residues were incorporated in viral NAs, synthesized at the surface. Two amplification steps were thus included in these surface processes: (a) DNA or RNA polymerization, (b) introduction of biotin tags into the dsDNA (or RNA) molecules, providing numerous docking sites for enzyme binding. Biocatalyzed precipitation of an insoluble indigo product was detected by electrochemical or gravimetric methods. A similar approach was used to generate ferrocene-labeled DNA [143]. Ferrocene-dUTP was incorporated in viral DNA, replicated at the surface by DNA polymerase. Redox-active DNA was then coupled to glucose oxidase and oxidation of glucose mediated by this enzyme provided biocatalytic amplification of the DNA detection by means of DPV.

5.3.3.7 Biocatalysis

Electrode processes in experiments with redox indicators or direct DNA electrooxidation (Figure 5.5C) involve only one or few electrons per indicator or G residue, being therefore inherently low-yield reactions. Recently catalytic processes have been used in DNA analysis to collect higher amounts of electrons [12, 13, 16, 17, 82, 90, 97, 205–210]. In addition to processes mentioned in Section 5.2.2.4, enzymatic reactions were utilized to enhance the sensitivity of the detection of the DNA hybridization event (Figure 5.5Cii). For example, horse radish peroxidase (HRP) was coupled to tDNA and used to detect the hybridization by electrocatalytic reduction of the hydrogen peroxide (Eq. 3):

$$H_2O_2 + 2e^- \longrightarrow 2OH^- \tag{3}$$

Peroxidase molecule turned about 1800 times per second, involving thus about 3600 electrons in 1 s [205]. Using a redox polymer-coated microelectrode and thermostable soybean peroxidase-labeled tDNA, a single-base mismatch in an 18-mer ODN was detected [206]. The discrimination between a perfectly matched duplex and a single-base mismatch was obtained at elevated temperature, while at room temperature such discrimination was not possible. Heller's group further developed their conception on the basis of a conducting osmium redox polymer and enzyme amplified amperometric detection [97, 205–210]. Recently they reported the detection of the 38-base tDNA at 20 pM concentration [210] in 30 µl volume. They used a carbon electrode with an electrodeposited redox polymer, comprising a DNA capture sequence. After capturing the tDNA, HRP-labeled reporter probe was used, producing an amperometric signal because of H_2O_2 electroreduction. Recently the same group reported a >10 000-fold improvement of the detection limit of their assay (down to 0.5 fM) when a 3.6-mm-diameter carbon electrode was replaced by a 10-µm-diameter microelectrode. The radial diffusion of electrons through the microelectrode film allowed the electrodeposition of a thicker layer of the redox polymer, higher capture sequence loading and increased collection efficiency of the electron vacancies originating from the electroreduced H_2O_2. 3000 copies of tDNA were detected in a 10-µl volume. In addition to peroxidases, phosphatases [211] and other enzymes were used both in single-surface and double-surface DNA hybridization sensors by different authors (Section 5.3.4). Various catalytic processes were also utilized in some DNA hybridization systems mentioned in Section 5.3.3.9.

5.3.3.8 Detection of Point Mutations

Many diseases are connected with point mutations (single-base mutation) at specific locations of the genome. Detection of a change in a single nucleotide per DNA sequence is a difficult task requiring highly specific and sensitive methods. First successful electrochemical analysis of point mutation (detection of single-base mismatch in the probe-tDNA duplex) was achieved in 1996 by applying PNA probe instead of DNA [30]. Mutation hot spot of the tumor suppressor *p53* gene was detected by this method using ODN as a target [212]. A relatively simple technique based on a carbon electrode and a redox indicator was used to obtain these results. Caruana and Heller [206] detected in 1999 a single-base mismatch in an 18-mer ODN using a redox polymer-coated microelectrode and thermostable soybean peroxidase-labeled tDNA. They obtained good discrimination between a perfectly matched duplex and a single-base mismatch at elevated temperature but not at room temperature. Since that time, a number of techniques have been developed (e.g., [13, 30, 54, 90, 119, 141, 147, 156, 157, 160, 161, 170, 206, 212–214]) suitable for the detection of point mutations. Some of them will be mentioned below in greater detail.

5.3.3.9 Recent Electrochemical Detection Systems

In the past decade, new principles and approaches were proposed and applied in DNA hybridization sensors. Most of them were demonstrated with synthetic ODNs, but only in some cases their applicability to real DNA samples was tested. Several promising techniques shown in Figure 5.7 have been developed and recently reviewed by Popovich and Thorp [90] and Drummond and coworkers [170]. Elaborated systems and new interesting principles will be discussed below (Figure 5.7).

Sandwich Assay with Signaling Probes

The first commercially available electrochemical DNA hybridization device was introduced by Motorola [141, 215, 216]. It was based on a three-component sandwich assay involving the ferrocene-labeled signaling (reporter) probe complementary to an overhang portion of the probe-target complex. Capture probes were immobilized on gold electrodes via thiol-end-labeled phenyl acetylene bridges allowing for strong electronic coupling between the ferrocene labels and the electrode. Binding of the signaling probe to the capture probe-tDNA complex resulted in an electrochemical signal because of oxidation of ferrocene (Figure 5.7B). Application of a second signaling probe with a different oxidation potential allowed monitoring of two targets simultaneously. Detection of 50-nM target by a.c.V was reported. These principles were used in the eSensor™ DNA Biochip, produced and sold by Motorola. The critical components (target, capture and signaling probes) are in a cartridge, and the current produced by this system is measured and interpreted by the eSensor™ DNA Detection Reader and Software. More details can be obtained from http://www.motorola.com/lifesciences/esensor/.

DNA-mediated Charge Transport

J. Barton and coworkers developed a new approach based on the electronic structure of the double-helical DNA and intercalated planar redox probe molecules to report on base stacking perturbations in DNA forming a film on the gold electrode [170, 171]. In their assay, the intercalator is not used to bind preferentially to dsDNA (see Sections 5.2.3.1 and 5.3.3.4) or to help determine the amount of DNA bound to the electrode. Instead, bound at the top of the film, it reports on the charge transport from the electrode inducing its reduction (Figure 5.7A). In duplexes with perfectly matched base pairs, an appreciable current flow can be observed.

In a typical assay, DNA duplexes end-labeled with a thiol are formed in solution and self-assembled into densely packed films on gold electrodes. The DNA-modified electrodes are treated with micromolar concentrations of a redox-active intercalator [120]. Upon intercalation to the perfectly matched DNA duplex, the intercalator is reduced while no reduction is observed with DNA duplexes containing imperfections in the base stacking because of single-base mismatches, DNA kinking, and so forth [118, 170]. These results were explained by a DNA-mediated charge transport through the perfectly matched DNA duplexes but not through the duplexes where the base stacking is disturbed [170] (see Chapter 2 for more detailed discussion of the electron transfer in DNA). Because of its remarkable specificity to the electronic structure of the π-stacked DNA, this assay is well suited for the detection of point mutations [213]. To increase

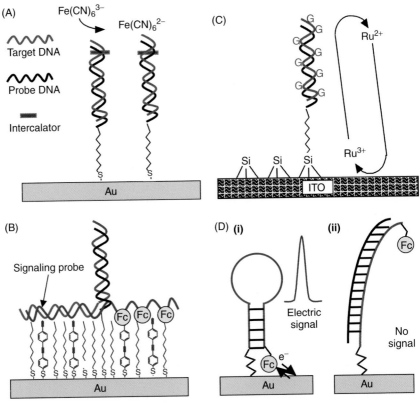

Fig. 5.7 Recent single-surface DNA hybridization sensors (Section 5.3.3.9). (A) DNA-mediated charge transport [213] Boon EM et al., (2000) *Nature Biotechnology* 18, 1096. Alkanethiol-modified probes are attached to gold electrodes and hybridized to target DNA. In the presence of redox-active intercalator, reduction of ferricyanide in solution is observed; (B) Sandwich assay with signaling probes as described in [215] Umek RM et al. (2001) *Journal of Molecular Diagnosis* 3, 74. Alkanethiol-modified capture probes are attached to gold surfaces in the presence of phenylacetylene thiols. Hybridization of target DNA and signaling probes position ferrocene labels (Fc) adjacent to the phenylacetylene thiols, which communicate with the gold electrode. Diluent thiols are terminated with polyethylene glycol (not shown); (C) Electrocatalytic oxidation of guanine with mediators. Silane-modified capture probes are attached to ITO electrodes. Hybridization of target allows for electrocatalytic oxidation of guanine by $Ru(bpy)_3^{3+/2+}$. Adapted from [90] with permission. (D) Electrochemical interrogation of DNA conformational changes [203]. Scheme of surface-attached electrochemical molecular beacon (E-DNA sensor). (i) A stem-loop ODN) with a terminal thiol is immobilized at a gold surface (Au) forming a SAM (Section 5.3.3.1). The other end of the ODN contains electroactive label (ferrocene, Fc). The stem-loop structure holds the Fc label in close proximity with the gold electrode, ensuring efficient electron transfer, manifested by an electrochemical signal. (ii) On hybridization with the complementary tDNA, the stem-loop structure is lost and a rigid rodlike duplex DNA is formed. As a result of duplex DNA formation, Fc label is moved away from the electrode, and the electrochemical signal is decreased or eliminated.

the sensitivity of the assay, a coulometric readout strategy based on the electrocatalytic reduction of ferricyanide by methylene blue (MB) was developed [119]. In this assay, electron flows from the electrode to the intercalated MB^+ to reduce it to leukomethylene blue (LB) in a reaction mediated by the surface-attached DNA. On reducing ferricyanide in solution, LB is oxidized back to MB, which can then participate in more electron transfer in an electrocatalytic cycle, involving repeated interrogation of the surface bound DNA (Figure 5.7A). Longer interrogation times yield greater absolute signals and better discrimination between mismatched and fully matched DNA. All of the possible DNA single-base mismatches (including the thermodynamically stable GA mismatch) were detected by this assay [213]. The total charge accumulated at the electrode was linearly proportional to the electrode area ranging from 30 to 500 μm. At a 30-μm electrode, as few as 10^8 DNA molecules were detectable. The reaction was performed on an addressable 18-gold electrode array. This new approach, based on the DNA-mediated charge transport, provides sensitivity through electrocatalysis, specificity in mutation analysis and a facile access to low-density array format.

Electrocatalytic Oxidation of Guanine with Mediators
The electrocatalytic strategy using mediators (see Section 5.2.2.4) was applied to develop a new type of a DNA sensor (Figure 5.7C). Several immobilization methods were used, including irreversible adsorption from dimethylformamide solution [86], covalent attachment of NAs to the dicarboxylic acid SAMs [81] and covalent coupling to microporous polymer membranes. In these experiments, ITO electrodes were used. At ITO electrodes nonspecific adsorption of DNA was minimized and at the potentials of the electrocatalytic reactions electrolysis of water was rather low not interfering with the analysis.

The rapid electron transfer between G and $[Ru(bipy)_3]^{3+}$ was primarily explained by nearly the same standard redox potentials of $[Ru(bipy)_3]^{3+}$ and G and the low reorganization energy of the Ru(II/III) couple [87]. High sensitivity of this method is based not only on its electrocatalytic nature of the processes yielding multiple electrons per G residue but also on the possibility of oxidizing all G residues in tDNA. About 100 electrons per hybridized G could be collected under favorable conditions [82]. The sensitivity thus increases with the length of the tDNA similarly as it is the case with the double-surface label-free detection (all purines yield the electrochemical signal, Section 5.3.4.3) or detection of osmium-modified DNA (all pyrimidines can produce the catalytic signal, Section 5.3.4.4) [130, 131, 133, 202]. To remove the background signal resulting from G oxidation of the immobilized probe DNA, all guanosine residues in the probe were substituted by inosines, which have similar base-pairing properties as guanines but do not produce any electrochemical signal comparable to that of guanine.

In a typical experiment, silane-modified probes are attached to ITO electrodes and immersed in a solution, containing tDNA. After the DNA hybridization, the electrodes are washed to remove unhybridized molecules and then exposed to $[Ru(bipy)]^{3/2+}$ mediator. Electrochemical measurements are usually performed by means of CV, chronoamperometry or chronopotentiometry. mRNA target composed of

973 nucleotides (containing 305 guanines) was detected by means of CP at a 200-μm ITO in a microtiter plate with a detection limit < 1 fmol of target mRNA in a 50-μL sample [90].

Electron transfer rates for oxidation of G by metal mediators depend on G solvent accessibility [80]. This makes it possible to differentiate not only between ss and dsDNA but also to detect all possible mismatches (see Section 5.2.2.4). A more reliable mismatch detection can be accomplished by using a G derivative (such as 8-oxoguanine, 8-OG) in the probe, which is selectively oxidized with $[Os(bipy)_3]^{3+}$ at about 0.85 V [85]. Current enhancement increases in the order 8-OG·C < 8-OG·T < 8-OG·G < 8-OG·A < 8-OG·G, that is, in the same order observed for guanine and for $[Ru(bipy)_3]^{3+}$ [80]. A DNA probe containing 8-OG was used to detect TTT deletion in cystic fibrosis [85].

Electrochemical Interrogation of DNA Conformational Changes
The so-called molecular beacons attracted great attention (reviewed in [217]) as a tool for the solution of DNA hybridization analysis. They are composed of an ODN forming a hairpin-like stem-loop structure, with a fluorescent label and its quencher at either of its ends. In its folded stem-loop configuration, the ODN termini are held in close proximity, and fluorescence is quenched. On binding of a complementary tDNA strand, the stem-loop structure is converted in a rigid linear DNA duplex, thus moving the fluorophore far from the quencher and inducing strong light emission. To create a molecular beacon, potentially useful for arrayable chip-based optical detection, its solid-state version was recently introduced [218]. Almost at the same time Fan and coworkers [203] proposed an electrochemical variant of molecular beacon and developed a new strategy for the sensitive, reagentless transduction of DNA hybridization into an electrochemical signal (Figure 5.7D).

They used a 27-mer ss ODN labeled by ferrocene at its 5'-end to immobilize it at the gold electrode via its – SH group at the other end. The sequence of this ODN was chosen to form readily the stem-loop structure at the electrode surface. Within this structure ferrocene was located close to the electrode surface, producing a distinguished cyclic voltammetric redox pair in the absence of the complementary tDNA. These voltammetric signals were due to electron tunneling between ferrocene and the electrode. Hybridization of tDNA with the 17-base loop of the 27-mer ODN resulted in the formation of a rigid rodlike duplex DNA, and the ferrocene label was moved away from the electrode surface (Figure 5.7D). This change in the ferrocene distance from the electrode surface resulted in a decrease or elimination of the ferrocene electrochemical signal (efficiency of electron tunneling decreases exponentially with increasing distance). In this way, a new electrochemical sensor (E-DNA sensor) was created, which is reusable, reagentless, sensitive and simple, thus offering new possibilities in the development of electrochemical DNA sensors. The 10-pM detection limit of the E-DNA sensor was comparable with other electrochemical sensors [16, 17, 170, 219] and substantially better than the 10-nM sensitivity reported in [218] for the surface-attached optical molecular beacon. The E-DNA sensor is potentially arrayable, thus providing a promising alternative to traditional fluorescence-based gene arrays. A similar approach was used by Youdong Mao and coworkers [142], but instead of ferrocene, thionine was used as an electroactive marker at the 3'-end of the DNA forming the hairpin loop structure. The thermal gradient detection method allowed discrimination of single-base mismatches from perfectly matched tDNAs.

Recently, Thorp and coworkers used $[Os(bpy)_3]^{2+}$ covalently tethered to one end of an ODN to detect electrocatalytic 8-OG oxidation in a DNA secondary structure-controlled manner [91]. In a hairpin form of the ODN (but not in a linear ds conformation), the osmium complex was situated near the 8-OG, being able to mediate its electrocatalytic oxidation.

Earlier described reagentless sensors [219], based on G oxidation at carbon electrodes, required the probe DNA without guanine and their sensitivity was substantially lower than that of the E-sensor [203]. The E-sensor is a signal-off device, being thus susceptible to false-positive responses due to various sample contaminants degrading DNA. Simultaneous monitoring, at a single electrode of both sensor elements and control sequences with "multicolor" redox markers (producing signals at different potentials) [141, 220, 221] will be necessary to detect the false-positive signals.

A different ferrocene-based signal-off device, called EDDA, is advertised by a German company FRIZ, Munich (http://www.frizbiochem.com/en/0830_idx.htm). Capture probe DNA in this device is unlabeled and DNA hybridization results in the removal of weakly bound ferrocene-labeled signaling probes from the capture probe, thus decreasing their ferrocene electrochemical signal.

5.3.4
Double-surface Techniques

For almost a decade, only single-surface strategy was considered for the electrochemical DNA sensors (reviewed in [16, 17, 90, 166]). After the application of the first simple electrodes with adsorbed or covalently attached DNA probes (reviewed in [167, 177]), it became clear that the electrode should be engineered to optimize the hybridization conditions and to minimize nonspecific adsorption of DNA at the electrode surface. Perhaps the most successful in this respect were the electrodes covered with conducting polymers (Sections 5.2.2.5 and 5.3.3.7) and gold electrodes with DNA probe and alkanethiol SAMs (Sections 5.3.3.1 and 5.3.3.2). In the recent two to three years, new technologies that differ from the classical single-surface strategies have been developed. They use two different surfaces, one for the DNA hybridization and the other one for the electrochemical detection. Application of these new approaches to DNA chips relies on microfluidic systems [222], which should be included into the chip to manipulate the analyte at different surfaces.

Conditions that are optimum for the DNA hybridization at surfaces greatly differ from those necessary for the electrochemical detection, thus making optimization of both hybridization and detection at a single surface difficult. For example, (a) strong nonspecific adsorption of DNA at the surface during DNA hybridization can decrease or even eliminate the specific DNA signals. On the other hand, strong adsorption of

DNA at the detection electrode (DE) can be exploited to accumulate DNA at DE to increase sensitivity of the DNA hybridization; (b) the DE should be small to be able to work with small analyte volumes, while the surface used for hybridization (surface H) should be sufficiently large to capture appreciable amount of target NA; (c) surface H should be suitable for the probe immobilization, while DE should be best suited for the given electrode process (probe immobilization may partially interfere with the electrochemical analysis); (d) DE, but not surface H, has to be conductive, and so on.

Recently we proposed a new method [12, 13, 16, 168, 223, 224] in which DNA hybridization has been performed at surface H and electrochemical detection at DE (Figure 5.8). We started our work with commercially available superparamagnetic Dynabeads oligo(dT) (DBT), with the covalently attached DNA probe (dT)$_{25}$, which we used as surface H, and mercury, SAE or carbon electrodes as the DE. We call this arrangement Double-Surface Technique (DST).

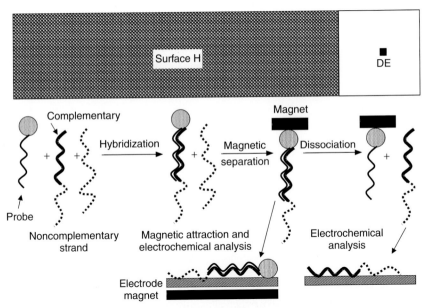

Fig. 5.8 Principles of double-surface (DS) DNA hybridization sensors. In variance to the single-surface techniques (SST, Figures 5.5–7) (Section 5.3.3) in the DS sensors, the DNA hybridization takes place at one surface, and electrochemical detection of the hybridization event at another surface (Section 5.3.4). Hybridization is performed at a relatively large surface H best suited for this purpose [e.g. magnetic beads with covalently (or avidin-biotin) bound DNA probe shown in this Figure]. Then the target DNA is released from the surface H and determined at a small detection electrode (DE, micro- or nanoelectrodes can be used for this purpose). Alternatively, the magnetic beads with the bound target DNA can be attracted to the electrode (under which a magnet is placed), and DNA can be determined without dissociating it from the beads. Various electroactive DNA labels were designed, such as metal nanoparticles, osmium complexes, enzymes and enzyme-linked immunoassays (Section 5.3.4.1). Carbon, solid dental amalgam and mercury electrodes were mainly applied.

Because of minimum nonspecific DNA and RNA adsorption at the beads, we achieved very high specificity of hybridization. Optimum DE was chosen only with respect to the given electrode process. By means of DST, detection of relatively long tDNAs is possible, and various electrochemical detection methods can be used without considering the suitability of the given DE for the immobilization of the DNA probe. We have been able to apply for the first time the mercury drop electrodes as well as solid dental amalgam electrodes in the DNA hybridization studies [67, 225, 226]. In addition to DBT, other types of magnetic beads such as streptavidin-coated magnetic beads were used to bind biotinylated DNAs [163, 219, 227]. Magnetic beads are very convenient, but also other materials can be used as surface H in electrochemical DST. Such surface H can be completely separated for the DE (e.g. wells of polystyrene microtiter plates) or can be somehow connected with the DE (e.g. nitrocellulose membranes attached to the solid electrode surface [228, 229]). We have recently used nitrocellulose membrane-covered carbon electrodes in hybridization experiments, and our preliminary results suggest that such electrodes may be of use in DNA sensors [230].

Willner and coworkers developed a number of methods of magneto-switchable electrocatalytic and biocatalytic transformations (reviewed in [163]). Recently they used rotating magnetic particles to enhance biocatalytic processes by convection-controlled rather than diffusion-controlled interactions of the substrate with the enzyme in DNA sensors [55, 161, 162].

5.3.4.1 New Detection Methods

Application of DST made it possible to propose several new highly sensitive assays, such as label-free DNA and RNA detection [67], DNA osmium-modification yielding catalytic signals at mercury electrodes and SAE [131, 133, 134, 201, 202], differently coded reporter probes [202, 231, 232], enzyme-linked immunoassays [211, 233, 234], nanoparticle labels (reviewed in [172]), including electroactive beads [220] and indium microrod tags [235] enabling a very high sensitivity of DNA detection (see Chapter 8). Most of these methods would be difficult or impossible to apply in combination with the single-surface techniques. Particularly interesting appears the use of nanoparticles (reviewed in [172]) and especially the electroactive beads [220], containing a large amount of small redox molecules (see Chapter 8) and the possibility of determination of the length of the DNA repetitive sequences without using any nonelectrochemical method. Some of the above methods will be briefly discussed below.

5.3.4.2 Nanoparticles in DNA Detection

Application of nanoparticles in optical detection resulted in enhanced sensitivity and selectivity. Methods of multicolor optical coding based on different sized quantum dots [236] and "bar-coded" metallic microrods [237] have been developed. Inspired by the success of nanoparticles in optical detection, analogous techniques for electrochemical detection were developed [172] by J. Wang and coworkers (see Chapter 8). Newly designed protocols are based mainly on magnetic beads (used for separation) combined with reporter probes labeled with nanoparticles such as gold and silver tags [238–240], semiconductor quantum dot tracers or with larger polymeric "electroactive" beads (as carriers of a huge amount of electroactive molecules). Acid dissolution followed by anodic stripping determination of the metal tracers was used to detect picomolar concentrations of the DNA captured at magnetic beads [240] or at some other surface H (e.g. chitosan or polypyrrole layers) [238, 239]. Nanoparticle-promoted precipitation of gold [240] or silver [241–243] allowed further sensitivity enhancement.

In addition to capturing and separating the hybridized assembly from the mixtures, magnetic beads have been also utilized for controlling and triggering the electrochemical DNA detection [244–246]. Magnetic attraction of the bead–DNA–metal tracer assembly to the mercury film-electrode, which allowed direct electrical contact of the precipitate with the electrode surface, was used to simplify the electrochemical detection by eliminating the acid-dissolution step [245].

Unique properties of semiconductor nanocrystals (quantum dots) ([172] and references therein) have been utilized for electrochemical detection [158, 232, 247]. Cadmium sulfide nanoparticles were used to label the reporter probes complementary to tDNA captured at magnetic beads [247]. ZnS, CdS and PbS nanocrystals (with different redox potentials) were applied for simultaneous detection of multiple tDNA sequences [232] with femtomolar detection limits.

ODN-functionalized gold nanoparticles and closely spaced interdigitated electrodes were used by Mirkin's group to develop a new array-based electrical DNA hybridization device [248]. The ODN probe was attached in the gap between two microelectrodes followed by the DNA hybridization, which brought the gold nanoparticles in the gap. The subsequent silver deposition resulted in a conductivity signal only in samples where DNA hybridization took place. Point mutations were detected with high selectivity at controlled salt conditions. Monitoring of the hybridization-induced resistance changes across the electrode gap provided high sensitivity with a detection limit of 500 fM.

5.3.4.3 Label-free DNA and RNA Detection

Label-free DNA hybridization assay is based on the ability of diluted acids to release purine bases from DNA and RNA. tDNA captured at the beads (Figure 5.8) is treated with acid, and the released bases are directly determined by cathodic stripping voltammetry (Section 5.2.1.2). Subnanomolar concentrations of bases can be determined using CuSAE or HMDE in the presence of copper [67]. Analysis can be performed in 3–10 µL volumes, which makes it possible to determine picogram or subpicogram quantities of DNA, corresponding to subattomole amounts of 1000-bp DNA.

5.3.4.4 Osmium-modified DNA

Earlier we showed that ssDNA can be easily modified by osmium tetroxide complexes with nitrogen ligands (Os,L) (Sections 5.2.2.4 and 5.2.3.2). Os,bipy-modified DNA produces three redox couples and a high-electron yield catalytic signal at about −1.2 V (Figure 5.4). Some Os,L complexes (such as Os,bipy) show a high selectivity for DNA structure (only ssDNA is modified) and for pyrimidine bases (purine bases do not react); with Os,bipy, thymine reacts 10 times faster than cytosine [130]. Combined with DST (Figure 5.8) Os,bipy modification of DNA provides a sensitive detection mode of the DNA hybridization event. Because of its specific reactivity, some of the Os,L complexes can be applied for an easy end-labeling of reporter probes [202] and for pre- or postmodification of tDNA [131]. Signals of Os,L-modified DNA can be measured at mercury and amalgam electrodes [134] while at carbon electrodes only the Os,L redox couples are detectable [201]. Using the catalytic signal, picomolar DNA concentrations were detected [20, 132].

Chromosomal DNA is nonimmunogenic or only very weakly immunogenic, but after Os,L modification it becomes a strong immunogen. We generated polyclonal and monoclonal antibodies against DNA modified by Os,bipy [135, 136, 249] and developed an enzyme-linked immunoassay of Os,bipy-modified tDNA at carbon DEs [233].

5.3.4.5 Determination of Lengths of Repetitive Sequences

Determination of lengths of repetitive sequences is important in various areas of molecular biology. Moreover, genomic expansions of trinucleotide repeats are connected with neurodegenerative diseases such as fragile X syndrome (CGG triplet), myotonic dystrophy (CTG), Huntington's disease (CAG) or Friedreich ataxia (FRDA; GAA), and determination of the triplet length expansion is used in diagnosing these diseases [250, 251]. Fast electrochemical DNA sensor capable of detecting the triplet lengths may thus be very useful. In SST, combination of electrochemistry with DNA radioactive labeling was, however, used to determine the length of the CGG or CTG triplet repeats [252]. Reporter probes are well suited for this purpose, but as we already pointed out in SST (see Section 5.3.3.5), they usually have to be bound to tDNA sequences that are close to the electrode surface. In contrast, in DST a reporter probe can bind to any site in tDNA. This made it possible for us to propose a novel method of the determination of the triplet repeat lengths (Figure 5.9). In principle, two RP with different labels can be used, one complementary to the repetitive sequence (RP1) and the other one (RP2), complementary to a nonrepetitive sequence present constantly in the genomic DNA (Figure 5.9). Length of the triplet expansion is calculated from the ratio of the intensities of RP1/RP2. To detect the homopurine $(GAA)_n$ repeat in FRDA, we applied [231] a slightly different strategy based on reactivity of Os,L complexes (see Section 5.2.3.2 and above) with pyrimidine but not purine bases. We used biotin-labeled RP1 (binding to the repeat), which was detected via an enzyme-linked assay. Pyrimidine residues within sequences flanking the homopurine $(GAA)_n$ repeat in tDNA were premodified with Os,bipy introducing electroactive labels in tDNA. Length of the triplet expansion was calculated from the ratio of the intensities of signals of hybridized RP/tDNA-Os,bipy (measured at carbon and/or mercury electrodes) [231]. Os,bipy-premodification of tDNA prevented reassociation of the pyrimidine and purine repeat strands [131, 231], strongly increasing efficiency of hybridization of homopurine DNA stretches with the RP1. Principles of this DST assay can be used in determination of expansion of different DNA nucleotide triplets (related to other neurodegenerative diseases [250–252]) as well as in determination of lengths of any repetitive sequence.

5 Electrochemical DNA Sensors

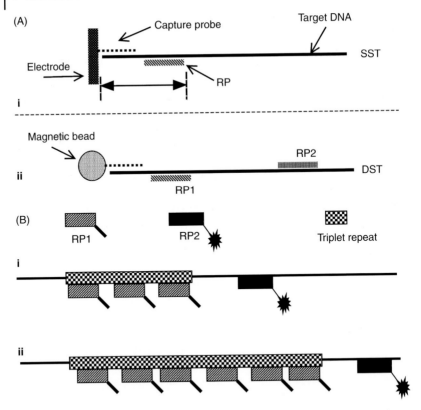

Fig. 5.9 Application of reporter probes (RP) in single-surface (SST) (Section 5.3.3) and double-surface techniques (DST) (Section 5.3.4). (A,i) In SST, the RP redox label has to communicate with the electrode and must be therefore bound to the tDNA sequence, which is close to the electrode surface. This limitation can be overcome by using enzyme–labeled RP that catalyze formation or disappearance of a low-molecular mass compound. (A,ii) In DST, the RP can be bound to any site at tDNA because it interacts with tDNA captured at surface H (e.g. magnetic beads), and the electrochemical analysis is performed in the next step where the RP is attached to the electrode (see Figure 5.8). (B) In DST, multiple RPs labeled with various electroactive markers producing electrochemical signals at different potentials can be used (Sections 5.3.4.1 and 5.3.4.2). To determine the lengths of the triplet expansion in neurodegenerative diseases, two RPs are used: RP1, complementary to a set of triplets and RP2, complementary to a nonrepetitive sequence present constantly in the genomic DNA. The length of the triplet expansion is calculated from the ratio of the intensities of RP1/RP2.

5.3.5
Concluding Remarks to DNA Hybridization Sensors

This review shows that PCR-amplified tDNAs (and RNAs) can be sensitively detected by a number of the current electrochemical DNA hybridization sensors. By these methods, detection of various kinds of point mutations and determination of lengths of long repetitive sequences is possible, in addition to the detection of relatively short specific DNA sequences.

On the other hand, electrochemical detection of nonamplified tDNA (for which the picomolar sensitivity is too low) has not yet been reported. Further improvement of sensitivity and selectivity of the current methods will be necessary to reach this goal. The ultrasensitive DST assay, based on polystyrene beads impregnated with a huge number of redox marker molecules (ferrocenecarboxaldehyde), was recently reported by Wang and coworkers [220]. By this assay, about 30 000 tDNA molecules were detectable. Shortly afterwards, this record sensitivity was surmounted by Heller's group [210], who achieved detection of about 3000 tDNA molecules by combining their biocatalytic system (Section 5.3.3.7) with carbon microelectrodes. There is no doubt that it is the signal amplification involving beads, nanoparticles and catalytic reactions on one hand and miniaturization of the electrode systems on the other hand that represent the ways toward reaching high sensitivities that are necessary for the detection of unamplified genomic DNA sequences in an electrochemical DNA sensor. Quite recently J. Wang et al. [423] used DST in combination with carbon nanotubes to increase the sensitivity of the DNA sensor down to z-mol (about 800 tDNA molecules).

5.4
Sensors for DNA Damage

5.4.1
DNA Damage

DNA in the cells is permanently subjected to a number of physical or chemical agents that induce chemical alterations in its molecules, representing damage to the genetic material. It has been estimated that 10^4 to 10^6 DNA damage events take place per cell per day (reviewed in [214, 253–255]). Prevailing DNA lesions (Figure 5.10) involve apurinic sites arising from hydrolysis of the purine N-glycosidic bonds, strand breaks (sb), 8-OG and thymine glycol induced by reactive oxygen species, base deamination products (uracil, hypoxanthine or xanthine), and so on. Some of these lesions originate from DNA interactions with intermediates or by-products of normal cellular metabolism. Other adducts are generated by genotoxic agents occurring in the environment. For example, rather abundant pyrimidine dimers are induced by sunlight, while a variety of bulky adducts are formed upon DNA interactions with metabolically activated carcinogens such as aromatic amines or polycyclic aromatic hydrocarbons (PAH). Persisting DNA damage may result in changes of the genetic information (mutations) and subsequently may have severe impacts on the cellular functions and life of the organisms. To maintain integrity of their genome, cells possess enzymatic machineries capable of recognizing and repairing the DNA damage [214, 253–255]. Different lesions are repaired via different pathways. Damaged bases or base mismatches are excised by action of N-glycosylases and/or nucleases (in base excision repair, nucleotide excision repair or mismatch repair) while highly cytotoxic double-strand breaks (dsb) are repaired through a mechanism involving DNA recombination. Cells with defects in either of

160 | 5 Electrochemical DNA Sensors

the DNA repair pathways are usually extremely prone to malignant transformation. Extensive DNA damage that would be hardly repairable frequently induces programmed cell death (apoptosis).

5.4.1.1 Techniques Used to Detect DNA Damage

Analysis of DNA damage induced by environmental mutagens and carcinogens, including industrial pollutants and their metabolically activated products, is of great

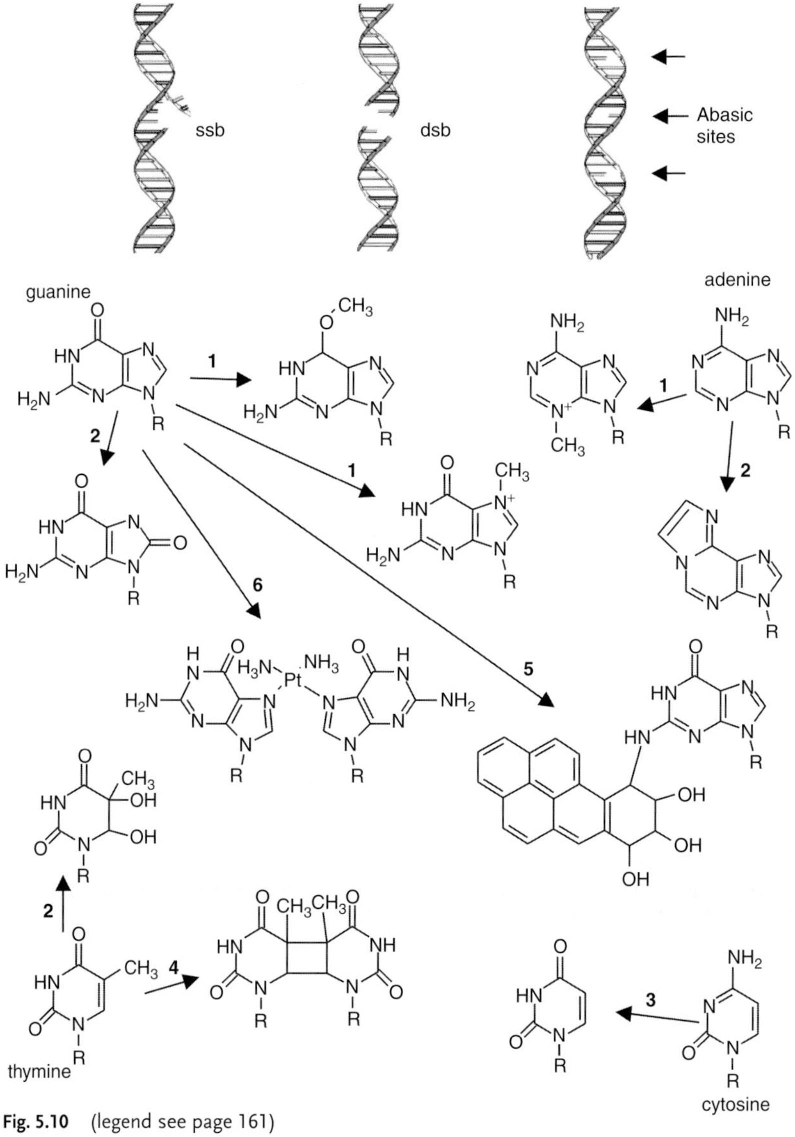

Fig. 5.10 (legend see page 161)

importance for human health protection. In addition, a number of chemotherapeutics act through formation of specific DNA lesions. Studies of DNA damage by these drugs and its repair are important steps in the development of novel drugs. Some DNA adducts have been chosen as biomarkers of exposure of an organism to reactive oxygen (8-OG) [256, 257] or nitrogen (8-nitroguanine) [258] species, alkylating agents (8-hydroxyethyl guanine, 6-O-methylguanine) [259, 260], and so forth. To analyze DNA damage, different analytical techniques have been applied, among which two basic approaches can be distinguished. In the first, the analyzed DNA is hydrolyzed and products of DNA damage are determined usually by chromatographic techniques. Products of oxidative DNA damage have been determined in DNA hydrolysates by HPLC [261–266], gas chromatography [267] or capillary electrophoresis [268] combined with different detection techniques. These methods are highly sensitive, allowing for detection of one damaged base among 10^7 normal ones [264], but may suffer from false-positives arising from additional DNA oxidation during the sample preparation [262, 264, 267, 269]. Different DNA adducts have been analyzed via postlabeling techniques [270–272]. These methods involve enzymatic hydrolysis of damaged DNA, followed by introduction of fluorescent or radioactive (^{32}P, ^{3}H) tags to the hydrolysis products and their chromatographic separation and detection. Differential cleavage of chemically modified DNA sites by some nucleases (such as nuclease P1 [273, 274]) may be utilized for a better resolution of the damaged entities. Identification of DNA adducts has been achieved also by using mass spectrometry techniques [262, 266, 267, 272, 275].

The other approaches are based on tracking changes of the features of whole DNA molecules (without hydrolysis) upon their damage. Most of these techniques involve detection of DNA sb by agarose gel electrophoresis. Individual ssb or dsb can be detected via relaxation or linearization of sc plasmid DNA molecules ([26, 276] and references therein). Analysis of multiple DNA strand breaking in individual cells is often performed via "comet" or alkaline elution assays [261, 269, 277–279]. These methods are based on electrophoretic estimation of mean length of ds or ssDNA fragments formed upon introduction of dsb or ssb, respectively. Combination of the comet assay with DNA cleavage by specific DNA repair endonucleases has been utilized for detection

---

Fig. 5.10 Most common products of DNA damage. Upper panel shows lesions originating from interruptions of the DNA sugar-phosphate backbone (single- or double-strand breaks, ssb or dsb) or from hydrolysis of the N-glycosidic bonds between the backbone and base residues (abasic sites). The most frequent modifications of the nucleobases guanine (G), adenine (A), thymine (T) and cytosine (C) induced by genotoxic agents are shown in the bottom panel. Alkylating agents (**1**) form a variety of base lesions, including 3-alkyl adenine, O^6- and 7-alkyl guanine derivatives shown in the Figure. Oxidative DNA damage (**2**) involves formation of 8-oxo guanine, thymine glycol and 1,N^6-ethenoadenine (formed because of A reaction with acrolein, a product of polyunsaturated fatty acids oxidation). Some mutagenic chemicals such as nitrous acid or bisulfite may cause base deamination, thus converting, for example, C to uracil (**3**). Photochemical reactions induced by ultraviolet light (**4**) yield several products, with the cyclobutane thymine dimers being the most abundant ones. Metabolically activated carcinogens such as polycyclic aromatic hydrocarbons (**5**) form bulky DNA adducts, predominantly at the G residues. Many anticancer drugs, including platinum complexes (**6**), act by inducing specific DNA damage.

of certain kinds of DNA base damage [256, 261]. Number of ends of DNA molecules in cell nuclei may be estimated by the "TUNEL" test [280, 281], a technique based on labeling of free 3'-OH polynucleotide termini via a reaction catalyzed by a terminal nucleotide transferase. Antigenic features of some DNA adducts have been utilized in immunochemical methods of their detection [249, 263, 266, 269, 282, 283].

5.4.2
Relations Between DNA Damage and its Electrochemical Features

It has been recently proposed [214, 284] that electrochemical and electronic features of DNA may play roles in processes resulting in DNA damage, recognition of DNA lesions and DNA repair *in vivo*. DNA double helix may behave as a "molecular wire" mediating electron (hole) transfer over long distances [118, 171, 285–290] (see also Sections 5.2.3.1 and 5.3.3.9). In addition, DNA contains sites of anomalous redox potential (such as GG doublets in undamaged DNA and some base lesions such as pyrimidine dimers or 8-OG), which can be oxidatively attacked and/or recognized by the DNA repair machinery. Considering the long-range charge transfer, recognition of these sites by specific proteins may be attained from remote positions [214, 285]. Perturbations in the DNA double helix such as single-base mismatches, insertion/deletion lesions or abasic sites dramatically decrease the charge transfer efficacy [171, 214] (see Section 5.3.3.9). Some proteins involved in recognition of DNA lesions (e.g., MutY) contain redox-active prosthetic groups that have been shown to take part in DNA-mediated redox reactions [214, 284].

Above and beyond their proposed biological consequences, changes of electrochemical properties of DNA upon its damage can be exploited analytically. Electrochemical detectors have been employed in techniques of damaged DNA analysis on the basis of hydrolysis DNA and separation of the product by HPLC [258, 261, 263, 264, 266], mass spectroscopy [262, 267] or capillary electrophoresis [263, 268]. On the other hand, it has been shown that electrochemical measurements can provide information about the properties of nonhydrolyzed, highly polymeric DNA molecules (reviewed in [7–14, 16, 90, 111, 166, 167, 291–293]). Relatively small changes in DNA structure (including those induced by DNA damaging or otherwise biologically important agents) may significantly affect behavior of the DNA at electrodes (see Section 5.2). Damage to DNA may lead to altered availability of its electroactive sites, resulting in changes in the measured electrochemical signals. Interruptions of the sugar-phosphate backbone of dsDNA confer increased accessibility to base moieties next to the strand ends, significantly influencing behavior of the DNA at mercury electrodes (Figure 5.11A; see Sections 5.4.2 and 5.4.4) [26, 111, 294–301]. Local changes of DNA conformation arising from chemical damage of DNA bases may also influence electrochemical features of the polynucleotide molecules [296, 302–306]). As a result of modification of electroactive base residues, their intrinsic oxidation or reduction response may be lost [29, 38, 175, 307–314] (Figure 5.11B; see Section 5.4.5.1). Some products of DNA base damage (e.g., 8-OG [85, 91, 315–318] or bulky DNA adducts with electroactive compounds [20, 129, 130, 132, 201, 309–311]) yield new electrochemical signals not observed in the

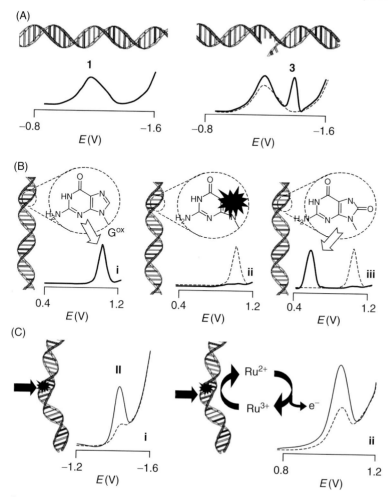

Fig. 5.11 Principles of electrochemical detection of covalent DNA damages.
(A) DNA strand breaks can be sensitively detected at mercury electrodes. Specific signals, including the tensammetric peak **3**, are yielded by dsDNA only in the presence of the sb [26]. (B) Chemical modifications of nucleobases may alter their electrochemical features. For example, intrinsic electroactivity of the G residues (yielding peak G^{ox} at carbon electrodes) (**i**) may be lost (**ii**) owing to disruption of the G moiety (e.g. [307, 313]). (**iii**) Some adducts, including 8-oxo guanine [315, 316], produce specific electrochemical response at potentials differing from the parent base signals. (C) Base lesions may cause distortions of the DNA double helix, connected with increased accessibility of adjacent electroactive base moieties. (**i**) For example, UV light- [296] or cisplatin- [335] induced DNA damage can be detected via measurements of DNA DPP peak **II**. (**ii**) Formation of DNA adducts with styrene oxide [303, 304, 336] facilitates electrocatalytic oxidation of the DNA G residues.

unmodified DNA (Figure 5.11B; Section 5.4.5.2). Noncovalent DNA interactions with potentially genotoxic agents can be monitored via changes of electrochemical signals of either DNA [110, 319, 320] or the binders (e.g., [110, 321–330]) (Section 5.4.6). Structure selectivity of some electroactive noncovalent [326, 331–334] or covalent [68, 129, 130, 201] DNA binders can be utilized in redox marker–based methods of electrochemical DNA structural probing (see Sections 5.2.3 and 5.4.4.2). A number of techniques of electrochemical DNA damage detection and sensors (detectors) for DNA damage have been proposed to date (reviewed in [13, 16, 111, 167, 291, 292]). In the following sections, various electrochemical approaches suitable for the detection of DNA damage and of DNA-damaging substances will be described and their applications discussed.

5.4.2.1 Studies of DNA Damage by Polarographic Techniques

The seminal electrochemical studies of DNA were connected with the dropping mercury electrode (DME) and polarographic methods (see Sections 5.1 and 5.2). Oscillographic polarography at controlled a.c. and later DPP revealed striking differences between native (ds) and denatured (ss) DNA at the DME [7–10, 37, 294, 337, 338] (Figure 5.1; see Section 5.2.2.1). Importantly, intensity of the dsDNA-specific peak II responded sensitively to subtle dsDNA structural transitions, including premelting changes of the DNA double helix [9] and perturbations related to DNA damage. Height of the DPP peak II increased with the number of ssb and/or dsb resulting from DNA treatment with ultrasound, DNase I or ionizing radiation [294, 295]. Covalently closed circular DNAs (possessing no strand ends) did not yield any peak II under the same conditions [291, 339]. Formation of DNA lesions involving conformational distortions of the DNA double helix such as pyrimidine dimers in UV-irradiated DNA [296] or DNA adduct with chemical agents (such as some platinum [14, 335, 340] or osmium tetroxide [22] complexes) led to increase of the peak II height as well (Figure 5.11C). Moreover, DPP measurements could be used to recognize specific structural features of different DNA lesions. Presence of unpaired bases in denaturation lesions induced by transplatin or some other platinum complexes [14, 335, 340] resulted in formation of the ssDNA-specific peak III. On the other hand, nondenaturation lesions that induce distortions of the DNA double helix without disruption of the Watson–Crick base pairs were connected with increasing intensity of the peak II [14, 335]. DNA covalently modified with osmium tetroxide complexes (electroactive DNA markers and structural probes, see Section 5.2.3.2, Figure 5.4) was analyzed [20] by DPP using a catalytic osmium signal that provides a high sensitivity of the DNA detection.

Analogous information about DNA damage can be obtained through measurements of tensammetric DNA response in weakly alkaline background electrolytes (reviewed in [10, 13, 14]). Specific a.c. polarographic signal (denoted as peak 2) responds to DNA damage by various agents (including ultrasound, γ-rays, etc.). This peak has been attributed to distorted segments of dsDNA, in contrast to peak 3 occurring at more negative potential, which is specific for denatured DNA with freely accessible bases [10, 13, 14]. Helix opening transitions induced by negative DNA superhelicity (involving unpaired bases within covalently closed DNA chains) are manifested by a formation of another tensammetric peak 3* (occurring at intermediate potential, compared to peak 2 and peak 3) detectable by means of a.c. polarographic or voltammetric measurements [68].

5.4.2.2 Voltammetric Approaches

Although polarographic techniques (using the DME) proved excellent for probing DNA structure in solution (reviewed in [9, 10]), they require rather large quantities of the analyte and are not compatible with construction of DNA biosensors. Development of such sensors, based on DNA immobilization at surfaces of stationary electrodes (see Section 5.2.2.2, Figure 5.2), was preceded by application of HMDE and various kinds of solid electrodes in connection with a variety of electrochemical techniques, including voltammetry (CV, DPV, SWV), impedance spectroscopy or CP (reviewed in [11–13, 16, 111, 167, 293]). Linear sweep [15, 341] and AC voltammetry [26, 298–301, 342] measurements of DNA tensammetric peaks at the HMDE were employed in investigations of electrochemical behavior of DNA containing sb (Figure 5.11A, Figure 5.12A) induced by γ-rays, ultrasound, chemical or enzymatic nucleases, and so on. Formation of DNA sb facilitates potential-induced unwinding of DNA duplex at the mercury surface, resulting in formation of ssDNA-specific signals at the mercury electrodes under certain conditions [26, 111] (see Section 5.2.2.3). Changes in intensity of CV, SWV, DPV or chronopotentiometric stripping analysis (CPSA) peaks yielded by DNA G residues on either HMDE or carbon electrodes (Figure 5.11B) have been utilized to monitor DNA damage caused by a variety of chemicals, including alkylating agents [38], hydrazine derivatives [307], anticancer drugs [309–312, 343], environmental pollutants [175, 308, 344–346], ultraviolet light [313], and so forth. Voltammetric signals yielded by specific products of DNA damage such as 8-OG [315, 316, 318], adenine ethenoderivatives [347] or DNA adducts with electroactive compounds (mitomycin C [309–311], osmium tetroxide complexes [21, 129, 130, 132, 201, 348]) at mercury or carbon electrodes have been utilized (for more details see Sections 5.4.4–5.4.8). Electrocatalytic oxidation of DNA G residues by redox mediators (see Section 5.2.2.4) (ruthenium or rhenium complexes in solution [79, 80, 82, 83, 85] or bound in polymeric assemblies at the electrode surface [84, 302–306, 349, 350]) at carbon or ITO electrodes has been used to enhance the electron yields and to probe accessibility of the G moieties in mismatched [85] or chemically damaged [302–306, 349, 350] DNA molecules. Voltammetric and CPSA measurements were utilized in a variety of studies of noncovalent DNA interactions with small molecules (reviewed in [13, 16, 111, 292]), including toxic metal ions and their complexes, natural and man-made carcinogens, drugs and so forth (see Section 5.4.6).

Other techniques involve application of indirect, redox indicator–based approaches. A metallointercalator $[Co(phen)_3]^{3+}$ binding selectively to intact dsDNA responded to DNA degradation by diminution of its voltammetric signal [326, 331–334] (Figure 5.13). Analogous system was used to probe DNA interactions with noncovalently binding species in a competitive mode [351]. Another technique for studies of DNA–drug binding was based on a restricted communication of anionic depolarizer ferro/ferricyanide with dsDNA-coated gold electrodes due to electrostatic repulsion [352]. In the presence

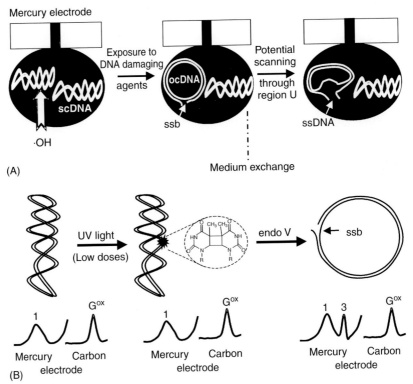

Fig. 5.12 (A) Scheme of electrochemical sensor for DNA strand breaks. Covalently closed circular (supercoiled, sc) DNA is adsorbed at surface of HMDE, MFE or AgSAE, thus forming the DNA sensor. The latter is immersed into a solution containing DNA-damaging agents (such as hydroxyl radicals) inducing the sb. Then, the electrode is washed and transferred into blank background electrolyte. During slow potential scanning through the "region U" (Section 5.2.2.3), the nicked ocDNA, but not the intact scDNA, is partially unwound (denatured) at the electrode surface. The resulting ssDNA regions are then detected using the a.c.V peak 3. (B) An example of detection of DNA base damage via its conversion to strand breaks. Upon DNA UV irradiation, thymine dimers are formed as one of the major products. These lesions are recognized by a specific enzyme, T4 endonuclease V (endo V) that introduces ssb next to the thymine dimer. The ssb can be detected using the principle shown in (A). Although deeper UV DNA damage can be detected directly by DPP [296], combination of the enzymatic and electrochemical approach makes it possible to increase considerably the sensitivity of the base damage detection when mercury electrodes are used. On the other hand, measurements of peak G^{ox} at carbon electrodes does not allow for discrimination between the sc and ocDNA (see curves on the bottom).

of cationic DNA-binding drugs, reduction/oxidation signals of ferro/ferricyanide increased. Electrochemistry of ferricyanide was utilized [213, 353, 354] also in systems proposed for mismatch-sensitive analysis of DNA hybridization involving long-range electron transfer through the DNA double helix (see Sections 5.2.3.1 and 5.3.3.9).

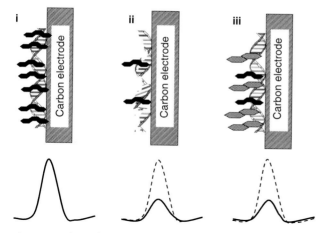

Fig. 5.13 Indicator-based sensor for DNA damage developed by Labuda et al. [326, 331, 333, 351]. (i) A metallointercalator $[Co(phen)_3]^{3+}$ (black figures) binds efficiently to dsDNA immobilized at a carbon electrode surface, producing a specific redox signal. (ii) Upon exposure to genotoxic agents, the anchored DNA is degraded, losing its double-helical structure. This results in diminished binding of the redox indicator and decrease of its signal. (iii) The technique can be used also as a competitive assay for other DNA binders, including electroinactive ones (cross-hatched figures).

5.4.3
DNA-modified Electrodes as Sensors for DNA Damage

Invention of Adsorptive Transfer Stripping (AdTS) voltammetry of DNA [based on DNA immobilization (adsorption) onto the electrode surface followed by medium exchange and measurement in a blank background electrolyte (see Section 5.2.2.2, Figure 5.2)] was the first step toward development of electrochemical DNA biosensors (reviewed in [12, 13, 16, 29, 90, 111, 166, 175, 219, 292]). Upon exposure of the sensor to different chemical or physical agents (including species causing DNA damage or otherwise interacting with the anchored DNA molecules), changes in the anchored DNA recognition layer may take place, which can subsequently be electrochemically detected.

Various techniques have been applied to immobilize DNA at electrode surfaces (see Sections 5.2.2.2 and 5.3.3.1). The simplest one involves DNA adsorption that is applicable in connection with mercury or carbon electrodes. Structure of adsorbed dsDNA at the mercury electrode is stable within a relatively wide potential window, different salt concentrations and pHs (reviewed in [10, 11, 13, 14, 111, 291, 293]). The DNA-modified mercury electrode is therefore applicable for detection of DNA-damaging species under a broad range of conditions. Adsorption of DNA at carbon electrodes may be facilitated by application of positive electrode potential providing electrostatic attraction of negatively charged DNA [110, 307, 308, 346, 355, 356]. "Bulk modification" of the carbon pastes with DNA (i.e., addition of DNA to the carbon powder prior to mixing with the oil component) [33, 357] or modification of GCE surface with acid-hydrolyzed DNA [358–362] have been used by some authors to construct DNA biosensors, while others have prepared "thick DNA layers" by application of concentrated DNA solutions onto solid electrode surfaces and subsequent drying of the resulting DNA gels or films [336, 363–365]. Although some of these procedures may result in poorly defined DNA layers, which may, in addition, suffer from inherent instability (e.g., due to release of water-soluble DNA from the bulk-modified carbon paste or from dried thick DNA films, changes in hydration of the latter), they have been empirically optimized for given purposes. An interesting technique has been developed by Rusling and coworkers, who have prepared layered DNA films on cationic polymer-coated solid electrodes [302–306, 336, 349, 350]. Such DNA films can be assembled layer-by-layer with enzymes exhibiting catalytic activity mimicking metabolic carcinogen activation [303] (see Sections 5.4.5.3 and 5.4.8.2). Application of other electrode materials, including ITO [79–81, 83, 349], gold [118–120, 213, 327, 352, 354, 366–368], carbon composites [176] or chemically modified electrodes [79, 81, 369], usually involves covalent coupling of derivatized oligonucleotides to the surface (reviewed in [17]) (see Section 5.3.3.1).

5.4.4
Sensors for DNA Strand Breaks

Strand breaks (Figure 5.10) belong to the most abundant products of DNA damage [253–255]. These lesions are formed because of attacking the deoxyribose moiety by oxygen radicals, as a result of some kinds of DNA base damage, and also upon action of nucleases (including those involved in DNA repair or programmed cell death processes). Besides electrophoretic techniques [26, 276–278] and methods based on determination of the ends of polynucleotide chains (such as the TUNEL test [280, 281]), cleavage of DNA sugar-phosphate backbone can be detected by electrochemical techniques. The latter include highly sensitive detection of the sb formation using mercury electrodes [26, 111, 297–301, 370] (Section 5.4.4.1) and other techniques suitable for monitoring deeper DNA degradation employing solid (primarily carbon) electrodes [326, 331–334, 371] (Section 5.4.4.2). The latter methods are based either on measurements of intrinsic guanine oxidation signal (peak G^{ox}) or on application of redox indicators of DNA degradation.

5.4.4.1 Detection of DNA Strand Breaks with Mercury and Amalgam Electrodes
Behavior of DNA at the mercury electrodes is strongly dependent on accessibility of the DNA base residues. In ssDNA, the base moieties can freely communicate with the electrode surface, yielding specific reduction (peak CA, peak III) or capacitive (peak 3) signals (reviewed in [10, 11, 13, 111, 291]) (see Section 5.2.2.1). In intact dsDNA, the nucleobases are hidden in the double helix interior (Figure 5.1), resulting in a relatively weak electroactivity of intact native DNA and a remarkable dependence of the intensity of specific DNA signals on formation of the sb facilitating communication of the base residues with the mercury surface (reviewed in [9, 10, 13, 111, 291]).

In addition to the effects of transient DNA double helix opening around the ssb, detectable polarographically (see Section 5.4.2.1), ccc (lacking free chain ends) and open circular (oc, containing one or more ssb per molecule) or linear (lin) DNA molecules remarkably differ also in their susceptibilities to irreversible denaturation. A procedure involving differential thermal DNA denaturation in solution followed by AdTS voltammetric microanalysis at HMDE was proposed [276]. The technique was utilized for determination of ocDNA in samples of cccDNA (such as plasmid scDNA) and for quantification of ssb induced by ionizing radiation. Nicked or linear DNAs, but not the scDNA, were irreversibly denatured at 85 °C. Only the former DNAs (i.e., the products of DNA damage) contributed to the observed changes in the intensity of CV [276] or a.c. voltammetric [26] DNA signals.

DNA double helix can be denatured at the mercury electrode in a potential-dependent manner (see Section 5.2.2.3) [11, 13, 26, 92, 201, 291, 319, 372, 373]. Again, extensive unwinding at the mercury surface can undergo only DNA molecules possessing free strand ends (lin or ocDNA) [11, 13, 16, 26, 111, 291] (Figure 5.12A). Unwinding of cccDNA molecules lacking the ssb is prevented for topological reasons [137]. When the electrode potential is slowly scanned from positive to negative values across the region U [13, 26, 49, 111], the lin or ocDNA are partially unwound before potentials of the ssDNA-specific peaks are reached, allowing for detection of the sb in a single voltammetric scan without any preconditioning. Under such conditions, both tensammetric [26, 68] and reduction [25, 297] responses of intact cccDNAs (scDNAs) at mercury electrodes strongly differ from those of lin or ocDNAs (Figure 5.12): the scDNA produces no peak 3 [26] and only negligible cathodic peak CA [25, 297], while oc and linDNAs yield both of these signals well-developed.

A simple DNA biosensor for DNA sb consisting of intact scDNA adsorbed at the mercury electrode surface has been proposed [26, 111, 291, 297, 298, 300] (Figure 5.12A). Introduction of ssb or dsb into the surface-confined DNA results in formation of the peak 3, and its intensity increases with the number of strand scissions [26, 297, 298, 300]. This type of biosensor has been applied in experiments involving DNA cleaving species including reactive oxygen species generated in Fenton-type reactions [26, 297, 300, 301] or nucleases [298]. DNA-damaging substances were detected in model samples as well as in various "real" specimens of natural and industrial waters or food [299]. Effects of the electrode potential on damage to the surface-confined DNA were investigated using complexes or compounds of redox-active transition metals [300, 301] (see Section 5.4.7). Certain kinds of DNA base lesions can be converted into ssb by DNA repair enzymes, which offers a possibility of indirect electrochemical detection of damage to the nucleobases [16, 374] (Section 5.4.5.4). Measurements of the DNA strand scissions at the HMDE exhibit a high sensitivity, allowing for detection of an ssb per more than 2×10^5 nucleotides [26].

Attempts have recently been made to find alternatives to the classical HMDE that would possess similar electrochemical features but would be better suited for application in simple, inexpensive and environment-friendly biosensors. MFE [57, 348, 375–378] and SAE [67, 225, 226, 370, 379, 380] have been successfully applied in NA analysis, including development of sensors for DNA damage. Both redox and tensammetric signals of NAs could be detected with a mercury-coated GCE (MF/GCE) [376, 377].

The MF/GCE modified with scDNA provided analogous response to ssb formation as the scDNA-modified HMDE [377]. Similar results have been obtained with silver SAE (AgSAE) coated with either mercury meniscus (m-AgSAE) or mercury film (MF-AgSAE). Voltammetric behavior of sc, lin and ssDNA at the m-AgSAE [370] or MF-AgSAE [381] exhibited analogous differences as observed previously with the HMDE. Both variants of the AgSAE modified with scDNA were used as sensors for DNA cleaving species.

5.4.4.2 Use of Solid (Mercury-free) Electrodes to Detect DNA Cleavage

Guanine oxidation DNA signal (peak G^{ox}) usually used in electrochemical analysis of DNA at carbon electrodes exhibits no significant differences between the ccc (sc) DNA and dsDNA molecules possessing free ends (oc, linDNA) [355, 377]. This behavior results partly from relatively easy accessibility of the G oxidation sites via a major groove of the DNA double helix (providing substantial electroactivity of the undamaged dsDNA) (Figure 5.1, see Section 5.2.2.1) and partly from absence of extensive DNA surface denaturation at the carbon surfaces (reviewed in [11, 13, 111]). Intensity of the peak G^{ox} therefore does not respond to formation of individual ssb or dsb. Nevertheless, measurements of the latter signal at GCE [358], CPE or SPE [326, 331, 333, 351] have been used to detect deeper degradation of the DNA molecules, involving major changes of the DNA molecular mass, disruption of its double-helical structure and/or release of the DNA monomeric components (due to e.g., DNA acidic hydrolysis [358, 382] or oxidative damage [326]). Besides the measurements of intrinsic DNA oxidation response, a redox indicator–based sensor (see also Sections 5.2.3 and 5.3.3.4) for the detection of DNA damage with carbon electrodes was developed by Labuda and coworkers (Figure 5.13) [326, 331–334, 351, 357, 371]. This technique is based on selective binding of a metallointercalator [Co(phen)$_3$]$^{3+}$ to dsDNA at the carbon electrode surface, resulting in enhancement of its voltammetric signals. Since affinity of the cobalt redox marker to degraded DNA is much lower than to the intact dsDNA, decrease of its voltammetric peak represents the response to the DNA damage. This system was applied, for example, in detecting DNA damage by chemical nucleases such as copper, 1,10-phenanthroline complex [326] or in studies of antioxidative properties of yeast polysaccharides [331], flavonoids [332, 334] or plant extracts [333] (Section 5.4.8.2). Electrically heated CPEs were employed in these investigations to explore temperature effects on the processes of DNA damage [332, 371].

5.4.5
Detection of Covalent Damage to DNA Bases

DNA base residues possess a number of reactive sites that are susceptible to chemical modification [116, 253, 254] (see Figure 5.10 for examples). Chemical and photochemical attacks to the nucleobase moieties may lead to changes in the DNA's electrochemical features (Figure 5.11B). Modification of electroactive bases (such as G) may lead to a loss of their intrinsic electroactivity [29, 38, 175, 307–313]. Other lesions (e.g., 8-OG [315, 316, 318], Figure 5.11B) or some bulky adducts involving introduced electroactive moieties [20, 129, 132, 201, 309–311]) yield new, specific electrochemical signals. In addition, formation of some lesions may lead to conformational changes of the DNA double helix (local distortions, unwinding) exposing some base residues to the environment and facilitating their interactions with electrodes [110, 296, 302, 319, 350, 336] (Figure 5.11C).

5.4.5.1 Techniques Based on Guanine Redox Signals

Because of the susceptibility of the guanine base to damage by a broad range of genotoxic agents [116, 253, 254] and its well-defined electrochemistry at both carbon and mercury electrodes, guanine redox peaks have been frequently utilized as transduction signals in DNA biosensors [29, 38, 175, 307–313] (Figure 5.11B). Upon damage to the G moiety, its electrochemical features may be changed, and signals corresponding to the parent base lost. In addition, chemical alterations in the G residue (that is, within its imidazole ring) often lead to release of the base from the DNA chains. These depurination events can also result in decrease of the intensity of the DNA guanine peaks under certain conditions. Since natural DNA contains many G residues, partial decrease of the G peaks are usually observed, depending on the extent of the DNA damage.

Alkylation of N7 position within the guanine imidazole moiety resulted in a decrease of DNA peak G measured by cyclic or SWV at the HMDE. Using this signal, DNA modification by a methylation agent dimethyl sulfate (DMS) was studied both in solution and at the HMDE surface [38]. Decrease of the same signal was observed because of DNA modification by mitomycin C [309–311] and thiotepa [312], anticancer drugs known to primarily attack guanine. Peak G^{ox} at carbon electrodes responded to the G base damage in a similar way. PGE, GCE, CPE or SPE coated with DNA were applied as sensors for a variety of agents including antitumor platinum complexes [343], hydrazine derivatives [307], polychlorinated biphenyls, aflatoxins, anthracenes, acridines, phenol compounds [29, 308, 344–346, 351], ultraviolet light [313], arsenic oxide [383,] and so on.

These types of DNA sensors are relatively simple, do not require mercury electrodes and can work with relatively poorly defined, inexpensive commercially available DNAs. On the other hand, their sensitivity is inherently low: when a decrease of initially large signal is evaluated, change of the peak height has to exceed standard deviation of the measurement (usually 5–10%, depending on the electrode material, surface pretreatment and DNA immobilization procedure). This implies that a relatively large portion of the G residues has to be damaged to gain a reliable response. Another disadvantage of this approach is that the signal decrease may be caused by nonspecific destruction of the DNA recognition layer (e.g., in the presence of surfactants), which can hardly be distinguished from specific damage to the G residues [344].

5.4.5.2 DNA Adducts Yielding Specific Electrochemical Responses

Better results (regarding sensitivity, and obviously also specificity) can in principle be obtained when a DNA lesion yields a new, specific signal. This is the case of one of the most abundant products of oxidative DNA damage, the 8-OG (Figure 5.11B).

8-OG having lower redox potential than G yields an oxidation peak at significantly less positive potential [85, 91, 315, 316, 318]. Electrooxidation of 8-OG was recently studied at different solid electrodes, including carbon, gold, platinum and tin oxide ones [318]. Specific 8-OG signal appeared because of adriamycin-mediated oxidative damage of DNA deposited at the GCE [316, 365]. Differences in the 8-OG and G redox potentials have been utilized in redox mediator–based techniques of differential detection of both bases at ITO electrodes [85, 91]. While $[Os(bipy)_3]^{3+}$ chelate was capable of electrocatalytic oxidation of 8-OG, oxidation of G required a mediator with higher redox potential such as $[Ru(bipy)_3]^{3+}$ (see Section 5.2.2.4) 1,N^6-ethenoadenine (a reaction product of A with a chemical DNA probe chloroacetaldehyde [137], but also one of the products of indirect oxidative DNA damage in cells [253]) yielded a cathodic peak at a potential remarkably less negative than the unmodified parent base [347].

Chemically modified DNA may acquire specific electrochemical features because of substances forming bulky adducts with it. A clinically used antitumor agent, mitomycin C, covalently binds primarily to G residues, forming interstrand and/or intrastrand cross-links ([309–311] and references therein). The MC moiety involves reversibly electroreducible quinone moiety. Interactions of acid- or reductively activated MC with DNA were investigated by means of CV at HMDE [309–311], including studies of electrochemical MC activation [309, 310]. Osmium tetroxide complexes (Os,L) represent another class of species forming covalent DNA adducts displaying distinct electrochemical features. Although these substances belong neither to typical natural genotoxic agents nor to chemotherapeutics, their properties have been studied in detail because of their wide practical usability in DNA analysis (see Sections 5.2.3.2 and 5.3.4.4).

5.4.5.3 Techniques Sensitive to DNA Conformational Distortions

Chemical changes of the DNA base residues may be connected with perturbations of the DNA double helix structure, including its untwisting or bending, rupture of the Watson–Crick pairing of the damaged and/or adjacent bases, formation of base mismatches, and so forth [116, 253–255]. These events may influence behavior of the damaged DNA at electrodes. For example, irradiation of DNA with UV light (causing formation of a number of base photoproducts, including pyrimidine cyclobutane dimers [116, 253–255]), resulted in enhancement of reducibility of the dsDNA at the DME (Figure 5.11C) [296], (see Section 5.4.2.1). Conformational distortions of DNA arising from its UV-induced damage also influenced intensity of amperometric signals related to the DNA interaction with polypyrrole-modified electrodes in a flow-through device [369]. Accessibility of G residues for their oxidation at solid electrodes may be also improved because of perturbations in the dsDNA structure. This phenomenon was shown to be analytically useful mainly in systems involving electrocatalytic G oxidation mediated by ruthenium or rhenium chelates (see Sections 5.2.2.4 and 5.3.3.9) Ruthenium-mediated electrocatalytic guanine oxidation in dsDNA layers assembled with $poly[Ru(bipy)_2poly(4-vinylpyridine)_{10}Cl]^+$ films at carbon electrodes was enhanced upon exposure of the DNA to styrene oxide or methylating agents forming covalent DNA adducts primarily at G residues [302, 304–306, 349, 350, 336]. This phenomenon was ascribed to conformation changes ("unraveling") of the DNA double helix, facilitating interactions of other guanines with the redox mediator. Enhancement of electrocatalytic current due to the ruthenium chelate-mediated guanine oxidation (Figure 5.11C) or of electrogenerated chemiluminescence signal [336, 384] was observed as responses to the DNA damage.

5.4.5.4 Enzymatic Probing of DNA Base Damage

DNA repair enzymes have been used to convert specific base lesions to ssb in techniques involving electrophoretic analysis of the length of DNA strands [256, 261]. Electrochemical measurements at mercury electrodes provide highly sensitive detection of the ssb (see Section 5.4.4.1). It has been shown that abasic sites arising from spontaneous depurination of DMS-treated DNA can be detected by a.c. voltammetry at the HMDE after the DNA enzymatic digestion by *Escherichia coli* exonuclease III [137, 374], an enzyme introducing ssb next to abasic lesions, followed by exonucleolytic degradation of one strand of the DNA. Similarly, in UV-irradiated DNA, pyrimidine dimers were converted into the ssb by T4 endonuclease V [374] (Figure 5.12B). In both cases, these procedures allowed for detection of small extents of base modifications, not detectable without the enzymatic digestion, and was successfully applied to probe base lesions induced by DMS or UV light in living bacterial cells [374].

5.4.6
Genotoxic Substances Interacting with DNA Noncovalently

A broad range of biologically important substances interact with DNA noncovalently (reversibly) [116]. Depending on the chemical nature of the binder, the noncovalent interactions may involve electrostatic outer-sphere binding along the DNA chains, binding within major or minor grooves of the DNA double helix, or intercalation between adjacent DNA base pairs (see Section 5.2.3.1). Action of many carcinogens, cytostatics, environmental pollutants and so forth, involves reversible binding to DNA. Noncovalent interaction via an intercalative or groove-binding moiety leads to accumulation of the DNA-damaging substance at the DNA molecules, thus facilitating subsequent covalent attacks to the genetic material via reactive groups of the binder (e.g., epoxide [336] or aziridine [309]) or via mediating formation of other DNA-damaging species such as oxygen radicals ([301, 365] and references therein).

Formation of noncovalent DNA complexes with small molecules can in principle be detected via changes in electrochemical behavior of both the binders (Figure 5.14A) and the DNA (Figure 5.14B). Upon binding to large molecules of DNA, apparent diffusion coefficients of the binder decrease, and altered mass transport of these species influence its electrochemical signals [322, 323, 326, 328, 329, 385, 386]. This phenomenon has been used to determine association/dissociation constants of the binder-DNA complexes. Peak potentials of the binders may be also changed upon the complex formation, depending on the binding mode [110, 322, 323]. Association of DNA with transition metals (that can form coordination bonds with nitrogen atoms of

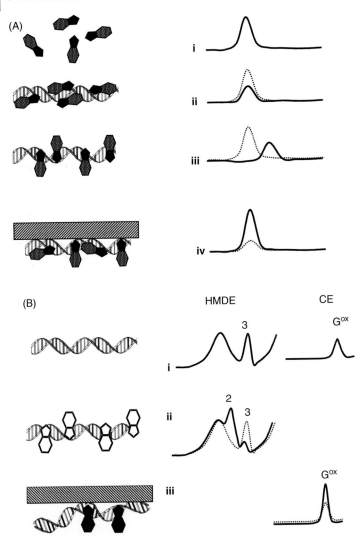

Fig. 5.14 Principles of electrochemical detection of noncovalent DNA interactions with small molecules. (A, i) Possible effects of the interactions on signals of an electroactive binder. (ii) Association of the binder with large DNA molecules may result in decrease of its signal because of reduced mass transport. (iii) When an electroactive moiety of the binder is hidden upon the complex formation (e.g., in intercalation complexes), decrease and/or potential shifts of the peaks may be observed. (iv) Accumulation of electroactive binders in electrode surface-confined DNA layer may result in enhancement of the binder signals. (B, i) Changes in the intrinsic DNA electrochemical signals. (ii) Owing to binding of DNA intercalators, adsorption/desorption behavior of dsDNA at mercury electrodes is remarkably changed, including formation of a.c.V peak 2 and decrease of the peak 3 height [319]. (iii) Untwisting and bending of dsDNA adsorbed at a carbon electrode surface upon binding of an intercalator (daunomycin [110]) results in enhancement of the peak G^{ox} intensity.

the bases) or their complexes influence adsorbability of both DNA and the metals at the electrodes. Voltammetric studies of complexes and associates of toxic heavy metals (lead, cadmium [387, 388], copper [67, 389–391], mercury [392], nickel [393]) and/or their complexes ([Co(NH$_3$)$_6$]$^{3+}$ [70], dimeric rhodium complexes [394]) with DNA and its components have been studied electrochemically at mercury as well as solid electrodes. Studies of DNA interactions with ruthenium, iron or osmium organic chelates [321, 322] (involving nitrogenous heterocyclic bidentate ligands such as 2,2′-bipyridine or 1,10-phenanthroline) by voltammetric and electrogenerated chemiluminescence [323, 325] measurements at carbon, ITO or gold electrodes provided basic information about relations between the interaction mode and behavior of the DNA complexes at electrodes. The metal chelates, exhibiting a well-defined electrochemistry, may undergo either electrostatic, minor groove or intercalative binding to dsDNA. The interaction mode depends on the valence (charge) of the central metal ion and/or on the heterocyclic ligand type. Different binding modes of these chelates specifically influenced their electrochemical signals, being manifested by either positive (for intercalative binding) or negative potential shifts (in the case of primarily electrostatic interaction) [321–323, 325]. DNA binding of metalloporphyrins (MPs) (containing copper, lead, zinc or nickel ions) was studied by CV at the HMDE [395]. In the absence of axial ligands, strong diminution of the MP signals was observed upon their DNA binding, suggesting a deep MP intercalation into the DNA double helix. Only weak signal decrease was detected with MPs containing the axial ligands. Intercalative transition-metal complexes can mediate oxidative DNA damage and may be involved in the DNA-mediated charge transfer reactions [118–120, 171, 213, 214, 285, 353] (see Sections 5.2.3.1, 5.3.3.9 and 5.4.2).

Electrodes modified with dsDNA have been used as sensors for various reversibly binding substances. Interactions of intercalators or cationic groove binders with the DNA recognition layer results in accumulation of these substances at the DNA-modified electrode (Figure 5.14A), resulting in enhancement of the binder electrochemical signals (redox mediator–based DNA hybridization sensors utilize the same phenomenon, see Section 5.3.3.4). Aromatic amines, important environmental carcinogens, were sensitively detected upon their accumulation within dsDNA anchored at the CPE [29, 345, 346, 356]. DNA interactions with clinically applied anticancer drug daunomycin was studied [110] using DNA-modified CPE. Electrooxidation of the daunomycin anthraquinone moiety intercalated into the DNA double helix was less feasible than in the free drug, resulting in potential shift toward more positive values. A similar compound, adriamycin, intercalated into a thick dsDNA layer at the GCE and mediated electrochemically induced oxidative DNA damage [316, 317, 365] (see Section 5.4.7). A bis-intercalator ECHI yielded well-developed reversible redox peaks on HMDE [123] or MFE [122] in the presence of ds, but not ssDNA (more details in Section 5.3.3.4). Labuda and coworkers proposed [351, 396] a competitive assay involving redox indicator [Co(phen)$_3$]$^{3+}$ for the detection of noncovalent DNA binders (Figure 5.13). The indicator signal measured at DNA-modified SPE or CPE decreased with increasing concentrations of quinazoline [351], acridine or catechin [396] derivatives whose molecules competed with the cobalt complex for binding sites in the DNA.

Noncovalent interactions (and especially intercalation) may induce remarkable changes in the dsDNA conformation, including DNA unwinding and increasing accessibility of the base residues to the environment. These phenomena are often connected with alterations in the intrinsic DNA redox and/or tensammetric signals at carbon or mercury electrodes (Figure 5.14B). Low (submicromolar) concentrations of daunomycin induced conformational changes in DNA anchored at CPE detectable via enhancement of the guanine CP signal (peak G^{ox}) [110]. Intercalative complexes of dsDNA with chloroquine, [Co(phen)$_3$]$^{3+}$, doxorubicin, 9-amino acridine [319], acridine orange or ethidium [320] were studied by a.c. voltammetry at HMDE. Conformational changes of the DNA double helix lead to changes of the intensity of the tensammetric DNA peak 2 and peak 3 (Figure 5.14B) [319]. It has been proposed that DNA adsorbed in the presence of the above-mentioned substances adopted a specific structure at the HMDE surface. After the intercalator removal, this structure involved untwisted regions adsorbed at the surface (contributing to increased intensity of DNA peak 2) and sc loops protruding to the solution. Such dsDNA structure was relatively resistant to potential-induced unwinding (resulting in decrease of the peak 3 intensity) (see Sections 5.2.2.3 and 5.4.4.1) unless ssb were introduced into it [319].

5.4.7
Electrochemically Induced DNA Damage

DNA-damaging species can be generated in electrochemically modulated reactions. For example, oxidation states of transition metals such as copper, iron or manganese can be controlled by the electrode potential. Lower-valency states of these metals undergo Fenton-type reactions, yielding reactive oxygen species that are vigorous DNA-damaging agents ([300, 301] and references therein). Supercoiled DNA was cleaved in solution in the presence of oxygen and electrochemically activated Mn(III) or Fe(III) porphine complexes [324]. Similarly, electrolysis of DNA solution (scDNA or an ODN) in the presence of $trans$-[Re(O)$_2$(4-OMe-py)$_4$]$^+$ resulted in formation of both ssb and piperidine-labile sites because of one-electron oxidation of G mediated by the electrochemically oxidized metal complex [79]. Electrolysis of an imidazoacridinone antitumor drug C-1311 at a platinum electrode led to formation of reactive intermediates forming covalent adduct with guanosine, mimicking behavior of the drug upon its metabolic oxidative activation [397]. Similarly tirapazamine, a cytotoxin of hypoxic cells, formed a radical anion because of its electrochemical one-electron reduction [398]. The lifetime of this species was remarkably decreased in the presence of DNA, suggesting that the DNA and the radical anion underwent a direct interaction.

Electrodes modified with DNA offer a unique possibility of electrochemical control of damage to the surface-confined DNA, followed by *in situ* electrochemical detection. ScDNA adsorbed at the surface of HMDE was cleaved in the presence of iron or copper complexes and hydrogen peroxide (or oxygen being electroreduced to the H_2O_2) in a distinctly electrode potential-dependent manner [300, 301]. In the presence of iron EDTA complex, the cleavage reaction takes place at potentials sufficiently negative to reduce iron(III) ion within the EDTA chelate to the iron(II) form undergoing the Fenton reaction [300]. Copper mediated DNA damage was seen only in a narrow potential region around the copper reduction peak, suggesting that stabilization of Cu(I) due to interactions with DNA bases and redox cycling of DNA-bound Cu(II)/Cu(I) were essential for formation of the DNA-damaging reactive oxygen species [301]. A Cu(I)-stabilizing ligand 1,10-phenanthroline increased the efficacy of DNA damage and shifted the maximum effect to more negative potentials. Chromate (which belongs to important carcinogens), but not Cr(III), salts caused strong electrochemically induced DNA cleavage upon its reduction at the scDNA-modified HMDE (T. Mozga and M. Fojta, unpublished), probably due to formation of highly reactive Cr(V) or Cr(IV) intermediates [399, 400]. A considerable extent of DNA cleavage at the HMDE charged to potentials ≤ -0.1 V was also observed in the absence of the metal complexes, suggesting a role of radical intermediates of oxygen electrochemical reduction [300, 301]. At GCE modified with thin or thick layers of calf thymus dsDNA, oxidative DNA damage (including formation of 8-OG) mediated by adriamycin (doxorubicin) intercalated into the DNA double helix was observed upon applying negative electrode potentials bringing about one-electron reduction of the drug [316, 365]. Electroreduced thiophene-S-oxide also induced *in situ* DNA damage in a thick dsDNA layer at the GCE [364]. Electrochemical activation of mitomycin C was studied by CV at the HMDE [309, 310].

5.4.8
Analytical Applications of Electrochemical Sensors for DNA Damage

Electrochemical methods have been widely applied in analysis of a broad spectrum of biologically active or toxic compounds. Analytical protocols involving polarographic, voltammetric, potentiometric or amperometric techniques for the determination of various analytes including drugs, carcinogens, pesticides, heavy metals, and so on, in water, soil, food or biological matrices have been developed (reviewed in [401–403]). These methods, together with techniques involving other detection principles (e.g., spectroscopic), are based on measurements of specific physico-chemical features of the substances of interest. They do not primarily provide information about the consequences of interactions the given compound had with components of living matter (although there have been attempts to correlate carcinogenic potentials of different substances with their electrochemical features [404–407]). On the other hand, responses of DNA sensors are based on specific interactions of the analyzed species with the genetic material, which are closely related to their toxicological, carcinogenic and/or pharmacological activity. Although the sensors for DNA damage may yield response that is inherently nonspecific regarding chemical nature of the genotoxic analyte, they readily provide information about the effects on the genetic material. Such information may be valuable as prompt alert to occurrence of dangerous substances and thus for preservation of humans' health.

5.4.8.1 Environmental Monitoring

Many pollutants occurring in the environment can interact with DNA noncovalently and/or covalently (usually attacking of the G moiety or inducing the DNA sb). ScDNA-modified mercury electrodes (Section 5.4.4.1) have been shown to be applicable as sensors for DNA cleaving agents in "real" samples of food, beverages, river or waste waters (including possible applications in water purification processes) [299]. DNA sensors based on carbon electrodes were applied to detect different toxic pollutants, including strongly carcinogenic aromatic amines [29, 345, 346, 356]. These species accumulate in the DNA recognition layer via intercalative DNA binding (Section 5.4.6), resulting in enhancement of their CPSA signals. Some of the aromatic amine derivatives also affected the DNA peak G^{ox} intensity (Section 5.4.5.1) in a concentration-dependent manner [345]. The latter guanine oxidation peak has been used as a transduction signal in techniques proposed for detection of species, inducing covalent base damage (e.g., hydrazine derivatives [307]). Height of the peak G^{ox} measured with DNA-modified CPE sensitively responded to these compounds at the ppb level. Analogous responses were obtained when the DNA-modified CPE or SPE was used to detect other pollutants, including polychlorinated biphenyls, aflatoxins, anthracene and acridine derivatives, phenol compounds or surfactants [308, 344, 345]. Carcinogenic PAH yielded a linear response at a DNA-modified SPE in model samples as well as in a biological matrix represented by bile of fish exposed to the PAH [408]. Brett and coworkers applied a GCE-based DNA biosensor to detect s-triazine derivatives used as herbicides [409] and a molluscicidic agent niclosamide [410].

5.4.8.2 Testing of Toxicity or Protection Capacity

DNA sensors offer the possibilities of rapid genotoxicity screening of a broad range of substances (including novel products of chemical and pharmaceutical industry) and of testing DNA-protective effects of other species like antioxidants. In contrast to detection of trace levels of genotoxic agents in environmental or biological specimens, these applications usually do not require extremely high sensitivities of the analytical techniques. Any of the above-mentioned methods can thus be utilized for these purposes. Mascini and coworkers compared responsibility of the DNA-modified carbon electrodes to a variety of substances, including those mentioned at the end of the previous paragraph, to a currently used bioluminescence toxicity test Toxalert® 100 [308, 344]. They concluded that the faster and less expensive electrochemical approach may be used for preliminary screening. A redox indicator–based system responding to oxidative DNA degradation was developed by Labuda and coworkers (Figure 5.13) and applied to evaluate antioxidant capacity of different yeast polysaccharides [331], flavonoids [332, 334] or plant extracts [333]. Rusling and coworkers developed a device for toxicity screening based on layer-by-layer assemblies of DNA and heme proteins on GCE surfaces modified with redox-active cationic polymer films (Figure 5.15) [302, 303, 305, 306, 349, 350, 384]. Layers of cytochrome P450, myoglobin or hemoglobin exhibited enzymatic activity mimicking metabolic carcinogen activation. For example, styrene was enzymatically converted to styrene oxide subsequently reacting with G moieties in the DNA layer. This DNA damage caused conformational changes of

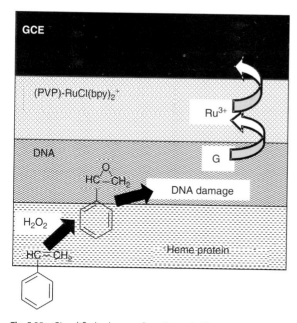

Fig. 5.15 Simplified scheme of an electrode-film assembly developed by Rusling et al. [302, 303, 305, 306, 349] as a device for toxicity screening. Cationic redox-active film poly[Ru(bipy)$_2$poly(4-vinylpyridine)$_{10}$Cl]$^+$ facilitates electrostatic immobilization of ultrathin DNA films and mediates electrostatic oxidation of the DNA guanine residues. The layer of a heme protein (cytochrome P450, myoglobin or hemoglobin) possesses enzymatic activity mimicking metabolic carcinogen activation. Such assembly may be composed of several layers of the cationic polymer, DNA and the protein. An analyzed substance (styrene) penetrates through the protein film, being enzymatically converted into the active form (styrene oxide), causing damage to the DNA. Formation of the DNA-styrene oxide adducts cause conformational changes of the DNA molecules, facilitating the electrocatalytic oxidation.

the DNA molecules (see Section 5.4.5.3), resulting in a better accessibility of G residues for electrocatalytic oxidation mediated by immobilized ruthenium complex (see Sections 5.2.2.4 and 5.3.3.9). Incubation of the sensor with styrene thus resulted in remarkable enhancement of electrocatalytic current. On the other hand, toluene, a similar compound, which is, however, not a substrate for the layered heme enzymes, did not yield any significant response [349, 350, 384].

5.4.8.3 DNA–drug Interactions

Actions of many drugs, namely, anticancer agents, involve specific interactions with DNA, including reversible binding and/or formation of covalent adducts. DNA biosensors, including electrochemical ones (reviewed in [111, 292]) thus appear to be well suited for the drug analysis. Some of these devices employ biologically relevant DNA-drug binding while in others the "recognition" DNA component of the sensor represents a nonspecific anionic binding substrate for positively charged molecules of the analyte rather than a pharmacologically specific target [357, 411, 412]. Some of these studies have been reviewed in previous paragraphs (Sections 5.4.5–5.4.7), including those dealing with noncovalent DNA binding of daunomycin [110, 413], doxorubicin [316, 317, 330, 365] or ECHI [122, 123] and those devoted to detecting covalent DNA damage induced by cisplatin [343], mitomycin C [309–311], or thiotepa [312]. Voltammetry at different carbon electrodes was utilized in studies of ds or ssDNA interaction with other chemotherapeutics, including epirubicin [414], mitoxantrone, [415], actinomycin D [329, 416] and lycorine [314]. GCE modified with acid-degraded DNA was applied by Brett and coworkers in analysis of a variety of clinically applied drugs, including metronidazole [361] or benznidazole [417]. Labuda and coworkers utilized [357, 411] CPE modified with DNA at its surface or in the bulk phase to detect phenothiazine or azepine derivatives used as antidepressive drugs [357] or of local anesthetics based on alkoxyphenylcarbamate esters [411]. ITO electrodes coated with DNA-modified gold nanoparticles selectively responded to mifepristone, a hormonal contraceptive agent [418]. Interactions between an antibacterial drug lumazine, cyclodextrins and DNA were studied by voltammetry at HMDE [419]. Inclusion cyclodextrin-lumazine complexes were decomposed in the presence of DNA to which the drug displays a high affinity. Interactions of dsDNA with another antibacterial agent, a fluoroquinolone derivative levofloxacin, were investigated by CV at the GCE [420]. Among other approaches, a redox indicator–based sensor providing response to DNA–drug interactions has been proposed [352, 421]. DNA layer (immobilized via thiol groups at one end of a dsODN) at a gold electrode strongly suppressed CV signals of the ferro/ferricyanide redox pair because of electrostatic repulsion between the charged DNA and the anionic depolarizer. Binding of cationic drug (e.g., quinacrine) to the surface-confined DNA decreased the negative charge density, resulting in enhancement of the ferro/ferricyanide CV current response. Another approach involves incorporation of ssDNA into surface-stabilized bilayer lipid membrane [412, 422]. Such a biosensor was applied [422] in analysis of atenolol, a β_1-blocker used to treat myocardial infarctions and hypertension.

5.4.9
Concluding Remarks to DNA Damage Sensors

Electrochemical analysis offers a variety of ways to detect DNA damage and its interactions with genotoxic agents, including carcinogens, mutagens, drugs, and so on. One of the major advantages of the DNA sensors is that their responses are based on specific interactions of the analyzed substances with genetic material, forming the sensor recognition component. These interactions are closely related to biological, toxicological or pharmacological activities of the analytes. Analytical approaches and biosensors so far applied in DNA damage assays involve techniques and devices differing in their specificity, sensitivity as well as practical usability. All of them are applicable in basic studies of DNA interactions with the damaging agents and in testing of toxicity or DNA-protective effects of various substances (Section 5.4.8.2). In these cases, the effects can easily be followed at relatively high concentrations in a dose-dependent manner. More strict requirements regarding sensitivity arise when a device should be used in trace assays, for example, in environmental analysis (Section 5.4.8.1), detection of drugs (Section 5.4.8.3) or their metabolites in clinical samples, and so forth. In such applications, mainly the "signal-off" approaches involving measurements of redox G responses or redox markers of DNA integrity may be insufficiently sensitive (their sensitivity can hardly exceed recognition of one lesion among few tens of intact G residues, see Sections 5.4.4.2 and 5.4.5.1). Using sufficiently long accumulation (exposure) times, these sensors may yield appreciable responses to, for example, relatively massive environmental pollution. However, the 10^{-1} to 10^{-2} sensitivities are far below the requirements for techniques of detecting biologically relevant levels of DNA damage (one lesion per 10^4 to 10^6 nucleotides). On the other hand, the "signal-on" methods, including detection of the DNA strand breaks with mercury or SAE (Section 5.4.4.1) and measurements of specific signals of DNA adduct or complexes (basically with all electrode types) (Sections 5.4.5.2 and 5.4.6), are more promising for these purposes. Sensitivity of the sb detection with the scDNA-modified HMDE (one lesion per about 2×10^5 nucleotides [26]) complies with the above-mentioned demands. It may be expected that improvement of the existing techniques and devices via, for example, appropriate signal amplification or utilization of DNA repair enzymes (Section 5.4.5.4), will result in attaining sensitivities currently reached with the DNA hydrolysis–based methods [264] (Section 5.4.1.1).

Acknowledgements

The authors express their gratitude to Drs Z. Pechan, F. Jelen and L. Havran for critical reading of the manuscript and for help with the preparation of the manuscript. This work was supported by grant Nos. 204/03/0566 from the Grant Agency of the Czech Republic (GACR), and IBS5004355 from the Academy of Sciences of the Czech Republic to Emil Palecek, and by grants 203/04/1325 from the GACR and A4004402 from the Grant Agency of the Academy of Sciences of the Czech Republic to Miroslav Fojta.

References

1 A. BENDICH, in *The Nucleic Acids* (Eds.: E. CHARGAFF, J.N. DAVIDSON), Vol. 1, Academic Press, New York, **1955**, 119–120.
2 H. BERG, *Biochem. Z.*, **1957**, *329*, 274–276.
3 E. PALECEK, *Naturwissenschaften*, **1958**, *45*, 186–187.
4 E. PALECEK, *Nature*, **1960**, *188*, 656–657.
5 E. PALECEK, *Biochim. Biophys. Acta*, **1961**, *51*, 1–8.
6 E. PALECEK, B. JANIK, *Arch. Biochem. Biophys.*, **1962**, *98*, 527.
7 E. PALECEK, in *Methods in Enzymology: Nucleic Acids, Part D* (Eds.: L. GROSSMAN,

K. Moldave), Vol. 21, Academic Press, New York, **1971**, 3–24.

8 E. Palecek, in *Progress in Nucleic Acid Research and Molecular Biology* (Eds.: J.N. Davidson, W.E. Cohn), Vol. 9, Academic Press, New York, **1969**, 31–73.

9 E. Palecek, in *Progress in Nucleic Acid Research and Molecular Biology* (Ed.: W.E. Cohn), Vol. 18, Academic Press, New York, **1976**, 151–213.

10 E. Palecek, in *Topics in Bioelectrochemistry and Bioenergetics* (Ed.: G. Milazzo), Vol. 5, John Wiley, Chichester, **1983**, 65–155.

11 E. Palecek, *Electroanalysis*, **1996**, *8*, 7–14.

12 E. Palecek, *Talanta*, **2002**, *56*, 809–819.

13 E. Palecek, M. Fojta, F. Jelen, V. Vetterl, in *The Encyclopedia of Electrochemistry*, Vol. 9, *Bioelectrochemistry* (Eds.: A.J. Bard, M. Stratsmann), Wiley-VCH, Weinheim, **2002**, 365–429.

14 V. Brabec, V. Vetterl, O. Vrana, in *Experimental Techniques in Bioelectrochemistry* (Eds.: V. Brabec, D. Walz, G. Milazzo), Vol. 3, Birkhauser Verlag, Basel, **1996**, 287–359.

15 J.-M. Sequaris, P. Valenta, H.W. Nurnberg, B. Malfoy, *Bioelectrochem. Bioenerg.*, **1978**, *5*, 483–503.

16 E. Palecek, M. Fojta, *Anal. Chem.*, **2001**, *73*, 74A–83A.

17 M.J. Tarlov, A.B. Steel, in *Biomolecular Films. Design, Function, and Applications* (Ed.: J.F. Rusling), Marcel Dekker, New York, **2003**, 545–608.

18 E. Palecek, E. Lukasova, F. Jelen, M. Vojtiskova, *Bioelectrochem. Bioenerg.*, **1981**, *8*, 497–506.

19 E. Palecek, M. Vojtiskova, F. Jelen, M. Kozinova, in *Charge and Field Effects in Biosystems* (Ed.: M.J. Allen), P.N.R. Usherwood, Abacus Press, Tonbridge, **1984**, 397–404.

20 E. Palecek, M.A. Hung, *Anal. Biochem.*, **1983**, *132*, 236–242.

21 E. Lukasova, F. Jelen, E. Palecek, *Gen. Physiol. Biophys.*, **1982**, *1*, 53–70.

22 E. Lukasova, M. Vojtiskova, F. Jelen, T. Sticzay, E. Palecek, *Gen. Physiol. Biophys.*, **1984**, *3*, 175–191.

23 E. Palecek, I. Postbieglova, *J. Electroanal. Chem.*, **1986**, *214*, 359–371.

24 E. Palecek, *Anal. Biochem.*, **1988**, *170*, 421–431.

25 C. Teijeiro, K. Nejedly, E. Palecek, *J. Biomol. Struct. Dyn.*, **1993**, *11*, 313–331.

26 M. Fojta, E. Palecek, *Anal. Chim. Acta*, **1997**, *342*, 1–12.

27 G. Marrazza, I. Chianella, M. Mascini, *Anal. Chim. Acta*, **1999**, *387*, 297–307.

28 G. Marrazza, I. Chianella, M. Mascini, *Biosens. Bioelectron.*, **1999**, *14*, 43–51.

29 M. Mascini, I. Palchetti, G. Marrazza, *Fresenius' J. Anal. Chem.*, **2001**, *369*, 15–22.

30 J. Wang, E. Palecek, P.E. Nielsen, G. Rivas, X.H. Cai, H. Shiraishi, N. Dontha, D. Luo, P.A.M. Farias, *J. Am. Chem. Soc.*, **1996**, *118*, 7667–7670.

31 J. Wang, X. Cai, J.Y. Wang, C. Jonsson, E. Palecek, *Anal. Chem.*, **1995**, *67*, 4065–4070.

32 J. Wang, G. Rivas, X.H. Cai, M. Chicharro, N. Dontha, D.B. Luo, E. Palecek, P.E. Nielsen, *Electroanalysis*, **1997**, *9*, 120–124.

33 J. Wang, J.R. Fernandes, L.T. Kubota, *Anal. Chem.*, **1998**, *70*, 3699–3702.

34 E. Palecek, F. Jelen, C. Teijeiro, V. Fucik, T.M. Jovin, *Anal. Chim. Acta*, **1993**, *273*, 175–186.

35 C. Prado, G.U. Flechsig, P. Grundler, J.S. Foord, F. Marken, R.G. Compton, *Analyst*, **2002**, *127*, 329–332.

36 M. Fojta, V. Vetterl, M. Tomschik, F. Jelen, P. Nielsen, J. Wang, E. Palecek, *Biophys. J.*, **1997**, *72*, 2285–2293.

37 E. Palecek, B.D. Frary, *Arch. Biochem. Biophys.*, **1966**, *115*, 431–436.

38 F. Jelen, M. Tomschik, E. Palecek, *J. Electroanal. Chem.*, **1997**, *423*, 141–148.

39 E. Palecek, F. Jelen, L. Trnkova, *Gen. Physiol. Biophys.*, **1986**, *5*, 315–329.

40 E. Palecek, *J. Mol. Biol.* **1966**, *20*, 263–281.

41 E. Palecek, V. Vetterl, *Biopolymers*, **1968**, *6*, 917–928.

42 H. Berg, D. Tresselt, J. Flemming, H. Bär, G. Horn, *J. Electroanal. Chem.*, **1969**, *21*, 181–186.

43 M. Tomschik, F. Jelen, L. Havran, L. Trnkova, P.E. Nielsen, E. Palecek, *J. Electroanal. Chem.*, **1999**, *476*, 71–80.

44 J. Wang, S. Bollo, J.L.L. Paz, E. Sahlin, B. Mukherjee, *Anal. Chem.*, **1999**, *71*, 1910–1913.
45 L. Trnkova, F. Jelen, I. Postbieglova, *Electroanalysis*, **2003**, *15*, 1529–1535.
46 L. Trnkova, R. Kizek, O. Dracka, *Electroanalysis*, **2000**, *12*, 905–911.
47 L. Trnkova, R. Kizek, O. Dracka, *Bioelectrochemistry*, **2002**, *55*, 131–133.
48 L. Trnkova, *Talanta*, **2002**, *56*, 887–894.
49 F. Jelen, V. Vetterl, P. Belusa, S. Hason, *Electroanalysis*, **2000**, *12*, 987–992.
50 V. Vetterl, N. Papadopoulos, V. Drazan, L. Strasak, S. Hason, J. Dvorak, *Electrochim. Acta*, **2000**, *45*, 2961–2971.
51 F. Patolsky, E. Katz, I. Willner, *Angew. Chem., Int. Ed.*, **2002**, *41*, 3398–3402.
52 F. Patolsky, A. Lichtenstein, I. Willner, *Chem.-Anlagen Eur. J.*, **2003**, *9*, 1137–1145.
53 F. Patolsky, E. Katz, A. Bardea, I. Willner, *Langmuir*, **1999**, *15*, 3703–3706.
54 F. Patolsky, A. Lichtenstein, I. Willner, *J. Am. Chem. Soc.* **2001**, *123*, 5194–5205.
55 E. Katz, L. Sheeney-Haj-Ichia, A.F. Buckmann, I. Willner, *Angew. Chem., Int. Ed.*, **2002**, *41*, 1343–1346.
56 A. Bardea, F. Patolsky, A. Dagan, I. Willner, *Chem. Commun.*, **1999**, *7*(1), 21–22.
57 S. Hason, V. Vetterl, *Bioelectrochemistry*, **2002**, *56*, 43–45.
58 V. Vetterl, *Experientia*, **1965**, *21*, 9–11.
59 V. Vetterl, J. Pokorny, *Bioelectrochem. Bioenerg.*, **1980**, *7*, 517–526.
60 V. Vetterl, R. de Levie, *J. Electroanal. Chem.*, **1991**, *310*, 305–315.
61 E.E. Ferapontova, E. Dominguez, *Electroanalysis*, **2003**, *15*, 629–634.
62 E. Palecek, F. Jelen, M. Hung, J. Lasovsky, *Bioelectrochem. Bioenerg.*, **1981**, *8*, 621–631.
63 E. Palecek, *Anal. Biochem.*, **1980**, *108*, 129–138.
64 E. Palecek, *Anal. Chim. Acta*, **1985**, *174*, 103–113.
65 E. Palecek, J. Osteryoung, R.A. Osteryoung, *Anal. Chem.*, **1982**, *54*, 1389–1394.
66 L. Trnkova, J. Friml, O. Dracka, *Bioelectrochemistry*, **2001**, *54*, 131–136.
67 F. Jelen, B. Yosypchuk, A. Kourilova, L. Novotny, E. Palecek, *Anal. Chem.*, **2002**, *74*, 4788–4793.
68 M. Fojta, R.P. Bowater, V. Stankova, L. Havran, D.M.J. Lilley, E. Palecek, *Biochemistry*, **1998**, *37*, 4853–4862.
69 M. Fojta, C. Teijeiro, E. Palecek, *Bioelectrochem. Bioenerg.*, **1994**, *34*, 69–76.
70 M. Fojta, R. Doffkova, E. Palecek, *Electroanalysis*, **1996**, *8*, 420–426.
71 E. Palecek, M. Fojta, *Anal. Chem.*, **1994**, *66*, 1566–1571.
72 E. Palecek, *Bioelectrochem. Bioenerg.* **1988**, *20*, 171–194.
73 R. Kizek, L. Havran, T. Kubicarova, B. Yosypchuk, M. Heyrovsky, *Talanta*, **2002**, *56*, 915–918.
74 E. Palecek, V. Vetterl, *Bioelectrochem. Bioenerg.*, **1977**, *4*, 361–368.
75 R.J. Heaton, A.W. Peterson, R.M. Georgiadis, *Proc. Natl. Acad. Sci. U.S.A.*, **2001**, *98*, 3701–3704.
76 F.G. Banica, A. Ion, in *Encyclopedia of Analytical Chemistry* (Ed.: R.A. Meyers), John Wiley & Sons, Chichester, **2000**, 11115–11144.
77 M. Tomschik, L. Havran, M. Fojta, E. Palecek, *Electroanalysis*, **1998**, *10*, 403–409.
78 P. de-Los-Santosalvarez, P. Rodriguez-Granda, M.J. Lobo-Castanon, A.J. Miranda-Ordieres, P. Tunon-Blanco, *Electrochem. Commun.*, **2003**, *5*, 267–271.
79 D.H. Johnston, C.-C. Cheng, K.J. Campbell, H.H. Thorp, *Inorg. Chem.*, **1994**, *33*, 6388–6390.
80 D.H. Johnston, K.C. Glasgow, H.H. Thorp, *J. Am. Chem. Soc.*, **1995**, *117*, 8933–8938.
81 M.E. Napier, H.H. Thorp, *Langmuir*, **1997**, *13*, 6342–6344.
82 H.H. Thorp, *Tibtech*, **1998**, *16*, 117–121.
83 M.E. Napier, H.H. Thorp, *J. Fluoresc.*, **1999**, *9*, 181–186.
84 A.C. Ontko, P.M. Armistead, S.R. Kircus, H.H. Thorp, *Inorg. Chem.*, **1999**, *38*, 1842–1846.
85 P.A. Ropp, H.H. Thorp, *Chem. Biol.*, **1999**, *6*, 599–605.

86 P.M. ARMISTEAD, H.H. THORP, *Anal. Chem.*, **2000**, *72*, 3764–3770.

87 V.A. SZALAI, H.H. THORP, *J. Phys. Chem. B*, **2000**, *104*, 6851–6859.

88 P.M. ARMISTEAD, H.H. THORP, *Anal. Chem.*, **2001**, *73*, 558–564.

89 E. ECKHARDT, M. ESPENHANH, N. NAPIER, N.D. POPOVICH, H.H. THORP, R. WITVER, in *DNA Arrays: Technologies and Experimental Strategies* (Ed.: E.V. GRIGONENKO), CRC Press, Boca Raton, **2001**, 39–60.

90 N. POPOVICH, H. THORP, *Interface*, **2002**, *11*, 30–34.

91 R.C. HOLMBERG, M.T. TIERNEY, P.A. ROPP, E.E. BERG, M.W. GRINSTAFF, H.H. THORP, *Inorg. Chem.*, **2003**, *42*, 6379–6387.

92 V. BRABEC, E. PALECEK, *Stud. Biophys.*, **1976**, *60*, 105–110.

93 F. GARNIER, *Biomed. Chem.*, **2000**, 349–368.

94 T.A. SKOTHEIM, *Handbook of Conducting Polymers*, Marcel Dekker, New York, **1986**.

95 T. de LUMLEY-WOODYEAR, P. ROCCA, J. LINDSAY, Y. DROR, A. FREEMAN, A. HELLER, *Anal. Chem.*, **1995**, *67*, 1332–1338.

96 T. de LUMLEY-WOODYEAR, C.N. CAMPBELL, A. HELLER, *J. Am. Chem. Soc.*, **1996**, *118*, 5504–5505.

97 T. de LUMLEY-WOODYEAR, C.N. CAMPBELL, E. FREEMAN, A. FREEMAN, G. GEORGIOU, A. HELLER, *Anal. Chem.*, **1999**, *71*, 535–538.

98 Y. DEGANI, A. HELLER, *J. Phys. Chem.*, **1987**, *91*, 1285–1289.

99 I. KATAKIS, A. HELLER, in *Frontiers in Bioelectronics I. Fundamental Aspects* (Eds.: F.W. SCHELLER, F. SCHUBERT, J. FEDROWITZ), Birkhauser Verlag, Basel, **1997**, 229–241.

100 A. HELLER, *Acc. Chem. Res.*, **1990**, *23*, 128–134.

101 J. WANG, M. JIANG, B. MUKHERJEE, *Anal. Chem.*, **1999**, *71*, 4095–4099.

102 J. WANG, B. TIAN, E. SAHLIN, *Anal. Chem.*, **1999**, *71*, 3901–3904.

103 H. KORRI-YOUSSOUFI, F. GARNIER, P. SRIVASTAVA, P. GODILLOT, A. YASSAR, *J. Am. Chem. Soc.*, **1997**, *119*, 7388–7389.

104 P. BAUERLE, A. EMGE, *Adv. Mater.*, **1998**, *3*, 324–330.

105 S. COSNIER, *Electroanalysis*, **1997**, *9*, 894–902.

106 S. COSNIER, *Biosens. Bioelectron.*, **1999**, *14*, 443–456.

107 F. PALMISANO, G. ZAMBONIN, D. CENTOZE, *Fresenius' J. Anal. Chem.*, **2000**, *366*, 586–601.

108 J. WANG, *Chem.-Anlagen Eur. J.*, **1999**, *5*, 1681–1685.

109 L.A. THOMPSON, J. KOWALIK, M. JOSOWICZ, J. JANATA, *J. Am. Chem. Soc.*, **2003**, *125*, 324–325.

110 J. WANG, M. OZSOZ, X.H. CAI, G. RIVAS, H. SHIRAISHI, D.H. GRANT, M. CHICHARRO, J. FERNANDES, E. PALECEK, *Bioelectrochem. Bioenerg.*, **1998**, *45*, 33–40.

111 M. FOJTA, *Electroanalysis*, **2002**, *14*, 1449–1463.

112 H. BERG, K. ECKHARDT, *Z. Naturforsch.*, **1970**, *25b*, 362–367.

113 H. BERG, G. HORN, U. LUTHARDT, W. IHN, *Bioelectrochem. Bioenerg.* **1981**, *8*, 537–553.

114 E. BAUER, H. BERG, H. SCHÜTZ, *Faserfosrchung Textiltechnik*, **1973**, *24*, 58–64.

115 K. HASHIMOTO, K. ITO, Y. ISHIMORI, *Anal. Chem.*, **1994**, *66*, 3830–3833.

116 M.G. BLACKBURN, M.J. GAIT, *Nucleic Acids in Chemistry and Biology*, IRL Press, New York, **1990**.

117 S. PALANTI, G. MARRAZZA, M. MASCINI, *Anal. Lett.*, **1996**, *29*, 2309–2331.

118 S.O. KELLEY, N.M. JACKSON, M.G. HILL, J.K. BARTON, *Angew. Chem., Int. Ed.*, **1999**, *38*, 941–945.

119 S.O. KELLEY, E.M. BOON, J.K. BARTON, N.M. JACKSON, M.G. HILL, *Nucleic Acids Res.*, **1999**, *27*, 4830–4837.

120 S.O. KELLEY, J.K. BARTON, N.M. JACKSON, M.G. HILL, *Bioconjugate Chem.*, **1997**, *8*, 31–37.

121 S. TAKENAKA, T. IHARA, M. TAKAGI, *J. Chem. Soc., Chem. Commun.*, **1990**, 1485–1487.

122 S. HASON, J. DVORAK, F. JELEN, V. VETTERL, *Talanta*, **2002**, *56*, 905–913.

123 F. JELEN, A. ERDEM, E. PALECEK, *Bioelectrochemistry*, **2002**, *55*, 165–167.

124 S. Takenaka, Y. Uto, H. Saita, M. Yokoyama, H. Kondo, W.D. Wilson, *Chem. Commun.*, **1998**, *10*, 1111–1112.
125 S. Takenaka, K. Yamashita, Y. Uto, M. Takagi, H. Kondo, *Denki Kagaku*, **1998**, *12*, 1329–1334.
126 S. Takenaka, Y. Uto, M. Takagi, H. Kondo, *Chem. Lett.*, **1998**, *27*, 989–990.
127 H. Miyahara, K. Yamashita, M. Kanai, K. Uchida, M. Takagi, H. Kondo, S. Takenaka, *Talanta*, **2002**, *56*, 829–835.
128 E. Palecek, M. Vojtiskova, F. Jelen, E. Lukasova, *Bioelectrochem. Bioenerg.*, **1984**, *12*, 135–136.
129 F. Jelen, P. Karlovsky, P. Pecinka, E. Makaturova, E. Palecek, *Gen. Physiol. Biophys.*, **1991**, *10*, 461–473.
130 E. Palecek, in *Methods in Enzymology* (Ed.: J.N. Abelson, M.I. Simon), Vol. 212, Academic Press, New York, **1992**, 139–155.
131 M. Fojta, L. Havran, S. Billova, P. Kostecka, M. Masarik, R. Kizek, *Electroanalysis*, **2003**, *15*, 431–440.
132 R. Kizek, L. Havran, M. Fojta, E. Palecek, *Bioelectrochemistry*, **2002**, *55*, 119–121.
133 L. Havran, M. Fojta, E. Palecek, *Bioelectrochemistry*, **2004**, *63*, 239–243.
134 B. Yosypchuk, M. Fojta, L. Havran, E. Palecek, **2004**, in preparation.
135 E. Palecek, F. Jelen, E. Minarova, K. Nejedly, J. Palecek, P. Pecinka, J. Ricicova, M. Vojtiskova, D. Zachova, In: *Structural Tools for the Analysis of Protein-Nucleic Acid Complexes.* (Eds.: D. M. J. Lilley, H. Heumann and D. Suck), Birghäuser Verlag, Basel, **1992**, 1–22.
136 E. Palecek, in *Nucleic Acids and Molecular Biology* (Eds.: F. Eckstein, D.M.J. Lilley), Vol. 8, Springer-Verlag, Berlin, **1994**, 1–13.
137 E. Palecek, *Crit. Rev. Biochem. Mol. Biol.*, **1991**, *26*, 151–226.
138 T. Ihara, Y. Maruo, S. Takenaka, M. Takagi, *Nucleic Acids Res.*, **1996**, *24*, 4273–4280.
139 T. Ihara, M. Nakayama, M. Murata, K. Nakano, M. Maeda, *Chem. Commun.*, **1997**, *17*, 1609–1610.
140 C.J. Yu, H. Yowanto, Y.J. Wan, T.J. Meade, Y. Chong, M. Strong, L.H. Donilon, J.F. Kayyem, M. Gozin, G.F. Blackburn, *J. Am. Chem. Soc.*, **2000**, *122*, 6767–6768.
141 C.J. Yu, Y.J. Wan, H. Yowanto, J. Li, C.L. Tao, M.D. James, C.L. Tan, G.F. Blackburn, T.J. Meade, *J. Am. Chem. Soc.*, **2001**, *123*, 11155–11161.
142 Y.D. Mao, C.X. Luo, Q. Ouyang, *Nucleic Acids Res.*, **2003**, *31*.
143 F. Patolsky, Y. Weizmann, I. Willner, *J. Am. Chem. Soc.*, **2002**, *124*, 770–772.
144 E. Uhlmann, A. Peyman, G. Breipohl, D.W. Will, *Angew. Chem., Int. Ed.*, **1998**, *37*, 2797–2823.
145 P.E. Nielsen, *Acc. Chem. Res.*, **1999**, *32*, 624–630.
146 J. Wang, *Biosens. Bioelectron.*, **1998**, *13*, 757–762.
147 O. Lioubashevski, F. Patolsky, I. Willner, *Langmuir*, **2001**, *17*, 5134–5136.
148 I. Willner, S. Rubin, *Angew. Chem., Int. Ed.*, **1996**, *35*, 367–385.
149 F. Patolsky, A. Lichtenstein, I. Willner, *Angew. Chem., Int. Ed.*, **2000**, *39*, 940–943.
150 F. Patolsky, A. Lichtenstein, I. Willner, *J. Am. Chem. Soc.*, **2000**, *122*, 418–419.
151 F. Patolsky, K.T. Ranjit, A. Lichtenstein, I. Willner, *Chem. Commun.*, **2000**, *12*, 1025–1026.
152 L. Alfonta, A. Bardea, O. Khersonsky, E. Katz, I. Willner, *Biosens. Bioelectron.*, **2001**, *16*, 675–687.
153 L. Alfonta, I. Willner, *Chem. Commun.*, **2001**, 1492–1493.
154 L. Alfonta, A.K. Singh, I. Willner, *Anal. Chem.*, **2001**, *73*, 91–102.
155 F. Patolsky, A. Lichtenstein, M. Kotler, I. Willner, *Angew. Chem., Int. Ed.*, **2001**, *40*, 2261–2265.
156 F. Patolsky, A. Lichtenstein, I. Willner, *Nat. Biotechnol.*, **2001**, *19*, 253–257.
157 Y. Weizmann, F. Patolsky, I. Willner, *Analyst*, **2001**, *126*, 1502–1504.
158 I. Willner, F. Patolsky, J. Wasserman, *Angew. Chem., Int. Ed.*, **2001**, *40*, 1861–1864.
159 F. Patolsky, Y. Weizmann, O. Lioubashevski, I. Willner, *Angew. Chem., Int. Ed.*, **2002**, *41*, 2323–2327.
160 I. Willner, F. Patolsky, Y. Weimann, B. Willner, *Talanta*, **2002**, *56*, 847–856.

161 F. Patolsky, Y. Weizmann, E. Katz, I. Willner, *Angew. Chem., Int. Ed.*, **2003**, *42*, 2372–2376.
162 Y. Weizmann, F. Patolsky, E. Katz, I. Willner, *J. Am. Chem. Soc.*, **2003**, *125*, 3452–3454.
163 I. Willner, E. Katz, *Angew. Chem., Int. Ed.*, **2003**, *42*, 4576–4588.
164 Y. Xiao, A.B. Kharitonov, F. Patolsky, Y. Weizmann, I. Willner, *Chem. Commun.*, **2003**, *3*(13) 1540–1541.
165 M.J. Heller, *Annu. Rev. Biomed. Eng.*, **2002**, *4*, 129–153.
166 J. Wang, *Nucleic Acids Res.*, **2000**, *28*, 3011–3016.
167 E. Palecek, M. Fojta, M. Tomschik, J. Wang, *Biosens. Bioelectron.*, **1998**, *16*, 621–628.
168 E. Palecek, F. Jelen, *Crit. Rev. Anal. Chem.*, **2002**, *32*, 73–83.
169 J.J. Gooding, *Electroanalysis*, **2002**, *14*, 1149–1156.
170 T.G. Drummond, M.G. Hill, J.K. Barton, *Nat. Biotechnol.*, **2003**, *21*, 1192–1199.
171 E.M. Boon, J.K. Barton, *Curr. Opin. Struct. Biol.*, **2002**, *12*, 320–329.
172 J. Wang, *Anal. Chim. Acta*, **2003**, *500*, 247–257.
173 J. Marmur, R. Rownd, C.L. Schildkraut, in *Progress in Nucleic Acid Research* (Eds.: J.N. Davidson, W.E. Cohn), Vol. 1, Academic Press, New York, **1963**, 232–300.
174 J. Janata, *Principles of Chemical Sensors*, Plenum Press, New York, **1989**.
175 J. Wang, G. Rivas, X. Cai, E. Palecek, P. Nielsen, H. Shiraishi, N. Dontha, D. Luo, C. Parrado, M. Chicharro, P.A.M. Farias, F.S. Valera, D.H. Grant, M. Ozsoz, M.N. Flair, *Anal. Chim. Acta*, **1997**, *347*, 1–8.
176 J. Schülein, B. Graßl, J. Krause, C. Schulze, C. Kugler, P. Müller, W.M. Bertling, J. Hassmann, *Talanta* **2002**, *56*, 875–885.
177 S.R. Mikkelsen, *Electroanalysis*, **1996**, *8*, 15–19.
178 M. Masarik, R. Kizek, K.J. Kramer, S. Billova, M. Brazdova, J. Vacek, M. Bailey, F. Jelen, J.A. Howard, *Anal. Chem.*, **2003**, *75*, 2663–2669.
179 Y. Okahata, Y. Matsunobu, K. Ijiro, M. Mukae, A. Murakami, K. Makino, *J. Am. Chem. Soc.*, **1992**, *114*, 8299–8300.
180 T.M. Herne, M.J. Tarlov, *J. Am. Chem. Soc.*, **1997**, *119*, 8916–8920.
181 R. Levicky, T.M. Herne, M.J. Tarlov, S.K. Satija, *J. Am. Chem. Soc.* **1998**, *120*, 9787–9792.
182 J.C. O'Brien, J.T. Stickney, M.D. Porter, *J. Am. Chem. Soc.*, **2000**, *122*, 5004–5005.
183 E. Huang, M. Satjapipat, S.B. Han, F.M. Zhou, *Langmuir*, **2001**, *17*, 1215–1224.
184 J.J. Gooding, F. Mearns, W.R. Yang, J.Q. Liu, *Electroanalysis*, **2003**, *15*, 81–96.
185 K. Slowinski, R.V. Chamberlain, R. Bilewicz, M. Majda, *J. Am. Chem. Soc.*, **1996**, *118*, 4709–4710.
186 K. Slowinski, R.V. Chamberlain, C.J. Miller, M. Majda, *J. Am. Chem. Soc.*, **1997**, *119*, 11910–11919.
187 D. Mandler, I. Turyan, *Electroanalysis*, **1996**, *8*, 207–213.
188 N. Muskal, I. Turyan, D. Mandler, *J. Electroanal. Chem.*, **1996**, *409*, 131–136.
189 N. Muskal, D. Mandler, *Electrochim. Acta*, **1999**, *45*, 537–548.
190 A.B. Steel, T.M. Herne, M.J. Tarlov, *Anal. Chem.*, **1998**, *70*, 4670–4677.
191 K.A. Peterlinz, R.M. Georgiadis, *J. Am. Chem. Soc.*, **1997**, *119*, 3401–3402.
192 J. Li, G. Cheng, S. Dong, *Electroanalysis*, **1997**, *9*, 834–837.
193 A.B. Steel, R.L. Levicky, T.M. Herne, M.J. Tarlov, *Biophys. J.*, **2000**, *79*, 975–981.
194 L.A. Chrisey, G.U. Lee, C.E. Oferrall, *Nucleic Acids Res.*, **1996**, *24*, 3031–3039.
195 D.E. Gray, S.C. Casegreen, T.S. Fell, P.J. Dobson, E.M. Southern, *Langmuir*, **1997**, *13*, 2833–2842.
196 M.S. Shchepinov, S.C. Casegreen, E.M. Southern, *Nucleic Acids Res.*, **1997**, *25*, 1155–1161.
197 U. Maskos, E.M. Southern, *Nucleic Acids Res.*, **1992**, *20*, 1675–1678.
198 U. Maskos, E.M. Southern, *Nucleic Acids Res.*, **1993**, *21*, 4663–4669.
199 K.U. Mir, E.M. Southern, *Nat. Biotechnol.*, **2000**, *18*, 1209.
200 G.E. Poirier, *Chem. Rev.*, **1997**, *97*, 1117–1127.

201 M. Fojta, L. Havran, R. Kizek, S. Billova, *Talanta*, 2002, 56, 867–874.
202 M. Fojta, L. Havran, R. Kizek, S. Billova, E. Palecek, *Biosens. Bioelectron.*, 2004, in press.
203 C. Fan, K.W. Plaxco, A.J. Heeger, *Proc. Natl. Acad. Sci. U.S.A.*, 2003, 100, 9134–9137.
204 S. Takenaka, K. Yamashita, M. Takagi, Y. Uto, H. Kondo, *Anal. Chem.*, 2000, 72, 1334–1341.
205 T. Lumley-Woodyear, D.J. Caruana, C.N. Campbell, A. Heller, *Anal. Chem.*, 1999, 71, 394–398.
206 D.J. Caruana, A. Heller, *J. Am. Chem. Soc.*, 1999, 121, 769–774.
207 C.N. Campbell, D. Gal, N. Cristler, C. Banditrat, A. Heller, *Anal. Chem.*, 2002, 74, 158–162.
208 M. Dequaire, A. Heller, *Anal. Chem.*, 2002, 74, 4370–4377.
209 Y.C. Zhang, H.H. Kim, N. Mano, M. Dequaire, A. Heller, *Anal. Bioanal. Chem.*, 2002, 374, 1050–1055.
210 Y.C. Zhang, H.H. Kim, A. Heller, *Anal. Chem.*, 2003, 75, 3267–3269.
211 O. Bagel, C. Degrand, B. Limoges, M. Joannes, F. Azek, P. Brossier, *Electroanalysis*, 2000, 12, 1447–1452.
212 J. Wang, G. Rivas, X.H. Cai, M. Chicharro, C. Parrado, N. Dontha, A. Begleiter, M. Mowat, E. Palecek, P.E. Nielsen, *Anal. Chim. Acta*, 1997, 344, 111–118.
213 E.M. Boon, D.M. Ceres, T.G. Drummond, M.G. Hill, J.K. Barton, *Nat. Biotechnol.*, 2000, 18, 1096–1100.
214 S.R. Rajski, B.A. Jackson, J.K. Barton, *Mutat. Res.*, 2000, 447, 49–72.
215 R.M. Umek, S.W. Lin, J. Vielmetter, R.H. Terbrueggen, B. Irvine, C.J. Yu, J.F. Kayyem, H. Yowanto, G.F. Blackburn, D.H. Farkas, Y.P. Chen, *J. Mol. Diagn.*, 2001, 3, 74–84.
216 S. Creager, C.J. Yu, C. Bamdad, S. O'Connor, T. Maclean, E. Lam, Y. Chong, G.T. Olsen, J.Y. Luo, M. Gozin, J.F. Kayyem, *J. Am. Chem.Soc.*, 1999, 121, 1059–1064.
217 N.E. Broude, *Trends Biotechnol.*, 2002, 20, 249–256.
218 H. Du, M.D. Disney, B.L. Miller, T.D. Krauss, *J. Am. Chem. Soc.*, 2003, 125, 4012–4013.
219 J. Wang, *Anal. Chim. Acta*, 2002, 469, 63–71.
220 J. Wang, R. Polsky, A. Merkoci, K.L. Turner, *Langmuir*, 2003, 19, 989–991.
221 S.A. Brazill, P.H. Kim, W.G. Kuhr, *Anal. Chem.*, 2001, 73, 4882–4890.
222 J. Wang, R. Polsky, B. Tian, M.P. Chatrathi, *Anal. Chem.*, 2000, 72, 5285–5289.
223 E. Palecek, M. Fojta, F. Jelen, *Bioelectrochemistry*, 2002, 56, 85–90.
224 E. Palecek, S. Billova, L. Havran, R. Kizek, A. Miculkova, F. Jelen, *Talanta*, 2002, 56, 919–930.
225 B. Yosypchuk, L. Novotny, *Crit. Rev. Anal. Chem.*, 2002, 32, 141–151.
226 B. Yosypchuk, M. Heyrovsky, E. Palecek, L. Novotny, *Electroanalysis*, 2002, 14, 1488–1493.
227 J. Wang, D.K. Xu, A. Erdem, R. Polsky, M.A. Salazar, *Talanta*, 2002, 56, 931–938.
228 S.S. Babkina, N.A. Ulakhovich, Y.I. Zyavkina, *Ind. Lab.*, 2000, 66, 779–781.
229 S.S. Babkina, N.A. Ulakhovich, Y.I. Zyavkina, E.N. Moiseeva, *Russ. J. Phys. Chem.*, 2003, 77, 797–800.
230 S.S. Babkina, F. Jelen, A. Kourilova, M. Fojta, E. Palecek, 2004, in preparation.
231 M. Fojta, L. Havran, M. Vojtiskova, E. Palecek, 2004, in preparation.
232 J. Wang, G. Liu, A. Merkoci, *J. Am. Chem. Soc.*, 2003, 125, 3214–3215.
233 E. Palecek, R. Kizek, L. Havran, S. Billova, M. Fojta, *Anal. Chim. Acta*, 2002, 469, 73–83.
234 F. Azek, C. Grossiord, M. Joannes, B. Limoges, P. Brossier, *Anal. Biochem.*, 2000, 284, 107–113.
235 J. Wang, G.D. Liu, Q.Y. Zhu, *Anal. Chem.*, 2003, 75, 6218–6222.
236 M. Han, X. Gao, J. Su, S. Nie, *Nat. Biotechnol.*, 2001, 19, 631.
237 R. Nicewarner-Pena, R. Freeman, B. Reiss, L. He, D. Pena, I. Walton, R. Cromer, C.D. Keating, M.J. Natan, *Science*, 2001, 293, 137.
238 H. Cai, D. Xu, N. Zhu, P. He, Y. Fang, *Analyst*, 2002, 127, 803.
239 L. Authier, G.C.P. Berssier, B. Limoges, *Anal. Chem.*, 2001, 73, 4450.
240 J. Wang, D. Xu, A.N. Kawde, R. Polsky, *Anal. Chem.*, 2001, 73, 5576.

241 H. Cai, Y. Wang, P. He, Y. Fang, *Anal. Chim. Acta*, **2002**, *469*, 165.
242 T.M.H. Lee, L.L. Li, I.M. Hsing, *Langmuir*, **2003**, *19*, 4338.
243 J. Wang, R. Polsky, D. Xu, *Langmuir*, **2001**, *17*, 5739–5741.
244 J. Wang, G. Liu, R. Polsky, A. Merkoci, *Electrochem. Commun.*, **2002**, *4*, 722–726.
245 J. Wang, D. Xu, R. Polsky, *J. Am. Chem. Soc.*, **2002**, *124*, 4208.
246 R. Hirsch, E. Katz, I. Willner, *J. Am. Chem. Soc.*, **2000**, *122*, 12053.
247 J. Wang, G. Liu, R. Polsky, *Electrochem. Commun.*, **2002**, *4*, 819.
248 S. Park, T.A. Taton, C.A. Mirkin, *Science*, **2002**, *295*, 1503.
249 A. Kuderova-Krejcova, A.M. Poverenny, E. Palecek, *Nucleic Acids Res.*, **1991**, *19*, 6811–6817.
250 V. Campuzano, L. Montermini, M.D. Molto, L. Pianese, M. Cossee, F. Cavalcanti, E. Monros, F. Rodius, F. Duclos, A. Monticelli et al., *Science*, **1996**, *271*, 1423–1427.
251 H.L. Paulson, K.H. Fischbeck, *Annu. Rev. Neurosci.*, **1996**, *19*, 79–107.
252 I.V. Yang, H.H. Thorp, *Anal. Chem.*, **2001**, *73*, 5316–5322.
253 O.D. Scharer, *Angew. Chem., Int. Ed.*, **2003**, *42*, 2946–2974.
254 E.C. Friedberg, *Nature*, **2003**, *421*, 436–440.
255 E.C. Friedberg, *Cold Spring Harbor Symp. Quant. Biol.* **2000**, *65*, 593–602.
256 C.M. Gedik, S.P. Boyle, S.G. Wood, N.J. Vaughan, A.R. Collins, *Carcinogenesis*, **2002**, *23*, 1441–1446.
257 B. Halliwell, *Free Radical Res.*, **1998**, *29*, 469–486.
258 M. Masuda, H. Nishino, H. Ohshima, *Chem.-Biol. Interact.*, **2002**, *139*, 187–197.
259 L.S. Nakao, E. Fonseca, O. Augusto, *Chem. Res. Toxicol.*, **2002**, *15*, 1248–1253.
260 H. Vasquez, W. Seifert, H. Strobel, *J. Chromatogr. B*, **2001**, *759*, 185–190.
261 A. Collins, C. Gedik, N. Vaughan, S. Wood, A. White, J. Dubois, J.F. Rees, S. Loft, P. Moller, J. Cadet, T. Douki, J.L. Ravanat, S. Sauvaigo, N. Faure, I. Morel, M. Morin, B. Epe, N. Phoa, A. Hartwig, T. Schwerdtle, P. Dolara, L. Giovannelli, M. Lodovici, R. Olinski, K. Bialkowski, M. Foksinski, D. Gackowski, Z. Durackova, L. Hlincikova, P. Korytar, M. Sivonova, M. Dusinska, C. Mislanova, J. Vina, L. Moller, T. Hofer, J. Nygren, E. Gremaud, K. Herbert, J. Lunec, C. Wild, L. Hardie, J. Olliver, E. Smith, *Free Radical Biol. Med.*, **2003**, *34*, 1089–1099.
262 T.G. England, A. Jenner, O.I. Aruoma, B. Halliwell, *Free Radical Res.*, **1998**, *29*, 321–330.
263 G. Guetens, G. De Boeck, M. Highley, A.T. van Oosterom, E.A. de Bruijn, *Crit. Rev. Clin. Lab. Sci.*, **2002**, *39*, 331–457.
264 H.J. Helbock, K.B. Beckman, M.K. Shigenaga, P.B. Walter, A.A. Woodall, H.C. Yeo, B.N. Ames, *Proc. Natl. Acad. Sci. U.S.A.*, **1998**, *95*, 288–293.
265 H. Kasai, *J. Radical Res.*, **2003**, *44*, 185–189.
266 A. Weimann, D. Belling, H.E. Poulsen, *Nucleic Acids Res.*, **2002**, *30*, e7 (8pp).
267 A. Jenner, T.G. England, O.I. Aruoma, B. Halliwell, *Biochem. J.*, **1998**, *331*, 365–369.
268 S. Inagaki, Y. Esaka, M. Sako, M. Goto, *Electrophoresis*, **2001**, *22*, 3408–3412.
269 J. Cadet, C. D'Ham, T. Douki, J.P. Pouget, J.L. Ravanat, S. Sauvaigo, *Free Radical Res.*, **1998**, *29*, 541–550.
270 M. Stiborova, V. Simanek, E. Frei, P. Hobza, J. Ulrichova, *Chem.-Biol. Interact.*, **2002**, *140*, 231–242.
271 M. Koskinen, P. Vodicka, L. Vodickova, K. Hemminki, *Biomarkers*, **2001**, *6*, 175–189.
272 S. Swaminathan, J.F. Hatcher, *Chem.-Biol. Interact.*, **2002**, *139*, 199–213.
273 J.M. Falcone, H.C. Box, *Biochim. Biophys. Acta*, **1997**, *1337*, 267–275.
274 G.D. Li, O. Shimelis, X.J. Zhou, R.W. Giese, *Biotechniques*, **2003**, *34*, 908–909.
275 Y.S. Wang, J.S. Taylor, M.L. Gross, *Chem. Res. Toxicol.*, **1999**, *12*, 1077–1082.
276 P. Boublikova, M. Vojtiskova, E. Palecek, *Anal. Lett.*, **1987**, *20*, 275–291.
277 J.P. Pouget, T. Douki, M.J. Richard, J. Cadet, *Chem. Res. Toxicol.*, **2000**, *13*, 541–549.

278 P.L. OLIVE, R.E. DURAND, J.A. RALEIGH, C. LUO, C. AQUINO-PARSONS, *Br. J. Cancer*, **2000**, *83*, 1525–1531.
279 B. MARCZYNSKI, H.P. RIHS, B. ROSSBACH, J. HOLZER, J. ANGERER, M. SCHERENBERG, G. HOFFMANN, T. BRUNING, M. WILHELM, *Carcinogenesis*, **2002**, *23*, 273–281.
280 A. MIGHELI, *Methods Mol. Biol.*, **2002**, *203*, 31–39.
281 D.T. LOO, *Methods Mol. Biol.*, **2002**, *203*, 21–30.
282 E. PALECEK, A. KREJCOVA, M. VOJTISKOVA, V. PODGORODNICHENKO, T. ILYINA, A. POVERENNYI, *Gen. Physiol. Biophys.*, **1989**, *8*, 491–504.
283 J. BUZEK, A. KUDEROVA, T. PEXA, V. STANKOVA, L. LAUEROVA, E. PALECEK, *J. Biomol. Struct. Dyn.*, **1999**, *17*, 41–50.
284 E.M. BOON, A.L. LIVINGSTON, N.H. CHMIEL, S.S. DAVID, J.K. BARTON, *Proc. Natl. Acad. Sci. U.S.A.*, **2003**, *100*, 12543–12547.
285 D.T. ODOM, J.K. BARTON, *Biochemistry*, **2001**, *40*, 8727–8737.
286 B. GIESE, S. WESSELY, M. SPORMANN, U. LINDEMANN, E. MEGGERS, M.E. MICHEL-BEYERLE, *Angew. Chem. Int. Ed.*, **1999**, *38*, 996–998.
287 G. HARTWICH, D.J. CARUANA, T. dELUMLEY-WOODYEAR, Y.B. WU, C.N. CAMPBELL, A. HELLER, *J. Am. Chem. Soc.*, **1999**, *121*, 10803–10812.
288 J. JORTNER, M. BIXON, T. LANGENBACHER, M.E. MICHELBEYERLE, *Proc. Natl. Acad. Sci. U.S.A.*, **1998**, *95*, 12759–12765.
289 E. MEGGERS, M.E. MICHEL-BEYERLE, B. GIESE, *J. Am. Chem. Soc. U.S.A.*, **1998**, *120*, 12950–12955.
290 Y. OKAHATA, T. KOBAYASHI, K. TANAKA, M. SHIMOMURA, *J. Am. Chem. Soc.*, **1998**, *120*, 6165–6166.
291 M. FOJTA, *Collect. Czech. Chem. Commun*, **2004**, *69*, 715–747.
292 A. ERDEM, M. OZSOZ, *Electroanalysis*, **2002**, *14*, 965–974.
293 E. PALECEK, in *Encyclopedia of Analytical Science* (Ed.: A. TOWNSHEND), Vol. 6, Academic Press, London, **1995**, 3600–3609.
294 E. PALECEK, *Biochim. Biophys. Acta*, **1967**, *145*, 410–417.
295 J. PURANEN, M. FORSS, *Strahlentherapie*, **1983**, *159*, 505–507.
296 M. VORLICKOVA, E. PALECEK, *Int. J. Radiat. Biol.*, **1974**, *26*, 363–372.
297 M. FOJTA, L. HAVRAN, E. PALECEK, *Electroanalysis*, **1997**, *9*, 1033–1034.
298 M. FOJTA, T. KUBICAROVA, E. PALECEK, *Electroanalysis*, **1999**, *11*, 1005–1012.
299 M. FOJTA, V. STANKOVA, E. PALECEK, J. MITAS, P. KOSCIELNIAK, *Talanta*, **1998**, *46*, 155–161.
300 M. FOJTA, T. KUBICAROVA, E. PALECEK, *Biosens. Bioelectron.*, **2000**, *15*, 107–115.
301 M. FOJTA, L. HAVRAN, T. KUBICAROVA, E. PALECEK, *Bioelectrochemistry*, **2002**, *55*, 25–27.
302 J.F. RUSLING, L.P. ZHOU, J. YANG, A. MUGWERU, *Abstr. Pap. Am. Chem. Soc.*, **2002**, *223*, U389–U389.
303 L.P. ZHOU, J. YANG, C. ESTAVILLO, J.D. STUART, J.B. SCHENKMAN, J.F. RUSLING, *J. Am. Chem. Soc.*, **2003**, *125*, 1431–1436.
304 A. MUGWERU, J.F. RUSLING, *Electrochem. Commun.*, **2001**, *3*, 406–409.
305 B.Q. WANG, J.F. RUSLING, *Anal. Chem.*, **2003**, *75*, 4229–4235.
306 J. YANG, J.F. RUSLING, *Chem. Res. Toxicol.*, **2002**, *15*, 1659–1660.
307 J. WANG, M. CHICHARRO, G. RIVAS, X.H. CAI, N. DONTHA, P.A.M. FARIAS, H. SHIRAISHI, *Anal. Chem.*, **1996**, *68*, 2251–2254.
308 F. LUCARELLI, A. KICELA, I. PALCHETTI, G. MARRAZZA, M. MASCINI, *Bioelectrochemistry*, **2002**, *58*, 113–118.
309 C. TEIJEIRO, P. PEREZ, D. MARIN, E. PALECEK, *Bioelectrochem. Bioenerg.*, **1995**, *38*, 77–83.
310 P. PEREZ, C. TEIJEIRO, D. MARIN, *Chem.-Biol. Interact.*, **1999**, *117*, 65–81.
311 D. MARIN, P. PEREZ, C. TEIJEIRO, E. PALECEK, *Biophys. Chem.*, **1998**, *75*, 87–95.
312 D. MARIN, R. VALERA, E. DE LA RED, C. TEIJEIRO, *Bioelectrochem. Bioenerg.*, **1997**, *44*, 51–56.
313 J. WANG, G. RIVAS, M. OZSOS, D.H. GRANT, X.H. CAI, C. PARRADO, *Anal. Chem.*, **1997**, *69*, 1457–1460.
314 H. KARADENIZ, B. GULMEZ, F. SAHINCI, A. ERDEM, G.I. KAYA, N. UNVER, B. KIVCAK, M. OZSOZ, *J. Pharm. Biomed. Anal.*, **2003**, *33*, 295–302.
315 A.M.O. BRETT, J.A.P. PIEDADE, S.H.P. SERRANO, *Electroanalysis*, **2000**, *12*, 969–973.

316 A.M.O.B. Rett, M. Vivan, I.R. Fernandes, J.A.P. Piedade, *Talanta*, **2002**, *56*, 959–970.

317 J.A.P. Piedade, I.R. Fernandes, A.M. Oliveira-Brett, *Bioelectrochemistry*, **2002**, *56*, 81–83.

318 J. Langmaier, Z. Samec, E. Samcova, *Electroanalysis*, **2003**, *15*, 1555–1560.

319 M. Fojta, L. Havran, J. Fulneckova, T. Kubicarova, *Electroanalysis*, **2000**, *12*, 926–934.

320 I.C. Gherghi, S.T. Girousi, A.N. Voulgaropoulos, R. Tzimou-Tsitouridou, *Talanta*, **2003**, *61*, 103–112.

321 M.T. Carter, A.J. Bard, *J. Am. Chem. Soc.*, **1987**, *109*, 7528–7530.

322 M.T. Carter, M. Rodriguez, A.J. Bard, *J. Am. Chem. Soc.*, **1989**, *111*, 8901–8911.

323 M. Rodriguez, A.J. Bard, *Anal. Chem.*, **1990**, *62*, 2658–2662.

324 M. Rodriguez, T. Kodadek, M. Torres, A.J. Bard, *Bioconjugate Chem.*, **1990**, *1*, 123–131.

325 X.H. Xu, A.J. Bard, *J. Am. Chem. Soc.*, **1995**, *117*, 2627–2631.

326 J. Labuda, M. Buckova, M. Vanickova, J. Mattusch, R. Wennrich, *Electroanalysis*, **1999**, *11*, 101–107.

327 K. Maruyama, Y. Mishima, K. Minagawa, J. Motonaka, *J. Electroanal. Chem.*, **2001**, *510*, 96–102.

328 W. Sufen, P. Tuzhi, C.F. Yang, *Electroanalysis*, **2002**, *14*, 1648–1653.

329 S. Wang, T.Z. Peng, C.F. Yang, *J. Electroanal. Chem.*, **2003**, *544*, 87–92.

330 H.M. Zhang, N.Q. Li, *J. Pharm. Biomed. Anal.*, **2000**, *22*, 67–73.

331 M. Bukova, J. Labuda, J. Sandula, L. Krizkova, I. Stepanek, Z. Durackova, *Talanta*, **2002**, *56*, 939–947.

332 O. Korbut, M. Buckova, J. Labuda, P. Grundler, *Sensors*, **2003**, *3*, 1–10.

333 J. Labuda, M. Buckova, L. Heilerova, A. Caniova-Ziakova, E. Brandsteterova, J. Mattusch, R. Wennrich, *Sensors*, **2002**, *2*, 1–10.

334 J. Labuda, M. Buckova, L. Heilerova, S. Silhar, I. Stepanek, *Anal. Bioanal. Chem.*, **2003**, *376*, 168–173.

335 V. Brabec, V. Kleinwachter, J.L. Butour, N.P. Johnson, *Biophys. Chem.*, **1990**, *35*, 129–141.

336 J. Mbindyo, L. Zhou, Z. Zhang, J.D. Stuart, J.F. Rusling, *Anal. Chem.*, **2000**, *72*, 2059–2065.

337 E. Palecek, *Abhandlungen Der DAW*, Berlin, **1964**, *4*, 270–274.

338 E. Palecek, in *Proceedings of Electroanalysis in Hygiene, Environmental, Clinical and Pharmacological Chemistry* (Ed.: W.F. Smyth), Elsevier, Amsterdam, **1980**, 79–99.

339 M. Vojtiskova, E. Lukasova, F. Jelen, E. Palecek, *Bioelectrochem. Bioenerg.*, **1981**, *8*, 487–496.

340 V. Marini, J. Kasparkova, O. Novakova, L. Monsu scolaro, R. Romeo, V. Brabec, *J. Biol. Inorg. Chem.*, **2002**, *7*, 725–734.

341 J.M. Sequaris, P. Valenta, H.W. Nurnberg, *Int. J Rad. Biol. Relat. Stud. Phys. Chem. Med.*, **1982**, *42*, 407–415.

342 D. Krznaric, B. Cosovic, J. Stuber, R.K. Zahn, *Chem.-Biol. Interact.*, **1990**, *76*, 111–128.

343 V. Brabec, *Electrochim. Acta* **2000**, *45*, 2929–2932.

344 F. Lucarelli, I. Palchetti, G. Marrazza, M. Mascini, *Talanta*, **2002**, *56*, 949–957.

345 G. Chiti, G. Marrazza, M. Mascini, *Anal. Chim. Acta*, **2001**, *427*, 155–164.

346 M. Mascini, *Pure Appl. Chem.*, **2001**, *73*, 23–30.

347 E. Palecek, *Bioelectrochem. Bioenerg.*, **1986**, *15*, 275–295.

348 P. Kostecka, L. Havran, H. Pivonkova, M. Fojta, *Bioelectrochemistry*, **2004**, *63*, 245–248.

349 J. Yang, Z. Zhang, J.F. Rusling, *Electroanalysis*, **2002**, *14*, 1494–1500.

350 A. Mugweru, J.F. Rusling, *Anal. Chem.*, **2002**, *74*, 4044–4049.

351 J. Labuda, M. Buckova, S. Jantova, I. Stepanek, I. Surugiu, B. Danielsson, M. Mascini, *Fresenius' J. Anal. Chem.*, **2000**, *367*, 364–368.

352 M. Maeda, Y. Mitsuhashi, K. Nakano, M. Takagi, *Anal. Sci.*, **1992**, *8*, 83–84.

353 E.M. Boon, J.K. Barton, *Langmuir*, **2003**, *19*, 9255–9259.

354 E.M. Boon, J.E. Salas, J.K. Barton, *Nat. Biotechnol.*, **2002**, *20*, 282–286.

355 X. Cai, G. Rivas, P.A.M. Farias, H. Shirashi, J. Wang, M. Fojta, E. Palecek, *Bioelectrochem. Bioenerg.*, **1996**, *401*, 41–47.

356 J. WANG, G. RIVAS, D.B. LUO, X.H. CAI, F.S. VALERA, N. DONTHA, *Anal. Chem.*, **1996**, *68*, 4365–4369.

357 M. VANICKOVA, M. BUCKOVA, J. LABUDA, *Chem. Anal.*, **2000**, *45*, 125–133.

358 C.M.A. BRETT, A.M.O. BRETT, S.H.P. SERRANO, *J. Electroanal. Chem.* **1994**, *366*, 225–231.

359 A.M.O. BRETT, T.R.A. MACEDO, D. RAIMUNDO, M.H. MARQUES, S.H.P. SERRANO, *Biosens. Bioelectron.*, **1998**, *13*, 861–867.

360 A.M.O. BRETT, S.H.P. SERRANO, T.R.A. MACEDO, D. RAIMUNDO, M.H. MARQUES, M.A. LASCALEA, *Electroanalysis*, **1996**, *8*, 992–995.

361 A.M.O. BRETT, S.H.P. SERRANO, I. GUTZ, M.A. LASCALEA, *Bioelectrochem. Bioenerg.*, **1997**, *42*, 175–178.

362 A.M.O. BRETT, S.H.P. SERRANO, I.G.R. GUTZ, *Electroanalysis*, **1997**, *9*, 110–114.

363 D.W. PANG, M. ZHANG, Z.L. WANG, Y.P. QI, J.K. CHENG, Z.Y. LIU, *J. Electroanal. Chem.*, **1996**, *403*, 183–188.

364 A.M.O. BRETT, L.A. dA SILVA, H. FUJII, S. MATAKA, T. THIEMANN, *J. Electroanal. Chem.*, **2003**, *549*, 91–99.

365 A.M.O. BRETT, M. VIVAN, I.R. FERNANDES, J.A.P. PIEDADE, *Talanta*, **2002**, *56*, 959–970.

366 J. WANG, M. JIANG, E. PALECEK, *Bioelectrochem. Bioenerg.*, **1999**, *48*, 477–480.

367 D.W. PANG, Q. LU, Y.D. ZHAO, M. ZHANG, *Acta Chim. Sin.*, **2000**, *58*, 524–528.

368 D.W. PANG, Y.D. ZHAO, M. ZHANG, Y.P. QI, Z.L. WANG, J.K. CHENG, *Anal. Sci.*, **1999**, *15*, 471–475.

369 J. WANG, M. JIANG, A.N. KAWDE, *Electroanalysis*, **2001**, *13*, 537–540.

370 K. KUCHARIKOVA, L. NOVOTNY, B. YOSYPCHUK, M. FOJTA, *Electroanalysis*, **2004**, in press.

371 O. KORBUT, M. BUCKOVA, P. TARAPCIK, J. LABUDA, P. GRUNDLER, *J. Electroanal. Chem.*, **2001**, *506*, 143–148.

372 F. JELEN, E. PALECEK, *Gen. Physiol. Biophys.*, **1985**, *4*, 219–237.

373 E. PALECEK, *Bioelectrochem. Bioenerg.*, **1992**, *28*, 71–83.

374 K. CAHOVA-KUCHARIKOVA, M. FOJTA, E. PALECEK, **2004**, in preparation.

375 S. HASON, V. VETTERL, *Bioelectrochemistry*, **2002**, *57*, 23–32.

376 T. KUBICAROVA, M. FOJTA, J. VIDIC, D. SUZNJEVIC, M. TOMSCHIK, E. PALECEK, *Electroanalysis*, **2000**, *12*, 1390–1396.

377 T. KUBICAROVA, M. FOJTA, J. VIDIC, L. HAVRAN, E. PALECEK, *Electroanalysis*, **2000**, *12*, 1422–1425.

378 J. WU, Y. HUANG, J. ZHOU, J. LUO, Z. LIN, *Bioelectrochem. Bioenerg.*, **1997**, *44*, 151–154.

379 R. FADRNA, B. YOSYPCHUK, M. FOJTA, T. NAVRÁTIL, L. NOVOTNY, *Anal. Lett.*, **2004**; in press.

380 B. YOSYPCHUK, L. NOVOTNY, *Electroanalysis*, **2002**, *14*, 1733–1738.

381 M. FOJTA, R. FADRNA, L. HAVRAN, P. KOSTECKA, B. YOSYPCHUK, **2004**; in preparation.

382 F. JELEN, M. FOJTA, E. PALECEK, *J. Electroanal. Chem.*, **1997**, *427*, 49–56.

383 M. OZSOZ, A. ERDEM, P. KARA, K. KERMAN, D. OZKAN, *Electroanalysis*, **2003**, *15*, 613–619.

384 L. DENNANY, R.J. FORSTER, J.F. RUSLING, *J. Am. Chem. Soc.*, **2003**, *125*, 5213–5218.

385 S. WANG, T. PENG, C.F. YANG, *J. Biochem. Biophys. Methods*, **2003**, *55*, 191–204.

386 P.C. PANDEY, H.H. WEETALL, *Anal. Chem.*, **1994**, *66*, 1236–1241.

387 J.-M. SEQUARIS, M. ESTEBAN, *Electroanalysis*, **1990**, *2*, 35–41.

388 J.-M. SEQUARIS, J. SWIATEK, *Bioelectrochem. Bioenerg.*, **1991**, *26*, 15–28.

389 M.M. CORREIA DOS SANTOS, C.M.L.F. LOPES, M.L. SIMOES-GONCALVES, *Bioelectrochem. Bioenerg.*, **1996**, *39*, 55–60.

390 M.M. CORREIA DOS SANTOS, P.M.P. SOUSA, A.M.M. MODESTO, M.L. SIMOES GONCALVES, *Bioelectrochem. Bioenerg.*, **1998**, *45*, 267–273.

391 P.A.M. FARIAS, A.D. WAGENER, A.A. CASTRO, *Talanta*, **2001**, *55*, 281–290.

392 R.F. JOHNSTON, D.M. LEWIS, J.Q. CHAMBERS, *J. Electroanal. Chem.*, **1999**, *466*, 2–7.

393 J.L.M. ALVAREZ, J.A.G. CALZON, J.M.L. FONSECA, *J. Electroanal. Chem.*, **1998**, *457*, 53–59.

394 E.D. Gil, S.H.P. Serrano, E.I. Ferreira, L.T. Kubota, *J. Pharm. Biomed. Anal.*, **2002**, *29*, 579–584.
395 F. Qu, N.-Q. Li, *Electroanalysis*, **1997**, *9*, 1348–1352.
396 M. Vanickova, J. Labuda, M. Buckova, I. Surugiu, M. Mecklenburg, B. Danielsson, *Collect. Czech. Chem. Commun.*, **2000**, *65*, 1055–1066.
397 Z. Mazerska, A. Zon, Z. Stojek, *Electrochem. Commun.*, **2003**, *5*, 770–775.
398 J.H. Tocher, *Free Radical Res.*, **2001**, *35*, 159–166.
399 D.M. Stearns, L.J. Kennedy, K.D. Courtney, P.H. Giangrande, L.S. Phieffer, K.E. Wetterhahn, *Biochemistry* **1995**, *34*, 910–919.
400 D.M. Stearns, K.E. Wetterhahn, *Chem. Res. Toxicol.*, **1997**, *10*, 271–278.
401 R. Kalvoda, *Crit. Rev. Anal. Chem.* **2000**, *30*, 31–35.
402 K.Z. Brainina, N.A. Malakhova, N.Y. Stojko, *Fresenius' J. Anal. Chem.*, **2000**, *368*, 307–325.
403 C. Locatelli, *Electroanalysis*, **1997**, *9*, 1014–1017.
404 D. Romanova, A. Vachalkova, K. Horvathova, A. Krutosikova, *Collect. Czech. Chem. Commun.*, **2001**, *66*, 1615–1622.
405 H.L. Lin, P.F. Hollenberg, *Chem. Res. Toxicol.*, **2001**, *14*, 562–566.
406 L. Novotny, A. Vachalkova, T. Al-Nakib, N. Mohanna, D. Vesela, V. Suchy, *Neoplasma*, **1999**, *46*, 231–236.
407 C. Cojocel, L. Novotny, A. Vachalkova, B. Knauf, *Neoplasma*, **2003**, *50*, 110–116.
408 F. Lucarelli, L. Authier, G. Bagni, G. Marrazza, T. Baussant, E. Aas, M. Mascini, *Anal. Lett.*, **2003**, *36*, 1887–1901.
409 A.M. Oliveira-Brett, L.A. da Silva, *Anal. Bioanal. Chem.*, **2002**, *373*, 717–723.
410 F.C. Abreu, M.O.F. Goulart, A.M.O. Brett, *Biosens. Bioelectron.*, **2002**, *17*, 913–919.
411 M. Vanickova, J. Labuda, J. Lehotay, J. Cizmarik, *Pharmazie*, **2003**, *58*, 570–572.
412 D.P. Nikolelis, S.S.E. Petropoulou, M.V. Mitrokotsa, *Bioelectrochemistry*, **2002**, *58*, 107–112.
413 X. Chu, G.L. Shen, J.H. Jiang, T.F. Kang, B. Xiong, R.Q. Yu, *Anal. Chim. Acta*, **1998**, *373*, 29–38.
414 A. Erdem, M. Ozsoz, *Anal. Chim. Acta*, **2001**, *437*, 107–114.
415 A. Erdem, M. Ozsoz, *Turkish J. Chem.*, **2001**, *25*, 469–475.
416 I.C. Gherghi, S.T. Girousi, A.N. Voulgaropoulos, R. Tzimou-Tsitouridou, *J. Pharm. Biomed. Anal.*, **2003**, *31*, 1065–1078.
417 M.A. La-Scalea, S.H.P. Serrano, E.I. Ferreira, A.M.O. Brett, *J. Pharm. Biomed. Anal.*, **2002**, *29*, 561–568.
418 J.H. Xu, J.J. Zhu, Y.L. Zhu, K. Gu, H.Y. Chen, *Anal. Lett.*, **2001**, *34*, 503–512.
419 M.S. Ibrahim, I.S. Shehatta, A.A. Al-Nayeli, *J. Pharm. Biomed. Anal.*, **2002**, *28*, 217–225.
420 A. Radi, M.A. El ries, S. Kandil, *Anal. Chim. Acta*, **2003**, *495*, 61–67.
421 K. Nakano, S. Uchida, Y. Mitsuhashi, Y. Fujita, H. Taira, M. Maeda, in *Polymer in Sensors. Theory and Practise* (Eds.: N. Akmal, A.M. Usmani), American Chemical Society, Washington, **1998**, 34–45.
422 C.G. Siontorou, A.M.O. Brett, D.P. Nikolelis, *Talanta*, **1996**, *43*, 1137–1144.

6
Probing Biomaterials on Surfaces at the Single Molecule Level for Bioelectronics

Barry D. Fleming, Shamus J. O'Reilly, and H. Allen O. Hill

There has been much effort over the last thirty years or so in examining the electrochemistry of proteins and enzymes. The procedures for doing so are now well established and, by and large, there is little trouble in achieving good, reproducible and fast electrochemistry. Of course, in not every case examined have the authors been able to ensure that the proteins or enzymes are in their "native" state. (Denaturation[1] may involve, when binding to an electrode surface, minor changes only in the conformation of the protein or enzyme: though changes to the latter may be detected by an alteration in, or loss of enzymatic activity, changes to the former may be more difficult to witness). However, the battle to achieve the electrochemistry of proteins[2] has been solved, although much remains to be done on using the method to investigate their properties or exploitation.

The challenge now is to see if it is possible to investigate the electrochemistry of *single* molecules of proteins. Although the properties of such assemblies have been studied for a number of years, for example, by Scanning Probe Microscopy, electron transfer to individual molecules has only recently become an achievable goal. This is particularly true if the aim is to examine systems in which the proteins are in their native state, and this is much more likely to be the case when they are examined in a biologically relevant medium. Though such studies of proteins in air, or even more riskily in vacuum, have long yielded dramatic images by electron microscopy, there is no guarantee that they bear a close resemblance to proteins in their native state (this is one of the main driving forces behind the development of cryo-electron microscopy where the proteins are retained, presumably, in their native states, albeit in ice). The hunt is now on for methods that make possible efficient contact between a conducting surface and a single protein molecule, many of which are highlighted in this article.

---

[1] Denaturation. *Any* change in the native conformation of a protein. Denaturation may involve the breaking of noncovalent bonds such as hydrogen bonds and the breaking of covalent bonds such as disulfide bonds: it may be partial or complete, reversible or irreversible. Denaturation leads to changes in one or more of the characteristic chemical, biological, or physical properties of the protein [1].

[2] In this article, unless otherwise stated, the term "protein" can be taken to include enzyme.

Bioelectronics. Edited by Itamar Willner and Eugenii Katz
Copyright © 2005 WILEY-VCH Verlag GmbH & Co. KGaA, Weinheim
ISBN: 3-527-30690-0

6.1
Methods for Achieving Controlled Adsorption of Biomolecules

It is relatively easy to achieve the adsorption of proteins or enzymes onto a surface, indeed it is almost difficult to avoid. It requires some degree of ingenuity to control the process of adsorption such that a surface modified with protein is ordered, biologically active and fully functional. When considering bioelectrochemical functionalities, it is fundamental that there is electrical contact between the surface and protein. We will address some of the approaches that have been employed to obtain such modified surfaces, giving relevant examples from the literature and work performed in our laboratory.

A variety of methods have been used to anchor biological molecules to surfaces but as with all molecular systems the mechanism for binding is either chemical or physical in nature. The chemical approach involves strong covalent bonds that can be difficult to produce without damage to the enzyme, but offer the greatest resilience and stability. Physical interactions such as electrostatic, hydrophobic, hydrophilic and affinity associations are experimentally easier to achieve but may not be as stable over long periods.

Protein adsorption to bare or unmodified surfaces is often uncontrollable and unforgiving, causing denaturation and loss of reactivity and function. Cytochrome P450 enzymes, such as $P450_{cam}$ and $P450_{BM3}$, adsorb quite readily and strongly to gold and graphitic electrodes when in their wild-type form. Yet it is almost impossible to obtain any electrochemistry indicative of electron transfer to these enzymes, as a result of the random adsorption and the denaturative effect upon immobilization on the electrode. The solution to this intrinsic problem with adsorbed biomolecules has been achieved by modification of the *biomolecule* via genetic engineering or chemical cross-linking and/or by modification of the *surface* with self-assembled monolayers (SAMs).

The application of genetic engineering methods allows the introduction of single immobilization sites on to the surface of an enzyme. This can lead to reproducible adsorption in a controlled orientation. Examples of such site-directed mutagenesis can be found with the copper proteins azurin [2] and poplar plastocyanin [3] and the heme containing monooxygenase $P450_{cam}$ [4–6]. In each case, an outer surface residue was replaced with a cysteine residue; this sulfur-containing group interacts strongly with gold via covalent interactions. Regularly arranged arrays of electrochemically addressable molecules were observed.

Modification of biomolecules by chemical methods has also proven a successful approach for organized binding to surfaces. An amine-containing functionality on cytochrome *c* was modified with N-succinimidyl-3-(2-pyridyldithio)propionate (SPDP) [7] (SPDP contains a disulfide bond, which, when cleaved with dithiothreitol, forms a thiol end-group for covalent binding to gold substrates).

A greater degree of interest has been shown in immobilizing proteins using SAMs. The latter typically consist of organothiols on metals such as Au, Ag or Pt, and organosilanes on silicon. Methyl-terminated alkane thiols can interact with hydrophobic regions on protein molecules. A useful example of this was shown for azurin immobilized on a series of alkane thiol monolayers of various lengths [8]. The hydrophobic region of azurin was near the copper center, providing a very suitable orientation for electron transfer to the gold electrode. The alkane chains can be functionalized at their solution terminus to create a desired specific interaction with the protein. For example, the electrostatic interaction between a carboxylic acid–functionalized organo-thiol and an amine-containing group on the exterior of a protein causes presumed complex formation on the electrode surface. However, this physical interaction is generally reversible and normally too unstable for long-term binding and electrochemical behavior [9]. In addition, it also lacks specificity, leading to heterogeneity in linkage and protein orientation [10].

To alleviate this concern, covalent linkage of proteins to functionalized SAMs has been performed with much success, particularly with reference to biosensor applications [10, 11], as a result of the stronger bonding and electrical connection between the biomolecule and surface. Numerous examples can be found in the literature, and the reader is directed to review works by Willner and Katz [10–12].

Another reportedly successful process of biomolecular adsorption is via a laser-assisted deposition technique [13, 14]. In this procedure, biomolecules are gently vaporized and ionized by a laser source and then guided to the substrate by an electric field. The enzyme horse radish peroxidase (HRP) was deposited onto silicon as a uniform layer by this process, with 10 to 30% activity being retained [14].

6.2
Methods for Investigating Adsorbed Biomolecules

In this section, we shall briefly describe some methods for the investigation of proteins and enzymes when adsorbed to a surface. Of particular interest is the probing of single molecular structures and electronic function. Two approaches stand out in this regard: Scanning Probe Microscopy and fluorescence methods.

Scanning Probe Microscopy (SPM) techniques, such as Scanning Tunneling Microscopy (STM) and Atomic Force Microscopy (AFM), have become increasingly useful tools for characterization of adsorbed biomolecular layers because of their powerful resolution, resulting in molecular and submolecular structural detail being regularly observed. This resolution has improved with advancements in the instrumentation and in the immobilization methods for controlling the attachment of the biomolecules to the surface. In addition, other significant benefits of these techniques are that the imaging process is capable of operating *in situ*, and with surfaces under electrochemical control. The latter is of significance for studying the redox properties of adsorbed proteins.

The STM generates a real space image of a surface on the basis of changes in the electron tunneling behavior, that is, electronic conduction, between a scanned conducting tip and conducting surface. It is therefore able to provide topographical and electronic information about the surface or material adsorbed to it. Despite their apparent insulating behavior, a number of adsorbed protein layers have been successfully imaged *in situ* by STM. These include azurin [2, 8, 15–19], catalase [20–22], cytochrome P450$_{cam}$ [6, 23], cytochrome *c* [24], glucose oxidase [25, 26], nitrite reductase [27], poplar plastocyanin [3], and a *de novo* designed carboprotein [28].

A number of models based on an electron tunneling mechanism have been proposed to explain these images [29, 30], although there is still no widely accepted mechanism that supports electrons tunneling through molecules that are several nanometers thick.

Another feature of STM is that the tip can be positioned at an area of interest on the surface and the tunneling current, I_{tunnel}, measured as a function of the potential difference between the surface and tip, V_{bias} (effectively measuring the resistance at this location). This mode, referred to as *Scanning Tunneling Spectroscopy* (STS), is useful for studying the tunneling properties with subnanometer spatial resolution. The application of this technique to single biological molecules will be discussed later.

In AFM, the image is based on the changes in force experienced by the sensing probe as it is scanned over the surface. The most common use has been topographical imaging, with or without adsorbed material, but it can also measure the forces of interaction between the tip and underlying substrate at a designated position. These experiments have been applied to the study of intermolecular forces between a range of surfaces, including those that have been modified with biological material.

A relatively recent advance in AFM is coating the scanning probe in conductive material such as Ag, Au or Pt [31]. When scanning, the probe not only senses force and so measures topography but also detects a current due to the applied bias between the tip and sample, and so the image produced is also a map of the resistance of the sample. As with STS, conducting AFM can also be used to measure current–voltage (I–V) properties at precise areas on a sample but in addition to this, it can measure current–force spectra.

There are advantages and disadvantages in using either conducting AFM spectroscopy or STS [31]. The vertical distance over which STS is practical is no more than 10 nm, yet the lateral resolution is an order of magnitude better than with the AFM approach. STM tips remain a finite distance away from the surface, whereas an AFM tip initially impacts the surface and can be damaged. The experimental system to be studied will determine the best approach to be used. As will be more fully described in the subsequent sections, both STM and AFM have been utilized as tools not only for imaging but also for the construction of patterned samples.

Fluorescence microscopy techniques, such as Confocal Microscopy and Total Internal Reflection Fluorescence Microscopy (TIRFM), have recently become a powerful source of information on the properties of immobilized biomolecules [32]. These techniques utilize the fluorescence properties of certain molecules, such as fluorophores, which, on absorbing light at a particular wavelength, emit light of longer wavelength after a small period of time, which is referred to as the *fluorescence lifetime*. The fluorescent species in protein systems can be native to the protein, such as a tryptophan residue [33], covalently attached probes at specific sites [34], or green fluorescent proteins (GFP) fused to the molecule [32, 35]. The distribution of a single molecular species over a surface can be imaged using the fluorescence microscopy techniques. Given the sensitive nature of fluorescence to localized environmental changes and that fluorophores can be located essentially anywhere on a protein, a wealth of data related to conformational changes, substrate binding, effect of immobilization and changing redox state can be acquired [36].

6.3
Surfaces Patterned with Biomolecules

The formation of patterned surfaces of biomolecules is highly desirable from both a fundamental and applied viewpoint. One fundamental area utilizing biomolecules arranged in predetermined surface locations is the study of biomolecular interactions such as those between enzymes and substrates, ligands and receptors, antigens and antibodies, cells and surfaces. There are many current and developing technological interests in this area, for example, in molecular electronic devices, biosensors, proteomics and cell research, and pharmaceutical screening procedures.

A degree of ordering within a monolayer of protein occurs naturally, directed by the mode of adsorption and the physical constraints of molecular packing onto native or SAM-modified surfaces. A good example of this was shown for cytochrome P450$_{cam}$ immobilized onto a single crystalline Au surface [6]. Reproducibly ordered adsorption to the Au(111) surface was facilitated by a site-specifically engineered cysteine residue. Figure 6.1 shows an STM image of the P450 layer under ambient and *in situ* conditions. Similar highly ordered arrays have also been observed with azurin adsorbed directly to a Au(111) surface [16, 17, 25, 37], Au surfaces modified with alkanethiol [8] and 11-mercaptoundecanoic [18] SAMs, and a 3-mercaptopropyltrimethylsilane-treated silicon oxide surface [38]. As for P450, the binding to the untreated Au surface occurred via strong covalent bonds with sulfide groups on the azurin. In addition to the ordered arrangement on the surface, both the P450 and azurin were shown to be electrochemically addressable when immobilized on the modified and unmodified Au surfaces. A further example is the ordered monolayers on Au formed by chemically

 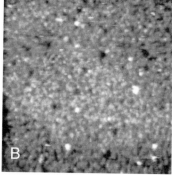

Fig. 6.1 STM images highlighting the natural ordering within a cytochrome P450$_{cam}$ monolayer adsorbed to a Au(111) surface when measured under (A) ambient and (B) fluid (deionized water) conditions. For (A), scan size is 290 × 290 nm, bias 1.1 V, tunnel current 190 pA, scan rate 6.8 Hz, z-scale 0–4.2 nm. For (B), scan size is 120 × 120 nm, bias 1.5 V, tunnel current 200 pA, scan rate 5 Hz, z-scale 0–3 nm. (J. J. Davis, D. Djuricic, K. K. W. Lo, E. N. K. Wallace, L.-L. Wong, H. A. O. Hill, *Faraday Discuss.* **2000**, *116*, 15–22; reproduced with permission of The Royal Society of Chemistry).

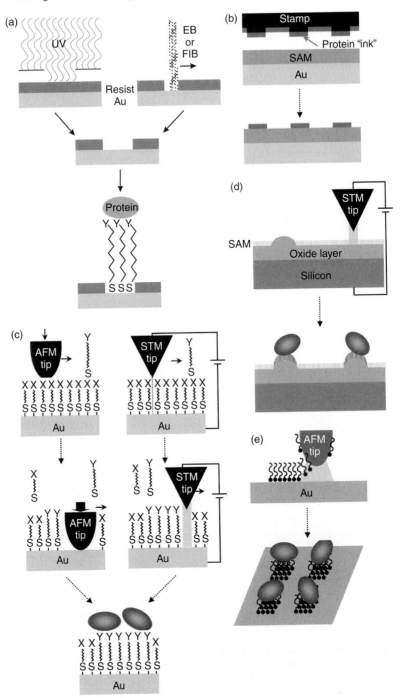

Fig. 6.2 (legend see page 199)

modified cytochrome c [7]. The presence of the monolayer was confirmed by AFM imaging and by adsorbing GFP onto the cytochrome by electrostatic forces and measuring the fluorescence emission spectra.

Several techniques have been applied to generate user-defined patterned surfaces of biomolecules. Some of those to be discussed here are microfabrication techniques such as photolithography, electron beam (EB) lithography and focused ion beam (FIB) lithography, microcontact printing (MCP) and scanning probe lithographic (SPL) techniques such as dip-pen lithography (DPL) and STM lithography. A schematic describing each of these procedures is shown in Figure 6.2.

The microfabrication techniques (Figure 6.2a) generate their patterns by removal of a polymeric resist material or insulative organic SAM from an underlying conducting or semiconducting substrate. Photolithographic procedures utilize a mask that shields some areas from the UV radiation sources. EB and FIB lithography are maskless procedures that directly mill or remove the resist material by way of a highly energetic focused beam of electrons or ions.

Oh et al. [39] used UV lithography to create a patterned monolayer of cytochrome c. First they coated a gold surface with a polyimide resist and UV lithography was then used to remove certain areas of resist, exposing the underlying Au substrate. A mixed SAM of alkanethiolates was then applied to the Au, and the remaining resist removed. The protein, cytochrome c, was then immobilized onto the SAM. More recently the same group selectively removed areas from an insulating SAM of 11-mercaptoundecanoic acid (11-MUA) by photochemical oxidation [40]. The remaining SAM was used to bind cytochrome c. A similar photolithographic procedure was performed by Veiseh et al [41]. In this instance, areas of photoresist were removed by UV irradiation to expose the underlying silicon surface. A titanium layer was deposited upon the silicon, and then a Au layer was deposited upon this. The remaining resist was removed with hot acetone. The Au areas were modified with a mixed COOH-terminated SAM for selective protein/peptide adsorption, while the silicon was modified with polyethylene glycol to inhibit protein adsorption. In both cases, AFM was used to image the protein-patterned surfaces. The latter also incorporated GFP into the protein matrix, highlighting the adsorbed protein areas by fluorescence emission imaging.

Gerardino et al. [14] incorporated EB lithography into their process for preparing silicon surfaces patterned with layers of the enzyme HRP. The polymethylmethacrylate resist-coated silicon was first patterned with the EB exposing the underlying silicon.

Fig. 6.2 Schematic descriptions of (a) UV and EB lithography, (b) microcontact printing, (c) nanografting and electrografting, (d) STM lithography, (e) dip-pen nanolithography. In all cases described, the final SAM is selective toward protein immobilization. In (a), after the resist has been selectively removed by use of UV irradiation, electron beam or FIB, the exposed gold surface is coated with an SAM. In (b), the protein is directly printed onto an SAM. In (c), an AFM (high force) or STM (high potential) tip is used to selectively remove an area of SAM (S-X). Another SAM (S-Y), present in solution, rapidly replaces the original SAM. In (d), an STM tip (high potential) is used to grow an isolated oxide area, upon which another SAM is formed. In (e), an AFM tip coated with "ink" molecules is used to directly "write" onto the surface. In a sufficiently humid environment, a meniscus between the tip and surface is formed, which permits the transport of the molecules to the surface.

HRP was then fixed to the silicon by laser-assisted deposition. EB lithography in this case enabled construction of nano- to micrometer-sized features of HRP.

Microcontact printing (Figure 6.2b) was recently used by Kwak et al. [42] to form patterned surfaces of cytochrome c. First, a stamp was prepared out of polydimethylsilane (PDMS) from a silicon master. The elastomeric PDMS stamp was used to deliver the cytochrome c "ink" molecules to the 16-mercaptohexadecanoic acid SAM on the Au surface where they were immobilized by electrostatic attractive forces. Protein dots of 1-μm radius with uniform spacing of 10 μm were formed by this method. Hong et al. [43] used a similar procedure to prepare arrays of fluorescent labeled avidin, antibiotin antibody and anti-BSA antibody. A poly(amidoamine) dendrimer was the patterned "ink" in this case, which, when functionalized, facilitated the biomolecular immobilization. Squares of 50 and 100 μm were fashioned in this case and imaged by AFM and fluorescence methods.

Recently, a series of lithographic techniques utilizing scanning probe technologies has been exploited for patterning surfaces. These include STM lithography and AFM-based approaches such as DPL, nanografting, current-sensing AFM lithography and anodization AFM lithography. AFM-based applications, particularly DPL, are an important process as they can pattern molecular layers that may be too thick to pass current or too insulating [44].

The majority of examples whereby STM lithography (Figure 6.2c) was used relate to the electrochemical removal of molecules from a SAM. The gaps formed in the monolayer can subsequently be filled with other self-assembling molecules bearing different functionalities for protein immobilization [45], or left bare and used in fine grating [46] or lateral nanostructure manufacture [47]. Alternatively, the STM has been used to locally anodize a thin insulating monolayer to form oxides of the underlying substrate, such as Ti or Si [48]. A schematic detailing this process is shown in Figure 6.2d. This is achieved by electrochemistry performed in a thin water column at the tip/surface junction. These areas can be subsequently modified with a second monolayer that facilitates the selective assembly of a protein. Sugimura et al. [48] performed this STM-based procedure to prepare linear patterns of HRP. This procedure is similar to anodization AFM lithography [49].

One of the more potentially useful AFM-based lithographic techniques is DPL (Figure 6.2e). Here, the AFM tip behaves analogous to a pen and at high-resolution nanofeatures can be created [44]. First reported for the formation of octadecanethiol surface structures [50], it has since been applied to generate patterns of proteins [34, 42, 51, 52]. Work by Mirkin's group has produced lysozome nanoarrays with dots as small as 100 nm and 1 to 2 molecules thick [51]. Kwak et al. [42] used a similar procedure to produce cytochrome c arrays having a lateral scale of ~200 nm. The protein nanoarrays are achieved by writing with "ink" molecules, which, when adsorbed to the substrate, have some affinity for the protein. The majority of examples have reported using a –COOH functionalized SAM. The unwritten areas are then passivated with an insulating SAM such as a long chain alkanethiol. This helps avoid any unintended binding to, and fouling of the underlying substrate. Mirkin's group has also developed an approach whereby the biomolecule is itself used as the "ink" [34]. They showed that, with the use of a suitably modified AFM tip, proteins such as rabbit IgG and antirabbit IgG could be directly written onto a negatively charged or aldehyde-modified silicon oxide surface. The key development in this approach was the modification of the AFM tip, which made it biocompatible and hydrophilic, suitable for protein protection and delivery.

A combined electrochemical and DPL approach was used by Agarwal et al. [35] to immobilize and pattern histidine-tagged proteins on to nickel. An applied potential between the conducting AFM probe and the Ni substrate facilitated by the small water meniscus between the two resulted in local ionization of the Ni substrate. The His-tag proteins that coat the tip, such as IgG, then attach to the ionized Ni sites. Deposition by this process was confirmed by using fluorescence microscopy imaging of GFP patterned in a similar way.

Another AFM-based approach to protein patterning is by nanografting (Figure 6.2c). A sufficiently high contact force is applied to an AFM tip while it scans an SAM-coated surface. This results in removal of the SAM from where the tip has scanned. If this procedure is performed in a solution containing a different self-assembling molecule, these will spontaneously adsorb to the vacancies in the SAM. Wadu-Mesthridge et al. [53] used this approach to pattern rabbit IgG and lysozome. They grafted away an area of alkanethiol and replaced it with a −CHO terminated alkanethiol. The proteins were bound to these areas of the SAM by imine bonds. Square areas as small as 40 nm in width were prepared and showed by AFM to contain a small number of proteins.

An electrografting procedure using conductive probe (CP) AFM was recently shown to have significant potential to create spatially defined arrays of different surface functionalities [54]. The procedure resulted in alkynes being electrografted onto a Si−H surface. The alkynes can have a variety of end functionalities and chain lengths. Lines of 40-nm width have been drawn by this approach in a rapid and reproducible manner. A further CP-AFM approach is the voltage-dependent oxidation of thiolated silicon surfaces resulting in localized activated areas whereby protein immobilization can take place [55]. The resolution of this process was dependent on the size of the cathodic AFM tip, with line widths of 70 to 200 nm being reported.

6.4
Attempts at Addressing Single Biomolecules

Several groups have addressed single protein molecules by using STM imaging. In Oxford, we have investigated the adsorption of cytochrome $P450_{cam}$, in addition to its redox partners putidaredoxin (PdX) and putidaredoxin reductase (PdR), to Au(111) and highly oriented pyrolytic graphite (HOPG) surfaces [6, 23, 56]. The high-resolution images obtained (Figure 6.3) revealed the trigonal prismatic structure of the chemisorbed $P450_{cam}$ molecules, each having molecular dimensions comparable to the crystallographically derived values. The complex formed between individual $P450_{cam}$ and PdX molecules was also observed.

The morphological properties and functionality of the metalloprotein yeast cytochrome *c* adsorbed to a gold substrate was recently reported [24]. Single molecule level characterization was performed using a variety of SPM techniques and under

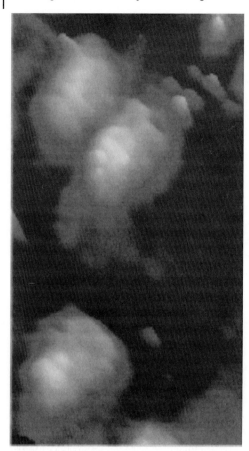

Fig. 6.3 *In situ* STM image of individual molecules of P450$_{cam}$ K344C mutant. Scan size is 10 × 20 nm, bias 1.2 V, tunnel current 215 pA, scan rate 9.4 Hz, z-scale 0–1.4 nm. (J. J. Davis, D. Djuricic, K. K. W. Lo, E. N. K. Wallace, L.-L. Wong, H. A. O. Hill, *Faraday Discuss.* **2000**, *116*, 15–22; reproduced with permission of The Royal Society of Chemistry).

different conditions (ambient, buffered solution, electrochemical control). More extensive single-molecule level studies have been undertaken with the redox metalloprotein azurin. Initial STM work revealed significant ordering within azurin monolayers when adsorbed on bare Au or Pt surfaces or those covered with an SAM [15, 57]. Imaging with submolecular resolution indicated a particular role being played by the copper redox level. Facci [19] showed how STM images of azurin bound to Au(111), as shown in Figure 6.4, were clearly dependent on tuning the Au substrate potential to that of the azurin redox potential. At −25 mV (Figure 6.4a), the bright spots represent the single azurin molecules. When the potential was reduced to −125 mV (Figure 6.4b), the same spots were no longer visible. Reverting back to −25 mV (Figure 6.4c) saw the return of the spots. This was a very important result, giving further insight into the mechanism involved in STM imaging of biomolecules.

In addition to the topographical and electronic information on single protein molecules obtained by STM imaging, the I-V properties derived from STS and conducting AFM are important. Investigating the I-V characteristics of single or groups

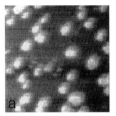  

Fig. 6.4 Sequence of *in situ* STM images of azurin-adsorbed Au(111) in 50 mM NH₄Ac (pH 4.6) obtained at (a) −25 mV (vs SCE), (b) −125 mV and (c) −25 mV. (Reprinted from P. Facci, D. Alliata, S. Cannistraro, *Ultramicroscopy*. 2001, *89*, 291–298, with permission from Elsevier Science).

of molecules can give information on the electron transfer mechanism(s) within a protein system and can also be used in the development of molecular-scale bioelectronic devices. STS and c-AFM had been the preferred method of obtaining such data. More recently, the use of nanogap electrodes [58], where a single or a group of molecules can be placed between two closely spaced planar electrodes and their conducting properties measured directly, has created a new pathway for this study. Some examples of using the above techniques will be discussed.

One of the first applications of STS to biological systems was with a single photosystem I (PSI) reaction center [59, 60]. The initial work [59] reported I-V data for untreated, and lightly and heavily platinized PSI when adsorbed to a Au electrode surface. The I-V curve for native PSI resembled that of a band gap semiconductor or insulator, and was symmetric in nature. The lightly platinized PSI I-V curve was diode-like, typical of an n-type doped semiconductor, with large currents observed with increasing negative bias potential. The changing contrast between negative and positive bias was also reflected in the electrochemically controlled STM images. The heavily platinized PSI behavior was close to that of platinum metal. Lee later used STS to further examine PSI electronic properties in addition to its orientation when immobilized to various chemically modified Au surfaces [60]. The orientation of the PSI molecules could be deduced by analysis of the shape of the I-V curve. A different I-V curve was produced, depending on the orientation of the electron transport vector with respect to the electrode surface. For instance, when parallel, a semiconductor-like I-V profile was observed. When perpendicular to the Au surface, a diode-like I-V curve was observed, with the asymmetry of the I-V curve dependent on whether the direction of the electron transport vector was toward or away from the Au surface. Each chemically modified surface did not show complete preference for one distinct orientation, but an increased or decreased preference for a certain orientation. For example, mercaptoacetic acid–treated Au surface showed high affinity for a parallel orientation (83%), cf. 33% and 28% for 2-dimethylaminoethanol and 2-mercaptoethanol–treated surfaces respectively.

Khomutov [61] provided an informative discussion on the I-V curves produced by cytochrome *c* molecules immobilized on HOPG. One important consideration presented was the strong dependence of the STM tip position over the protein on the I-V curve shape – with different I-V profiles observed at different STM tip positions over a cytochrome *c* molecule. They also described the fine structure observed in I-V curves such as steps, troughs or peaks, as the result of the interplay between the double-tunnel junction (STM tip/protein and HOPG/protein) and the electron tunneling features of the protein system, for example, the resonant tunneling through the discrete levels of cytochrome c along with charging effects.

Andolfi [3] incorporated STS in their study of two mutants of poplar plastocyanin, PCSS and PCSH. PCSS has a disulphide (S–S) bridge and PCSH has a single sulphide (S–H) covalently bound to the Au substrate. Different I-V characteristics were observed for each mutant, with PCSS showing rectifying-like behavior and PCSH showing semiconductor or insulating behavior. These differences were confirmed by electrochemical STM, with individual PCSS molecules showing some contrast bleaching (as in Figure 6.4) when the substrate potential was far from the redox potential of the mutant protein. These differences were partly explained by the differences in conformational flexibility imparted to each mutant molecule by its binding to the Au substrate. Though both mutants were well coupled to the Au electrode by covalent linkage, the single cysteine C-terminal extension of PCSH gave that mutant greater molecular flexibility, compared to the disulphide bridge in PCSS. This meant that PCSH was less likely to align its molecular redox levels with the tip and substrate Fermi levels. An important point, in respect to both interpretation of all STS data and explanation of the differences seen between PCSS and PCSH, concerns the difficulties in obtaining accurate and consistent tip-sample separations. This can result in unequal stresses being applied to each molecule, possibly affecting their conductive behavior.

Nam [62] recently showed that the heterolayer consisting of ferredoxin adsorbed onto a polylysine/11-MUA SAM on Au had rectifying properties. The asymmetric I-V curve exhibited higher current flow at positive bias potentials, that is, with a forward bias. This unsymmetrical behavior was explained in terms of the difference between the redox potential of the ferredoxin (~500 mV) and 11-MUA (−15 mV) layer that is, at highly negative potentials a counter current would flow between the ferredoxin (electron donor) and 11-MUA (electron acceptor).

The development of nanodevices incorporating single biomolecules, such as rectifiers and transistors, requires the manufacture of metal electrodes separated by the size of one molecule. This has proven to be a challenge, but recent reports [63, 64] suggest that the realization of such a bioelectronic construct may not be too far away. A typical device setup consisting of two triangular-shaped metallic electrodes is shown in Figure 6.5.

The early biology related work using nanogap electrodes used DNA. This involved tethering single or multiple strands of DNA across the gap between the two electrodes and measuring its conductive properties. Contrasting results demonstrating conducting, semiconducting and insulating behavior have been reported [58, 65, 66]. As the gap between the electrodes has decreased, it has permitted the study of smaller-sized molecules such as proteins.

Another prototype solid-state rectifying device was constructed with azurin [63]. On the basis of the same two-electrode design, the azurin layer was bound to the functionalized SiO_2 in the 50 to 100 nm gap. Two types of immobilization were employed: covalent bonding via a disulphide bridge resulting in an oriented layer, and with physical interactions using amino groups on the protein resulting in a nonoriented layer. The effect of azurin orientation was pertinent to the conductive

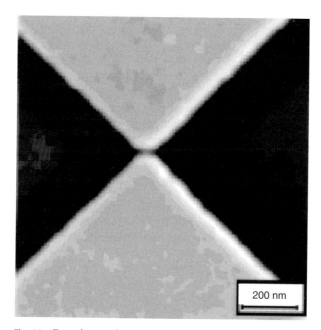

Fig. 6.5 Typical triangular-shaped metallic electrodes separated by a nanometer-sized gap. High magnification SEM image of two Cr/Au nanotips with a separation of 20 nm is shown. (Reprinted in part with permission from G. Maruccio, P. Visconti, V. Arima, S. D'Amico, A. Biasco, E. D'Amore, R. Cingolani, R. Rinaldi, S. Masiero, T. Giorgi, G. Gottarelli, *Nano. Lett.* **2003**, *3*, 479–483. Copyright 2003, American Chemical Society).

and rectifying behavior. Only oriented, purified, Cu-containing azurin showed any significant rectifying behavior. This was explained in terms of the dipole alignment and intermolecular electron transfer fostered by the uniform orientation.

To date, the scope of these nanogap devices has been limited to solid-state, humid or vacuum conditions. Recently, work in our laboratory gave rise to a convenient method to fabricate nanogap electrodes for use in aqueous environments (Figure 6.6). This was achieved by using the FIB to create an electrode device, with nanogaps as small as 14 nm, which was then insulated with photoresist. The FIB was then used to remove the insulating material from the gap to produce two well-defined opposing electrode surfaces, which are suitable for single molecule interrogation as shown schematically in Figure 6.6.

6.5
Conclusions

It must be apparent that, for devices whose use is envisaged where the protein or enzyme is in its dry state, there are sufficient exemplifications to warrant attempts toward

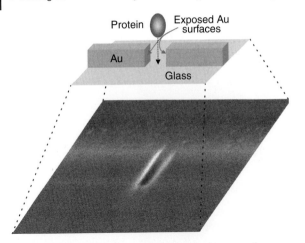

Fig. 6.6 High magnification FIB image and schematic of an insulated nanogap electrode. The schematic shows in 3-D how an individual protein molecule could be localized within the nanometer-sized gap for aqueous electrochemical interrogation.

manufacture. How feasible would it be to employ a biologically derived substance in such a device? It is most unlikely that a protein or an enzyme would retain its particular characteristic *when dehydrated*. This is not to say that some may not survive the process of dehydration and subsequent hydration: irreversible denaturation need not accompany this seemingly dramatic event. Do proteins or enzymes still retain some useful property when dried? Are the redox potentials of proteins still valuable in some device? Is it possible that, even if catalysis is impeded, enzymes or antibodies can still act as recognizing entities for interesting compounds? The latter would appear more likely, though the myriad attempts to mimic enzymes indicate that recognition can be particularly sensitive to the state of the enzyme. In our opinion, it would be fortuitous if proteins, enzymes, antibodies, and so on, retained valuable properties when dry.

This is far from the case when considering such materials in solution: the overwhelming evidence is that the properties of proteins, enzymes or antibodies are retained even when present at low concentration *except* if the properties of interest depend on, for example, a protein–protein complex for which dilution leads to dissociation. What happens if the desire is to achieve the adsorption of a protein/enzyme/antibody, either physically or chemically, on an electrode? That depends on the extent of denaturation of the material when adsorbed, either directly to the surface or to an SAM: many materials *are* denatured; some are not. It is imperative that checks are made to ensure that the biological material is still intact. If that is so, then there are now a number of methods highlighted in this article, which should allow the design and production of useful devices, certainly in air and even in the more relevant physiological solutions.

Acknowledgments

We would like to thank European Union for the grant, ECEnzymes, the BBSRC and Abgene, and Dr R. M. Langford.

References

1 J. Stenesh, *Dictionary of Biochemistry*, John Wiley & Sons, New York, **1975**.
2 J.J. Davis, H.A.O. Hill, A.M. Bond, *Coord. Chem. Rev.*, **2000**, 200(202), 411–442.
3 L. Andolfi, B. Bonanni, G.W. Canters, M.P. Verbeet, S. Cannistraro, *Surf. Sci.*, **2003**, 530, 181–194.
4 J.J. Davis, C.M. Halliwell, H.A.O. Hill, G.W. Canters, M.C.v. Amsterdam, M.P. Verbeet, *New J. Chem.*, **1998**, 22, 1119–1123.
5 K.K.-W. Lo, L.-L. Wong, H.A.O. Hill, *FEBS Lett.*, **1999**, 451, 342–346.
6 J.J. Davis, D. Djuricic, K.K.W. Lo, E.N.K. Wallace, L.-L. Wong, H.A.O. Hill, *Faraday Discuss.*, **2000**, 116, 15–22.
7 J.W. Choi, S.-J. Park, Y.-S. Nam, H.W. Lee, M. Fujihira, *Collids Surf. B*, **2002**, 23, 295–303.
8 Q. Chi, J. Zhang, J.E.T. Andersen, J. Ulstrup, *J. Phys. Chem. B*, **2001**, 105, 4669–4679.
9 L.-H. Guo, H.A.O. Hill, *Adv. Inorg. Chem.*, **1991**, 36, 341–375.
10 I. Willner, E. Katz, *Angew. Chem., Int. Ed.*, **2000**, 39, 1180–1218.
11 I. Willner, E. Katz, B. Willner, *Electroanalysis*, **1997**, 9, 965–977.
12 I. Willner, B. Willner, *Trends Biotechnol.*, **2001**, 19, 222–230.
13 P. Morales, A. Pavone, M. Sperandei, G. Leter, L. Mosiello, L. Nencini, *Biosens. Bioelectron.*, **1995**, 10, 847–852.
14 A. Gerardino, A. Notargiacomo, P. Morales, *Microelectron. Eng.*, **2003**, 67(68), 923–929.
15 E.P. Friis, J.E.T. Andersen, L.L. Madsen, P. Moller, R.J. Nichols, K.G. Olesen, J. Ulstrup, *Electrochim. Acta*, **1998**, 43, 2889–2897.
16 E.P. Friis, J.E.T. Andersen, Y.I. Kharkats, A.M. Kuznetsov, R.J. Nichols, J.-D. Zhang, J. Ulstrup, *Proc. Natl. Acad. Sci. U.S.A.*, **1999**, 96, 1379–1384.
17 Q. Chi, J. Zhang, J.U. Nielsen, E.P. Friis, I. Chorkendorff, G.W. Canters, J.E.T. Andersen, J. Ulstrup, *J. Am. Chem. Soc.*, **2000**, 122, 4047–4055.
18 O. Cavalleri, C. Natale, M.E. Stroppolo, A. Relini, E. Cosulich, S. Thea, M. Novi, A. Gliozzi, *Phys. Chem. Chem. Phys.*, **2000**, 2, 4630–4635.
19 P. Facci, D. Alliata, S. Cannistraro, *Ultramicroscopy*, **2001**, 89, 291–298.
20 G.J. Legget, C.J. Roberts, P.M. Williams, M.C. Davies, D.E. Jackson, S.J.B. Tendler, *Langmuir*, **1993**, 9, 2356–2362.
21 N. Patel, M.C. Davies, R.J. Heaton, C.J. Roberts, S.J.B. Tendler, P.M. Williams, *Appl. Phys. A*, **1998**, 66, S569–S574.
22 J. Zhang, Q. Chi, B. Zhang, S. Dong, E. Wang, *Electroanalysis*, **1998**, 10, 738–746.
23 D. Djuricic, H.A.O. Hill, K.K.-W. Lo, L.-L. Wong, *J. Inorg. Biochem.*, **2002**, 88, 362–367.
24 B. Bonanni, D. Alliata, A.R. Bizzarri, S. Cannistraro, *Chem. Phys. Chem.*, **2003**, 5, 1183–1188.
25 J.J. Davis, H.A.O. Hill, *Chem. Commun.*, **2002**, 393–401.
26 D. Losic, J.G. Shapter, J.J. Gooding, *Langmuir*, **2002**, 18, 5422–5428.
27 S.A. Contera, H. Iwasaki, I. Suzuki, *Ultramicroscopy*, **2003**, 97, 65–72.
28 J. Brask, H. Wackerbarth, K.J. Jensen, J. Zhang, I. Chorkendorff, J. Ulstrup, *J. Am. Chem. Soc.*, **2003**, 125, 94–104.
29 S.M. Lindsay, T.W. Jing, J. Pan, A. Vaught, D. Rekesh, *Nanobiology* **1994**, 3, 17–27.
30 E.P. Friis, Y.I. Kharkats, A.M. Kuznetsov, J. Ulstrup, *J. Phys. Chem. A*, **1998**, 102, 7851–7859.
31 T.W. Kelley, E.L. Granstrom, C.D. Frisbie, *Adv. Mater.*, **1999**, 11, 261–264.
32 S. Weiss, *Science*, **1999**, 283, 1676–1683.

33 P.P. Pompa, L. Blasi, R. Longo, R. Cingolani, G. Ciccarella, G. Vasapollo, R. Rinaldi, A. Rizzello, C. Storelli, M. Maffia, *Phys. Rev. E*, **2003**, *67*, 1–8.

34 J.-H. Lim, D.S. Ginger, K.-B. Lee, J. Heo, J.-M. Nam, C.A. Mirkin, *Angew. Chem., Int. Ed.*, **2003**, *42*, 2309–2312.

35 G. Agarwal, R.R. Naik, M.O. Stone, *J. Am. Chem. Soc.* **2003**, *125*, 7408–7412.

36 P. Schwille, U. Kettling, *Curr. Opin. Biotechnol.*, **2001**, *12*, 382–386.

37 J.J. Davis, D. Bruce, G.W. Canters, J. Crozier, H.A.O. Hill, *Chem. Commun.*, **2003**, *5*, 576–577.

38 A. Alessandrini, M. Gerunda, P. Facci, B. Schnyder, R. Kotz, *Surf. Sci.*, **2003**, *542*, 64–71.

39 S.Y. Oh, J.-K. Park, J.W. Choi, C.-M. Chung, *Mol. Cryst. Liq. Cryst.*, **2002**, *377*, 241–244.

40 S.Y. Oh, H.S. Choi, H.S. Jie, J.K. Park, *Mater. Sci. Eng.*, **2004**, *C24*, 91–94.

41 M. Veiseh, M.H. Zareie, M. Zhang, *Langmuir*, **2002**, *18*, 6671–6678.

42 S.K. Kwak, G.S. Lee, D.J. Ahn, J.W. Choi, *Mater. Sci. Eng.*, **2004**, *C24*, 151–155.

43 M.-Y. Hong, D. Lee, H.C. Yoon, H.-S. Kim, *Bull. Korean Chem. Soc.*, **2003**, *24*, 1197–1202.

44 P. Gould, *Mater. Today*, **2003**, *6*(5), 34–39.

45 C.B. Gorman, R.L. Carroll, Y. He, F. Tian, R. Fuierer, *Langmuir*, **2000**, *16*, 6312–6316.

46 U. Kleineberg, A. Brechling, M. Sundermann, U. Heinzmann, *Adv. Funct. Mater.*, **2001**, *11*, 208–212.

47 J. Hartwich, M. Sundermann, U. Kleineberg, U. Heinzmann, *Appl. Surf. Sci.*, **1999**, *144*(145), 538–542.

48 H. Sugimura, N. Nakagiri, *J. Am. Chem. Soc.*, **1997**, *119*, 9226–9229.

49 S. Lee, J. Kim, W.S. Shin, H.-J. Lee, S. Koo, H. Lee, *Mater. Sci. Eng.*, **2004**, *C24*, 3–9.

50 R.D. Piner, J. Zhu, F. Xu, S. Hong, C.A. Mirkin, *Science*, **1999**, *283*, 661–663.

51 K.-B. Lee, S.-J. Park, C.A. Mirkin, J.C. Smith, M. Mrksich, *Science*, **2002**, *295*, 1702–1705.

52 J. Hyun, S.J. Ahn, W.K. Lee, A. Chikoti, S. Zauscher, *Nano Lett.*, **2002**, *2*, 1203–1207.

53 K. Wadu-Mesthrige, S. Xu, N.A. Amro, G.-Y. Liu, *Langmuir*, **1999**, *15*, 8580–8583.

54 P.T. Hurley, A.E. Ribbe, J.M. Buriak, *J. Am. Chem. Soc.*, **2003**, *125*, 11334–11339.

55 E. Pavlovic, S. Oscarsson, A.P. Quist, *Nano Lett.*, **2003**, *3*, 779–781.

56 W. McGee, D. Djuricic, K. Lorimer, L.-L. Wong, H.A.O. Hill, *J. Electroanal. Chem.*, **2002**, *538*(539), 261–265.

57 E.P. Friis, J.E.T. Andersen, L.L. Madsen, P. Moller, M.H. Thuesen, N.H. Andersen, J. Ulstrup, *Comprehensive Chemistry Kinetics*, Elsevier, Amsterdam, **1999**, pp. 133–160.

58 D. Porath, A. Bezryadin, S. de Vries, C. Dekker, *Nature*, **2000**, *403*, 635–638.

59 I. Lee, J.W. Lee, R.J. Warmack, D.P. Allison, E. Greenbaum, *Proc. Natl. Acad. Sci. U.S.A.*, **1995**, *92*, 1965–1969.

60 I. Lee, J.W. Lee, E. Greenbaum, *Phys. Rev. Lett.*, **1997**, *79* 3294–3297.

61 G.B. Khomutov, L.V. Belovolova, S.P. Gubin, V.V. Khanin, A.Y. Obydenov, A.N. Sergeev-Cherenkov, E.S. Soldatov, A.S. Trifonov, *Bioelectrochemistry*, **2002**, *55*, 177–181.

62 Y.S. Nam, K.-W. Park, W.H. Lee, J.-W. Choi, *Mater. Sci. Eng.*, **2004**, *C24*, 95–98.

63 R. Rinaldi, A. Biasco, G. Maruccio, V. Arima, P. Visconti, R. Cingolani, P. Facci, F. De Rienzo, R. Di Felice, A. Molinari, M.P. Verbeet, G.W. Canters, *Appl. Phys. Lett.*, **2003**, *82*, 472–474.

64 P. Visconti, G. Maruccio, E. D'Amore, A. Della Torre, A. Bramanti, R. Cingolani, R. Rinaldi, *Mater. Sci. Eng.*, **2003**, *C23*, 889–892.

65 H.-W. Fink, C. Schonenberger, *Nature*, **1999**, *398*, 407–410.

66 R. Rinaldi, E. Branca, R. Cingolani, R. Di Felice, A. Calzolari, E. Molinari, S. Masiero, G. Spada, G. Gottarelli, A. Garbesi, *Ann. N.Y. Acad. Sci.*, **2002**, *960*, 184–192.

7
Interfacing Biological Molecules with Group IV Semiconductors for Bioelectronic Sensing

Robert J. Hamers

7.1
Introduction

Recent advances in areas such as molecular electronics and biological sensing have placed increased emphasis on understanding the electrical properties of organic and biological molecules at surfaces [1–5]. Gold [6–12] and silicon [13–17] have been especially widely used as model systems for understanding structural and electrical properties of monolayer systems. Silicon has been studied in depth because of the possibility of fabricating field-effect devices, analogous to field-effect transistors but with chemical and/or biochemical sensitivity [18, 19]. Field-effect devices are also particularly attractive for converting biological information directly into electrical signals because these devices are essentially sensitive to the presence of electrical charge. Since the vast majority of biomolecules are charged at physiological pH values, this means that the use of the field effect can be a very general method for detecting a wide variety of biomolecules.

The use of silicon is especially attractive because it can be easily made atomically flat and homogeneous and because of the possibility of integrating the sensing element with conventional silicon electronics for amplification and signal-processing electronics to fabricate completely integrated bioelectronic sensing systems. Consequently, successful implementation of silicon as a basis for bioelectronic sensing system places very high demands on the chemical modification process, in order to have the needed stability of the interface. Field-effect biomolecular devices on silicon have typically been fabricated by chemically modifying the SiO_2 insulator that is typically present at the gate electrode [18–21].

However, the widespread use of silicon as a basis for bioelectronic devices has been plagued by the fact that silicon surfaces oxidize in aqueous environments, and the resulting SiO_2 layers are sufficiently soluble so that when silicon surfaces modified with biological molecules are exposed to aqueous media, the resulting layers degrade because of hydrolysis of the interfacial bonds.

Here, I will review some of the approaches to the integration of biological molecules with silicon and diamond and demonstrate the use of electrical impedance spectroscopy to characterize the electrical properties of the interfaces.

Bioelectronics. Edited by Itamar Willner and Eugenii Katz
Copyright © 2005 WILEY-VCH Verlag GmbH & Co. KGaA, Weinheim
ISBN: 3-527-30690-0

7.2
Semiconductor Substrates for Bioelectronics

7.2.1
Silicon

The use of semiconductors such as silicon and/or diamond for bioelectronics provides the opportunity to integrate the sensing with signal transduction via field-effect bioelectronic devices. Each of these materials has its own advantages and disadvantages.

Silicon is very attractive as a substrate for bioelectronic sensing not only because of its obvious semiconducting properties but also because it can be obtained in single-crystal form in extraordinarily high purity and flatness. Single-crystal silicon is readily available in a number of different crystallographic orientations, but the most common are the (001) orientation (sometimes also written as the (100) orientation), and the (111) orientation. Most microelectronic devices such as microprocessors and memory chips use silicon with a (001) crystallographic orientation. This particular crystal face is used because the interatomic spacing of silicon atoms of this crystal plane is a good match to the lattice constant of silicon dioxide, SiO_2 [22]. Hence, silicon dioxide

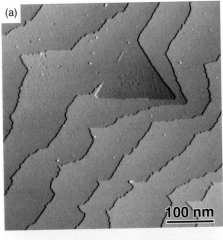

Fig. 7.1 (a) Atomic force microscope image of a Si(111) surface after etching in a 40% NH_4F solution. The steps arise from a 0.35 degree miscut. From Ref. [23]. (b) Scanning electron microscope image of a nanocrystalline diamond surface after cleaning in a hydrogen plasma and functionalization with DNA [25].

films grown on the Si(001) surfaces have a very low concentration of defects at the Si–SiO$_2$ interface. This low density of defect sites is important for fabrication of high-performance electronic devices that use the oxide as an electrical insulator. However, for bioelectronics the use of the Si(111) surface has one very important advantage, which is that it is possible to prepare the (111) surface atomically flat via wet-chemical processing alone.

This ability arises because of wet-chemical etching preferentially etching particular crystal planes and atomic step configurations [23]. When etched in concentrated NH$_4$F solution, Si(111) surface that is initially rough will spontaneously smoothen as the sample is etched; however, to achieve this result it is important to eliminate dissolved oxygen [23, 24]. With a moderate amount of care, it is possible to prepare Si(111) surfaces that are atomically flat except for a small number of atomic steps that arise from the fact that the samples are often not cut precisely along the crystallographic plane. Figure 7.1 shows an atomic force microscope image of such a surface; the image shows atomically flat terraces that are separated by atomic steps that are only a single atom high [23].

The high reactivity of silicon is largely associated with the disruption of the chemical bonding at the interface. The etching procedure described above leaves all silicon atoms with nearly ideal coordination by terminating the "dangling bonds" with hydrogen atoms [26]. The resulting hydrogen-terminated surfaces are sufficiently stable that they can be handled in air for short periods of time (approximately 30 min). Thus, the use of the Si(111) surface as a starting point for biological modification brings with it an extraordinarily high degree of crystalline perfection that, in principle, should translate to improved homogeneity of biomolecular structures constructed on this surface.

7.2.2 Diamond

As will be discussed in more detail below, silicon surfaces are subject to degradation in aqueous solutions because of hydrolysis of the interfacial bonds. Also, silicon surfaces undergo facile electrochemical oxidation and reduction reactions, limiting the utility of silicon in some electrical measurements. In contrast, diamond is one of the most stable materials known. In addition to having good electrical [27] and chemical properties [28, 29], it is biocompatible [30–32], and can be deposited as a robust thin film on silicon and other substrates [33]. Its large band gap of 5.5 eV also makes it transparent in visible light.

Diamond deposition occurs under highly nonequilibrium conditions. Since graphite is the thermodynamically stable form of carbon at all reasonably accessible conditions, the ability to grow diamond films relies on controlling the kinetics of growth to favor the diamond allotrope. Because diamond crystals act as a good template for further diamond growth, growth of thin films is often initiated by embedding a high density of nanometer-sized diamond crystallites in the underlying substrate. Growth times are typically a few hours for growing high-quality films of ~1-μ thickness. Faster growth rates are possible, but result in increased incorporation of graphite into the lattice. For use in biological applications, pinholes (discontinuities in the film) must be strictly avoided, because the chemicals involved in the chemical modification process are sometimes fairly harsh. While these reagents typically do not harm the diamond film, leakage through any pinholes in the film can initiate an attack of the underlying substrate and delaminate the diamond film. With care, high-quality films as thin as ~0.5-μ thick can be made continuous. Films with slightly higher thickness, on the order of 1 to 2 μ, are commonly free from pinholes.

The highly nonequilibrium growth process of diamond makes the structure and electrical properties more complicated and potentially dependent on all the variables associated with the nucleation and growth. As with any semiconductor, the doping is extremely important. While most diamond samples are p-type due to the introduction of boron during growth [33], it is also possible to prepare n-type diamond samples with nitrogen as a dopant using substantially different growth conditions [34]. As will be discussed below, our results show that both p-type and n-type diamond samples in a variety of crystallite sizes [25, 35–37] and even diamond single crystals [38] exhibit similar chemical properties, despite differing substantially in their electrical properties. Figure 7.2 shows a scanning electron microscope image of a diamond thin film used in our studies [25]. The sample shows small crystallites with well-defined crystal planes; while these crystallites are smooth on atomic length scales, the surfaces exhibit significant roughness on longer length scales.

Like silicon, diamond surfaces will undergo partial oxidation when exposed to the environment. Additionally, the growth of diamond is also accompanied by small amounts of graphite. To prepare a clean diamond surface, it is necessary to remove the oxidized carbon atoms and to etch away the graphite. Both of these can be accomplished by exposing the sample to a hydrogen plasma [39]. The plasma etches both diamond and graphite, but because the etch rate for graphite is much faster than that of diamond, the hydrogen plasma eliminates virtually all graphite from the surface. The net result is a diamond sample with all exposed "dangling bonds" terminated with hydrogen atoms.

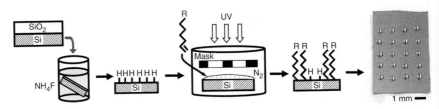

Fig. 7.2 Schematic illustration depicting photochemical modification of silicon surfaces. The oxidized surface is H-terminated in NH_4F solution; and then covered with a reactive alkene in a N_2-purged cell. Illumination with ultraviolet light links molecules to the surface. If illuminated through a mask, then the sample can be patterned with specific chemical groups. The photograph at right shows a Si sample that was functionalized to produce a square array of small dots that were functionalized with a hydrophilic molecule, and the surrounding regions were then modified with a hydrophobic molecule (dodecene). Water beads up in the hydrophilic regions but washes away from the hydrophobic regions, making the pattern clearly visible to the eye.

7.3
Chemical Functionalization

7.3.1
Covalent Attachment of Biomolecules to Silicon Surfaces

Upon exposure to ambient environments, silicon surfaces will eventually form a thin "native" oxide, approximately 1 nm in thickness. Past studies have typically used this oxide as an anchor point for subsequent covalent attachment of biomolecules. In these approaches, oxidized silicon samples are typically exposed to alkoxysilanes consisting of a long-chain hydrocarbon terminating at one end with a reactive group of interest (often an amine) and terminating at the other end with a group such as $-Si(OCH_3)_3$ that is reactive toward silicon dioxide [20]. Upon heating, the silane groups react with the SiO_2 surface and cross-link together. While silane chemistry has been widely used, it suffers from several disadvantages, especially poor reproducibility. This poor reproducibility stems from a combination of two factors. First is the fact that SiO_2 surfaces are terminated with both Si-O-Si and by Si-OH (silanol) species, in relative amounts that are dependent on the past history of the surface. Second, the cross-linking polymerization process is difficult to precisely control. Similarly, the degradation of chemically modified surfaces involving Si-O linkages, especially under basic conditions, is well known [40, 41]. Since oxidation of silicon is catalyzed by amines [42], both crystalline silicon and SiO_2 surfaces are susceptible to amine-induced degradation in the presence of water [40, 41, 43].

Because of the difficulties associated with silane and siloxane chemistry, a number of schemes have been recently developed for chemically modifying surfaces of silicon starting with the H-terminated surfaces, and fabricating direct Si-C linkages without intervening oxides [13, 14, 44–48]. Modification of the surfaces relies principally on the reaction of organic molecules containing one or more vinyl (alkene) groups, C=C, with the H-terminated surfaces. This process must be activated in some manner, to remove H from some of the surface sites to create reactive "dangling bonds" that will, in turn, react with the vinyl groups. This reaction can be initiated in several ways. Thermal activation was used in some early studies [14, 44–46, 48]. The principal disadvantage of the thermal method is that in order to proceed at an appreciable rate, rather higher temperatures and long reaction times are typically needed. To prevent evaporation of the reactant, it is necessary to perform the reaction under reflux and to continually exclude air. A second method is to initiate the reaction by use of some type of radical scavenger – a molecule that will pluck an H atom from the surface, leading to a reactive dangling bond. Diacyl peroxides [14] and atomic chlorine [47] (produced by illumination of Cl_2 gas) have been used to initiate the reactions.

More recently, several groups have shown that the reactions can be photo-initiated [13, 49–51]. Figure 7.3 outlines the procedure followed while Figure 7.3 shows the chemistry more completely. In this method, the H-terminated surface is covered with a thin film of the reactant, covered with quartz window, and then illuminated with a small ultraviolet light source, typically a low-pressure mercury lamp emitting at a wavelength of 254 nm. It is critical to eliminate all water during this process, and oxygen must be excluded by

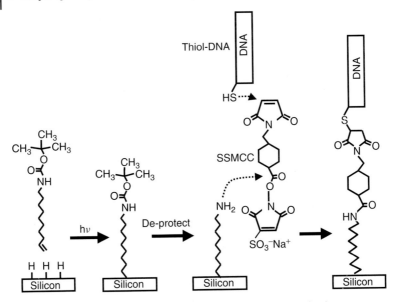

Fig. 7.3 Schematic illustration of the chemistry associated with covalently linking DNA to silicon surfaces.

flushing with dry nitrogen or argon while under UV illumination. The ultraviolet light initiates the removal of hydrogen and the attachment of the molecule to the surface through the vinyl (C=C) group. Using 254-nm light with an intensity of 0.35 mW cm^{-2}, the reaction is complete within approximately 2 h. One advantage in using this method is that by illuminating through a mask, it is possible to directly photopattern specific chemical functional groups onto the surface. For example, Figure 7.2(b) shows a silicon sample that was modified at selected locations with an amine and at all other locations with a hydrophobic molecule. When the sample is immersed in water, water is retained at the hydrophilic locations but not at the hydrophobic locations, making the pattern visible to the eye.

Although the process of chemically modifying silicon is simple in principle, in practice there are several other factors that must be taken into consideration. The important guiding principle is that because SiO_2 has a finite solubility in water, the stability of all modified silicon surfaces depends principally on the exclusion of water from the silicon-organic interface. If water gets to the interface, it hydrolyzes the Si–C bonds, and the layers become removed from the surface just as they will on an SiO_2 surface. The primary role of the alkyl chains at the interface is to provide a hydrophobic environment that is not readily penetrated by water molecules [17]. Consequently, in order to achieve good stability in aqueous solutions, it is important to have the densest possible packing of molecular layers.

The usual method for linking biomolecules to the surface is by first functionalizing the surface with specific reactive groups, such as amines, hydroxyls, carboxylic acids, or aldehydes, and to link these groups to thiol-terminated DNA by using a heterobifunctional cross-linker. The challenge here is that these chemical groups will also react directly with the H-terminated silicon surface. Hence, to get good reactivity, it is necessary to use molecules in which the reactive groups are modified with specialized "protecting groups", linked to the surface via a terminal vinyl group, and then deprotected to restore the reactive group [35, 48–50]. The choice of protecting group is critical to the final stability of the modified surfaces. While surfaces modified with simple alkyl groups are quite stable, [14, 46] surfaces modified with alkyl chains bearing reactive groups at their terminus are less stable [17]. This decrease in stability presumably arises from disruption of the packing and the resulting enhancement of hydrolysis of the Si–C bonds, releasing the molecular layers into the aqueous phase.

Because the stability of the surface relies on the close packing of the alkyl chains, it is very important to choose a protecting group that is small so that it does not inhibit this packing. For example, in organic chemistry the trityl (triphenylmethyl) group is often used as a protecting group; however, the three benzene rings inhibit the packing and produce less dense molecular layers [48]. For amines, we have found the trifluoroacetamide (TFA) group and the tert-butyloxycarbamate (t-BOC) groups to work well [50, 52], and for carboxylic acids we have found that the use of ethyl or methyl esters leads to good-quality monolayers that are effective attachment points for DNA.

A second approach in stabilizing the functionalized semiconductor surfaces is to use mixed layers, in which the molecules bearing both vinyl and protected amine groups are mixed in with other suitable molecules (such as dodecene) that have better packing. Mixing in even relatively small amounts of molecules such as dodecene can significantly improve the quality of the modified surfaces. Presumably, these smaller molecules can efficiently pack between the large ones, thereby facilitating the close packing of the alkyl chains.

Once the amine-modified surfaces are prepared, the remaining chemistry for attachment of DNA and other biomolecules is fairly straightforward. For DNA, it is usually most convenient to use oligonucleotides modified with a thiol group (usually at the 5′ end) and to use a heterobifunctional cross-linker such as SSMCC (depicted below) to link the amine-modified surface to the thiol-modified DNA. The net result is a DNA-modified surface with completely covalent linkages.

7.3.2
Hybridization of DNA at DNA-modified Silicon Surfaces

To test the biomolecular recognition properties of DNA-modified silicon surfaces, we tethered 31-mer oligonucleotides to silicon surfaces and then explored their selectivity for hybridization with complementary versus noncomplementary sequences. While a number of different sequences have been investigated, Figure 7.4 shows fluorescence data for surfaces that were covalently linked to several different sequences. The top panels show a Si(111) sample that was covalently linked to DNA oligonucleotides with the sequences 5′-HS-C_6-T_{15}AA CGA TCG AGC TGC AA-3′ (S1) in a spot in the upper half of the sample, and with 5′-HSC6-T_{15}AA CGA TGC AGG AGC AA-3′

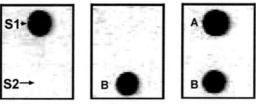

F1→Denature→F2→Denature→F1 + F2

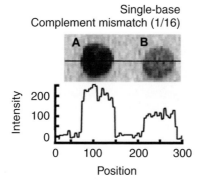

Fig. 7.4 Fluorescence images (black = high intensity) of the hybridization of fluorescently labeled DNA oligonucleotides with DN molecules covalently linked to silicon surfaces. The top panel shows results from perfect matches and four-base mismatches. The bottom panel compares a perfect match with a single-base mismatch. These data were obtained on a Si(001) surface [43].

(S2) on the lower half. Complementary oligonucleotides used for hybridization were modified with fluorescein on the 5′ end using 6-FAM phosphoramidite (Perkin-Elmer Biosystems). The three sequences employed were 5′-FAM-TT GCA GCT CGA TCG TT-3′ (F1, matches with S1), 5′-FAM-TT GCT CCT GCA TCG TT-3′ (F2, matches with S2). The fluorescence images show a high degree of specificity, with no significant nonspecific binding.

While the upper panels show that there is little binding between oligonucleotides with four mismatched bases, the bottom panel shows results for a single-base mismatch. In this case, the surface was modified with sequences 5′-HSC6-T_{15}AG GAA TGC CGG TTA T-3′ (S3) and 5′-HSC6-T_{15}AG GAT TGC CGG TTA T-3′ (S4). The fluorescence image shows the interaction with an oligonucleotide with the sequence 5′-FAM-AT AAC CGG CAT TCC T-3′(F3, matches with S3, single-base mismatches with S4). While the single-base mismatch shows some fluorescence, it is only ∼40% that of the perfect match. The important result from these studies was to show that covalent attachment of DNA to silicon surfaces leads to surfaces with good selectivity.

For many applications, the stability of the surfaces is also extremely important. While there is no generally accepted method for evaluating stability, one method

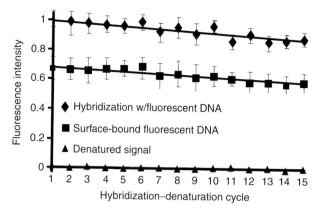

Fig. 7.5 Stability of DNA-modified silicon surfaces to repeated cycles of hybridization and denaturation. The plot shows the intensity of fluorescence observed when DNA-modified silicon surfaces were put through repeated cycles of exposure to their fluorescently labeled complements followed by denaturation in 8.3 M urea. Also shown is fluorescence data from fluorescently labeled DNA molecules that were directly linked to the surface [49].

is to investigate how well the surface responds to successive cycles of hybridization and denaturation. Figure 7.5 shows results from a Si(111) surface that was modified with DNA and then placed through repeated cycles of hybridization with fluorescently labeled, complementary DNA and denaturation in 8.3 M urea. Also shown is data from fluorescently modified DNA molecules that were directly linked to the surface. In both cases, the data shows that the fluorescence signal decreases with time, indicating some degradation of the interface.

7.3.3
Covalent Attachment and Hybridization of DNA at Diamond Surfaces

Chemical modification of diamond can be achieved in a manner similar to that of silicon. Although the wavelength of the 254-nm mercury line is below the band gap of diamond, diamond samples do exhibit significant absorption for lower-energy photons because of defects and impurities in the bulk and also because of localized electronics state on the surface atoms. The photochemical reaction efficiency on H-terminated diamond is approximately 5 to 6 times lower than that observed on H-terminated silicon under identical conditions [35–37, 53]. However, studies on p-type and n-type nanocrystalline diamond films, p-type microcrystalline diamond films, and single-crystal diamond (111) samples all show similar reaction efficiency [38].

Figure 7.6 shows the best pathway we have found for covalently linking DNA to diamond thin films [36]. DNA hybridization at modified diamond surfaces was investigated using two sequences: 5′HS-C_6H_{12}-T_{15}-GC TTA TCG AGC TTT CG3′ (**S1**) and 5′HS-C_6H_{12}-T_{15}-GC TTA AGG AGC AAT CG3′ (**S2**) and the complements

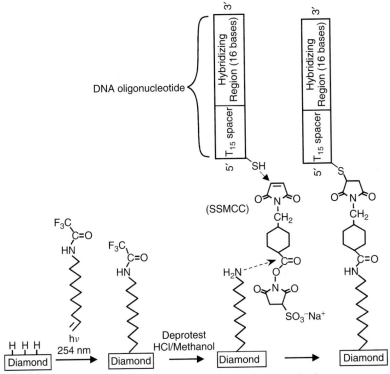

Fig. 7.6 Schematic illustration of the chemistry associated with covalently linking DNA to diamond surfaces. The procedure for linking other amine-bearing molecules such as antibodies is similar.

5′-FAM-CG AAA GCT CGA TAA GC-3′ (**F1**, 16 bases complementary to S1 and with a four-base mismatch to S2) and 5′-FAM-CG ATT GCT CCT TAA GC-3′ (**F2**, 16 bases complementary to S2 and with a four-base mismatch to S1). Fluorescence measurements looking at the interaction of the surface-bound S1 and S2 sequences with the solution-phase F1 and F2 sequences showed very high selectivity.

One of the potential advantages of diamond over alternative materials is that it provides extremely good stability. To demonstrate this, we performed stability studies of DNA-modified surfaces of diamond and a number of other materials including silicon, gold, a commercially available amine-terminated glass, and two types of glassy carbon [36]. In each case, the surfaces were terminated with amine groups, were reacted with SSMCC, and were then reacted with thiol-terminated DNA. Thus, each sample involves nearly the same covalent chemistry.

Each DNA-modified sample was then subjected to 30 successive cycles in which it was hybridized as above, the fluorescence image was obtained, the sample was denatured in 8.3 M urea, rinsed, and then the fluorescence image was measured again. Figure 7.7 shows a subset of this data, depicting the results for silicon, diamond, and for a gold sample functionalized with DNA using a self-assembled monolayer of

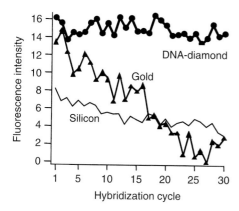

Fig. 7.7 Stability of DNA-modified surfaces to repeated cycles of hybridization and denaturation.

an amine-terminated thiol to provide a monolayer of amine groups, and then using chemistry identical to that used for silicon and diamond. These repetitive measurements show that DNA-modified diamond surfaces exhibit no measurable decrease in signal intensity even after 30 cycles. DNA-modified silicon surfaces show a small but steady decrease, consistent with previous studies [43, 49, 50]. The data show that the surface modified by gold-thiol chemistry degraded rapidly; this rapid degradation is due in part to the use of surfactants to minimize nonspecific binding, but is typical of DNA-modified gold surfaces used in biological buffer solutions.

Since all of the DNA-modified surfaces were prepared with nearly identical chemistry, the differences in stability can be ascribed to differences in the intrinsic stability of the starting surfaces to degradation. For example, it is widely recognized that gold–thiol bonds are very susceptible to oxidation, leading to solubilization of the attached layers [54]. In contrast, the DNA-modified diamond surfaces clearly exhibit extremely good stability. Moreover, the improved stability afforded by the diamond substrate is accomplished without loss of selectivity: the selectivity of binding to perfectly matched and mismatched sequences was also investigated at the end of the 30-cycle stability test and showed no decrease in intensity from a complementary sequence and no increased background after exposure to a noncomplementary sequence.

The stability of diamond persists even at elevated temperatures so that the modified surfaces can be kept near the melting temperature of the DNA duplex for extended periods of time with virtually no degradation [55].

7.4 Electrical Characterization of DNA-modified Surfaces

7.4.1 Silicon

While semiconductors have the attractive features of being atomically flat (silicon) and are ultrastable (diamond), the potential of using their semiconducting electrical properties makes them especially interesting. We have conducted several investigations aimed at understanding the interfacial electrical properties of biologically modified surfaces of silicon and diamond [25, 52, 53].

While for biomolecules such as DNA there are significant changes in the electrical impedance of the biomolecules themselves, the more general class of biomolecules is not necessarily conductive. Nevertheless, the impact of biomolecular binding on the interfacial conductivity can still be measured by applying a small AC modulation to the potential and measuring the amplitude and phase of the current conducted across the interface [19, 21].

Most previous studies have characterized the response at a fixed frequency and represented the data as an equivalent capacitance of the interface [19, 21, 56]. However, the complex nature of the semiconductor–biomolecular interface leads to a very frequency-dependent variation in the sensitivity to biomolecular binding [25, 52, 53, 57, 58]. By characterizing the electrical response as a function of frequency, it is possible to extract much more information about the physical nature of the interface and its bioelectronic response and to use this information to optimize the sensitivity for detection.

To extract the frequency-dependent information, the most common method is electrochemical impedance spectroscopy (EIS). In EIS, a small, sinusoidally varying potential (with a possible additional constant value), of the form $V = V_0 + V_{mod}\cos(2\pi ft)$, is applied across the interface, and the in-phase and out-of-phase components of the current at the modulation frequency are measured [1, 59]. The results are typically represented as the complex impedance $Z = Z' + jZ''$, where $j = \sqrt{-1}$. EIS has most commonly been performed on metallic substrates under Faradaic conditions, where a relatively large potential is applied and the impedance associated with various electron-transfer steps is probed [1]. However, one potential drawback to the use of any Faradaic process is that the applied potentials can alter the biological binding processes; for example, potentials of approximately 0.5 V can denature DNA hybridized at a surface [60]. Consequently, we typically perform measurements at or near the open-circuit potential, using a very small potential (typically 5–10 mV, although sometimes larger). Under these conditions, the electrical measurement process does not significantly perturb the biological interface. To understand the physical transduction process, however, it is necessary to also perform measurements as a function of potential, which can be very useful in identifying whether the semiconductor, the molecular layers, or the solution is controlling the overall impedance [25].

Electrochemical measurements were performed in a thin three-electrode electrochemical flow cell. In this cell, the DNA-modified diamond sample acts as a working electrode and a Pt foil acts as a counter electrode. These planar electrodes are separated by a thin sheet of polydimethylsiloxane (PDMS) with a 3 × 1.5 mm opening that forms a fluid cavity with a volume of 4.5 µL. The Pt foil (counter electrode) and the DNA-modified diamond (working electrode) press against the PDMS to seal the top and bottom of the cell. A reference electrode (Ag/AgCl) was made using a 25-µm diameter Ag wire coated with AgCl, which was then embedded into the PDMS with approximately 3 mm exposed to the solution. Microfluidic inlet and outlets were also

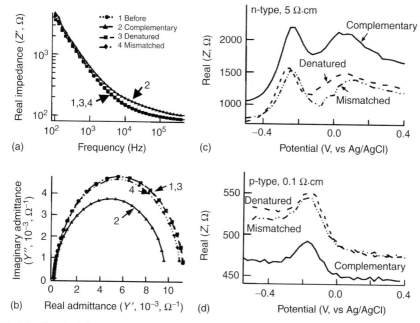

Fig. 7.8 Electrical characterization of DNA-modified silicon surfaces. (a) Impedance spectrum of DNA-modified silicon, showing the electrical response of the starting surface, the same surface after exposure to complementary DNA, the same surface after denaturing, and then after exposure to a four-base mismatched sequence. (b) Cole–Cole plot depicting the real and imaginary admittance as a function of frequency, from the same sample shown in (a). (c) Real impedance as a function of potential on an n-type silicon sample. (d) Real impedance as a function of potential on a p-type silicon sample.

embedded into the PDMS to permit solutions to be flowed continuously through the cell using a syringe pump.

Figure 7.8 shows the electrochemical response of a DNA-modified Si electrode measured at 0 V (vs Ag/AgCl) with a 5-mV root-mean-square sinusoidal modulation [52]. Figure 7.8a shows a logarithmic plot of the real part of the impedance, Z', versus frequency. While the impedance varies over many orders of magnitude, the sensitivity to DNA hybridization is most apparent at frequencies higher than ~1 kHz, and is most clear at frequencies of >10 kHz. As the frequency approaches 100 kHz, the sensitivity is again reduced. The sensitivity to DNA hybridization can be observed more clearly in linear plots. Since the sensitivity to DNA hybridization occurs at high frequencies where the overall impedance is small, the sensitivity is more clearly observed by plotting the admittance instead of the impedance. One common representation is a plot of the real versus imaginary components of the complex admittance, $Y' = \mathrm{Re}(1/Z) = Z'/|Z|^2$, $Y'' = \mathrm{Im}(1/Z) = -Z''/|Z|^2$ as a function of frequency [61, 62]. Such a "Cole–Cole" plot is shown in Figure 7.8b.

The overall impedance of the interface can be controlled by the molecular layers, by the semiconductor space-charge region, and/or by the solution resistance. By analyzing the impedance spectra and using circuit modeling methods, it is possible to extract out the contributions of each component. Because the molecular layers and the electrical double-layer have comparable dimensions and significant finite spatial extent, the use of discrete elements to model impedance spectra must be approached with caution. However, the modeling does give a very good qualitative, if not quantitative, understanding of the electrical response. At low frequencies of <1 kHz, the molecular layer dominates, and there is only a minor change in impedance upon hybridization with complementary DNA. At frequencies higher than ~10^3 Hz, where sensitivity to DNA hybridization is observed, the total impedance is dominated by the silicon space-charge layer. However, as the frequency approaches 100 kHz, the solution resistance becomes significant. Thus, the optimal frequency for detection lies over a range of intermediate frequencies that depend on the properties of the semiconductor and the molecular layer.

Further confirmation that the space-charge region of the silicon controls the impedance can be obtained in several ways. First, our studies show that DNA hybridization causes the impedance to increase on n-type substrates and to decrease on p-type substrates. While the frequency-dependent spectra shown above were obtained at the open-circuit potential, measuring the response at one particular frequency as a function of potential provides complementary information. Figure 7.8c and d shows plots of the measured impedance as a function of potential for n-type and p-type silicon substrates, measured at a frequency of 100 kHz. On n-type substrates, the impedance increases at more positive potentials as the surface is brought into depletion, and on p-type substrates, the impedance is higher at negative potentials. In both cases, a peak is

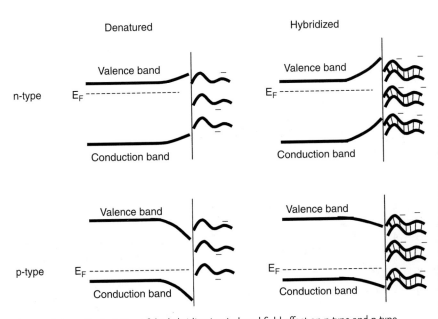

Fig. 7.9 Schematic depiction of the hybridization-induced field effect on n-type and p-type semiconductors.

observed near −0.2 V. Such a peak is commonly attributed to electrically active interface states that tend to lie near the middle of the silicon band gap. Even though the interface is modified by a layer of biological molecules, the behavior as a function of potential is that typically observed for semiconductor-electrolyte interfaces.

Additional measurements on multiple samples and as a function of doping and frequency confirm that the ability to detect DNA hybridization at higher frequencies (typically in the 10 kHz–1 MHz range) can be attributed to a field effect, in which the negative charge on the DNA molecules creates an electric field that penetrates into the silicon space-charge layer. On n-type samples, the valence band conduction bands are typically bent slightly upward near the surface so that at the surface the Fermi level lies closer to the middle of the band gap. In this case, the hybridizing DNA molecules push the bands upward even more, making the surface go further into depletion and increasing the interfacial resistance. Conversely, on p-type samples, the bands are typically bent downward; here, the negative charge on the hybridizing DNA molecules pushes the bands upward, but on n-type material, this makes the silicon come closer to

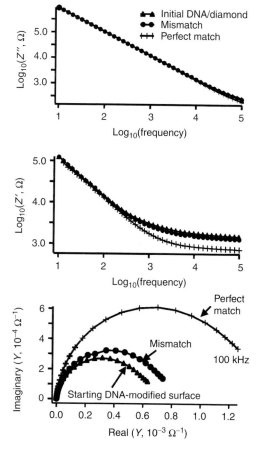

Fig. 7.10 Impedance spectroscopy data for DNA hybridization to p-type nanocrystalline diamond film, showing the impedance spectrum after exposure to a mismatched sequence and after exposure to a perfectly matched (complementary) sequence.

its flat-band condition. Figure 7.9 depicts the bands on n-type and p-type material and how they change in response to hybridization.

There are several important consequences to the above measurements. First, they show that DNA hybridization and other biological processes can induce a significant field effect in n-type and p-type silicon that has been covalently modified without any intervening oxide. Second, the fact that the DNA-induced hybridization changes are only observed over a selected range of frequencies demonstrates that an understanding of the frequency dependence is critical to achieving good sensitivity and understanding the physical basis of the electronic response. Finally, we note that the frequency dependence observed here is very different from that observed on conducting substrates such as gold and/or glassy carbon substrates [1, 63, 64]. On metallic substrates, most studies have found that biomolecular binding influences electron-transfer reactions and diffusion processes in the electrolyte solution. These are most easily observed at very low frequencies, where the impedance is high and is dominated by the overall resistance of the interface. In contrast, our results show that on silicon (and diamond, as described below) the field effect provides increased sensitivity that can be obtained by measuring the response at frequencies where the impedance is dominated by the semiconductor space-charge layer.

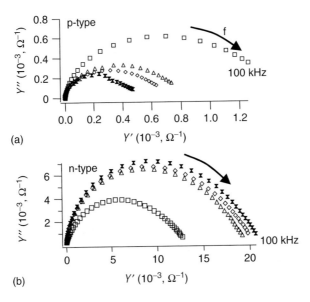

Fig. 7.11 Cole–Cole plots showing the changes in admittance on n-type and p-type samples due to DNA hybridization [25].

7.4.2
Impedance Spectroscopy of DNA-modified Diamond Surfaces

Impedance spectroscopy of DNA-modified diamond films has been performed in a manner identical to that described above for silicon. Figure 7.10 shows impedance data for a p-type nanocrystalline diamond film that was covalently modified with DNA and was then exposed to mismatched and matched sequences of DNA in solution. As in the case of silicon, the data show that hybridization induces a significant decrease in impedance when measured at frequencies greater than ∼1 kHz. At the highest frequency measured (100 kHz), the change in impedance is nearly a factor of 2. Figure 7.11 shows a comparison of results for a p-type nanocrystalline diamond [25] and an n-type ultrananocrystalline diamond film [53]. As with silicon, hybridization induces a decrease in impedance on p-type samples and an increase in impedance on n-type samples. Detailed analysis of the spectra, in conjunction with Mott–Schottky analysis of the spectra as a function of potential, shows that in a manner similar to that observed on silicon, the total impedance of the DNA-modified diamond films is controlled by the molecular layers at low frequencies (below ∼1 kHz). As the frequency increases beyond 1 kHz, the impedance of the system becomes dominated by the diamond space-charge layer, and the sensitivity to hybridization is observed. At the highest frequencies (near 1 MHz, not shown) the solution resistance again becomes important. Thus, sensitivity to hybridization is observed over a limited range of frequencies where the diamond space-charge layer dominates the total impedance. By measuring the impedance as a function of time at a fixed frequency, it is possible to achieve real-time detection of biomolecular binding processes, as depicted in Figure 7.12.

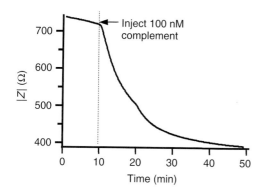

Fig. 7.12 Real-time measurement of DNA hybridization at DNA-modified surface of p-type diamond [25].

7.5
Extension to Antibody–Antigen Detection

Since the "field effect" is associated with the charge on the molecules binding to the surface, it can be used for a wide range of biomolecular binding processes. One example

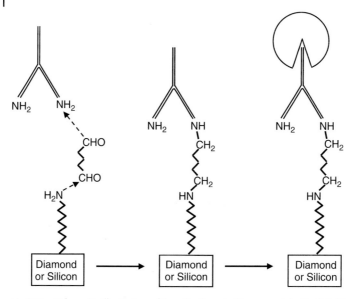

Fig. 7.13 Schematic illustration of the attachment of immunoglobulin G (IgG) to a diamond surface and its interaction with the anti-IgG antibody.

consists of antibody–antigen interactions at surfaces. For example, immunoglobulins often contain primary amine groups that can be used to link the antibodies to silicon or diamond surfaces by reacting the amine-terminated surfaces. We have found that one good way of doing this is to prepare an amine-terminated silicon or diamond surface as described above and then to use glutaraldehyde to react with the amine groups and provide a high density of aldehyde sites, as depicted in Figure 7.13. These can then react with the primary amines of the Fab fragment of the immunoglobulins to link them to the surface, leaving the Fc fragment exposed and relatively unperturbed.

Figure 7.14 shows impedance plots obtained for a diamond thin film that was modified with IgG as described above. The specificity of the binding can be assessed by comparing the ability of the IgG-modified surface to react with different antibodies, such as anti-IgG and anti-IgM. While exposure to anti-IgM elicits no significant response, exposure to anti-IgG produces a significant increase in impedance (decrease in admittance).

As noted earlier, one of the unusual features of electrical detection is that it is possible to directly observe binding events in real time. Figure 7.14(b) shows the results of such a real-time measurement, in which an IgG-modified surface was exposed to anti-IgG. The bottom panel shows measurements of the fluorescence intensity on samples exposed to IgG for varying lengths of time. While both fluorescence and electrical measurements show identical response time, only the electrical measurement is truly a real-time measurement. Thus, the use of impedance spectroscopy, when combined with diamond as a substrate, provides a good method for real-time detection of antibodies.

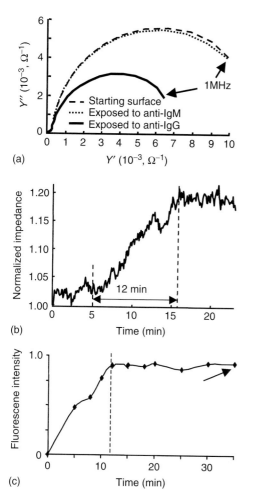

Fig. 7.14 Impedance data showing the use of impedance spectroscopy to detect antibody–antigen binding on p-type diamond. (a) Cole–Cole plot of the real and imaginary parts of the admittance for a p-type diamond surface that was covalently linked to IgG as shown in Figure 7.13, and was then exposed to anti-IgM and to anti-IgG. A large change in admittance is observed at higher frequencies because of anti-IgG, but no change in response is observed in response to anti-IgM, which is not expected to bind to the IgG-modified surface. (b) Real-time measurement showing the fractional change in impedance as a function of time after injection of anti-IgG to an IgG-modified diamond surface. (c) Fluorescence measurements obtained under conditions identical to those shown in (b). The fluorescence data yields the same effective temporal response as the electrical data.

7.6 Summary

The use of semiconductor substrates provides a good method for converting biological information directly into electrical signals, by virtue of the fact that nearly all biomolecules of interest are charged, and this charge can interact with the space-charge region of the semiconductors to induce changes in impedance via a "field effect" that is analogous to that used in field-effect transistors. However, the complexity of the biologically modified surfaces necessitates an understanding of the frequency response in order to understand the bioelectronic signal transduction process and to optimize factors such as sensitivity. Ultimately, the ability to achieve direct electronic sensing in this way hinges directly on the ability to prepare stable, reproducible semiconductor surfaces, modified with biomolecules in a way that preserves the specific biomolecular recognition properties. The use of direct covalent attachment chemistry and the use of ultrastable materials such as diamond thin films provides a pathway for fabrication of integrated bioelectronic signal transduction devices that provide unparalleled sensitivity, selectivity and stability. This stability may be particularly crucial for emerging applications such as antiterrorism and environmental sensing, in which it is desired to achieve continuous, unattended monitoring and integration with silicon-based microelectronics. While silicon itself is a good material, its stability is surpassed by that of diamond. Because diamond can be grown as a thin film on a variety of substrates (including silicon), it can be used as a thin protective coating that provides sensitivity and stability, while still being able to be integrated with silicon-based microelectronics technology.

Acknowledgments

The author wishes to acknowledge the students and postdocs who have performed most of the work described here, including Wei Cai, Tanya Knickerbocker, Tami Lasseter, Zhang Lin, Beth Nichols, Todd Strother, and Wensha Yang, along with a number of senior collaborators, including Lloyd Smith, John Russell, Jim Butler, John Carlisle, and Dieter Gruen.

References

1 E. Katz, I. Willner, *Electroanalysis*, **2003**, 15, 913–947.
2 W.-L. Xing, J. Cheng, Eds., *Biochips: Technology and Applications*, Springer-Verlag, Berlin, Heidelberg, **2003**.
3 A. Brajter-Toth, J.Q. Chamber, Eds., *Electroanalytical Methods for Biological Materials*, Marcel Dekker, New York, **2002**.
4 E. Souteyrand, J.P. Cloarec, J.R. Martin, M. Cabrera, M. Bras, J.P. Chauvet, V. Dugas, F. Bessueille, *Appl. Surf. Sci.*, **2000**, 164, 246–251.
5 J. Wang, *Chem. Eur. J.*, **1999**, 5, 1681–1685.
6 P.D. Laibinis, G.M. Whitesides, D.L. Allara, Y.-T. Tao, A.N. Parikh, R.G.J. Nuzzo, *Am. Chem. Soc.*, **1991**, 113, 7152–7167.
7 H.O. Finklea, D.D.J. Hanshew, *Am. Chem. Soc.*, **1992**, 114, 3173–3181.
8 C. Miller, M.J. Graetzel, *Phys. Chem.*, **1991**, 95, 5225–5233.
9 C. Miller, P. Cuendet, M.J. Graetzel, *Phys. Chem.*, **1991**, 95, 877–886.
10 M.-T. Lee, C.-C. Hsueh, M.S. Freund, G.S. Ferguson, *Langmuir*, **1998**, 14, 6419–6423.
11 N.T. Flynn, T.N.T. Tran, M.J. Cima, R. Langer, *Langmuir*, **2003**, 10909–10915.
12 F. Lucarelli, G. Marrazza, A.P.F. Turner, M. Mascini, *Biosens. Bioelectron.*, **2004**, 19, 515–530.
13 J.M. Buriak, *Chem. Rev.*, **2002**, 102, 1271–1308.
14 M.R. Linford, P. Fenter, P.M. Eisenberger, C.E.D. Chidsey, *J. Am. Chem. Soc.*, **1995**, 117, 3145–3155.
15 L.J. Webb, N.S.J. Lewis, *Phys. Chem. B*, **2003**, 107, 5404–5412.
16 P. Allongue, C.H. de Villeneuve, J. Pinson, *Electrochim. Acta*, **2000**, 45, 3241–3248.
17 C.J. Barrelet, D.B. Robinson, J. Cheng, T.P. Hunt, C.F. Quate,

C.E.D. Chidsey, *Langmuir*, **2001**, *17*, 3460–3465.
18 J. Fritz, E.B. Cooper, S. Gaudet, P.K. Sorger, S.R. Manalis, *Proc. Natl. Acad. Sci.*, **2002**, *99*, 14142–14146.
19 E. Souteyrand, J.P. Cloarec, J.R. Martin, C. Wilson, I. Lawrence, S. Mikkelsen, M.F. Lawrence, *J. Phys. Chem. B*, **1997**, *101*, 2980–2985.
20 L.A. Chrisey, G.U. Lee, E.O. O'Ferrall, *Nucleic Acids Res.*, **1996**, *24*, 3031–3039.
21 E. Souteyrand, J.R. Martin, C. Martelet, *Sens. Actuators, B*, **1994**, *20*, 63–69.
22 C.R. Helms, B.E., Eds., *The Physics and Chemistry of SiO2 and the Si-SiO2 Interface*, Plenum Press, **1989**.
23 Hines, M.A. *Annu. Rev. Phys. Chem.*, **2003**, *54*, 5429–5456.
24 C.P. Wade, C.E.D. Chidsey, *Appl. Phys. Lett.*, **1997**, *71*, 1679–1681.
25 W. Yang, J.E. Butler, J.N. Russell, Jr., R.J. Hamers, *Langmuir*, **2004**; submitted for publication.
26 G.S. Higashi, Y.J. Chabal, G.W. Trucks, K. Raghavachari, *Appl. Phys. Lett.*, **1990**, *56*, 656–658.
27 G.M. Swain, M. Ramesham, *Anal. Chem.*, **1993**, *65*, 345–351.
28 J. Wei, J.T., Yates, *J. Crit. Rev. Surf. Chem.*, **1995**, *5*, 1–71.
29 J. Wei, V.S. Smentkowski, J.T. Yates, Jr., *Crit. Rev. Surf. Chem.*, **1995**, *5*, 73–248.
30 F.Z. Cui, D.J. Li, *Surf. Coat. Technol.*, **2000**, *131*, 481–487.
31 H.J. Mathieu, *Surf. Interface Anal.*, **2001**, *32*, 3–9.
32 L. Tang, C. Tasi, W.W. Gerberich, L. Kruckeberg, D.R. Kanie, *Biomaterials*, **1995**, *16*, 483–488.
33 J.E. Butler, H. Windischmann, *MRS Bull.*, **1998**, *23*, 22–27.
34 M.D. Gruen, *Annu. Rev. Mater. Sci.*, **1999**, *29*, 211–259.
35 T. Strother, T. Knickerbocker, J.N. Russell, Jr., J.E. Butler, L.M. Smith, R.J. Hamers, *Langmuir*, **2002**, *18*, 968–971.
36 W. Yang, J.E. Butler, W. Cai, J. Carlisle, D. Gruen, T. Knickerbocker, J.N. Russell, Jr., L.M. Smith, R.J. Hamers, *Nat. Mater.*, **2002**, *1*, 253–257.
37 T. Knickerbocker, T. Strother, M.P. Schwartz, J.N. Russell, Jr., J.E. Butler, L.E. Smith, R.J. Hamers, *Langmuir*, **2003**, *19*, 1938–1942.

38 B. Nichols, W. Yang, R.J. Hamers, **2003**, in preparation.
39 B.D. Thoms, M.S. Owens, J.E. Butler, C. Spiro, *Appl. Phys. Lett.*, **1994**, *65*, 2957.
40 D.E. Gray, S.C. Case-Green, T.S. Fell, P.J. Dobson, E.M. Southern, *Langmuir*, **1997**, *13*, 2833–2842.
41 R.C. Major, X.-Y. Zhu, *Langmuir*, **2001**, *17*, 5576–5580.
42 R.M. Finne, D.L.J. Klein, *Electrochem. Soc.*, **1967**, *114*, 965–970.
43 Z. Lin, T. Strother, W. Cai, X. Cao, L.M. Smith, R.J. Hamers, *Langmuir*, **2002**, *18*, 788–796.
44 A.B. Sieval, A.L. Demirel, J.W.M. Nissink, M.R. Linford, J.H. van der Maas, W.H. de Jeu, H. Zuilhof, E.R.J. Sudhölter, *Langmuir*, **1998**, *14*, 1759–1768.
45 A.B. Sieval, R. Linke, G. Heij, G. Meijer, H. Zuilhof, E.J.R. Sudholter, *Langmuir*, **2001**, *17*, 7554–7559.
46 M.R. Linford, C.E.D.J. Chidsey, *Am. Chem. Soc.*, **1993**, *115*, 12631–12632.
47 M.R. Linford, C.E.D. Chidsey, *Langmuir*, **2002**, *18*, 6217–6221.
48 A.R. Pike, L.H. Lie, R.A. Eagling, L.C. Ryder, S.N. Patole, B.A. Connolly, B.R. Horrocks, A. Houlton, *Angew. Chem., Int. Ed.*, **2002**, *41*, 615–617.
49 T. Strother, W. Cai, X. Zhao, R.J. Hamers, L.M. Smith, *J. Am. Chem. Soc.*, **2000**, *122*, 1205–1209.
50 T. Strother, R.J. Hamers, L.M. Smith, *Nucleic Acids Res.*, **2000**, *28*, 3535–3541.
51 R.L. Cicero, M.R. Linford, C.E.D. Chidsey, *Langmuir*, **2000**, *16*, 5688–5695.
52 W. Cai, J. Peck, D. van der Weide, R.J. Hamers, *Biosens. Bioelectron.*, **2004**, *19*, 1013–1019.
53 W. Yang, O. Auciello, J.E. Butler, W. Cai, J. Carlisle, J.E. Gerbi, D.M. Gruen, T. Knickerbocker, T.L. Lasseter, J.N. Russell, Jr., L.M. Smith, R.J. Hamers, *Mater. Res. Soc. Symp. Proc.*, **2003**, *737*, F4.4.1–F4.4.6.
54 W.R. Everett, I. Fritschfaules, *Anal. Chim. Acta*, **1995**, *307*, 253–268.
55 M. Lu, T. Knickerbocker, W. Cai, W. Yang, R.J. Hamers, L.M. Smith, *Biopolymers* **2004**, *73*, 606–613.
56 F. Wei, B. Sun, Y. Guo, X.S. Zhao, *Biosens. Bioelectron.*, **2003**, *18*, 1157–1163.

57 W. Yang, R.J. Hamers, **2004**, in press.
58 A.B. Kharatinov, J. Wasserman, E. Katz, I.J. Willner, *Phys. Chem. B*, **2001**, *105*, 4205–4213.
59 S.O. Kelley, E.M. Boon, J.K. Barton, N.M. Jackson, M.G. Hill, *Nucleic Acids Res.*, **1999**, *27*, 4830–4837.
60 R.J. Heaton, A.W. Peterson, R.M. Georgiadis, *Proc. Natl. Acad. Sci. U.S.A.*, **2001**, *98*, 3701–3704.
61 K.S. Cole, R.H.J. Cole, *Chem. Phys.*, **1941**, *9*, 341.
62 J.R. Macdonald, W.R. Kenan, *Impedance Spectroscopy – Emphasizing Solid Materials and Systems*, John Wiley, New York, **1987**.
63 T.L. Lasseter, W. Cai, R.J. Hamers, *Analyst* **2004**, *128*, 3–8.
64 E. Katz, L. Alfonta, I. Willner, *Sens. Actuators, B*, **2001**, *76*, 134–141.

8
Biomaterial-nanoparticle Hybrid Systems for Sensing and Electronic Devices

Joseph Wang, Eugenii Katz, and Itamar Willner

8.1
Introduction

The emergence of nanotechnology is opening new horizons for the application of nanomaterials in electroanalytical chemistry. The unique electronic [1, 2], optical [3–6] and catalytic [7, 8] properties of metal and semiconductor nanoparticles (1 – 200 nm) have paved the way for new generations of devices [9–13] and materials [14–16] exhibiting novel properties and functions. A variety of synthetic methods for the preparation of metal or semiconductor nanoparticles and their stabilization by functional monolayers [17–22], thin films or polymers [23–25] are available. The functionalized metal or semiconductor nanoparticles provide useful building blocks for the assembly of functionalized nanostructures. The chemical functionalities associated with nanoparticles enable the assembly of 2D and 3D nanoparticle architectures on surfaces [26–29], resulting in the novel design of bioelectronic devices [30]. Several reviews have addressed the synthesis, properties and functions of nanoparticles [31–34] and the progress in the integration of composite nanoparticle systems on surfaces and their use as functional devices [34–36]. Functionalized nanoparticles have found novel applications for extremely efficient wiring of redox enzymes opening the way to novel electrochemical biosensors, bioelectronic devices and biofuel cells. Labeling of biorecognition elements such as oligonucleotides or antigens/antibodies with metal or semiconductor nanoparticle paves the way for new electrochemical immunosensors and DNA sensors [37–39].

It is the aim of this chapter to review recent advances in the use of functionalized nanoparticles for biosensing applications. Specifically, we will discuss: (1) The electrical wiring of redox enzymes via metal nanoparticles and the integration of semiconductor nanoparticles with enzyme-driven biocatalytic transformations for photoelectrochemical sensing; (2) The labeling of biomolecules with nanoparticle tracers and the use of the biomaterial–nanoparticle conjugates in electrochemical and photoelectrochemical sensing; (3) The application of magnetic particles for the separation of analyte molecules during bioelectrochemical analyses, and the triggering and amplification of bioelectroanalytical processes by means of magnetic particles.

Bioelectronics. Edited by Itamar Willner and Eugenii Katz
Copyright © 2005 WILEY-VCH Verlag GmbH & Co. KGaA, Weinheim
ISBN: 3-527-30690-0

8.2
Biomaterial–nanoparticle Systems for Bioelectrochemical Applications

Metal nanoparticles, such as gold or silver nanoparticles, exhibit plasmon absorbance bands in the visible spectral region that are controlled by the size of the respective particles. Numerous studies reported on the labeling of biomaterials and the staining of biological tissues by metal particles as a means to image and visualize biological processes [40–42]. The spectral shifts originating from adjacent or aggregated metal nanoparticles, for example, Au nanoparticles [43–46], find increasing interest in the development of optical biosensors based on biomaterial–nanoparticle hybrid systems. Similarly, semiconductor nanoparticles exhibit size-dependent tunable absorbance and fluorescence. The high fluorescence quantum yields of semiconductor nanoparticles, their photostability and their tunable fluorescence bands attracted substantial research efforts directed to the use of semiconductor nanoparticles as fluorescence labels for biorecognition processes [47]. The extensive use of knowledge in the preparation of metal and semiconductor nanoparticles functionalized with biomaterials suggests that the unique catalytic or photoelectrochemical properties of the nanoparticles could be used to develop electrochemical and photoelectrochemical biosensors. For example, the catalytic electroless deposition of metals on nanoparticle hybrid labels could be used to generate conductive domains on functionalized or patterned surfaces, and the conductivity properties of the systems may then transduce the biosensing processes [48]. The metal and semiconductor nanoparticles could be enlarged and then detected by stripping voltammetry, leading to greatly amplified bioelectronic assays. The conducting properties of metal nanoparticles could be used for electrical wiring of redox enzymes, whereas photoelectrochemical reactions of semiconductive nanoparticles could be coupled to biocatalytic enzymatic reactions, leading to a new generation of electrochemical and photoelectrochemical enzyme biosensors.

The following sections will address recent advances in the application of nanoparticle–biomaterial conjugates as active components in electrochemical biosensing systems.

8.2.1
Bioelectrochemical Systems Based on Nanoparticle-enzyme Hybrids

Electrical contacting of redox enzymes with electrodes is a key process in the tailoring of enzyme electrodes for bioelectronic applications such as biosensors [49–55] or biofuel cell elements [56–59]. While redox enzymes usually lack direct electrical communication with electrodes, the application of diffusional electron mediators [60], the tethering of redox-relay groups to the protein [61–65], or the immobilization of the enzymes in redox-active polymers [66–69] were applied to establish electrical communication between the redox proteins and the electrodes. Nonetheless, relatively inefficient electrical contacting of the enzymes with the electrode is achieved because of the nonoptimal modification of the enzymes by the redox units [70] or the lack of appropriate alignment of the enzymes in respect to the electrode. Efficient electrical communication between redox proteins and electrodes was achieved by the reconstitution of apoenzymes on relay-cofactor monolayers associated with electrodes [71–75]. For example, apo-glucose oxidase was reconstituted on a relay-FAD monolayer [71–74], and apo-glucose dehydrogenase was reconstituted on a pyrroloquinoline quinone (PQQ)–modified polyaniline film associated with an electrode [75]. Effective electrical communication between the redox centers of the biocatalysts and the different electrodes was observed and reflected by high turnover electron transfer rates from the redox sites to the electrode. The effective electrical contacting of these redox enzymes was attributed to the alignment of the proteins on the electrodes and to optimal positioning of the intermediary electron-relay units between the enzyme redox centers and the electrode.

The availability of metal nanoparticles that exhibit conductivity properties at nanoscale dimensions enabled the generation of nanoparticle–enzyme hybrid systems for controlled electron transfer. A few biocatalytic electrodes have been prepared for biosensor applications by codeposition of redox enzymes and Au nanoparticles on electrode supports [76–79]. The biocatalytic electrodes were reported to operate without electron transfer mediators, but the random and nonoptimized positioning of the redox proteins on the conductive nanoparticles did not allow efficient electron transfer between the enzyme active sites and the electrode support. Highly efficient electrical contacting of the redox enzyme glucose oxidase through a single Au nanoparticle was accomplished by the reconstitution of the apo-flavoenzyme, apo-glucose oxidase (apo-GOx), with a 1.4-nm Au_{55}-nanoparticle functionalized with N^6-(2-aminoethyl)-flavin adenine dinucleotide (FAD cofactor amino-derivative) **1**. The conjugate produced was assembled on a thiolated monolayer using different dithiols **2–4** as linkers, Figure 8.1A [80]. Alternatively, the FAD-functionalized Au nanoparticle could be assembled on a thiolated monolayer associated with an electrode, and apo-GOx was subsequently reconstituted on the functional nanoparticles, Figure 8.1B. The enzyme electrodes prepared by these two routes reveal similar protein surface coverages of ca 1×10^{-12} mol cm^{-2}. The nanoparticle-reconstituted glucose oxidase layer was found to be electrically contacted with the electrode without any additional mediators, and the enzyme assembly stimulates the bioelectrocatalyzed oxidation of glucose, Figure 8.1C. The resulting nanoparticle-reconstituted enzyme electrodes revealed a remarkably efficient electrical communication with the electrode (electron transfer turnover rate ca 5000 s^{-1}). This electrical contacting makes the enzyme electrode insensitive to oxygen or to common electroactive interferences such as ascorbic acid. The electron transfer from the enzyme redox center to the bulk electrode is mediated by the Au nanoparticle, and the rate-limiting step in the electron transfer was found to be the charge transport across the dithiol molecular linker that bridges the particle to the electrode. The conjugated benzene dithiol **4** was found as the most efficient electron-transporting unit among the linkers **2–4**. The future application of conjugated molecular wires such as oligophenylacetylene units could further improve and enhance the electrical contacting efficiency.

While the previous system employed the metal nanoparticle as a nano-electrode that electronically communicates the enzyme redox-site with the macroscopic electrode, one may use enzyme–nanoparticle hybrid systems where the product generated by the biocatalytic process activates the functions of the nanoparticle. This has recently been demonstrated by tailoring an acetylcholine esterase (AChE)–CdS nanoparticle hybrid

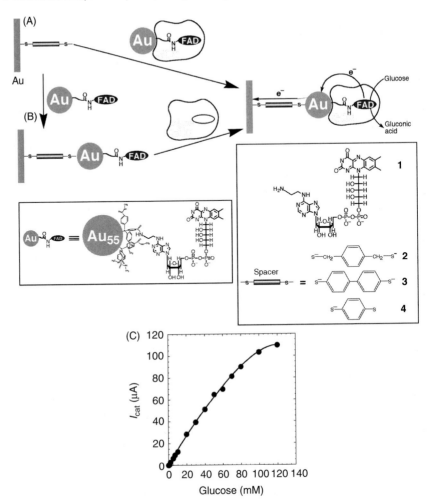

Fig. 8.1 Electrical "wiring" of glucose oxidase (GOx) by the apoenzyme reconstitution with a Au-nanoparticle- functionalized with a single FAD cofactor unit. (A) The reconstitution process performed in a solution followed by the assembly adsorption onto a dithiol-modified Au electrode. (B) The Au-FAD conjugate adsorption onto a dithiol-modified Au electrode, followed by the reconstitution of the apo-GOx at the interface. (C) Calibration plot of the electrocatalytic current developed by the reconstituted GOx electrode in the presence of different concentrations of glucose. (Adapted from reference [80], Figures 1(A) and 2(B), with permission.)

monolayer on a Au electrode, and the activation of the photoelectrochemical functions of the nanoparticles by the biocatalytic process [81]. The CdS–AChE hybrid interface was assembled on the Au electrode by the stepwise coupling of cystamine-functionalized CdS (thiol-sulfonate capping molecules were added to the Au nanoparticles in order to provide their solubility in water) to the electrode, and the subsequent covalent linkage of the enzyme AChE to the particles, Figure 8.2A. In the presence of acetylthiocholine

5 as substrate, the enzyme catalyzes the hydrolysis of **5** to thiocholine **6** and acetate. Photoexcitation of the CdS semiconductor yields the electron-hole pair in the conduction band and the valence band, respectively. The enzyme-generated thiocholine **6** acts as an electron donor for the valence-band holes. The scavenging of the valence-band holes results in the accumulation of the electrons in the conduction band and their transfer to the electrode with the generation of a photocurrent, Figure 8.2B. The addition of enzyme inhibitors such as 1,5-*bis*(4-allyldimethylammoniumphenyl)pentane-3-one dibromide **7** blocks the biocatalytic functions of the enzyme and as a result inhibits the photocurrent formation in the system, Figure 8.2C. Thus, the hybrid CdS/AChE system provides a functional interface for sensing of the AChE inhibitors (e.g. chemical warfare) by means of photocurrent measurements.

A similar system composed of a photoactivated CdS nanoparticle and co-immobilized formaldehyde dehydrogenase that utilizes formaldehyde as an electron donor has been reported [82]. In this hybrid system, the direct electron transfer from the enzyme active center to the CdS photogenerated holes was achieved, and the steady state photocurrent signal in the system was reported to be directly related to the substrate concentration.

8.2.2
Electroanalytical Systems for Sensing of Biorecognition Events Based on Nanoparticles

The unique optical [3–6], photophysical [34], electronic [1, 2, 83–87] and catalytic [7, 8] properties of metal and semiconductor nanoparticles turn them into ideal labels for biorecognition and biosensing processes. For example, the unique plasmon absorbance features of Au nanoparticles and specifically the interparticle-coupled plasmon absorbance of conjugated particles have been widely used for DNA [88] and antibody–antigen [89–91] analyses. Similarly, the tunable fluorescence properties of semiconductor nanoparticles were used for the photonic detection of biorecognition processes [47, 92].

Inspired by the novel use of nanoparticles in optical bioassays, recent studies have focused at developing analogous particle-based electrical routes for gene detection and immunosensing [93]. Such new protocols are based on the use of colloidal gold tags, semiconductor quantum dot tracers, metallic microrods, polymeric carrier (amplification) beads, or magnetic (separation) beads. These nanoparticle materials offer elegant ways for interfacing DNA recognition and antigen/antibody binding events with electrochemical signal transduction, and for amplifying the resulting electrical response.

Most of these schemes have commonly relied on a highly sensitive electrochemical stripping transduction/measurement of the metal tracer. Stripping voltammetry is a powerful electroanalytical technique for trace metal measurements [94]. Its remarkable sensitivity is attributed to the "built-in" preconcentration step, during which the target metals are accumulated (plated) onto the working electrode. The detection limits are thus lowered by 3 to 4 orders of magnitude, compared to pulse-voltammetric techniques, used earlier for monitoring DNA hybridization. Such ultrasensitive electrical detection of metal tags has been accomplished in connection with a variety of new and novel DNA-linked particle nanostructure networks.

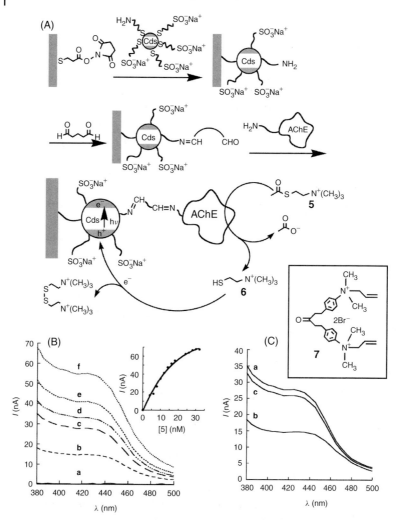

Fig. 8.2 (A) Assembly of the CdS nanoparticle/AChE hybrid system on a Au electrode for the photoelectrochemical detection of the enzyme activity. (B) Photocurrent action spectra observed in the presence of different concentrations of acetylthiocholine (**5**): (a) 0 mM, (b) 6 mM, (c) 10 mM, (d) 12 mM, (e) 16 mM, (f) 30 mM. Inset: Calibration curve corresponding to the photocurrent generated upon the excitation of the assembly at $\lambda = 380$ nm in the presence of variable concentrations of **5**. Spectra were recorded in 0.1 M phosphate buffer, pH = 8.1, under argon. (C) Photocurrent spectra corresponding to the CdS/AChE system in the presence of **5**, 10 mM, (a) without the inhibitor, (b) upon addition of **7**, 1×10^{-6} M, (c) after rinsing the system and excluding of the inhibitor. (Adapted from reference [81], Scheme 1, Figures 1 and 2(A), with permission.)

Metal and semiconductor nanoparticles coupled to biomaterials generate solubilized entities. Nonetheless, even nanoscale particulate–clustered systems include many atoms/molecules in the clusters. Thus, the clustered systems could be loaded with additional markers, including redox-active moieties. For example, silica nanoparticles loaded with *tris*(2,2′-bipyridyl)cobalt(III) [95] or Au nanoparticles functionalized with tethered ferrocene units [96] were applied for labeling of DNA and further electrochemical DNA detection based on the redox process of the redox-active complex units.

The solubility of the nanoparticle biomaterial structures allows the application of washing procedures on surfaces that include a sensing interface, and thus nonspecific adsorption processes are eliminated. On the other hand, the specific capturing of biomaterial nanoparticles on the respective sensing interfaces allows the secondary dissolution of the captured nanoparticles and thus enables the amplified detection of the respective analyte by the release of many ions/molecules as a result of a single recognition event.

8.2.2.1 Gold and Silver Metal Tracers for Electrical DNA Detection and Immunosensing

Powerful nanoparticle-based electrochemical DNA hybridization assays were developed using Au, Ag, Cu, or In metal tracers [97–101]. Such protocols have relied on capturing the gold [97, 98], silver [99], Cu_{core}-Au_{shell} [100] nanoparticles, or In nanorods [101] to the hybridized target, followed by anodic stripping electrochemical measurement of the metal tracer. The probe or target immobilization has been accomplished directly on carbon or indium-tin oxide (ITO) electrodes [102, 103]. Alternatively, the DNA probe was linked to streptavidin-coated magnetic beads [97] or adsorbed onto the walls of polystyrene microwells [98]. The DNA-functionalized beads were collected on an electrode surface, and picomolar levels of the DNA target have thus been electrochemically detected. For example, an electrochemical method was employed for the Au nanoparticle–based quantitative detection of the 406-base human cytomegalovirus DNA sequence (HCMV DNA) [98]. The HCMV DNA was immobilized on a microwell surface and hybridized with the complementary oligonucleotide-modified Au nanoparticle. The resulting surface-immobilized Au nanoparticle double-stranded assembly was treated with HBr/Br_2, resulting in the oxidative dissolution of the gold particles. The solubilized Au^{3+}-ions were then electrochemically reduced and accumulated on the electrode and subsequently determined by anodic stripping voltammetry using a sandwich-type screen-printed microband electrode (SPMBE). The combination of the sensitive detection of Au^{3+}-ions at the SPMBE due to nonlinear mass transport of the ions and the release of a large number of Au^{3+} ions upon the dissolution of the particle associated with a single recognition event provide an amplification path that enabled the detection of the HCMV DNA at a concentration of 5×10^{-12} M.

Further sensitivity enhancement can be obtained by catalytic enlargement of the gold tracer in connection with nanoparticle-promoted precipitation of gold [97] or silver [104]. Combining such enlargement of the metal-particle tags with the effective "built-in" amplification of electrochemical stripping analysis paved the way for sub-picomolar detection limits. Instead of enlarging nanometer-size metal particles, it is possible to use micrometer-long wire tracers for attaining ultrasensitive DNA measurements [101]. In particular, indium microrods – prepared by a porous membrane template electrical deposition – resulted in greatly enhanced sensitivity (compared to commonly used spherical nanoparticle tags) and with fM detection limits. The coupling of gold nanoparticle–based electrical measurements with on-chip PCR reactions has also been reported [105].

An electrochemical protocol for detecting DNA hybridization based on preparing the metal marker along the DNA backbone (instead of capturing it at the end of the duplex) was described [106]. This procedure relies on DNA template–induced generation of conducting nanowires as a mode of capturing the metal tag. The use of DNA as a metallization template [107] has evoked substantial research activity directed to the generation of conductive nanowires and the construction of functional circuits [108–111]. Such an approach was applied to growing silver [108, 109], palladium [111] or platinum [110] clusters on DNA templates. Elements from the methods used for the generation of metal nanocircuitry based on DNA templates were adapted to develop DNA detection schemes as outlined in Figure 8.3. The short DNA primer **8** attached to the electrode hybridizes with the target DNA **9** (Step a). The phosphate groups associated with the long target DNA **9** collect Ag^+ ions from the solution by electrostatic interaction (Step b). The bound Ag^+ ions are then reduced by hydroquinone, resulting in the formation of metallic silver aggregates along the DNA (Step c). The subsequent dissolution and stripping electrochemical detection of the nanoscale silver clusters (Step d) then provides the route to detect the hybridized DNA. It should, however, be noted that the short DNA primer could also bind some Ag^+ ions that yield a background response. The background signal could be avoided, and the sensitivity provided by this method could be improved upon application of peptide nucleic acid (PNA) that lacks phosphate groups and thus does not bind Ag^+ ions as the primer for hybridization of the target DNA.

The catalytic features of metal nanoparticles that enable the electroless deposition of metals on the nanoparticle clusters allow the enlargement of the particles to conductive interparticle-connected entities. The formation of conductive domains as a result of biorecognition events then provides an alternative path for the electrical transduction of biorecognition events. This was exemplified by the design of a miniaturized immunosensor based on Au nanoparticles and their catalytic properties [48], Figure 8.4A. Latex particles stabilized by an anionic protective layer were attracted to a gap between micron-sized Au electrodes by the application of a nonuniform alternating electric field between the electrodes (dielectrophoresis). Removal of the protective layer from the latex particles by an oppositely charged polyelectrolyte resulted in the aggregation of the latex particles and their fixation in the gap domain. Adsorption of protein A on the latex surface yielded a sensing interface for the specific association of the human immunoglobulin (IgG) antigen. The association of the human immunoglobulin on the surface was probed by the binding of the secondary Au-labeled anti-human IgG antibodies to the surface, followed by the catalytic deposition of a silver layer on the Au nanoparticles. The silver layer bridged the gap between the two microelectrodes, resulting in a conductive "wire". Typical resistances between the

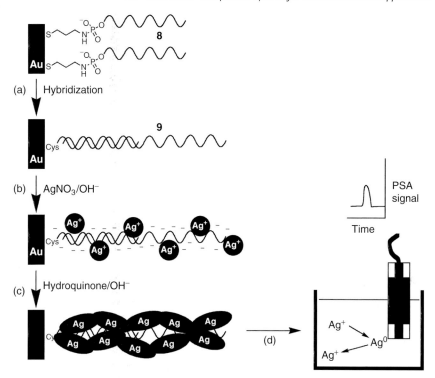

Fig. 8.3 Schematic outline of the steps involved in the amplified electrochemical detection of DNA by the catalytic silver cluster deposition on the DNA strand: (a) hybridization of the complementary target DNA **9** with the DNA probe **8** covalently linked to the electrode surface through a cystamine monolayer; (b) "loading" of the silver ions onto the immobilized DNA; (c) reduction of silver ions by hydroquinone to form silver aggregates on the DNA backbone; (d) dissolution of the silver aggregates in an acid solution and transfer to the detection cell followed by stripping potentiometric detection (PSA = potentiometric stripping analysis).

microelectrodes were 50 to 70 Ω, whereas control experiments that lack the specific catalytic enlargement of the domain by the Au-nanoparticle-antibody conjugate yielded resistances >10^3 Ω. The method enabled the analysis of human IgG with a detection limit of ca 2×10^{-13} M.

A related DNA detection scheme was developed using microelectrodes fabricated on a silicon chip [112], Figure 8.4B. A probe nucleic acid **10** was immobilized on the SiO_2 interface in the gap separating the microelectrodes. The target 27-mer-nucleotide **11** was then hybridized with the probe interface, and subsequently nucleic acid **12**–functionalized Au nanoparticles were hybridized with the free 3'-end of the target DNA. The Au nanoparticle catalyzed hydroquinone-mediated reduction of Ag^+-ions results in the deposition of silver on the Au nanoparticle assembly, lowering the resistance between the electrodes. Single-base mutants of the analyte oligonucleotide **11** were washed off from the capture-nucleic acid **10** by the use of a buffer with the appropriate ionic strength. A difference of 10^6 in the gap resistance was observed

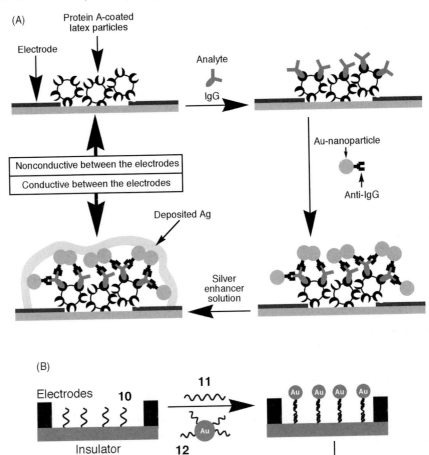

Fig. 8.4 (A) Immunosensing at micro-sized Au electrodes based on the change of conductivity between the Au strips upon binding of Au nanoparticles followed by silver deposition. (B) The use of a DNA-nanoparticle conjugate and subsequent silver deposition to connect two microelectrodes, as a means of sensing a DNA analyte. (Adapted from reference [34], Scheme 15(A), with permission).

upon analyzing the target DNA and its mutant by this method. The low resistances between the microelectrodes were found to be controlled by the concentration of the target DNA, and the detection limit for the analysis was estimated to be ca 5×10^{-13} M. This sensitivity translates to ca 1 µg µL^{-1} of human genomic DNA or ca 0.3 ng µL^{-1} of DNA from a small bacterium. These concentrations suggest that the DNA may be analyzed with no pre-PCR amplification. The simultaneous analysis of a collection of DNA targets was accomplished with a chip socket that included 42 electrode gaps and appropriate different nucleic acid sensing probes between the electrode gaps [113].

The immobilization of nanoparticles on surfaces may also be used to yield high surface area electrodes [114] and hence for increasing the hybridization capacity of the surface [115]. Such use of nanoparticle-supporting films relied on the self assembly on 16-nm diameter colloidal gold onto a cystamine-modified gold electrode and resulted in surface densities of oligonucleotides as high as 4×10^{14} molecules cm^{-2}. The detection of the ferrocenecarboxaldehyde tag (conjugated to the target DNA) resulted in a detection limit of 500 pM.

8.2.2.2 Use of Magnetic Beads for DNA Analysis

Biomaterial-functionalized magnetic particles (e.g. Fe_3O_4) have been extensively applied in a broad variety of bioelectronic applications [116]. Application of external magnetic fields to control bioelectrochemical processes, such as biocatalytic transformations of redox enzymes, was recently reported [117, 118]. Separation and purification steps prior to the DNA electrochemical analysis were simplified by the DNA analyte attachment to magnetic particles [119]. The reversible magnetically controlled oxidation of DNA was accomplished in the presence of nucleic acid-modified magnetic particles [120]. Avidin-modified magnetic particles were functionalized with the biotinylated probe nucleic acid and subsequently hybridized with the complementary DNA. Two carbon-paste electrodes were patterned on a surface and applied as working electrodes. Spatial deposition of the functionalized magnetic particles on the right (**R**) or left (**L**) electrode enabled the magneto-controlled oxidation of the DNA by chronopotentiometric experiments (potential pulse from 0.6 to 1.2 V), Figure 8.5. Changing the position of the magnet (below planar printed electrodes) was thus used for "ON" and "OFF" switching of the DNA oxidation (through attraction and removal of DNA-functionalized magnetic particles). The process was reversed and repeated upon switching the position of the magnet, leading to "ON" and "OFF" oxidation signals in the presence and absence of the magnetic field, respectively. Such magnetic triggering of the DNA oxidation holds great promise for the analysis of DNA on arrays.

Several of the protocols for DNA detection have coupled the inherent signal amplification of stripping analysis with an effective discrimination against nonhybridized DNA [121, 122]. In addition to efficient isolation of the duplex, magnetic spheres can open the door for elegant ways for triggering and controlling electrical DNA detection [120, 123]. For example, an attractive magnetic switching of the electrical DNA detection has been realized by the "magnetic" collection of the magnetic-bead/DNA-hybrid/metal-tracer assembly onto a thick-film electrode transducer that allowed direct electrical contact of the silver precipitate to facilitate a solid-state detection (without

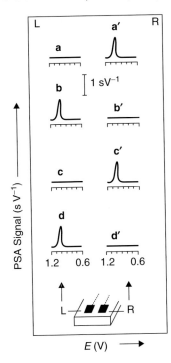

Fig. 8.5 Chronopotentiometric signals for the DNA oligomer-functionalized magnetic particles (100 μg) using the dual-carbon paste electrode assembly: **a–d** are potentiograms at the "left" (L) electrode, while **a′–d′** are potentiograms obtained at the "right" (R) electrode. **a, b′, c, d′** are potentiograms obtained in the absence of the magnet, while **a′, b, c′, d** are potentiograms recorded in the presence of the magnet. (Adapted from reference [120], Figure 2, with permission).

dissolving the metal tracer) [123]. By this approach, magnetic spheres functionalized with a biotinylated nucleic acid by an avidin bridge act as the capturing particles. Hybridization of the analyzed DNA with the capturing nucleic acid is followed by the secondary association of metal or semiconductor nanoparticles functionalized with a nucleic acid that is complementary to a free segment of the analyzed DNA. The binding of the nanoparticle labels to the biorecognition assay then provides amplifying cluster tags that by dissolution enable the release of numerous ion/molecule units. Also, the metal nanoparticles associated with the sensing interface may act as catalytic sites for the electroless deposition of other metals, thus leading to the amplified detection of DNA by the intermediary accumulation of metals that are stripped off, or by generating an enhanced amount of dissolved product that can be electrochemically analyzed [106, 124]. Figure 8.6A displays the amplified detection of DNA by the application of nucleic acid–functionalized magnetic beads and Au nanoparticles as catalytic seeds for the deposition of silver [123]. A biotin-labeled nucleic acid **13** was immobilized on the avidin-functionalized magnetic particles and hybridized with the complementary biotinylated nucleic acid **14**. The hybridized assembly was then reacted with a Au nanoparticle–avidin conjugate. The magnetic separation of the particles by an external magnet concentrated the hybridized assembly from the analyzed sample. Treatment of the magnetic particles-DNA-Au-nanoparticle conjugate with silver ions (Ag^+) in the presence of hydroquinone results in the electroless catalytic deposition of silver on the Au nanoparticles, acting as catalyst. The latter process provided the amplification route since the catalytic accumulation of silver on the Au nanoparticle originates from a single DNA recognition event. The current originating from the potential stripping-off of the accumulated silver then provided the electronic signal that transduced the analysis of the target DNA.

In a related system [97], the electrochemical detection of the DNA was accomplished by the use of Au nanoparticles as electroactive and catalytic tags, Figure 8.6B. The primer biotinylated nucleic acid **15** was linked to magnetic beads through an avidin bridge. The hybridization of the nucleic acid **16** functionalized with Au nanoparticle was then

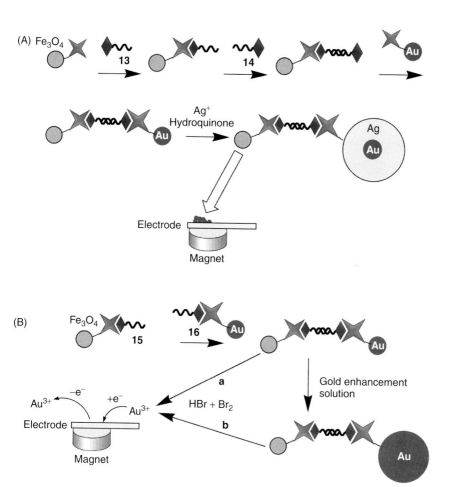

Fig. 8.6 Electrochemical analysis of DNA upon the assembly of DNA molecules at magnetic particles followed by their association with Au nanoparticles. (A) The Au nanoparticles are used for silver deposition and the DNA analysis is performed by electrochemical Ag stripping. (B) The Au nanoparticles are chemically dissolved, the resulting Au^{3+} ions electrochemically reduced, and the deposited gold is electrochemically stripped (Route **a**). The intermediate enlargement of the Au nanoparticles (Route **b**), results in the further amplification of the signal.

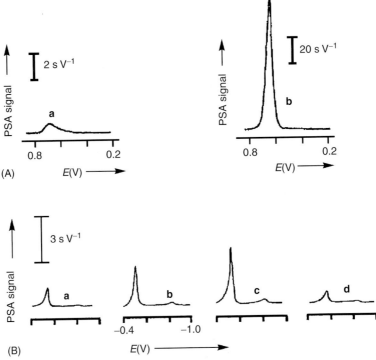

Fig. 8.7 (A) Effect of the gold enhancement upon the stripping response for the analyte DNA (**16**), 10 μg mL^{-1}: (**a**) Stripping signal of the Au nanoparticle label. (**b**) Stripping signal after 10 min of the system reaction in the gold enhancement solution. Hybridization time, 25 min; amount of magnetic beads, 90 μg; amount of 5 nm avidin-coated gold particles, 7.6 × 10^{10}; gold oxidation time, 5 min. (B) Stripping potentiograms measured upon the sensing of different concentrations of DNA, which is bound to magnetic particles and labeled with CdS nanoparticles: (**a**) 0.2 mg L^{-1} (**b**) 0.4 mg L^{-1} (**c**) 0.6 mg·L^{-1} (**d**) Control experiment with noncomplementary DNA, 0.6 mg L^{-1}. Amount of the magnetic particles, 20 μg; concentration of the DNA-functionalized CdS nanoparticles, 0.01 mg mL^{-1}; hybridization time, 10 min; accumulation potential, −0.9 V; accumulation time, 2 min; stripping current, 1 μA. (Adapted from reference [97], Figure 9, and reference [125], Figure 2, with permission).

detected by the dissolution of the Au nanoparticles with a HBr/Br$_2$ solution, followed by the electrochemical reduction of the generated Au^{3+} ions onto the electrode and the subsequent electrochemical stripping-off of the surface generated gold, Figure 8.6B, route (a). This analytical procedure was further amplified by the intermediary deposition of gold on the Au nanoparticles, Figure 8.6B, route (b). The higher gold content after the catalytic deposition leads to a higher chronopotentiometric signal. Figure 8.7A displays the chronopotentiograms corresponding to the stripping-off of the gold generated on the electrode upon the analysis of the nucleic acid associated with 5 nm Au nanoparticles, route (a), and after a 10-min deposition of gold on the Au nanoparticles, route (b).

It is also possible to use magnetic beads as reporters (tags) for DNA hybridization detection coupled to cathodic-stripping voltammetric measurements of their iron content [126]. A related protocol, developed in the same study, involved probes labeled with gold-coated iron core-shell nanoparticles. In both cases, the captured iron-containing particles were dissolved following the hybridization, and the released iron was quantified by stripping voltammetry in the presence of the 1-nitroso-2-naphthol complexing ligand and a bromate catalyst.

8.2.2.3 Semiconductive Nanoparticle Tags: Toward Electrical Coding

Efficient methods for the preparation of semiconductive nanocrystal particles (e.g. CdS, PbS, CdSe, ZnS) and their functionalization with biomaterials were recently developed [127]. These nanoparticles were applied for labeling of biomaterials in biorecognition processes (e.g. DNA sensing). For example, CdS semiconductor nanoparticles labeled oligonucleotides were employed as tags for the detection of hybridization events of DNA [125]. Dissolution of the CdS (in the presence of 1 M HNO_3), followed by the electrochemical reduction of the Cd^{2+} to Cd^0 that accumulates on the electrode, and the stripping-off of the generated Cd^0 (to Cd^{2+}) provided the electrical signal for the DNA analysis. Figure 8.7B displays the chronopotentiograms resulting in the analysis of different concentrations of the complementary target DNA using the CdS nanoparticles as tags. A nanoparticle-promoted cadmium precipitation was used to enlarge the nanoparticle tag and amplify the stripping DNA hybridization signal. In addition to measurements of the dissolved cadmium ion, solid-state measurements following a "magnetic" collection of the magnetic-bead/DNA-hybrid/CdS-tracer assembly onto a thick-film electrode transducer were demonstrated. Such protocol combines the amplification features of nanoparticle/polynucleotides assemblies and the highly sensitive potentiometric stripping detection of cadmium, with an effective magnetic isolation of the duplex. A low detection limit of 100 fmol was obtained along with a good reproducibility (RSD = 6%).

An interesting aspect of these systems is, however, the future possibility of using a combination of different metal or semiconductor tags linked to different nucleic acids for the simultaneous analysis of multiple DNA targets. By this approach [128], different nucleic acid probes complementary to series of DNA targets are linked to different magnetic particles. Similarly, different semiconductor or metallic nanoparticle tags complementary to segments of the series of target DNAs are used as amplifying detection units for the primary hybridization process. The hybridization of the nucleic acid–functionalized semiconductor or metal particle to the specific DNA targets, followed by the dissolution of the nanoparticles and the electrochemical accumulation and stripping-off of the metal, enables the determination of the specific DNA targets present in the sample, that is, the characteristic potentials needed to strip off the metal provide electrochemical indicators for the nature of the analyzed DNA. A model system that follows this principle was developed [128] in which three different kinds of magnetic particles modified by three different nucleic acids were hybridized with three different kinds of semiconductor nanoparticles, ZnS, CdS, PbS, that were functionalized with oligonucleotides complementary to the probes associated with the magnetic

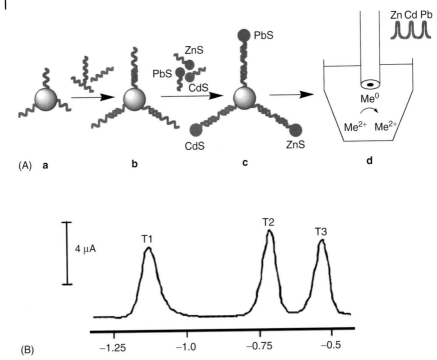

Fig. 8.8 (A) Multitarget electrochemical DNA detection protocol based on different nanocrystal tracers: (**a**) Introduction of probe-modified magnetic beads. (**b**) Hybridization with the DNA targets. (**c**) Second hybridization with the nanoparticle-labeled probes. (**d**) Dissolution of the nanoparticles and the electrochemical detection. (B) Stripping voltammogram recorded upon the simultaneous analysis of three different 60-mer DNA targets related to the *BRCA1* breast-cancer gene (54 nM each) labeled: (T1) with ZnS nanoparticles, (T2) with CdS nanoparticles, and (T3) with PbS nanoparticles. The measurements were performed on a mercury-coated glassy-carbon electrode, with 1 min pretreatment at 0.6 V; 2 min accumulation at −1.4 V; 15 s rest period (without stirring); square-wave voltammetric scan with a step potential of 50 mV; amplitude, 20 mV; frequency, 25 Hz. (Adapted from reference [128], Figure 2(E), with permission).

particles, Figure 8.8A. The magnetic particles allow easy transportation and purification of the analyte sample, whereas the semiconductor particles provide nonoverlapping electrochemical signals that transduce the specific kind of hybridized DNA. Stripping voltammetry of the semiconductive nanoparticles yields well-defined and resolved stripping peaks, for example, at −1.12 V (Zn), −0.68 V (Cd), and −0.53 V (Pb) (vs Ag/AgCl reference), thus allowing simultaneous electrochemical analysis of several DNA analytes tagged with the labeling semiconductive nanoparticles. For example, Figure 8.8B displays stripping voltammograms for a solution containing three DNA samples labeled with the ZnS, CdS and PbS nanoparticle tracers. The functionalization of the nanocrystal tags with thiolated oligonucleotide probes offered the voltammetric signature with distinct electrical hybridization signals for the corresponding DNA targets. The position and size of the resulting stripping peaks provided the desired identification and quantitative information, respectively, on a given target DNA. The multitarget DNA detection capability was coupled to the amplification feature of stripping voltammetry (to yield fmol detection limits) and with an efficient magnetic removal of nonhybridized nucleic acids to offer high sensitivity and selectivity. Up to five to six targets can thus be measured simultaneously in a single run-in connection to ZnS, PbS, CdS, InAs and GaAs semiconductor particles. Conducting massively parallel assays (in microwells of microtiter plates or using multichannel microchips, with each microwell or channel carrying out multiple measurements) could thus lead to a high-throughput analysis of DNA. Analogous coding of proteins, in connection with sandwich immunoassays of multiple antigens, is anticipated.

Further efforts should be aimed at developing large particle-based libraries for electrical coding, based on the judicious design of encoded "identification" beads or striped metal rods. Submicrometer metallic barcodes for fluorescence analysis of biomaterials such as DNA or antigens have recently been reported [129]. By incorporating different predetermined levels (or lengths) of multiple heavy-metal markers, such beads or microrods can lead to a large number of recognizable stripping-voltammetric fingerprints and hence to a reliable identification of a large number of DNA targets [130]. In addition to powerful bioassays, such "identification beads" hold great promise for the identification of counterfeit products and related authenticity testing.

8.2.2.4 Ultrasensitive Particle-based Assays for DNA Based on Multiple Amplification Routes

The amplification paths for electrochemical analyzing DNA that were discussed in the previous sections have employed a single reporter unit, for example, a metal or semiconductor nanoparticle, per one hybridization event. For further enhancing the sensitivity of DNA detection, it is possible to load multiple tags per binding event [131, 132]. This can be accomplished by linking the biorecognition units to polymeric microbeads carrying multiple redox tracers in external positions (on their surface) or internal positions (via encapsulation). A triple-amplification bioassay that couples the carrier-sphere amplifying units (loaded with numerous gold nanoparticles tags) with the "built-in" preconcentration feature of the electrochemical stripping detection and the catalytic enlargement of the multiple gold-particle tags was demonstrated [131], Figure 8.9. The gold-tagged beads were prepared by binding biotinylated Au nanoparticles to streptavidin-coated polystyrene spheres. These beads were functionalized with a single-stranded oligonucleotide, which was further hybridized with a complementary oligonucleotide linked to a magnetic particle, Figure 8.9, Step "a". The numerous Au nanoparticle labels associated with one ds-oligonucleotide pair were enlarged by electroless deposition of gold, Figure 8.9, Step "b", and transported to the electrode array with the use of the magnetic particle. Then the Au assembly was dissolved upon the reaction with HBr/Br_2 and electrochemically analyzed using electrochemical deposition/stripping procedure, Figure 8.9, steps "c" and "d". Such a triple-amplification route offered a dramatic enhancement of the sensitivity. In another approach, carbon nanotubes loaded with many CdS nanoparticles were employed as labels for DNA hybridization [132]. Dissolution of the bound CdS

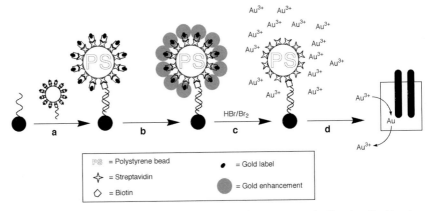

Fig. 8.9 The amplified DNA detection using nucleic acid/Au nanoparticle–functionalized beads as labels and electroless catalytic deposition of gold on the nanoparticles as amplification path. The analysis includes: (**a**) hybridization of the nucleic acid/Au nanoparticle–functionalized beads with the target DNA associated with a magnetic bead; (**b**) the enhanced catalytic deposition of gold on the nanoparticles; (**c**) dissolution of the gold clusters; (**d**) the detection of the Au^{3+} ions by stripping voltammetry.

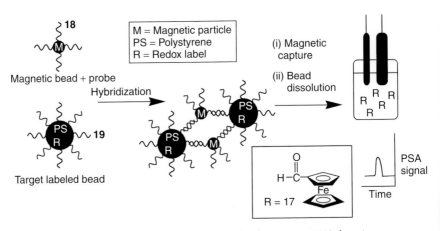

Fig. 8.10 The application of "electroactive" particles for ultrasensitive DNA detection, based on polystyrene beads impregnated with a redox marker.

nanoparticles in 1 M HNO_3 followed by the electrochemical detection of the released Cd^{2+} ions provided an amplified signal for the hybridization event.

Internal encapsulation of electroactive tags within carrier beads offers an alternative means to label the probes, and it might reveal some advantages as compared to the external labeling of the probes by nanoparticles. For example, ultrasensitive electrical DNA detection was recently reported on the basis of polystyrene beads impregnated with ferrocenecarboxaldehyde as a redox marker **17** [122], Figure 8.10. The capturing

DNA **18** was linked to magnetic particles, and the polystyrene beads functionalized with the complementary nucleic acid **19** were hybridized with the nucleic acid–modified magnetic particles. Collection of the hybridized system and the dissolution of the beads in an organic solvent released the beads-immobilized redox label **17**. This allowed the chronopotentiometric detection of the target DNA with a sensitivity that corresponds to the 5.1×10^{-21} mol (~31 000 molecules) under experimental conditions that involved 20-min hybridization and the "release" of the marker by dissolution of the modified beads in an organic medium. The amplified electrochemical readout signal was observed with the remarkable discrimination of a large excess (10^7-fold) of noncomplementary nucleic acids, revealing the analytical advantages of this sensing process. Further efforts should be directed for encapsulating different redox markers in different polystyrene host beads that could allow parallel multitarget DNA detection. Other marker encapsulation routes hold great promise for electrical DNA detection. Particularly attractive are the recently developed nanoencapsulated microcrystalline particles, prepared by the layer-by-layer technique, that offer large marker/biomolecule ratios and superamplified bioassays [133]. Related analytical procedures that combined multiple amplification pathways on the basis of enzyme-functionalized liposomes and the accumulation of the biocatalytic-reaction product were reported for the ultrasensitive DNA assays [134]. Such bioassay relied on the large surface area of the liposomes that carry a large number of enzyme molecules. Sensing of the accumulated product was accomplished by means of chronopotentiometry.

8.2.2.5 Photoelectrochemical Transduction of DNA Recognition Based on Semiconductive Nanoparticles

Photoelectrochemical transduction of DNA recognition processes has been demonstrated by using semiconductor (CdS) nanoparticles modified with nucleic acids [135]. Semiconductor CdS nanoparticles (2.6 ± 0.4 nm) were functionalized with one of the two thiolated nucleic acids **20** or **21** that are complementary to the 5′ and 3′ ends of a target DNA **22**. An array of CdS nanoparticle layers was then constructed on a Au electrode by a layer-by-layer hybridization process, Figure 8.11. A primary thiolated DNA **20** monolayer was assembled on a Au electrode, and the target DNA **22** hybridized to the interface acted as a cross-linking unit for the association of the **21**-functionalized CdS nanoparticles to the electrode by the hybridization of the ends of the **22**-units to the **21**-functionalized CdS particles. The subsequent association of the **20**-modified CdS particles prehybridized to **22** was then linked to the first generation of the CdS particles, resulting in the second generation of CdS particles. By the stepwise application of the two different kinds of nucleic acid–functionalized CdS- nanoparticles hybridized with **22**, an array with a controlled number of nanoparticle generations could be assembled on the electrode. This array was characterized by spectroscopic means (absorption, fluorescence) upon the assembly of the array on glass supports and by microgravimetric quartz-crystal microbalance analyses on Au-quartz piezoelectric crystals. Illumination of the array resulted in the generation of a photocurrent. The photocurrents increased with the number of CdS nanoparticle generations associated with the electrode, and

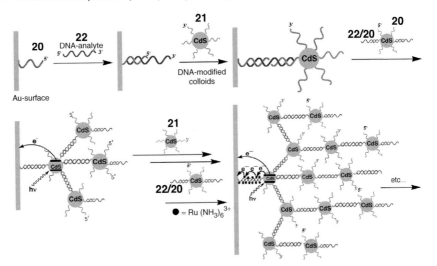

Fig. 8.11 The layer-by-layer assembly of CdS nanoparticle/DNA arrays on an electrode, and their use in the generation of photocurrents.

the photocurrent action spectra followed the absorbance features of the CdS nanoparticles, implying that the photocurrents originated from the photoexcitation of the CdS nanoparticles, that is, photoexcitation of the semiconductor induced the transfer of electrons to the conduction band and the formation of an electron-hole pair. Transfer of the conduction-band electrons to the bulk electrode, and the concomitant transfer of electrons from a sacrificial electron donor to the valence-band holes, yielded the steady state photocurrent in the system. The ejection of the conduction-band electrons into the electrode occurred from nanoparticles positioned in intimate contact with the electrode support, rather than being transported through the DNA strands. This was supported by the fact that $Ru(NH_3)_3^{6+}$ units ($E° = -0.16$ V vs SCE) that are electrostatically bound to the DNA enhanced the photocurrent from the DNA–CdS array, that is, the $Ru(NH_3)_6^{3+}$ units acted as electron wiring elements that facilitated electron hopping of conduction-band electrons of CdS particles that lack direct contact with the electrode through the DNA tether. The system is important not only because it demonstrates the use of photoelectrochemistry as a transduction method for DNA sensing but also since the system reveals the nano-engineering of organized DNA-tethered semiconductor nanoparticles on conductive supports.

8.3
Application of Redox-functionalized Magnetic Particles for Triggering and Enhancement of Electrocatalytic and Bioelectrocatalytic Processes

Microsize and nanosize magnetic particles functionalized with redox-active entities were applied for controlling bioelectrocatalytic reactions particularly important for bioelectroanalysis. A series of electroactive relay units consisting of 2,3-dichloro-1,4-naphthoquinone **23**, N-(ferrocenylmethyl)aminohexanoic acid **24**, N-methyl-N'-(dodecanoic acid)-4,4'-bipyridinium **25**, pyrroloquinoline quinone (PQQ, **26**), or microperoxidase-11 (MP-11, **27**), were covalently linked [117, 118] to the magnetite particles [136], Figure 8.12-B. Electrochemical activation of the redox units associated with the magnetic particles enables the secondary magneto-controlled switchable activation and deactivation of bioelectrocatalytic processes. For example, pyrroloquinoline quinone (PQQ, **26**), $E° = -0.16$ V (pH = 8.0; vs SCE), acts as an electrocatalyst for the oxidation of 1,4-dihydro-β-nicotineamide adenine dinucleotide, NADH [137, 138]. Accordingly, the **26**-functionalized magnetic particles (average surface coverage of 7.5×10^3 PQQ units per particle) were employed for controlling the electrocatalyzed oxidation of NADH by means of the external magnet [118]. Figure 8.13A depicts the cyclic switchable activation and deactivation of the electrocatalytic oxidation of NADH by the PQQ-functionalized particles upon their attraction to the electrode and their retraction from the electrode by means of the external magnet, respectively. Similarly, the magnetic particles functionalized with microperoxidase-11 **27** or with aminonaphthoquinone **28** were employed for magnetically switched electrocatalytic reduction of hydrogen peroxide or oxygen, respectively [118].

Electron-relay units act as electron transfer mediators that electrically communicate with redox enzyme and electrode supports. For example, ferrocene or bipyridinium redox-relay units were reported to electrically contact oxidative redox enzymes, for example glucose oxidase, or reductive biocatalysts, for example nitrate reductase, respectively [60]. Thus, the activation of the redox functionalities of the relay-modified magnetic particles enables the subsequent switchable activation and deactivation of bioelectrocatalytic transformations by means of the external magnet [117, 118]. Figure 8.14 shows the schematic magneto-switchable activation of redox enzymes in the presence of electron-relay–functionalized magnetic particles, for example, the bioelectrocatalyzed oxidation of glucose by glucose oxidase in the presence of the ferrocene **24**-functionalized magnetic particles. The magnetic attraction of the **24**-functionalized magnetic particles to the electrode results in the oxidation of the ferrocene units to the ferrocenyl cation components. The latter ferrocenyl cation relay units oxidize the redox center of glucose oxidase and activate the bioelectrocatalyzed oxidation of glucose. Retraction of the magnetic particles from the electrode support by means of the external magnet blocks the electrical communication between the electrode and the ferrocene units and inhibits the secondary bioelectrocatalyzed oxidation of glucose. Figure 8.13B displays the magnetic activation of the bioelectrocatalyzed oxidation of glucose, reflected by the electrocatalytic anodic current, as a result of the attraction of the functional magnetic particles to the electrode. Similarly, the retraction of the magnetic particles from the electrode blocks the bioelectrocatalytic oxidation of glucose. By the reversible attraction and removal of the magnetic particles onto the electrode and from the electrode, respectively, cyclic "ON" and "OFF" activation and deactivation of the bioelectrocatalytic process is accomplished. A similar approach was applied for the magnetic switching of the bioelectrocatalyzed reduction of nitrate to nitrite using nitrate reductase as a biocatalyst and the bipyridinium **25**-functionalized magnetic particles, $E° = -0.57$ V (vs SCE), as mediator units [118]. Upon attraction of the **25**-functionalized

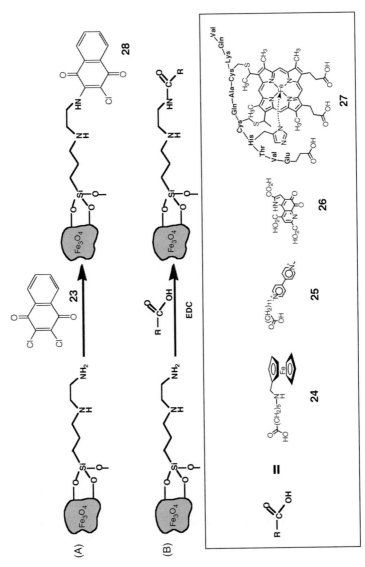

Fig. 8.12 Synthesis of relay-functionalized magnetic particles by the covalent linkage of redox-active units to magnetic particles functionalized with [3-(2-aminoethyl)aminopropyl]siloxane film: (A) Linkage of 2,3-dichloro-1,4-naphthoquinone **23** to the functionalized particles to yield the aminonaphthoquinone **28** functionalized magnetic particles. (B) Carbodiimide coupling of electron-relay carboxylic derivatives (**24**–**27**) to the amino groups of the siloxane layer. (C) Magnetic particles functionalized with the PQQ-NAD$^+$ dyad for the electrochemical activation of NAD$^+$-dependent enzymes. (Adapted from reference [118], Scheme 2, with permission).

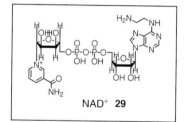

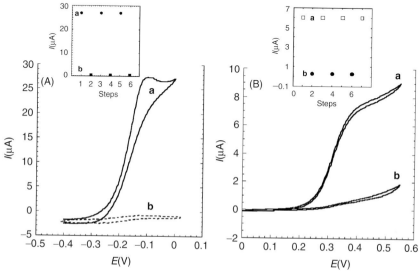

Fig. 8.12 (Continued)

Fig. 8.13 Bioelectrocatalysis in the presence of functionalized magnetic particles controlled by means of the external magnet: (A) Cyclic voltammograms at a Au electrode in the presence of the PQQ 26 functionalized magnetic particles (10 mg) and NADH (50 mM): (a) When the magnetic particles are attracted to the electrode. (b) When the magnetic particles are retracted from the electrode. Inset: Reversible changes of the electrocatalytic current (at $E = 0.05$ V vs SCE) upon attraction of the magnetic particles to the electrode (Points a) and retraction of the magnetic particles from the electrode (Points b). The data were obtained in 0.1 M Tris-buffer, pH = 8.0, in the presence of $CaCl_2$, 10 mM, under argon. Potential scan rate is 2 mV s^{-1}. (B) Cyclic voltammograms at a Au electrode in the presence of the ferrocene (24)-functionalized magnetic particles (10 mg), glucose (10 mM) and GOx (1 mg mL^{-1}): (a) The redox functions of the particles are switched "ON" by their attraction to the electrode. (b) The redox functions of the particles are switched "OFF" by their retraction from the electrode. Inset: Reversible changes of the electrocatalytic current (measured at $E = 0.45$ V vs SCE) upon attraction of the magnetic particles to the electrode (Points a) and retraction (Points b) of the magnetic particles from the electrode. The data were recorded in 0.1 M phosphate buffer, pH = 7.0) under argon. Potential scan rate is 5 mV s^{-1}. (Adapted from ref. [118], Figures 2(C) and 4(B), with permission).

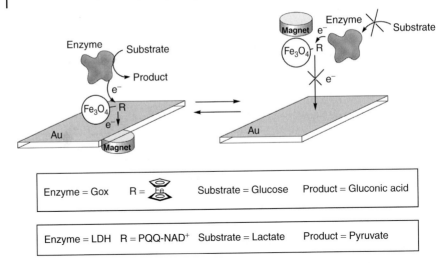

Fig. 8.14 Bioelectrocatalytic reactions of functional magnetic particles controlled by means of the external magnet: switching "ON" and "OFF" the bioelectrocatalytic oxidation of glucose in the presence of glucose oxidase (GOx) and ferrocene (**24**) functionalized magnetic particles or oxidation of lactate in the presence of lactate dehydrogenase (LDH) and PQQ-NAD$^+$-functionalized magnetic particles. (Adapted from reference [118], Schemes 1 and 3, with permission).

magnetic particles to the electrode and application of the potential of $E = -0.7$ V to the electrode, the bipyridinium radical cation-mediated bioelectrocatalyzed reduction of nitrate to nitrite occurred at a rate corresponding to 6.5×10^{-3} mM·min^{-1}, whereas retraction of the particles from the electrode by positioning the magnet above the electrochemical cell blocked the bioelectrocatalytic process. By switching the position of the external magnet below the electrode surface and above the cell, the electrocatalyzed reduction of NO$_3^-$ could be reversibly switched between "ON" and "OFF" states, respectively.

Functional magnetic particles were also employed for the cyclic activation and deactivation of NAD$^+$-dependent enzymes by means of an external magnet. It was reported [139] that a covalently linked pyrroloquinoline quinone (PQQ)-NAD$^+$ monolayer associated with an electrode provides an active interface for the activation of NAD$^+$-dependent enzymes, for example, lactate dehydrogenase. The biocatalyzed oxidation of the substrate yields NADH that is oxidized to NAD$^+$ by PQQ. The resulting PQQH$_2$ is electro-oxidized to PQQ, enabling the electrochemically driven biocatalyzed oxidation of the substrate, for example, lactate. Accordingly, the PQQ **26**-functionalized magnetic particles were reacted with N^6-(2-aminoethyl)-β-nicotinamide dinucleotide **29** [140] to yield the PQQ-NAD$^+$-modified particles, Figure 8.12C. The resulting particles were then applied for the magnetoswitching of the bioelectrocatalytic functions of the NAD$^+$-dependent enzyme lactate dehydrogenase (LDH) by means of the external magnet, Figure 8.14. Attraction of the PQQ-NAD$^+$-functionalized magnetic particles to the electrode in the presence of LDH activates the bioelectrocatalyzed oxidation of lactate to pyruvate. The biocatalyzed reduction of the NAD$^+$-cofactor unit by lactate is followed by the oxidation of the resulting NADH by the PQQ electrocatalyst. The application of the appropriate potential ($E = 0.05$ V vs SCE) for oxidation of the resulting PQQH$_2$ regenerates the electrocatalyst. This enables the continuous bioelectrocatalyzed oxidation of lactate to pyruvate in the presence of LDH. The electrocatalytic anodic current developed by the system increases in magnitude as the lactate concentration is elevated, and it levels-off to the value of the current density of $i_{max} = 1.8$ µA cm^{-2} at a concentration of lactate corresponding to 50 mM. Analysis of the electrocatalyzed oxidation of lactate to pyruvate by LDH under steady state electrolysis indicated that pyruvate is generated at a rate corresponding to 0.13 mM min^{-1}. Retraction of the functionalized magnetic particles from the electrode by means of the external magnet prohibited the bioelectrocatalyzed oxidation of lactate, and by the cyclic positioning of the external magnet below and above the cell, the process could be reversibly switched between "ON" and "OFF" states, respectively.

The ability to control bioelectrocatalytic transformations by means of an external magnet was utilized to develop selective dual biosensing systems [141]. The specific simultaneous electrochemical sensing of two substrates in an overlapping redox-potential region is a challenging topic in bioelectronics. Figure 8.15 outlines the concept for the dual sensing of lactate and glucose in the presence of the two oxidative enzymes: LDH and glucose oxidase (GOx) using the NAD$^+$-PQQ-functionalized magnetic particles and an external magnet. The electrode is modified with a monolayer of the ferrocene **24** and the NAD$^+$-PQQ-functionalized magnetic particles, the enzymes LDH and GOx, and the two substrates, lactate and glucose, are included in the system. Application of a potential on the electrode $E > 0.32$ V (vs SCE), while retracting the magnetic particles from the electrode, Figure 8.15A, results in the

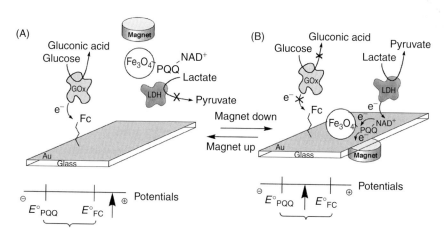

Fig. 8.15 Magneto-controlled dual biosensing of glucose (A) and lactate (B) in the presence of glucose oxidase (GOx), lactate dehydrogenase (LDH), magnetic particles functionalized with PQQ-NAD$^+$ and a Au electrode modified with a monolayer of ferrocene units. (Adapted from reference [118], Scheme 4, with permission).

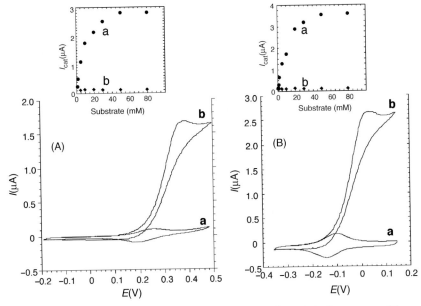

Fig. 8.16 Cyclic voltammograms of the ferrocene-modified Au electrode in the presence of the NAD$^+$-PQQ-functionalized magnetic particles: (A) When they are retracted from or (B) attracted to the electrode by the external magnet: (a) In the presence of GOx (1 mg mL^{-1}) and LDH (2 mg mL^{-1}). (b) In the presence of GOx (1 mg mL^{-1}), LDH (2 mg mL^{-1}), glucose (50 mM), and lactate (20 mM). Insets show the calibration plots of the amperometric responses of the system: (A) With the magnet in the up position and the applied potential $E = 0.50$ V. (a) At different concentrations of glucose. (b) At different concentrations of lactate. (B) With the magnet in the down position and the applied potential $E = 0.05$ V. (a) At different concentrations of lactate. (b) At different concentrations of glucose. The data were recorded against SCE under argon in the presence of the magnetic particles (20 mg) in 0.1 M Tris-HCl, pH = 7.0, CaCl$_2$ (10 mM). Potential scan rate, 5 mV s^{-1}. (Adapted from reference [118], Figure 9.8, with permission).

ferrocene-mediated bioelectrocatalyzed oxidation of glucose, Figure 8.16A, curve (b), whereas the bioelectrocatalyzed oxidation of lactate is blocked, Figure 8.16A, inset. Magnetic attraction of the NAD$^+$-PQQ-magnetic particles to the electrode, followed by sweeping the potential on the electrode in the range -0.13 V $< E < 0.32$ V (vs SCE), allows the PQQ-NAD$^+$-mediated oxidation of lactate in the presence of the NAD$^+$-dependent lactate dehydrogenase, Figure 8.15B. Since the ferrocene units are not oxidized in this potential range, the bioelectrocatalyzed oxidation of glucose is prohibited. Figure 8.16B, curve (b), depicts the electrocatalytic anodic current resulting in the biocatalyzed oxidation of lactate and the lack of electrical response of the system to glucose, Figure 8.16B, inset. The specific sensing of lactate is thus achieved by limiting the potential applied on the electrode. Note, however, that the specific analysis of glucose is accomplished only by retraction of the magnetic particles from the electrode by means of the external magnet, since both bioelectrocatalyzed transformations may proceed in the extended potential region.

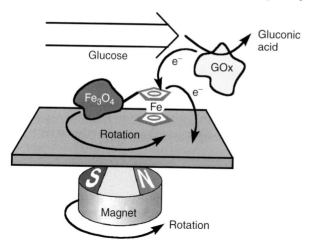

Fig. 8.17 Bioelectrocatalytic oxidation of glucose in the presence of glucose oxidase (GOx) and ferrocene-functionalized magnetic particles enhanced upon circular rotation of the particles by means of the external rotating magnet.

An important advance in the magnetic control of bioelectrocatalytic transformations was accomplished by magnetic attraction of the magnetic particles to the electrode support followed by the rotation of the magnetic particles on the electrode by means of an external rotating magnet [142]. The rotation of the magnetic particles turns the redox-functionalized magnetic particles into circularly rotating microelectrodes. As a result, redox-activated bioelectrocatalytic processes mediated by the functional particles are controlled by convection rather than by the diffusion of the respective substrate to the microelectrodes. Accordingly, enhanced amperometric responses of the particle-mediated bioelectrocatalytic processes are anticipated, and the resulting currents should be controlled by the rotation speed of the particles. Figure 8.17 displays schematically the amplified amperometric analysis of glucose by the rotation of ferrocene 24-functionalized magnetic particles on the electrode support. As the interaction of glucose oxidase (GOx) with the electron transfer mediator associated with the rotating magnetic particles is convection-controlled, the resulting electrocatalytic currents should relate to the square root of the rotation speed ($I_{cat} \propto \omega^{1/2}$). Figure 8.18 displays the bioelectrocatalytic currents obtained in the system at different rotation speeds of the external magnet. Clearly, the amperometric response increases as the rotation speed is elevated. Figure 8.18, inset, shows the analysis of variable concentrations of glucose by the particles rotating at a rotation speed of 400 rpm, curve (c), or 100 rpm, curve (b), as compared to the amperometric responses of the system to variable glucose concentrations where the particles are in a steel, nonrotating configuration, curve (a). The rotation of the particles (400 rpm) results in a ca. 15-fold increase in the amperometric responses of the bioelectrocatalytic process. The amperometric responses of the electrocatalytic transformations originating from the

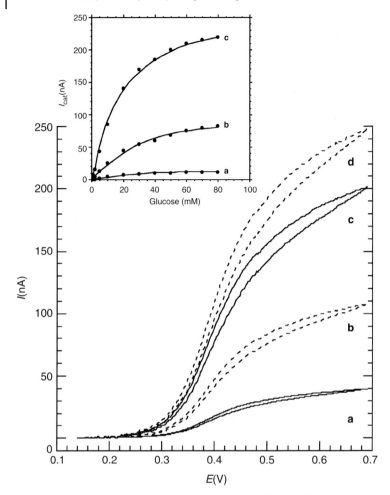

Fig. 8.18 Amplification of sensing events upon rotation of functionalized magnetic particles. Cyclic voltammograms of a Au electrode with the magnetically attracted 24-functionalized magnetic particles (6 mg) in the presence of GOx, 1×10^{-5} M, and glucose, 50 mM upon rotation of the magnet (rpm): **(a)** 0, **(b)** 10, **(c)** 100, **(d)** 400. Potential scan rate, 5 mV s^{-1}. Inset: Calibration plots for the amperometric detection of glucose ($E = 0.5$ V vs SCE) upon rotation of the magnet (rpm): **(a)** 0, **(b)** 100, **(c)** 400. The data were recorded in 0.1 M phosphate buffer, pH 7.0.
(Adapted from reference [142], Figure 9.1, with permission).

rotation of the redox-functionalized magnetic particles may be analyzed using the conventional rotating disc electrode (RDE) theory [143]. Analysis of the experimental currents obtained at different rotation speeds of the redox-functionalized magnetic particles in terms of Levich theory allows estimation of the interfacial electron transfer rate constant.

8.4
Conclusions and Perspectives

This chapter has addressed recent advances in the application of biomolecular–functionalized metal, semiconductor and magnetic particles for biosensing applications. We have described a variety of nanoparticle/biomaterial assemblies for advanced electrical biosensing. Metal nanoparticles reveal three general advantages in their functional application in electrochemical biosensing, and these provide great promise for further developments. (1) Their immobilization on surfaces generates a roughened conductive-high-surface area interface that enables the sensitive electrochemical detection of molecular and biomolecular analytes. The rapid progress in the synthetic functionalization of metal nanoparticles with molecular, polymer and biomaterial coatings suggests that many different biosensors could be tailored by such functional nanoparticles on surfaces. (2) The nanoparticles act as effective tags for the amplified bioaffinity assays of the respective bioanalytes. The amplification is accomplished by two general routes. By one method, the nanoparticle labels are dissolved upon sensing the analyte, and numerous ions are released as a result of a single recognition event. The electrochemical collection of the released ions and their subsequent stripping provides the indirect electrochemical readout of the analyte. The second amplification path utilizes the catalytic properties of the nanoparticles. The catalytic deposition of metals on the nanoparticle labels accompanying the sensing process accumulates new metals on the surface, and a high content of metal is formed as a result of the recognition process. The dissolution of the metal and its electrochemical stripping then provide the electrochemical readout signal for the analysis. Other amplification routes that employ metal nanoparticle labels (specifically, Au or Ag nanoparticles) for electroanalytical applications may be envisaged. The enhanced surface plasmon resonance shifts induced by nanoparticle labels [144] suggest that the functionalization of the particles by redox-active units may generate new sensing labels in coupled bioelectrochemical/surface plasmon resonance analysis systems [145]. (3) The conductivity properties of metal nanoparticles enable the design of biomaterial architectures with predesigned and controlled electrochemical functions. A recent report on the unprecedented effective electrical contacting of redox enzymes by means of reconstitution of the apoenzyme by its alignment on a single Au nanoparticle [80] paves the way for the construction of new semi-synthetic redox proteins. The extension of the approach to other metal nanoparticles, the application of metallic nanowires or other nanomaterials such as carbon nanotubes or semiconductive tubes for reconstitution and electrical contacting of redox proteins may provide interesting and challenging topics to follow. A great promise of metal nanoparticles in analytical applications, however, rests on the fundamental property of single-electron charging of quantum size particles [83]. The possibility of accumulating charges in metal nanoparticles suggests that catalytic and biocatalytic redox processes may control the charge associated with nanoparticles. As a result, the assembly of functional nanoparticles on ISFET devices may prove to be a new sensing configuration.

The nanoparticles add new dimension to electrical biosensing. The unique properties of nanoparticles lead to simple, highly sensitive, electroanalytical procedures that could not be accomplished by standard electrochemical methods. Also, nanoparticles that reveal high stability can substitute amplifying labels of limited stability such as enzymes or liposomes, with equivalent or improved sensitivities. In fact, many of the nanoparticle-based sensor devices discussed here reveal comparable or better sensitivities when compared to other sensing methods such as fluorescence, surface plasmon resonance (SPR), or microgravimetric quartz-crystal microbalance analyses.

While most of the biosensing applications of biomolecular-functionalized nanoparticles applied metal nanoparticles, the use of semiconductor nanoparticles and photoelectrochemistry as the readout signal holds great promise. Recent developments have revealed the practical utility of biomaterial–semiconductor nanoparticle hybrid systems for the photoelectrochemical detection of biorecognition events [81]. The size-controlled and tunable absorbance of semiconductor nanoparticles and the availability of semiconductor nanoparticles with different absorption properties indicate that arrays for the parallel photoelectrochemical analysis of biorecognition events could be achieved, that is, the functionalization of semiconductor nanoparticles of pre-designed size and composition with different biomaterials could enable the parallel detection of analytes, whereby the photocurrent action spectra act as the markers for the respective analytes.

To date, the bioelectroanalytical applications of functionalized nanoparticles rely on the collective properties of the nanoparticle ensemble associated with a conductive surface. A challenging topic is, however, the miniaturization of these assemblies with the ultimate goal of using a single functionalized nanoparticle for the electronic sensing events. Some initiatives directed to the application of biomaterial nanoelements (carbon nanotubes) as nanoscale electrical sensors, have recently been reported, yet the important challenging objectives are still ahead of us.

The introduction of biomolecular-functionalized nanoparticles as active components in bioelectroanalysis has a considerable impact on recent developments in electroanalytical and bioelectroanalytical sciences. The unique properties of nanoparticles and the nanoscale dimensions of the active sensing elements suggest that future interdisciplinary efforts of chemists, physicists, biologists and material scientists could yield new generations of miniaturized sensing devices. It is expected that future innovative research will lead to new particle-based sensing strategies that, coupled with other major technological advances, will result in effective, easy-to-use handheld analytical devices.

Acknowledgment

The research at New Mexico State University (Joseph Wang) is supported by the US National Science Foundation (Grant Number CHE 0209707). The research at the Hebrew University of Jerusalem (Itamar Wilnar and Eugenii Katz) on Biomaterial Nanoparticle Hybrid systems is supported by the German-Israeli Program (DIP).

References

1 R.F. Khairutdinov, *Colloid J.*, **1997**, *59*, 535–548.
2 *Single Charge Tunnelling and Coulomb Blockade Phenomena in Nanostructures*, H. Gzabezt, M.H. Devozet (Eds.), NATO ASI Ser. B., Vol. 294, Plenum Press, New York, 1996.
3 P. Mulvaney, *Langmuir*, **1996**, *12*, 788–800.
4 M.M. Alvarez, J.T. Khoury, T.G. Schaaff, M.N. Shafigullin, I. Vezmar, R.L. Whetten, *J. Phys. Chem. B*, **1997**, *101*, 3706–3712.
5 A.P. Alivisatos, *J. Phys. Chem.*, **1996**, *100*, 13226–13329.
6 L.E. Brus, *Appl. Phys. A*, **1991**, *53*, 465–474.
7 L.N. Lewis, *Chem. Rev.*, **1993**, *93*, 2693–2730.
8 V. Kesavan, P.S. Sivanand, S. Chandrasekaran, Y. Koltypin, A. Gedankin, *Angew. Chem., Int. Ed.*, **1999**, *38*, 3521–3523.
9 D.L. Klein, R. Roth, A.K.L. Kim, A.P. Alivisatos, P.L. McEuen, *Nature*, **1997**, *389*, 699–701.
10 T. Sato, H. Ahmed, D. Brown, B.F.G. Johnson, *J. Appl. Phys.*, **1997**, *82*, 696–701.
11 R.S. Ingram, M.J. Hostetler, R.W. Murray, T.G. Schaaff, J.T. Khoury, R.L. Whetten, T.P. Bigioni, D.K. Guthrie, P.N. First, *J. Am. Chem. Soc.*, **1997**, *119*, 9279–9280.
12 T. Sato, H. Ahmed, *Appl. Phys. Lett.*, **1997**, *70*, 2759–2761.
13 H. Weller, *Angew. Chem., Int. Ed.*, **1998**, *37*, 1658–1659.
14 J.H. Fendler, *Chem. Mater.*, **1996**, *8*, 1616–1624.
15 A.K. Boal, F. Ilhan, J.E. DeRouchey, T. Thurn-Albrecht, T.P. Russell, V.M. Rotello, *Nature*, **2000**, *404*, 746–748.
16 V.I. Chegel, O.A. Raitman, O. Lioubashevski, Y. Shirshov, E. Katz, I. Willner, *Adv. Mater.*, **2002**, *14*, 1549–1553.
17 A.C. Templeton, D.E. Cliffel, R.W. Murray, *J. Am. Chem. Soc.*, **1999**, *121*, 7081–7089.
18 M.J. Hostetler, S.J. Green, J.J. Stokes, R.W. Murray, *J. Am. Chem. Soc.*, **1996**, *118*, 4212–4213.
19 R.S. Ingram, R.W. Murray, *Langmuir*, **1998**, *14*, 4115–4121.
20 J.J. Pietron, R.W. Murray, *J. Phys. Chem. B*, **1999**, *103*, 4440–4446.
21 H. Imahori, S. Fukuzumi, *Adv. Mater.*, **2001**, *13*, 1197–1199.
22 H. Imahori, M. Arimura, T. Hanada, Y. Nishimura, I. Yamazaki, Y. Sakata, S. Fukuzumi, *J. Am. Chem. Soc.*, **2001**, *123*, 335–336.
23 N. Herron, D.L. Thorn, *Adv. Mater.*, **1998**, *10*, 1173–1184.
24 J.F. Ciebien, R.T. Clay, B.H. Sohn, R.E. Cohen, *New J. Chem.*, **1998**, *22*, 685–691.
25 R. Gangopadhyay, A. De, *Chem. Mater.*, **2000**, *12*, 608–622.
26 K.V. Sarathy, P.J. Thomas, G.U. Kulkarni, C.N.R. Rao, *J. Phys. Chem. B*, **1999**, *103*, 399–401.
27 R. Blonder, L. Sheeney, I. Willner, *Chem. Commun.*, **1998**, 1393–1394.
28 T. Zhu, X. Zhang, J. Wang, X. Fu, Z. Liu, *Thin Solid Films*, **1998**, *327–329*, 595–598.
29 K. Bandyopadhyay, V. Patil, K. Vijayamohanan, M. Sastry, *Langmuir*, **1997**, *13*, 5244–5248.
30 A.B. Kharitonov, A.N. Shipway, I. Willner, *Anal. Chem.*, **1999**, *71*, 5441–5443.
31 T. Trindade, P. O'Brien, N.L. Pickett, *Chem. Mater.*, **2001**, *13*, 3843–3858.
32 J.T. Lue, *J. Phys. Chem. Solids*, **2001**, *62*, 1599–1612.
33 K. Grieve, P. Mulvaney, F. Grieser, *Curr. Opin. Colloid Interface Sci.*, **2000**, *5*, 168–172.
34 A.N. Shipway, E. Katz, I. Willner, *Chem. Phys. Chem.*, **2000**, *1*, 18–52.
35 M. Brust, C.J. Kiely, *Colloids Surf., A*, **2002**, *202*, 175–186.
36 W.P. McConnell, J.P. Novak, L.C. Brousseau, R.R. Fuierer, R.C. Tenent, D.L. Feldheim, *J. Phys. Chem. B*, **2000**, *104*, 8925–8930.
37 E. Katz, I. Willner, J. Wang, *Electroanalysis*, **2004**, *16*, 19–44.

38 E. Katz, A.N. Shipway, I. Willner, in *Nanoparticles – From Theory to Applications* (Ed.: G. Schmid), Wiley-VCH, Weinheim, **2003**, Chapter 6, 368–421.

39 J. Wang, *Anal. Chim. Acta*, **2003**, *500*, 247–257.

40 L. Cognet, C. Tardin, D. Boyer, D. Choquet, P. Tamarat, B. Lounis, *Proc. Natl. Acad. Sci. U.S.A.*, **2003**, *100*, 11350–11355.

41 Y.F. Wang, D.W. Pang, Z.L. Zhang, H.Z. Zheng, J.P. Cao, J.T. Shen, *J. Med. Virol.*, **2003**, *70*, 205–211.

42 K. Sokolov, M. Follen, J. Aaron, I. Pavlova, A. Malpica, R. Lotan, R. Richards-Kortum, *Cancer Res.*, **2003**, *63*, 1999–2004.

43 R. Elghanian, J.J. Storhoff, R.C. Mucic, R.L. Letsinger, C.A. Mirkin, *Science*, **1997**, *277*, 1078–1081.

44 J.J. Storhoff, R. Elghanian, R.C. Mucic, C.A. Mirkin, R.L. Letsinger, *J. Am. Chem. Soc.*, **1998**, *120*, 1959–1964.

45 R.A. Reynolds, III, C.A. Mirkin, R.L. Letsinger, *J. Am. Chem. Soc.*, **2000**, *122*, 3795–3796.

46 G.R. Souza, J.H. Miller, *J. Am. Chem. Soc.*, **2001**, *123*, 6734–6735.

47 M. Bruchez, Jr., M. Moronne, P. Gin, S. Weiss, A.P. Alivisatos, *Science*, **1998**, *281*, 2013–2015.

48 O.D. Velev, E.W. Kaler, *Langmuir*, **1999**, *15*, 3693–3698.

49 I. Willner, E. Katz, *Angew. Chem., Int. Ed.*, **2000**, *39*, 1180–1218.

50 F.A. Armstrong, G.S. Wilson, *Electrochem. Acta*, **2000**, *45*, 2623–2645.

51 I. Willner, E. Katz, B. Willner, *Electroanalysis*, **1997**, *13*, 965–977.

52 L. Habermüller, M. Mosbach, W. Schuhmann, *Fresenius' J. Anal. Chem.*, **2000**, *366*, 560–568.

53 I. Willner, E. Katz, B. Willner, in *Biosensors and their Applications* (Eds.: V.C. Yang, T.T. Ngo), Kluwer Academic Publishers, New York, **2000**, Chapter 4, 47–98.

54 I. Willner, B. Willner, E. Katz, *Rev. Mol. Biotechnol.*, **2002**, *82*, 325–355.

55 F.A. Armstrong, H.A. Heering, J. Hirst, *Chem. Soc. Rev.*, **1997**, *26*, 169–179.

56 E. Katz, I. Willner, A.B. Kotlyar, *J. Electroanal. Chem.*, **1999**, *479*, 64–68.

57 E. Katz, A.N. Shipway, I. Willner, in *Handbook of Fuel Cells – Fundamentals, Technology, Applications* (Eds.: W. Vielstich, H. Gasteiger, A. Lamm), Vol. 1, Part 4, Wiley, Chichester, UK, **2003**, Chapter 21, 355–381.

58 E. Katz, I. Willner, *J. Am. Chem. Soc.*, **2003**, *125*, 6803–6813.

59 T. Chen, S.C. Barton, G. Binyamin, Z. Gao, Y. Zhang, H.-H. Kim, A. Heller, *J. Am. Chem. Soc.*, **2001**, *123*, 8630–8631.

60 P.N. Bartlett, P. Tebbutt, R.G. Whitaker, *Prog. React. Kinet.*, **1991**, *16*, 55–155.

61 I. Willner, A. Riklin, B. Shoham, D. Rivenzon, E. Katz, *Adv. Mater.*, **1993**, *5*, 912–915.

62 I. Willner, E. Katz, A. Riklin, R. Kasher, *J. Am. Chem. Soc.*, **1992**, *114*, 10965–10966.

63 Y. Degani, A. Heller, *J. Phys. Chem.*, **1987**, *91*, 1285–1289.

64 Y. Degani, A. Heller, *J. Am. Chem. Soc.*, **1988**, *110*, 2615–2620.

65 W. Schuhmann, T.J. Ohara, H.-L. Schmidt, A. Heller, *J. Am. Chem. Soc.*, **1991**, *113*, 1394–1397.

66 A. Heller, *Acc. Chem. Res.*, **1990**, *23*, 128–134.

67 S. Cosnier, *Electroanalysis*, **1997**, *9*, 894–902.

68 I. Willner, E. Katz, N. Lapidot, P. Bäuerle, *Bioelectrochem. Bioenerg.*, **1992**, *29*, 29–45.

69 S.A. Emr, A.M. Yacynych, *Electroanalysis*, **1995**, *7*, 913–923.

70 A. Badia, R. Carlini, A. Fernandez, F. Battaglini, S.R. Mikkelsen, A.M. English, *J. Am. Chem. Soc.*, **1993**, *115*, 7053–7060.

71 I. Willner, V. Heleg-Shabtai, R. Blonder, E. Katz, G. Tao, A.F. Bückmann, A. Heller, *J. Am. Chem. Soc.*, **1996**, *118*, 10321–10322.

72 E. Katz, A. Riklin, V. Heleg-Shabtai, I. Willner, A.F. Bückmann, *Anal. Chim. Acta*, **1999**, *385*, 45–58.

73 M. Zayats, E. Katz, I. Willner, *J. Am. Chem. Soc.*, **2002**, *124*, 14724–14735.

74 O.A. Raitman, E. Katz, A.F. Bückmann, I. Willner, *J. Am. Chem. Soc.*, **2002**, *124*, 6487–6496.
75 O.A. Raitman, F. Patolsky, E. Katz, I. Willner, *Chem. Commun.*, **2002**, 1936–1937.
76 J. Zhao, R.W. Henkens, J. Stonehurner, J.P. O'Daly, A.L. Crumbliss, *J. Electroanal. Chem.*, **1992**, *327*, 109–119.
77 A.L. Crumbliss, S.C. Perine, J. Stonehurner, K.R. Tubergen, J. Zhao, R.W. Henkens, J.P. O'Daly, *Biotechnol. Bioeng.*, **1992**, *40*, 483–490.
78 J. Zhao, J.P. O'Daly, R.W. Henkens, J. Stonehurner, A.L. Crumbliss, *Biosens. Bioelectron.*, **1996**, *11*, 493–502.
79 C.-X. Lei, S.-Q. Hu, G.-L. Shen, R.-Q. Yu, *Talanta*, **2003**, *59*, 981–988.
80 Y. Xiao, F. Patolsky, E. Katz, J.F. Hainfeld, I. Willner, *Science*, **2003**, *299*, 1877–1881.
81 V. Pardo-Yissar, E. Katz, J. Wasserman, I. Willner, *J. Am. Chem. Soc.*, **2003**, *125*, 622–623.
82 M.L. Curri, A. Agostiano, G. Leo, A. Mallardi, P. Cosma, M.D. Monica, *Mater. Sci. Eng. C*, **2002**, *22*, 449–452.
83 J.F. Hicks, D.T. Miles, R.W. Murray, *J. Am. Chem. Soc.*, **2002**, *124*, 13322–13328.
84 J.F. Hicks, F.P. Zamborini, A.J. Osisek, R.W. Murray, *J. Am. Chem. Soc.*, **2001**, *123*, 7048–7053.
85 S. Chen, R.W. Murray, *J. Phys. Chem. B*, **1999**, *103*, 9996–10000.
86 J.F. Hicks, F.P. Zamborini, R.W. Murray, *J. Phys. Chem. B*, **2002**, *106*, 7751–7757.
87 S. Chen, R.W. Murray, S.W. Feldberg, *J. Phys. Chem. B*, **1998**, *102*, 9898–9907.
88 L. He, M.D. Musick, S.R. Nicewarner, F.G. Salinas, S.J. Benkovic, M.J. Natan, C.D. Keating, *J. Am. Chem. Soc.*, **2000**, *122*, 9071–9077.
89 S. Kubitschko, J. Spinke, T. Brückner, S. Pohl, N. Oranth, *Anal. Biochem.*, **1997**, *253*, 112–122.
90 L.A. Lyon, M.D. Musick, M.J. Natan, *Anal. Chem.*, **1998**, *70*, 5177–5183.
91 P. Englebienne, A.V. Hoonacker, M. Verhas, *Analyst*, **2001**, *126*, 1645–1651.
92 C.M. Niemeyer, *Angew. Chem., Int. Ed.*, **2001**, *40*, 4128–4158.
93 J. Wang, *Anal. Chim. Acta*, **2003**, *500*, 247–257.
94 J. Wang, *Stripping Analysis*, VCH, New York, **1985**.
95 N. Zhu, H. Cai, P. He, Y. Fang, *Anal. Chim. Acta*, **2003**, *481*, 181–189.
96 J. Wang, J. Li, A.J. Baca, J. Hu, F. Zhou, W. Yan, D.-W. Pang, *Anal. Chem.*, **2003**, *75*, 3941–3945.
97 J. Wang, D. Xu, A.-N. Kawde, R. Polsky, *Anal. Chem.*, **2001**, *73*, 5576–5581.
98 L. Authier, C. Grossiord, P. Brossier, B. Limoges, *Anal. Chem.*, **2001**, *73*, 4450–4456.
99 H. Cai, Y. Xu, N. Zhu, P. He, Y. Fang, *Analyst*, **2002**, *127*, 803–808.
100 H. Cai, N. Zhu, Y. Jiang, P. He, Y. Fang, *Biosens. Bioelectron.*, **2003**, *18*, 1311–1319.
101 J. Wang, G. Liu, Q. Zhu, *Anal. Chem.*, **2003**, *75*, 6218–6222.
102 T.G. Drummond, M.G. Hill, J.K. Barton, *Nat. Biotechnol.*, **2003**, *21*, 1192–1199.
103 T.M.-H. Lee, L.-L. Li, I.-M. Hsing, *Langmuir*, **2003**, *19*, 4338–4343.
104 J. Wang, R. Polsky, D. Xu, *Langmuir*, **2001**, *17*, 5739–5741.
105 T. Lee, M. Carles, I. Hsing, *Lab Chip*, **2003**, *3*, 100–105.
106 J. Wang, O. Rincón, R. Polsky, E. Dominguez, *Electrochem. Commun.*, **2003**, *5*, 83–86.
107 J. Richter, *Physica E*, **2003**, *16*, 157–173.
108 E. Braun, Y. Eichen, U. Sivan, G. Ben-Yoseph, *Nature*, **1998**, *391*, 775–778.
109 Y. Eichen, E. Braun, U. Sivan, G. Ben-Yoseph, *Acta Polym.*, **1998**, *49*, 663–670.
110 M. Mertig, L.C. Ciacchi, R. Seidel, W. Pompe, A. De Vita, *Nano Lett.*, **2002**, *2*, 841–844.
111 J. Richter, R. Seidel, R. Kirsch, M. Mertig, W. Pompe, J. Plaschke, H.K. Schackert, *Adv. Mater.*, **2000**, *12*, 507–509.
112 S.-J. Park, T.A. Taton, C.A. Mirkin, *Science*, **2002**, *295*, 1503–1506.
113 M. Urban, R. Müller, W. Fritzsche, *Rev. Sci. Instrum.*, **2003**, *74*, 1077–1081.
114 A. Doron, E. Katz, I. Willner, *Langmuir*, **1995**, *11*, 1313–1317.

115 H. Cai, C. Xu, P. He, Y. Fang, *J. Electroanal. Chem.*, **2001**, *510*, 78–85.
116 I. Willner, E. Katz, *Angew. Chem., Int. Ed.*, **2003**, *42*, 4576–4588.
117 R. Hirsch, E. Katz, I. Willner, *J. Am. Chem. Soc.*, **2000**, *122*, 12053–12054.
118 E. Katz, L. Sheeney-Haj-Ichia, I. Willner, *Chem. Eur. J.*, **2002**, *8*, 4138–4148.
119 J. Wang, A.-N. Kawde, A. Erdem, M. Salazar, *Analyst*, **2001**, *126*, 2020–2024.
120 J. Wang, A.-N. Kawde, *Electrochem. Commun.*, **2002**, *4*, 349–352.
121 M. Ozsoz, A. Erdem, K. Kerman, D. Ozkan, B. Tugrul, N. Topcuoglu, H. Ekren, M. Taylan, *Anal. Chem.*, **2003**, *75*, 2181–2187.
122 J. Wang, R. Polsky, A. Merkoçi, K. Turner, *Langmuir*, **2003**, *19*, 989–991.
123 J. Wang, D. Xu, R. Polsky, *J. Am. Chem. Soc.*, **2002**, *124*, 4208–4209.
124 H. Cai, Y. Wang, P. He, Y. Fang, *Anal. Chim. Acta*, **2002**, *469*, 165–172.
125 J. Wang, G. Liu, R. Polsky, A. Merkoçi, *Electrochem. Commun.*, **2002**, *4*, 722–726.
126 J. Wang, G. Liu, A. Merkoçi, *Anal. Chim. Acta*, **2003**, *482*, 149–155.
127 E. Katz, A.N. Shipway, I. Willner, in *Nanoscale Materials* (Eds.: L.M. Liz-Marzan, P. Kamat), Kluwer, **2003**, Chapter 2, 5–78.
128 J. Wang, G. Liu, A. Merkoçi, *J. Am. Chem. Soc.*, **2003**, *125*, 3214–3215.
129 S.R. Nicewarner-Pena, R.G. Freeman, B.D. Reiss, L. He, D.J. Pena, I.D. Walton, R. Cromer, C.D. Keating, M.J. Natan, *Science*, **2001**, *294*, 137–141.
130 J. Wang, G. Liu, G. Rivas, *Anal. Chem.*, **2003**, *75*, 4667–4671.
131 A. Kawde, J. Wang, *Electroanalysis*, **2004**, *16*, 101–107.
132 J. Wang, G. Liu, M.R. Jan, Q. Zhu, *Electrochem. Commun.*, **2003**, *5*, 1000–1004.
133 D. Trau, W.J. Yang, M. Seydack, F. Carusu, N.-T. Yu, R. Renneberg, *Anal. Chem.*, **2002**, *74*, 5480–5486.
134 L. Alfonta, A. Singh, I. Willner, *Anal. Chem.*, **2001**, *73*, 91–102.
135 I. Willner, F. Patolsky, J. Wasserman, *Angew. Chem., Int. Ed.*, **2001**, *40*, 1861–1864.
136 L. Shen, P.E. Laibinis, T.A. Hatton, *Langmuir*, **1999**, *15*, 447–453.
137 E. Katz, T. Lötzbeyer, D.D. Schlereth, W. Schuhmann, H.-L. Schmidt, *J. Electroanal. Chem.*, **1994**, *373*, 189–200.
138 I. Willner, A. Riklin, *Anal. Chem.*, **1994**, *66*, 1535–1539.
139 A. Bardea, E. Katz, A.F. Bückmann, I. Willner, *J. Am. Chem. Soc.*, **1997**, *119*, 9114–9119.
140 A.F. Bückmann, V. Wray, *Biotech. Appl. Biochem.*, **1992**, *15*, 303–310.
141 E. Katz, L. Sheeney-Haj Ichia, A.F. Bückmann, I. Willner, *Angew. Chem., Int. Ed.*, **2002**, *41*, 1343–1346.
142 E. Katz, I. Willner, *J. Am. Chem. Soc.*, **2002**, *124*, 10290–10291.
143 A.J. Bard, L.R. Faulkner, *Electrochemical Methods: Fundamentals and Applications*, Wiley, New York, **1980**.
144 L.A. Lyon, M.D. Musick, M.J. Natan, *Anal. Chem.*, **1998**, *70*, 5177–5183.
145 M. Zayats, S.P. Pogorelova, A.B. Kharitonov, O. Lioubashevski, E. Katz, I. Willner, *Chem. Eur. J.*, **2003**, *9*, 6108–6114.

9
DNA-templated Electronics

Kinneret Keren, Uri Sivan, and Erez Braun

9.1
Introduction and Background

Molecular electronics deals with the electronic properties of molecules and their incorporation into devices and circuits. Technological advances in top-down nanofabrication technologies have facilitated contacting various single-molecule devices [1, 2] as well as simple circuits [3–5]. Notwithstanding this impressive progress, the integration of a large number of individual devices into a functional circuit remains an outstanding challenge. There is no paradigm for large-scale integration in molecular electronics analogous to integrated circuits in microelectronics. One of the promising routes toward integrated molecular electronics relies on self-assembly based on the recognition between molecular building blocks [2]. This approach is inspired by biology, which provides remarkable examples of functional systems assembled from molecular building blocks. The entire structural information must somehow be encoded into the building blocks, which then self-assemble, by virtue of their molecular recognition properties, into large-scale constructs.

Supramolecular chemistry aims at developing ways of designing and chemically synthesizing molecules that self-assemble into desired structures [6]. However, the design and synthesis of appropriate molecular building blocks is highly nontrivial, and the attained complexity is presently limited to relatively simple structures. Alternatively, one can borrow the remarkable molecular recognition properties of biological molecules to direct the self-assembly process. This approach has been successfully employed to assemble a broad range of nonbiological constructs as reviewed in [7]. Self-assembly of functional systems such as electronic circuits is, however, more demanding since the relationship between structure and function is, in general, complex. Minor changes in structure may dramatically influence its functionality. For example, minute changes in the layout of molecular objects can have a large effect on the electronic connectivity between them. The use of biological molecules to realize molecular-scale electronics is further complicated by the fact that all biological molecules are electrically insulating.

The present review describes our efforts to develop a framework for harnessing the biological machinery and its working principles for self-assembly of molecular electronics. Several years ago, our group proposed an approach combining the molecular recognition properties of DNA molecules and their related proteins with the

Bioelectronics. Edited by Itamar Willner and Eugenii Katz
Copyright © 2005 WILEY-VCH Verlag GmbH & Co. KGaA, Weinheim
ISBN: 3-527-30690-0

electronic functionality of semiconductors and metals [8, 9]. In this approach, termed *DNA-templated electronics*, the molecular recognition and self-assembly capabilities of the biomolecules serve first to construct an elaborate DNA scaffold with well-defined molecular addresses. Electronic functionality is then instilled to the scaffold by localizing electronic devices at desired addresses and interconnecting them by DNA-templated wires.

An outline of DNA-templated electronics is presented in the following section. DNA metallization, developed as a way to convert the insulating DNA molecules into conductive wires, is discussed in Section 9.3. Section 9.4 introduces the concept of "sequence-specific molecular lithography" on DNA templates [10], which provides a framework for defining the molecular circuit architecture, in analogy with photolithography in conventional microelectronics. Our molecular lithography relies on homologous genetic recombination processes carried out by the RecA protein from *Escherichia coli* bacteria. It facilitates sequence-specific patterning of DNA coating with metal, localization of labeled molecular objects at desired addresses, growing metal islands at specific sites along the DNA scaffold, and generating molecularly accurate stable DNA junctions. Relying on sequence-specific molecular lithography, we demonstrate in Section 9.5 self-assembly of a DNA-templated carbon nanotube field-effect transistor (FET) [11]. The development of sequence-specific molecular lithography and the realization of a DNA-templated transistor promote DNA-templated electronics as a realistic strategy for self-assembly of molecular electronics. The realization of complex DNA-templated circuits remains an outstanding challenge requiring additional new concepts and techniques. Our perspectives on DNA-templated electronics and more generally on biologically directed assembly of molecular electronics are discussed in Section 9.6.

9.2
DNA-templated Electronics

The assembly of DNA-templated electronics comprises two steps. First, the biological machinery is employed to assemble a DNA template or a scaffold with well-defined molecular addresses. Then, electronic functionality is instilled by localization of molecular devices at specific addresses along the scaffold and conversion of the DNA template into a conductive network, interconnecting devices to each other and to predefined electrode pads. DNA was chosen as the main building block because of its elaborate molecular recognition properties, the sophisticated enzymatic machinery allowing its manipulation and replication, and its mechanical strength and chemical stability. Unlike conventional semiconductor substrates, DNA scaffolds provide the high density of molecular addresses needed for self-assembly of elaborate molecular-scale electronics.

Figure 9.1 depicts a possible scheme for DNA-templated assembly of molecular-scale electronics. This scheme presents a heuristic solution to some of the major challenges faced by molecular electronics, namely, precise localization of a large number of molecular devices, interdevice wiring and electrically interfacing between the molecular

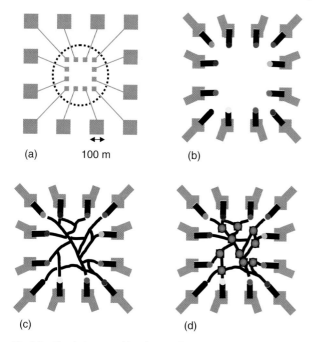

Fig. 9.1 Heuristic assembly scheme of a DNA-templated electronic circuit. (a) Gold pads are defined on an inert substrate. (b–d) correspond to the circle in (a) at different stages of circuit construction. (b) Oligonucleotides of different sequences are attached to the different pads. (c) DNA network is constructed and bound to the oligonucleotides on the gold electrodes by self-assembly. (d) Metal clusters or molecular electronic devices are localized on the DNA network. The DNA molecules are finally converted into metallic wires rendering the construct a functional electronic circuit. Note that the figures are not to scale – the metallic clusters are nanometer size, while the electrode pads measure micrometers. Typical spacing between scaffold junctions would be in the 10-nm range.

and the macroscopic worlds. Realization of this scheme depends on the developments of concepts and techniques dealing with the following aspects:

- Construction of elaborate DNA scaffolds with well-defined addresses and connectivities.
- Conversion of DNA molecules into conductive wires.
- Attachment of single DNA molecules to macroscopic electrodes.
- Localization and electrical contacting of molecular elements at specific addresses on the DNA network.

A prerequisite for DNA-templated electronics is the ability to transform the structural connectivity of the scaffold DNA network into electronic connectivity. The intrinsic conductivity of bare DNA is too low for using it directly as a molecular wire [8, 12–14].

DNA molecules can be converted into conductive wires by coating them with metal as demonstrated in [8]. Section 9.3 discusses DNA metallization in more detail.

Realization of DNA-templated devices and circuits requires tools for defining the circuit architecture analogous to photolithography in conventional microelectronics. This includes the formation of junctions, wire patterning at molecular resolutions, and molecularly accurate device localization. All these features are provided by our sequence-specific molecular lithography [10], described in Section 9.4. The glass masks used in conventional lithography are replaced by the information encoded in the DNA sequence, while the photoresist is replaced by the RecA protein. Relying on molecular lithography, we were able to demonstrate a fully self-assembled DNA-templated carbon nanotube FET (see Section 9.5). Thus, we have taken the concept of DNA-templated electronics all the way to the realization of a functional self-assembled device.

The compatibility between biomolecules and the nonbiological materials and processes used along the way presents a major hurdle. Processes adapted from biology, microelectronics and chemistry must be modified to match each other without losing biological or electronic functionality. The realization of the DNA-templated FET demonstrates that this demanding integration is nonetheless possible. The tools and concepts developed along the way were kept as general as possible so that they could in principle be extended to large-scale circuits. Yet, we have made little progress in the integration of numerous individual devices into functional circuits. Some ideas concerning the extension of DNA-templated electronics to large-scale circuits are discussed in Section 9.6.

9.3
DNA Metallization

The potential use of DNA in molecular-scale electronics led to the development of various DNA metallization schemes (see [15] and references therein). Continuous metallized DNA wires are fabricated in two steps. First, metallic nucleation sites are formed along the DNA molecule either by binding metal ions and subsequently reducing them [8, 10, 16] or by binding small metallic particles. The nucleation sites serve as catalysts for further selective deposition of metal from solution to form continuous conductive wires.

DNA metallization was first demonstrated in [8] using the process described schematically in Figure 9.2. Positive silver ions were localized along the poly-anion phosphate backbone of the DNA by ion exchange. A reducing agent, hydroquinone, was added after washing off excess silver, resulting in the formation of small metallic silver clusters along the DNA. These clusters served as seeds for subsequent silver deposition, leading to the formation of continuous silver wires, about 100 nm wide and 12 to 16 μm long (Figure 9.3a). The resistance of these wires was measured and found to be several MΩ (Figure 9.3). This work also demonstrated an approach to interfacing single DNA molecules with macroscopic electrodes. Electrode pads were imparted with DNA recognition properties by attaching oligonucleotides to them [8]. Individual electrodes could be assigned distinct recognition properties using oligonucleotides having different

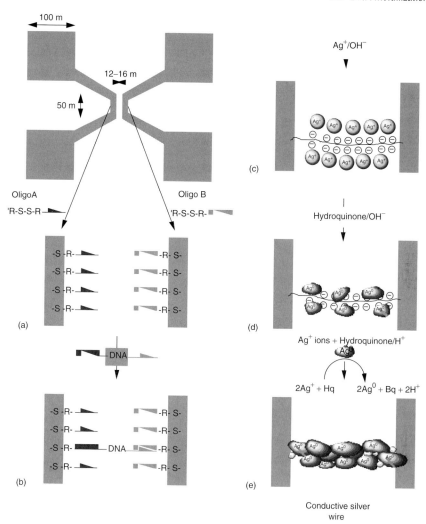

Fig. 9.2 Gold pattern, was defined on a passivated glass by microelectronics techniques. The pattern comprised four bonding pads, 100 μm in size, connected to two 50-μm long parallel gold electrodes, 12–16 μm apart. (a) The electrodes were each wet with a 10^{-4} μL droplet of disulfide-derivatized oligonucleotide solution of a given sequence (Oligos A and B). (b) After rinsing, the structure was covered with a 100-μΛ solution of λ-DNA having two sticky ends that are complementary to Oligos A and B. A flow was applied to stretch the λ-DNA molecules, allowing their hybridization to the bound oligos. (c) The DNA bridge was loaded with silver ions by Na^+/Ag^+ ion exchange. (d) The silver ion-DNA complex was reduced using a basic hydroquinone solution to form metallic silver aggregates bound to the DNA skeleton. (e) The DNA-templated wire was "developed" using an acidic solution of hydroquinone and silver ions. (Reprinted with permission from Ref. [8]. Copyright 1998 *Nature*).

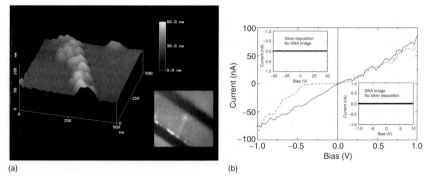

Fig. 9.3 (a) AFM image of a silver wire connecting two gold electrodes 12 μm apart, 0.5-μm field size. Inset: Fluorescently labeled λ-DNA molecule stretched between two gold electrodes (dark strips), 16 μm apart. The electrodes are connected to large bonding pads 0.25 mm away (see Figure 9.2). (b) Two-terminal *I-V* curves of the silver wire shown on left. Note the current plateau (dashed dotted line), on the order of 0.5 V. By applying 50 V to the wire, the plateau has been permanently eliminated to give an ohmic behavior (solid line), over the whole measurement range. *I-V* curves of a DNA bridge with no silver deposition and silver deposition without a DNA bridge are depicted at the bottom and top insets, respectively. Clearly, the sample is insulating in both cases. (Reprinted with permission from Ref. [8]. Copyright 1998 *Nature*).

sequences. A single λ-DNA molecule with two sticky ends complementary to the oligonucleotides on the electrode pads was hybridized to form a bridge between the two metal electrodes. The DNA bridge was then used as a template for growing a conductive silver wire connecting the two electrodes.

More recently, we have developed a gold metallization scheme that produces highly conductive DNA-templated wires [10]. In this scheme, the reducing agent, glutaraldehyde, is first localized on the DNA. Silver ions are then reduced selectively along the DNA, leading to the formation of small silver clusters. These clusters provide nucleation centers for electroless gold plating [17] until a continuous gold wire is formed. The localization of the reducing agent along the DNA, rather than being dispersed in the solution, results in a significantly lower metallization background. Moreover, the aldehyde-derivatized DNA retains its biological functionality, allowing the metallization pattern to be embedded in the scaffold DNA without compromising its molecular recognition properties [18]. The resistance of the resulting gold wire is 4 orders of magnitude lower compared to our previous silver wires. Figure 9.4 shows a scanning electron microscope (SEM) image of a DNA-templated gold wire produced this way. Direct electrical measurements reveal an $\sim 25\ \Omega$ resistance and ohmic characteristics up to currents on the order of 200 nA. The wire's resistivity, $\sim 1.5 \times 10^{-7}\ \Omega m$, is only seven times higher than that of polycrystalline gold. The DNA-templated gold wires are stable for extended periods of time.

The diameter of double-stranded DNA (dsDNA) measures only 2 nm. However, because of the inhomogeneous nature of the metallization processes, the width of microns-long continuous wires demonstrated thus far is ≥ 50 nm. This dimension has important consequences for the utilization of DNA-templated wires in molecular

(a)

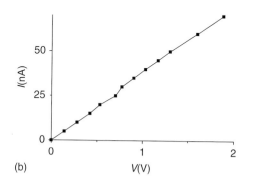

(b)

Fig. 9.4 (a) A DNA-templated gold wire and (b) its two-terminal I-V curve. SEM image of a typical DNA-templated gold wire stretched between two electrodes deposited by e-beam lithography. Bar 1 μm. The wire's resistivity, 1.5×10^{-7} Ωm, is only seven times higher than that of polycrystalline gold, 2.2×10^{-8} Ωm. (Reprinted with permission from Ref. [10]. Copyright 2002, The American association for the Advancement of Science).

electronics. Apart from the obvious disadvantage, the large diameter also makes the contact to molecular objects much harder.

The metallization process irreversibly eliminates the molecular recognition properties of DNA, as the molecule is buried under a large amount of metal coating. It therefore precludes further biological processes. Developing methods for using DNA as a template for conductive wires, while retaining its molecular recognition properties, would make DNA-templated electronics a much more powerful approach. For example, it could then be possible to design feedback loops from the electronic functionality to the biological assembly process, which is currently impossible (see also Section 9.6 below). The aldehyde derivatization method presents a step in this direction [18].

9.4
Sequence-specific Molecular Lithography

Sequence-specific molecular lithography enables elaborate manipulations of dsDNA molecules, including patterning the metal coating of DNA, localization of labeled molecular objects at arbitrary addresses on dsDNA, and generating molecularly accurate stable DNA junctions. It utilizes homologous genetic recombination processes carried out by the RecA protein from *E. coli*. The information guiding the lithography is encoded in the DNA substrate molecules and in short auxiliary probe DNA molecules.

The RecA protein provides the assembling capabilities as well as the resist function. Molecular lithography works with high resolution over a broad range of length scales, from nanometers to many micrometers.

Homologous genetic recombination is one of several mechanisms that cells use to manipulate their DNA [19]. In homologous recombination, two parental DNA molecules, possessing some sequence homology, crossover at equivalent sites. The reaction is based on protein-mediated sequence-specific DNA–DNA interaction. RecA is the major protein responsible for this process in *E. coli*. Moreover, it is able to carry out the essential steps of the recombination process *in vitro*. In our procedure, RecA proteins are polymerized on a probe DNA molecule to form a nucleoprotein filament (Figure 9.5a). The nucleoprotein filament binds to a substrate molecule at homologous probe-substrate location (Figure 9.5b). RecA allows addressing an arbitrary sequence, from as few as 15 bases [20] to many thousands of bases, by the same standard reaction. This is in contrast to DNA binding proteins, which are specific to particular DNA sequences. Unlike DNA hybridization, sequence-specific recognition by the RecA operates on dsDNA, rather than single-stranded DNA (ssDNA). Being chemically more inert and mechanically more rigid, the former provides a better substrate than the latter. The high efficiency and specificity of the recombination reaction, evidently essential for its biological roles, are also advantageous for molecular lithography.

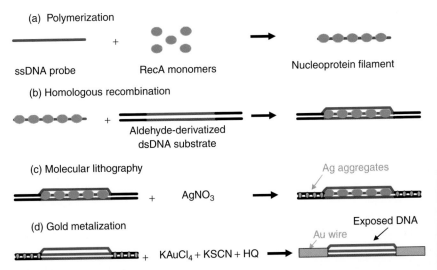

Fig. 9.5 Schematics of the homologous recombination reaction and molecular lithography. (a) RecA monomers polymerize on an ssDNA probe molecule to form a nucleoprotein filament. (b) The nucleoprotein filament binds to an aldehyde-derivatized dsDNA substrate molecule at a homologous sequence. (c) Incubation in AgNO$_3$ solution results in the formation of silver aggregates along the substrate molecule at regions unprotected by RecA. (d) The silver aggregates catalyze specific gold deposition on the unprotected regions. A highly conductive gold wire is formed with a gap in the protected segment. (Reprinted with permission from Ref. [10]. Copyright 2002, The American association for the Advancement of Science).

Homologous recombination can be harnessed for sequence-specific patterning of DNA metal coating (Figure 9.5). We employ the DNA metallization scheme described above in which DNA-bound glutaraldehyde serves as a localized reducing agent. An ssDNA probe is synthesized so that its sequence is identical to the designated unmetallized region on the dsDNA substrate molecule. RecA monomers polymerize on the ssDNA probe and form a nucleoprotein filament (Figure 9.5a), which locates and binds the homologous sequence on the dsDNA molecule (Figure 9.5b). Once bound, the RecA in the nucleoprotein filament acts as a sequence-specific resist, physically protecting the aldehyde-derivatized substrate DNA and preventing silver cluster formation in the bound region (Figure 9.5c). Subsequent gold metallization leads to the growth of two extended DNA-templated wires separated by the predesigned gap (Figure 9.5d).

The recombination reaction was carried out using a 2027-base long ssDNA probe molecule identical to a section in the middle of a 48 502 bp aldehyde-derivatized λ-DNA substrate molecule. Following the recombination reaction, the molecules were stretched on a passivated silicon wafer. Figure 9.6a depicts an atomic force microscope (AFM) image of the 2027-base long RecA nucleoprotein filament bound specifically to the homologous sequence on the DNA substrate. Next, the sample was incubated in the silver solution. The selective reduction of silver ions by the DNA-bound aldehyde in the unprotected segments of the substrate molecule led to the formation of tiny silver aggregates along the DNA skeleton. The localized RecA proteins, serving as a resist, prevented silver deposition on the protected aldehyde-derivatized DNA segment and created a gap of exposed sequence between the silver-loaded segments of the substrate molecule (Figure 9.6b). The silver aggregates catalyzed subsequent electroless gold deposition, which produced two continuous gold wires separated by the predesigned gap (Figures 9.6c and d). Extensive AFM and SEM imaging confirmed that the metallization gap was located where expected. The position and size of the insulating gap could be tailored by choosing the probe's sequence and length.

The ability to pattern DNA metallization facilitates circuit design and is therefore valuable for the realization of DNA-templated electronics. Insulating and conducting segments can be defined on the DNA scaffold according to the underlying sequence, thus determining the electrical connectivity in the circuit. In addition, patterning DNA metallization is useful for the integration of molecular objects into a circuit. Such objects can be localized within the exposed DNA sequences in the gaps and then contacted.

The DNA scaffold provides a substrate with a high density of molecular addresses. Homologous recombination provides direct access to any address on the scaffold. As described above, a nucleoprotein filament formed by polymerization of RecA on an ssDNA probe can locate and bind a homologous segment on any dsDNA molecule. We have shown that this process facilitates localization of molecular objects at arbitrary addresses along the substrate dsDNA molecule with no prior modification of the DNA [10]. The localization of a molecular object by strand exchange with a labeled strand is illustrated in Figure 9.7. The recombination reaction was carried out using a biotin-labeled 500-base long ssDNA probe and λ-DNA as a substrate. After the recombination reaction, the RecA

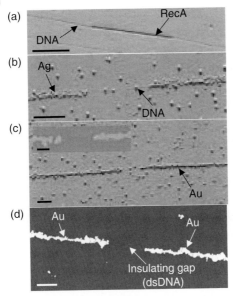

Fig. 9.6 Sequence-specific molecular lithography on a single DNA molecule. (a) AFM image of a 2027 base–long RecA nucleoprotein filament bound to an aldehyde-derivatized λ-DNA substrate molecule. (b) AFM image of the sample after silver nucleation. Note the exposed DNA at the gap between the silver-loaded sections. (c) AFM image of the sample after gold metallization. Inset: zoom on the gap. The height of the metallized sections is ~50 nm. (d) SEM image of the wire after gold metallization. All scale bars – 0.5 μm; inset to c – 0.25 μm. The variation in the gap length is mainly because of variability in DNA stretching on the solid support. The very low background metallization in the SEM image compared to the AFM ones indicates that most of the background in panels b, c is insulating. (Reprinted with permission from Ref. [10]. Copyright 2002, The American association for the Advancement of Science).

proteins were decomposed and the purified reaction products were incubated with streptavidin-conjugated 1.4-nm gold particles. The streptavidin-conjugated Nanogold bound specifically to the biotinilated probe. An AFM image (Figure 9.7a) shows the Nanogold-bound ssDNA probe localized on the substrate dsDNA, and Figure 9.7b shows that electroless gold deposition resulted in gold growth around the catalyzing Nanogold particles. Thus, metal islands can be defined in a sequence-specific manner.

The same recombination machinery can be utilized to realize branched DNA junctions. Such structures arise in nature as intermediates in the recombination process. RecA can generate stable three- or four-arm junctions between any two dsDNA molecules having homologous regions [21]. The boundary between the homologous and heterologous regions of the two molecules determines the junction position. The arms length can vary from a few tens to many thousands of bases. The realization of a molecularly accurate three-arm DNA junction is shown schematically in Figure 9.8a. Briefly, two types of DNA molecules, 15 and 4.3 kbp long, were synthesized by PCR

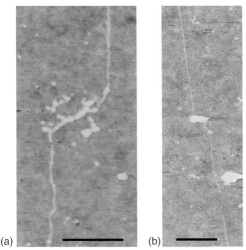

Fig. 9.7 Sequence-specific localization of labeled nanometer-scale objects on a dsDNA substrate. (a) AFM image of a nanogold-labeled 500-base long probe bound to a λ-DNA substrate molecule. The height of the central feature is 5 nm. Bar – 0.2 μm. (b) Sample after electroless gold metallization with nanogold particles serving as nucleation centers. The metallized object at the center of the DNA molecule is over 60-nm high. Bar – 1 μm. (Reprinted with permission from Ref. [10]. Copyright 2002, The American association for the Advancement of Science).

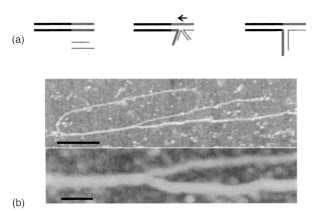

Fig. 9.8 Stable three-arm junction.
(a) A scheme for junction formation by homologous recombination. Two types of molecules are prepared by PCR. The short molecule is identical to a segment of the long one. RecA is first polymerized on the short molecule and the nucleoprotein filament is then reacted with the long molecule. The junction created by this reaction migrates until the end of sequence of the short molecule, resulting in a stable three-armed junction.
(b) AFM images of a three-arm junction, which can serve as a scaffold for a three-terminal device. Bars: upper panel 0.25 μm and lower panel, 50 nm. The arms lengths are consistent with the expected values considering the variations due to interaction with the substrate in the combing process. (Reprinted with permission from Ref. [10]. Copyright 2002, The American association for the Advancement of Science).

on a λ-DNA template. The short molecule was identical to a 4.3-kbp segment at one end of the long molecule. RecA was first polymerized on the short molecule and then reacted with the long molecule. The recombination reaction led to the formation of a stable three-arm junction with two 4.3 kbp–long arms and an 11 kbp–long third arm. AFM images of such a junction are shown in Figure 9.8b. Such a junction can be used to scaffold a three-terminal device such as a FET. In addition, DNA junctions may be used as building blocks to form complex geometries. Note that branched DNA structures have also been constructed by other means. Chemical DNA junctions have been constructed using multifunctional linkers that bind several DNA strands simultaneously [22]. Artificial DNA junctions have been demonstrated using hybridization of ssDNA [23]. Both these approaches rely on synthesized ssDNA and are hence limited to relatively short arms (<100 bases). Extended arms require hybridization of longer DNA molecules.

The various molecular lithography processes are compatible with each other. Thus, several lithographic processes can be carried out sequentially. For example, junction definition followed by specific metallization of the unprotected arms and colloid localization at the junction represent three levels of lithography.

9.5
Self-assembly of a DNA-templated Carbon Nanotube Field-effect Transistor

The superb electronic properties of carbon nanotubes [24], their large aspect ratio, and their inertness with respect to the DNA metallization process make them an ideal choice for active elements in DNA-templated electronics. The abilities to localize molecular objects at any desired address along a dsDNA molecule and to pattern DNA metallization facilitate the incorporation of carbon nanotubes into DNA-templated devices and circuits. This was demonstrated by the self-assembly of a DNA-templated carbon nanotube FET [11]. A DNA scaffold molecule provided the address for precise localization of a semiconducting single wall carbon nanotube (SWNT) and templated the extended wires contacting it. The conduction through the SWNT was controlled by a voltage applied to the substrate supporting the structure.

The assembly of the SWNT-FET, shown schematically in Figure 9.9, employed a three-strand homologous recombination reaction between a long dsDNA molecule serving as a scaffold and a short auxiliary ssDNA. The short ssDNA molecule was synthesized so that its sequence was identical to the dsDNA at the designated location of the FET. RecA proteins were first polymerized on the auxiliary ssDNA molecules to form nucleoprotein filaments (Figure 9.9a), which were then mixed with the scaffold dsDNA molecules. A nucleoprotein filament bound to the dsDNA molecule according to the sequence homology between the ssDNA and the designated address on the dsDNA (Figure 9.9b; AFM image: Figure 9.10a).

The SWNTs were solubilized in water by micellization in SDS [24] and functionalized with streptavidin by nonspecific adsorption [25–27] (Figure 9.9c). Primary anti-RecA antibodies were reacted with the product of the homologous recombination reaction,

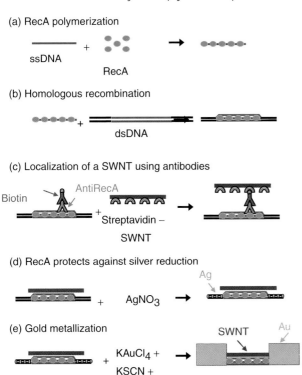

Fig. 9.9 Assembly of a DNA-templated FET and wires contacting it. (a) RecA monomers polymerize on an ssDNA molecule to form a nucleoprotein filament. (b) Homologous recombination reaction leads to binding of the nucleoprotein filament at the desired address on an aldehyde-derivatized scaffold dsDNA. (c) The DNA-bound RecA is used to localize a streptavidin-coated SWNT, utilizing a primary anti-RecA antibody and a biotin-conjugated secondary antibody. (d) Incubation in an $AgNO_3$ solution leads to the formation of silver clusters on the segments that are unprotected by RecA. (e) Electroless gold deposition, using the silver clusters as nucleation centers, results in the formation of two DNA-templated gold wires contacting the SWNT bound at the gap. (Reprinted with permission from Ref. [11]. Copyright 2003, The American association for the Advancement of Science).

resulting in specific binding of the antibodies to the RecA nucleoprotein filament. Next, biotin-conjugated secondary antibodies, having high affinity to the primary ones, were localized on the primary anti-RecA antibodies. Finally, the streptavidin-coated SWNTs were added, leading to their localization on the RecA via biotin-streptavidin specific binding (Figure 9.9c). The DNA/SWNT assembly was then stretched on a passivated oxidized silicon wafer. An AFM image of a SWNT bound to a RecA-coated 500-base long ssDNA localized at the homologous site in the middle of a scaffold λ-DNA molecule is shown in Figure 9.10b. Figure 9.10c depicts a scanning conductance microscopy image of the same area, in which the conducting carbon nanotube is clearly distinguished

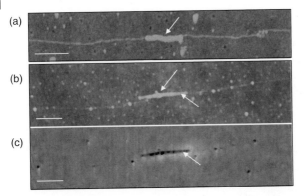

Fig. 9.10 Localization of an SWNT at a specific address on the scaffold dsDNA using RecA. (a) An AFM image of a 500-base long RecA nucleoprotein filament localized at a homologous sequence on a λ-DNA scaffold molecule. Scale bar, 200 nm. (b) An AFM image of a streptavidin-coated SWNT (white arrow) bound to a 500-base long (~250 nm) nucleoprotein filament (black arrow) localized on a λ-DNA scaffold molecule. Scale bar, 300 nm. (c) A scanning conductance image of the same region as in (b). The conductive SWNT (white arrow) yields a significant signal while the insulating DNA is hardly resolved. Scale bar, 300 nm. (Reprinted with permission from Ref. [11]. Copyright 2003, The American association for the Advancement of Science).

from the insulating DNA [13, 14]. Note that the carbon nanotube is aligned with the DNA. This is almost always the case due to the stiffness of the SWNT and the stretching process.

Following stretching on the substrate, the scaffold DNA molecule was metallized. The RecA, doubling as a sequence-specific resist, protected the active area of the transistor against metallization. We employed the metallization scheme described above in which aldehyde residues, acting as reducing agents, are bound to the scaffold DNA molecules by reacting the latter with glutaraldehyde. Highly conductive metallic wires were realized by silver reduction along the exposed parts of the aldehyde-derivatized DNA (Figure 9.9d) and subsequent electroless gold plating using the silver clusters as nucleation centers (Figure 9.9e). Since the SWNT was longer than the gap dictated by the RecA, the deposited metal covered the ends of the nanotube and contacted it. Figures 9.11a,b depict, respectively, SEM images of an individual SWNT and a rope contacted by two DNA-templated gold wires. The scaffold DNA molecule and the RecA were not resolved by the SEM.

The extended DNA-templated gold wires were contacted by e-beam lithography, and the device was characterized by direct electrical measurements under ambient conditions. Schematics of the measurement circuit are given in Figure 9.12a and the electronic characteristics of the device are shown in Figures 9.12b,c. The gating polarity indicates p-type conduction of the SWNT as is usually the case with semiconducting carbon nanotubes in air [28]. The saturation of the drain-source current for negative gate voltages indicates resistance in series with the SWNT. The resistance is attributed to the contacts between the gold wires and the SWNT since the resistance of the

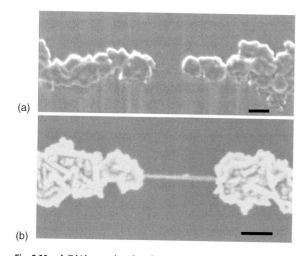

Fig. 9.11 A DNA-templated carbon nanotube FET and metallic wires connecting it. SEM images of SWNTs contacted by self-assembled DNA-templated gold wires. The FET is gated by the substrate. (a) An individual SWNT. (b) A rope of SWNTs. Scale bars, 100 nm. (Reprinted with permission from Ref. [11]. Copyright 2003, The American association for the Advancement of Science).

DNA-templated gold wires is typically smaller than 100 Ω (see Figure 9.4). Different devices have somewhat different turn-off voltages.

9.6
Summary and Perspective

The work described here presents our efforts to transform DNA-templated electronics from a heuristic scheme to a realistic approach for self-assembly of molecular electronics. Relying on molecular lithography, we were able to self-assemble a functional DNA-templated transistor, which represents the first realization of an entirely self-assembled electronic device. The transistor fabrication process demonstrates that functional devices can indeed be realized by self-assembly but at the same time highlights the overwhelming challenges expected along the way to complex circuits.

New metallization schemes for DNA were invented, enabling the fabrication of highly conductive DNA-templated gold wires. The conductivity of these wires is superior to earlier wires fabricated both by our group [8] and by others [16]. The minimal width of long, continuous DNA-templated wires is about 50 nm. This width is dictated by inhomogeneities in the metallization process. New methods that produce nanometer wide wires are certainly desirable. Apart from the obvious benefits of size reduction, thinner wires should contact nanometer-scale objects more easily without shorting them. The main drawback of DNA metallization is in the fact that metallized wires

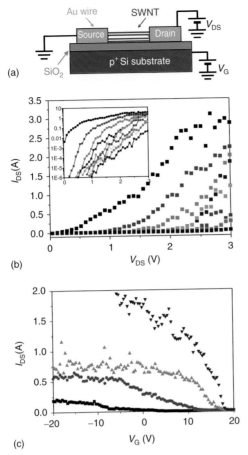

Fig. 9.12 Electrical characterization of the DNA-templated FET. (a) Schematics of the electrical measurement circuit. (b) Drain-source current versus drain-source bias for different values of gate bias. $V_G = -20$ V (black), -15 V (red), -10 V (green), -5 V (blue), 0 V (cyan), 5 V (magenta), 10 V (yellow), 15 V (olive), 20 V (navy). The inset depicts the same data on a logarithmic scale. (c) Drain-source current versus gate voltage for different values of drain-source bias (same device as (b)). $V_{DS} = 0.5$ V (black), 1 V (red), 1.5 V (green), 2 V (blue). (Reprinted with permission from Ref. [11]. Copyright 2003, The American association for the Advancement of Science).

are no longer accessible to the biological machinery. Converting DNA into conductive wires without interfering with their biological properties would open up entirely new opportunities.

The utilization of homologous recombination to realize sequence-specific molecular lithography demonstrates that biological processes can be successfully harnessed in a completely nonbiological context. Homologous recombination operates on any sequence, over scales varying from several bases to many thousands of bases with essentially single-base accuracy. These characteristics were instrumental for the development of our molecular lithography. It is hard to imagine how such a general, yet precise, lithographic process on DNA could be developed without the extraordinary recombination machinery.

We have shown that molecular lithography facilitates the self-assembly of a functional DNA-templated FET, which incorporates several types of molecular building blocks including DNA, proteins (RecA, antibodies, streptavidin), biotin, carbon nanotubes and metals. The different materials and techniques were made compatible with each other. The information is encoded into the molecular building blocks, and they assemble properly by mutual interaction. The fabrication process is devoid of any external molecular-scale operations, relying entirely on self-assembly. Ideally, the self-assembly process would comprise mixing of all the ingredients into a "transistor soup" and letting them self-assemble into numerous functional devices. In practice, the fabrication process is divided into several distinct steps. In each step, the assembly process proceeds spontaneously toward some equilibrium configuration.

Numerous functional devices are fabricated in parallel. Ideally, all devices of the same type should turn identical to each other structurally and, more importantly, functionally. In practice, the devices have a spectrum of different characteristics. The nanotube preparation contains a mixture of metallic and semiconducting SWNT as well as ropes, so that only a fraction of the devices comprise individual semiconducting SWNTs. This drawback can be mitigated by recently developed techniques for separating semiconducting SWNTs from metallic ones [29–31]. Sorting the different semiconductor SWNTs according to their characteristics is presently impossible. The variability in characteristics of nominally identical molecular-scale devices is, in fact, a major hurdle on the way to molecular-scale electronics. The assembly process adds uncertainties. Critical parameters such as contact resistance between the nanotube and the DNA-templated wires are poorly controlled. The way the streptavidin coats the nanotube, the nanotube's exact position on the DNA, the DNA metallization process and its contact with the nanotube, and the capacitive coupling between the gate and the nanotube combine to produce significant variability. These variations can probably be partially suppressed by process optimization but a certain degree of variability and finite yield is inevitable.

The construction of a three-terminal FET, using the three-arm DNA junction as a scaffold, suggests a natural extension of the present work. After assembly of the junction, a semiconducting SWNT is localized there. The junction is then stretched on a surface, with the nanotube aligned in the correct orientation. In addition to the drain and source contacts, the third arm serves to template a third, capacitively coupled, gate contact to the SWNT. Such a three-terminal layout provides individual gating of devices, which is essential for any practical application.

Other molecular devices can be assembled by the same approach. Molecular lithography enables incorporation of an arbitrary molecular object as an active element, the only requirement being its compatibility with the biological reactions and the metallization process. A rich arsenal of standard chemistries is readily available for modifications of both the electronic components and the DNA scaffold to facilitate such incorporation. For example, semiconductor quantum dots or nanorods could be used as active elements in a DNA-templated single-electron transistor. Localization of semiconductor quantum dots and nanorods is readily achieved. Contacting them electrically may prove challenging because of their small dimensions.

The approach developed here can be generalized, in principle, to form a functional circuit on a scaffold DNA network. Assuming an elaborate DNA network could be constructed, many DNA-templated devices could, in principle, be fabricated at desired locations on the scaffold. The extension of molecular lithography to networks may, however, prove difficult. For example, it is known that the RecA-promoted recombination reaction is sensitive to torsion and strain in the DNA, which can be imposed by the network structure. In addition, the efficiency and specificity of the recombination reaction may be affected when the reaction is carried out in parallel at many different sites.

The discussion above summarizes our work on DNA-templated electronics and outlines several conceptually straightforward (though experimentally, in some cases, rather difficult) extensions to it. In the remaining part of the article, the challenging task of assembling large-scale DNA-templated circuits is considered. Drawbacks and limitations of the current approach are discussed and potentially useful, mostly biologically inspired, new concepts are introduced. The assembly of the DNA-templated FET employs a deterministic engineering approach; the building blocks are designed to self-assemble into a well-defined structure as accurately as possible, and the conditions are set to optimize the yield. This approach can likely be extended to construction of a simple circuit comprising a few interconnected DNA-templated devices. Progressing beyond this point relying solely on free-running self-assembly may turn impossible. Even if individual devices could be made reliably, the inherent error-prone nature of the self-assembly process is likely to dramatically limit the complexity of realizable circuits. Self-assembly of a functional circuit is further complicated by the highly nontrivial relation between physical structure and electronic functionality. Bridging the gap between a single DNA-templated device and large-scale circuits will thus require new concepts.

Biology is the only paradigm for the assembly of complex functional systems from molecular building blocks [32]. Self-assembly, even in biology, is limited to relatively small systems such as proteins, protein complexes (e.g., nuclear pores or flagellar motors) or nucleoprotein complexes (e.g., ribosomes or viruses). Assembly of more complex systems such as cells or complex organs relies on self-organization rather than self-assembly. It is hard to imagine how the position of each neuron and its connections in a brain, or the entire repertoire of antibodies in a mammal, could be designed and precisely implemented. The information required to specify the structure with such detail would be overwhelming. Moreover, the exact implementation of such a design would be impossible because of inevitable variability in the assembly process. Self-organization facilitates assembly of complex systems without requiring a precise design. As in self-assembly, there is no supervisor; the information is encoded in the building blocks, and the interactions between them and with the environment direct the assembly process. However, unlike self-assembly, the rules that govern the interaction between building blocks are not rigid. The system explores various configurations, and different strategies are used to direct the assembly process toward a favorable configuration. This configuration is generated by trial and error and is only vaguely constrained by design, yet it displays the desired functionality. Self-organization makes the assembly of complex functional systems possible in spite of variability in the building blocks, in the assembly process, and in the external conditions.

Self-organization is a dynamic, out-of-equilibrium process generated by a rich and context-dependent repertoire of interactions between building blocks. Signals can be transduced between building blocks. Building blocks can respond to external signals according to their internal state. This facilitates feedback processes from the structural and/or functional state of the system to the assembly process. Such feedback processes are essential for directing the assembly process toward particular functionality or structure, without requiring precise design. A continuous supply of energy is required to maintain the dynamic state during the assembly of self-organized systems. Self-assembly, in contrast, relies on more rigid interactions essentially limited to affinity, and the system eventually reaches some well-defined "minimal free-energy" state. The structure of this state is hardwired by the design of the building blocks.

Selection is a very powerful concept to find solutions without a detailed knowledge of their nature. A mechanism for generating variation is required, and the solutions are found by exploring the various configurations. Selection from a large library of suitable configurations can take part in parallel. This strategy is used, for example, in the adaptive immune system. Mechanisms such as recombination create an enormous diversity of specific receptors and antibodies. Negative selection is employed to eliminate anti-self agents, while positive selection leads to a rapid increase in the number of T cells and antibodies that are useful in fighting an invading pathogen. The system is designed to respond to all encountered pathogens, despite the fact that their identity is not known in advance. By self-organization, the system can deal successfully with unforeseen tasks. An alternative variation and selection mechanism involves dynamic exploration of different configurations and once found, stabilization of a suitable configuration. This strategy is employed by cells to locate their chromosomes in order to correctly distribute them between daughter cells during mitosis. A cell generates many highly unstable microtubules that grow in random directions from a single organizing center. The microtubules probe the entire volume of the cell and are stabilized once they accidentally collide with a chromosome. This localization process operates rapidly and efficiently despite the uncertainty in the position of the chromosomes and the variable cell morphology.

Let us now return to DNA-templated electronics and consider the possibilities and difficulties in implementing the concepts underlying the inspiring examples discussed above. The capacity of a system to obtain a well-defined function through exploration depends on its ability to assess its state and direct the assembly process accordingly. Feedback loops and error-correction mechanisms are instrumental in this respect. Molecular biology provides such tools for certain operations but as soon as the assembly departs from the biological pathway, many of these tools prove useless. Developing analogous concepts for DNA-templated electronics is crucial. However, this is an incredibly challenging task given the fact that electronic functionality is alien to biology.

DNA metallization was used to convert DNA into conductive wires. The metallization eliminates the molecular recognition properties of the underlying DNA, and hence, the possibility of circuit reconfiguration due to feedback from the electronic functionality to the biological assembly process. This is a major drawback of the current approach. Overcoming this limitation by developing ways to convert DNA into conductive wires without destroying its biological features would open up much richer possibilities for DNA-templated integrated electronics. Alternatively, developing a scheme for biologically assisted self-assembly of functional circuits relying on ionic conduction (which is compatible with biology, and present in many biological systems) rather than electronic conduction may turn more productive.

Another powerful biological concept is replication. Biological replication is employed to produce the large number of building blocks used in DNA-templated electronics. It would be more useful if, rather than replicating individual building blocks, an entire DNA network or even a functionalized circuit could be replicated. Moreover, combining selection and replication, one can imagine evolving DNA networks or even DNA-templated circuits to obtain new desired functionalities. Biology, however, does not provide tools for replicating DNA networks or even branched DNA. A unique example of synthetic DNA junction replication is given in [33,] but replication of an entire DNA network is significantly harder. Replicating a functionalized DNA-templated circuit is just a dream at this point. In the process of electronic functionalization, a large amount of information is lost and, as discussed above, the system is forced out of the biological realm. Thus, there are no conceivable means to replicate a functionalized circuit.

In summary, DNA-templated electronics provides a framework for investigating self-assembly and, in particular, self-assembly of molecular electronics. It displays the power as well as the limitations of harnessing biology to realize nonbiological functionality. The assembly of a "transistor in a test tube", facilitated by the powerful tools of sequence-specific molecular lithography, is an important step toward DNA-templated electronics. Generalization of this approach to large-scale DNA-templated circuits presents outstanding and exciting challenges that require new concepts and tools.

References

1. C. Joachim, J.K. Gimzewski, A. Aviram, *Nature*, **2000**, *408*, 541–548.
2. J.R. Heath, M.A. Ratner, *Phys. Today*, May, **2003**, 43–49.
3. Y. Huang, X. Onan, Y. Cui, L.J. Lauhon, K.H. Kim, C.M. Lieber, *Science*, **2001**, *294*, 1313–1317.
4. A. Bachtold, P. Hadley, T. Nakanishi, C. Dekker, *Science*, **2001**, *294*, 1317–1320.
5. A. Javey, Q. Wang, A. Ural, Y. Li, H. Dai, *Nano Lett.*, **2002**, *2*(9), 929–932.
6. J.M. Lehn, *Proc. Natl. Acad. Sci.*, **2002**, *99*(8), 4763–4768.
7. N. Seeman, *Nature*, **2003**, *421*, 427–431.
8. E. Braun, Y. Eichen, U. Sivan, G. Ben-Yoseph, *Nature*, **1998**, *391*, 775–778.
9. Y. Eichen, E. Braun, U. Sivan, G. Ben-Yoseph, *Acta Polym.*, **1998**, *49*, 663–670.
10. K. Keren, M. Krueger, R. Gilad, G. Ben-Yoseph, U. Sivan, E. Braun, *Science*, **2002**, *297*, 72–75.

11 K. Keren, R.S. Berman, E. Buchstab, U. Sivan, E. Braun, *Science*, **2003**, *302*, 1380–1382.
12 C. Dekker, M. Ratner, *Phys. World*, **2001**, *14*(8), 29–33.
13 C. Gómez-Navarro, F. Moreno-Herrero, P.J. De Pablo, J. Colchero, J. Gómez-Herrero, A.M. Baró, *Proc. Natl. Acad. Sci. U.S.A.*, **2002**, *99*, 8484–8487.
14 M. Bockrath, N. Markovic, N. Shepard, M. Tinkham, L. Gurevich, L.P. Kouwenhoven, M.W. Wu, *Nano Lett.*, **2002**, *2*, 187–190.
15 J. Richter, *Physica E*, **2003**, *16*, 157–173.
16 J. Richter, M. Mertig, W. Pompe, I. Monch, H.K. Schackert, *Appl. Phys. Lett.*, **2001**, *78*(4), 536–538.
17 Y. Eichen, U. Sivan, E. Braun, PCT WO0025136, **1999**.
18 K. Keren, R.S. Berman, E. Braun, *Nano Lett.*, **2004**, *4*, 323.
19 M.M. Cox, *Prog. Nucl. Acid Res. Mol. Biol.*, **2000**, *63*, 311–366.
20 P. Hseih, C.S. Camerini-Otero, D. Camerini-Otero, *Proc. Natl. Acad. Sci.*, **1992**, *89*, 6492–6496.
21 B. Muller, I. Burdett, S.C. West, *Embo*, **1992**, *11*(7), 2685–2693.
22 M. Scheffler, A. Dorenback, S. Jordan, M. Wustefeld, G. von Kiedrowski, *Angew. Chem. Int. Ed.*, **1999**, *3*(22), 3312–3315.
23 N.C. Seeman, *Annu. Rev. Biophys. Biomol. Struct.*, **1998**, *27*, 225–248.
24 C. Dekker, *Phys. Today*, **May 1999**, 2–28.
25 J. Liu, A.G. Rinzler, H. Dai, J.H. Hafner, R.K. Bradley, P.J. Boul, A. Lu, T. Iverson, K. Shelimov, C.B. Huffman, F. Rodriguez-Macias; Y.-S. Shon, T. Randall Lee, D.T. Colbert, R.E. Smalley, *Science*, **1998**, *280*, 1253–1256.
26 F. Balavoine, P. Schultz, C. Richard, V. Mallouh, T.W. Ebbesen, C. Mioskowski, *Angew. Chem. Int. Ed.*, **1999**, *38*, 1912–1915.
27 M. Shim, N.W.S. Kam, R.J. Chen, Y. Li, H. Dai, *Nano Lett.*, **2002**, *2*(4), 285–288.
28 P. Avouris, *Acc. Chem. Res.*, **2002**, *35*, 1026.
29 R. Krupke, F. Hennrich, Hv. Lohneysen, M.M. Kappes, *Science*, **2003**, *301*, 344.
30 D. Chattopadhyay, I. Galeska, F. Papadimitrakopoulos, *J. Am. Chem. Soc.*, **2003**, *125*, 3370.
31 M. Zheng, A. Jagota, E.D. Semke, B.A. Diner, R.S. Mclean, S.R. Lustig, R.E. Richardson, N.G. Tassi, *Nat. Mater.*, **2003**, *2*, 338.
32 J. Gerhart, M. Kirschner, *Cells Embryos and Evolution,*, Blackwell Science, Inc., Massachusetts, USA **1997**, Chapter 4.
33 L.H. Eckardt, K. Naumann, W.M. Pankau, M. Rein, M. Schweitzer, N. Windhab, G. Von Kiedrowski, *Nature*, **2002**, *420*, 286.

10
Single Biomolecule Manipulation for Bioelectronics

Yoshiharu Ishii and Toshio Yanagida

Biomolecules of the order of nanometers are interacting in unique assemblies and networks. The functions of the biomolecules result from the dynamic changes in their interaction. To understand the mechanisms of their function, it is therefore critical to measure the dynamic interactions between biomolecules while they are working. However, the dynamic properties are obscured by averaging large numbers of molecules in conventional ensemble measurements. Recently, single molecule detection techniques have been developed to describe the behavior of single molecules and have thus overcome this problem [1]. These techniques include the manipulation and imaging of single biomolecules. The manipulation techniques allow single molecules to be captured and handled and their behavior to be monitored in a controlled manner.

Manipulation using microneedles and a laser trap are methods used to trap single molecules mechanically. These methods can also be used to impose forces on the system and to measure the mechanical response of the biomolecules. The techniques have been developed for the studies of molecular motors, and the field of molecular motors has made great progress over recent years [2]. In this chapter, we describe the manipulation techniques and their applications to the molecular motor systems.

10.1
Single Molecule Manipulation

In order to efficiently monitor reactions caused by single biomolecules, it is critical to manipulate biomolecules. Single molecule measurements are performed at low density of the molecules and the probability that the molecules happen to come to the place of the detector is low. Biomolecules are captured and brought to the place where the interaction is supposed to take place. Given that biomolecules are on the order of nanometers in size and that they are very fragile, special techniques are required to catch and hold them without damage. Biomolecules, either single assemblies of biomolecules or single molecules themselves were captured by large probes such as a microneedle and beads trapped by a laser (Figure 10.1). By manipulating these probes, biomolecules can be positioned to a desired place, brought into contact with other

Bioelectronics. Edited by Itamar Willner and Eugenii Katz
Copyright © 2005 WILEY-VCH Verlag GmbH & Co. KGaA, Weinheim
ISBN: 3-527-30690-0

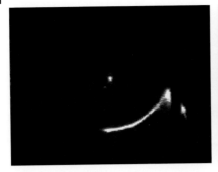 

Fig. 10.1 Manipulation of single actin filament. One end of an actin filament was caught by a bead (indicated by an arrow) held by a focused laser. The actin filament and bead are fluorescently labeled and monitored using fluorescence microscopy. The actin filament moves when the position of a laser moves.

molecules and made to react to monitor their function. It has been demonstrated that the single filaments or single molecules are moved, stretched and fixed under a microscope.

Manipulating biomolecules has been facilitated by visualizing them. Visualization of biomolecules was accomplished by attaching fluorescent dyes to them, because single fluorescent molecules can be successfully visualized in aqueous solution under a microscope. To monitor single molecules with higher signal-to-noise ratio, local illumination techniques have been employed. In total internal reflection fluorescence microscopy (TIRFM), an evanescent field is created with a depth of ~100 nm near the surface of the slide glass when incident laser light is totally reflected. The evanescent field has been used to illuminate only molecules that existed near the glass surface [3]. Visualization of biomolecules not only allows biomolecules to be captured easily but also allows chemical reactions and movement of single molecules to be visualized [3, 4]. Manipulation of biomolecules is required to visualize the chemical reactions and movement of single molecules occurring near the surface.

Microneedles and laser traps used for trapping biomolecules are also used for imposing force on biomolecules (Figure 10.2). Microneedles and laser traps have the nature of a spring; when the biomolecules attached to microneedles or beads are displaced from a stable position, a restoring force is exerted with a strength that is proportional to the distance required to pull them back to the original position. The behavior of the molecules depends on the stiffness of the spring. For stiff springs, the molecules do not move readily. In contrast, the molecules move sensitively to the applied forces for compliant springs. Thus, these manipulation systems provide tools to apply the mechanical perturbations on protein systems and to measure forces from the changes in the displacement.

In the first few sections, the typical manipulation techniques, microneedle and laser trap methods, are mentioned.

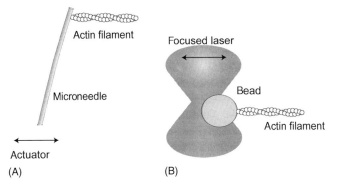

Fig. 10.2 Manipulation using (A) microneedle and (B) laser trap. (A) Biomolecules (actin filament) are attached to a glass microneedle that is connected to an actuator. The flexible microneedle bends when the actin filament is pulled. (B) For a laser trap, a bead is attached to protein molecules and held by a focused laser. The biomolecules are manipulated in a noncontact manner.

10.1.1
Glass Microneedle

A microneedle is made from glass needles [5]. Glass microneedles ~0.3 to 0.5 μm in diameter are made from 1.1-nm diameter glass rods using a glass electrode puller. The microneedle is then attached to a rigid glass rod with a tip diameter of ~100 μm with epoxy resin (and cut in 50–100 μm in length. The stiffness of a microneedle with a length L and radius r is given by $K_n = 3E\pi r^4/4L$, where E is Young's moduli (7.1×10^{10} Nm^2 for glass). For example, a glass microneedle 100 μm in length with a radius of 0.5 μm gives a value for K of 0.66 pN nm^{-1}, which has been used for these measurements. The stiffness of the microneedle is determined by cross-calibration against a standard needle of known stiffness. The stiffness of the standard needle is determined by measuring the vertical bending due to the weight of the 25-μm diameter steel wires, which is determined by measuring its length under a microscope. A stiff microneedle is useful for manipulating biomolecules. If a microneedle is flexible enough, it bends when the biomolecule is pulled by the external force or by the interaction with other molecules fixed on the glass surface. As the flexible microneedle bends, the restoring force the microneedle exerts against the force pulls the biomolecules. When these forces are balanced, the movement ceases. The strength of the restoring force by the microneedle is proportional to its displacement and stiffness, as long as the external force is not too large.

10.1.2
Laser Trap

Laser trap is a method to trap small dielectrics at a focal point of a laser utilizing the pressure of light [6]. The generation of the light pressure on dielectric spheres by a focused laser is explained as follows. An incident laser beam is reflected at the surface of dielectric materials of which the refractive index is higher than the medium solution. When the direction of a light ray changes, the momentum of the photons change. This results in the radiation pressure of light on the bead. The total forces integrated over light rays are directed toward the focal point of the laser regardless of the position of beads. The strength of the force is proportional to the distance from the focal points. Thus, the force is exerted on the beads to pull back toward the focal point of the laser, when they are displaced from the focal points. The beads are thus trapped by the focused laser. The trapped beads behave like a material connected to the focal point of the laser via a spring. Biomolecules are too small to be trapped directly. Dielectric spheres in the order of μm, which the biomolecules are attached to, are trapped. For beads 1 μm in radius, the force is in the order of tens of pN's for a laser several hundreds mW, which is usually used for the manipulation. The spring force of the laser trap is on the order of 0.1 pN nm^{-1}. When either the focal points of the laser or the stage of the microscope change, the position of the biomolecules attached on the beads moves.

10.1.3
Space and Time Resolution of Nanometry

The position of a bead and a microneedle is subject to thermal fluctuation. The mean square displacement $\langle x^2 \rangle$ of a bead and microneedle is related to the stiffness of the laser trap and the microneedle, respectively, by an equipartition law as $K\langle x^2\rangle/2 = k_B T/2$. According to this equation, for example, the root mean square distance is 6.43 nm for a spring of 0.1 pN nm^{-1} and 2.03 nm for a spring of 1 pN nm^{-1}. Using this equation, the stiffness can be obtained from the fluctuation of the microneedle. To reduce the thermal fluctuation and increase the spatial resolution, a stiff system should be used.

On the other hand, the time resolution is determined by the viscous drag of a probe. The relaxation time of the system τ can be written as the friction coefficient β divided by the stiffness K. The friction coefficient for a microneedle is given by $\beta = 4\pi \eta L/[\ln(L/R) + 1/2]$ and for a bead $\beta = 4\pi \eta R$. A glass microneedle with a diameter of 1 μm and a length of 1 cm gave a 100-nm displacement for a force of 1 pN. For such a long and flexible microneedle, however, the response to the applied force was slow owing to viscous drag and hence the temporal resolution was only 1 s. To increase the temporal resolution, a short and thin glass microneedle was needed, though it was not easy to make these needles. For a microneedle with a length of 50 μm and diameter of 0.3 μm (which gave a stiffness 1 pN nm^{-1}), the temporal resolution increased to be as fast as 1 ms. For a bead, the viscous drag is small because a bead is small compared to a microneedle. In contrast, a bead 1 μm in diameter has temporal resolution as good as 0.1 ms. In addition to this time resolution, the laser trap experiments are performed more easily than the experiments with the microneedle. For instance, it is easy to control the stiffness of the system, that is, the stiffness of the laser trap can be changed by changing the laser power. In the microneedle measurements, the microneedle must be individually handcrafted (or custom-made) and it is practically impossible to control the stiffness.

10.1.4
Molecular Glues

The displacement of the biomolecules can be determined by measuring the displacement of the probes. The existence of compliant linkages between the biomolecules and the probes results in inaccurate measurements, and these effects must be corrected. Needless to say, tight linkages are favorable. There are many kinds of glue to attach an actin filament to a glass surface. A strong bond, the avidin–biotin bond, has been used. Streptoavidin can be attached to a specific site of the protein and biotin then attached to the other side. Avidin and biotin are attached by mixing them. Thus, a strong and specific bond can be attained, and the strength of this bond is independent of the solution conditions or the addition of ligands such as ATP and Ca^{2+}. NEM-treated myosin heads, in which NEM is attached to the most reactive cysteine residue of the myosin head, were also used to attach actin filaments to a glass surface. Treatment with NEM prevents the myosin heads from hydrolyzing ATP and dissociating from actin filaments even in the presence of ATP, where myosin head binds to actin very weakly. These materials have been used to glue actin filaments to probes. However, a stronger glue still needs to be developed.

10.1.5
Comparisons of the Microneedle and Laser Trap Methods

Biomolecules are attached to large probes. The probes restrict the degree of freedom of the movement of the molecules, depending on the type of the probes. Microneedles are directly attached to manipulators. The motion is restricted on a 2-D plane, and rotational motion is prohibited. In contrast, beads are trapped by a laser in noncontact mode. The noncontact manipulation avoids topological restrictions. For example, it was impossible to make a knot using a string of long molecules such as DNA and actin filaments when microneedles were attached at both ends; it was impossible unless one microneedle was detached. However, it was possible to make a knot when both ends of the molecules were attached to beads trapped by a laser. The free motion of the beads allows 3-D movement, and the beads are also able to freely rotate. These differences are important considerations when designing an experiment. They should also be taken into consideration when the data is interpreted. Some examples of molecular motors are described below.

10.2
Mechanical Properties of Biomolecules

10.2.1
Protein Polymers

The first couple of examples of the measurements use manipulation techniques to measure the mechanical properties of proteins and protein assemblies by measuring the responses when mechanical modulation is applied.

Actin filaments are double-stranded helical polymers of actin molecules. Fluorescently labeled actin filaments were visualized under fluorescence microscopy, and the visualization of single actin filaments makes it easier to catch and manipulate them. The ends of the filaments were attached to microneedles or beads. Actin filaments serve as tracks for the molecular motor, myosin. The physical properties of the actin filaments are known to change when they interact with motor proteins. Actin may play an active role in the function of molecular motors. The physical parameters of actin filaments have been measured in ensemble measurements, which give averaged numbers over a large number of molecules. Using fluorescently labeled actin filaments, it has been demonstrated that single filaments were flexible in solution. The flexibility was affected by the interaction with myosin molecules in the presence of ATP, which is an energy source for the motile activity of myosin.

Given that single actin filaments were manipulated by microneedles and a laser trap, mechanical properties of single actin filaments were measured. Actin filaments are broken when they are pulled with large force. The force required to break actin–actin interaction has been determined (Figure 10.3) [7]. A stiff microneedle is attached at one end of the actin filaments to apply force and a flexible microneedle is attached at the other end to monitor the responses to the force. The maximum displacement of the microneedle immediately before the actin filament breaks gives the force that is required to break the interaction between the actin monomers in the filament. For a sinusoidal displacement at one end of the actin filament, the displacement at the other end is damped and delayed because of the extensibility of the filament. The tensile strength of the actin filament is approximately 100 pN, which is not much greater than the force applied on the filament in a contracting muscle.

The tensile strength and the stiffness of the actin filaments are measured by attaching the actin filament at one end to a very flexible microneedle and the other end to a stiff microneedle [8]. The stiffness of actin filaments was measured by monitoring the displacement of the tip of the microneedle, when the other end is moved with an actuator attached. The stiffness of the actin filament was approximately 65 pN nm^{-1}. The value can be compared with those obtained from other measurements. Dynamic light scattering in ensemble measurements gives the average motion of the actin filaments. From the imaging of single actin filaments, the dynamic changes of the shape of single actin filaments have been detected directly. The data suggested that ~50% of the compliance of the active muscle was due to the extensibility of the actin filaments. The value is also similar to that of other proteins such as collagen.

Torsional rigidity is another important parameter when considering actin filaments. Torsional motion of actin filaments was measured by monitoring the rotation of the end of the filaments when the other end was fixed on the glass surface (Figure 10.4A) [9]. The actin filaments were held taut vertically by trapping a bead attached at the end by a laser. The rotational Brownian motion of the end of actin filaments was measured by monitoring the motion of small fluorescent beads on a larger bead attached to the end of the actin filament. The bead trapped by laser was allowed to rotate around the axis of the actin filaments, while the rotational motion was stopped at the other end. The variance of the fluctuation of the angle was recorded. From the variance, the rotational rigidity κ is obtained using the relation $k\langle\theta^2\rangle/2L = k_B T/2$, where L is the length of

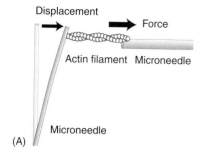

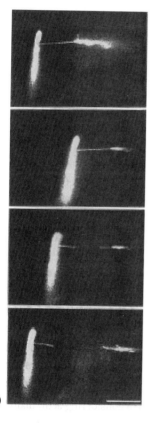

Fig. 10.3 Breaking force of actin filaments. (A) Schematic diagram of the measurements. (B) Time sequence of the measurements. The actin filament breaks when it is pulled strongly. In the photos, thick rods are microneedles, and a thin thread is a fluorescently labeled actin filament.

actin filaments. The torsional rigidity was 8.0×10^{-26} Nm^2, in agreement with the data obtained by other experiments.

This method to monitor the rotation of long filamentous molecules can also be applied to DNA molecules (Figure 10.4B) [10]. RNA polymerase was shown to move along the helical structure of DNA when it read the genetic information and synthesized RNA. The rotation of DNA molecules was demonstrated when it interacted with an RNA polymerase molecule immobilized on a glass surface. As in the case of actin

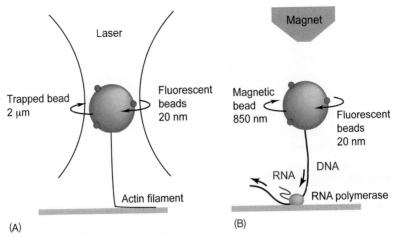

Fig. 10.4 Measurements of torsional motion of fibrous molecules. The torsional motion is monitored by rotation of a bead attached at the end of the molecule. The bead is trapped optically or magnetically, while it is allowed to rotate freely. The rotation of the bead is monitored by attaching a small fluorescent bead on it. (A) To measure the torsional rigidity of actin filaments, thermal rotational fluctuation of actin filaments was recorded when the other end was immobilized on the glass surface. (B) The rotational motion of a DNA molecule was visualized when the DNA molecule moves on an RNA polymerase immobilized on a slide glass.

filaments, the rotation of the DNA molecules was detected by monitoring the rotation of a bead attached to one end of a DNA molecule.

10.2.2
Mechanically Induced Unfolding of Single Protein Molecules

Biomolecules are chains connecting tens to hundreds of unit structures. Proteins are chains of amino acids, and DNA and RNA are composed of nucleotides. In the native structure, protein chains are folded in unique structures as a result of various types of interactions, including electrostatic interactions, hydrophobic interactions and hydrogen bonding. When both ends of the proteins are captured and pulled, the individual interactions are broken, and whole or parts of protein molecules are unfolded (Figure 10.5). These studies have provided information on the structural stability of proteins and insight into the mechanism of the folding of proteins.

The measurements of mechanically induced unfolding at the single molecule level was first carried out with titin [11–13]. Titin is a gigantic protein, which is responsible for the passive elasticity of the muscle. The mechanically induced unfolding of titin has been measured using a laser trap and atomic force microscopy, showing that this protein is extensible when it is stretched by external forces (Figure 10.5). The unfolding of titin occurs in a repeating unit of immunoglobulin (Ig) domain in an all-or-none fashion. The force-extension curve showed a characteristic sawtooth pattern when titin

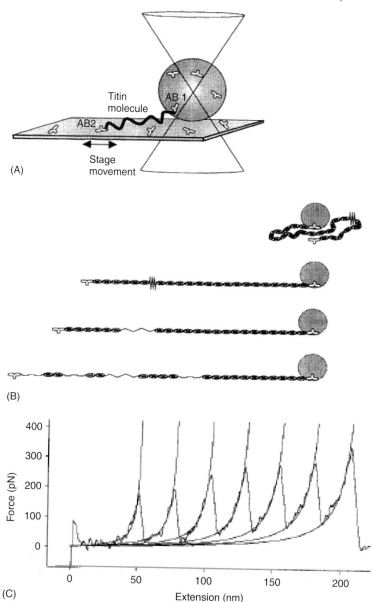

Fig. 10.5 Mechanically induced unfolding of titin. (A) One end of a titin molecule is pulled, while the other end is fixed using a bead trapped by a laser or a cantilever of AFM. (B) Compactly folded titin molecules are expanded and the Ig domain of titin unfolds in all-or-none fashion sequentially. (C) The force–extension relation obtained using AFM showed a characteristic sawtooth pattern due to the unfolding of individual Ig domains.

molecules were stretched between both ends. The force required for the unfolding of the Ig domain depends on the speed at which the filament is pulled.

The mechanically induced unfolding pathway has been demonstrated to be basically different from chemical and thermally induced unfolding pathway. This method provides information on another aspect of biomolecular structures. This method has been used for the unfolding of RNA and DNA [14].

10.2.3
Interacting Molecules

Protein molecules interact with other molecules, whether they are the same or different types of molecules, to perform their functions. When one of the interacting molecules is captured and pulled, the interaction is broken, resulting in unbinding. It is possible to determine the force that is required to break the interaction. This method was used to determine the mechanical properties of the unbinding of molecular motors from protein tracks. The effects of the external force are important in the case of the molecular motors. An increase in the force decreases the velocity. The maximum velocity is attained at no load, and the maximum force is attained at zero velocity. The relationship between velocity and force characterizes molecular motors. The force may affect local structures and interactions of molecular motors. The measurements of the effects of force on the interaction may be important in relation to the mechanism.

Myosin is a molecular motor that moves along acrin filaments. Myosin interacts with actin strongly in the absence of nucleotides. To study the binding of myosin to actin, an actin filament was pulled to break, when it interacts with myosin (Figure 10.6A) [15]. A bead trapped by a laser was attached to one end of an actin filament and the actin filament was brought into contact with myosin molecules immobilized on the glass surface. The unbinding force of myosin from actin filaments or the force required to break the binding was 7 pN, much less than the interaction force between actin molecules in the filament, and only a little greater than the force myosin generates when it works. According to biochemical studies of the interaction in which the binding affinity is measured by estimating how many molecules are bound on average, the interaction between myosin and actin is strong.

Kinesin is a molecular motor that moves along microtubules using the energy released from the ATP hydrolysis. Two-headed kinesin moves in a hand-over-hand manner, repeating one head attached state and two heads attached state coupled to the ATP hydrolysis reactions. The unbinding force in several nucleotide conditions was measured by pulling kinesin molecules along microtubules through beads attached to kinesin molecules using a laser trap (Figure 10.6B) [16]. The information is useful to interpret the data of the stepwise movement of kinesin. It is thought that the conformational changes in one head causes load imposed on the other head, which induces the step movement. The data of the load dependence are useful to understand the mechanism of the movement of kinesin.

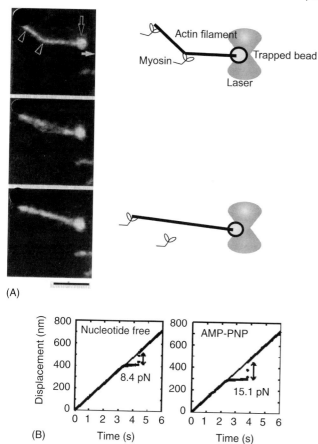

Fig. 10.6 Unbinding force of molecular motors from protein tracks. (A) It was demonstrated that unbinding of myosin immobilized on a slide glass from an actin filament occurred when the actin filament was pulled using a laser trap. In the photos, a bead and a fluorescently labeled actin filament are visualized, but myosin molecules immobilized on a slide glass (the positions are indicated by arrows) are not visualized. (B) Unbinding force of kinesin was measured using a laser trap. The displacement of the bead was recorded with time when the laser position was moved at a constant speed.

10.3
Manipulation and Molecular Motors

Manipulation techniques are also used for the measurements of the interaction of biomolecules at work. The techniques allow biomolecules to interact in a controlled manner and make the measurements efficient. These mechanical manipulation techniques are also used to impose forces on biomolecules while the biomolecules are working and to measure the effects of the force on their function. Thus, the mechanical properties of molecular motors have been measured, and the basic understanding of their mechanism has been greatly deepened.

Muscle contraction and the motile activity involving myosin motors are explained by the sliding movement of myosin along actin filaments using the energy released from the hydrolysis of ATP. The sliding movement of actin filaments was visualized under microscopy when fluorescently labeled actin filaments were added to myosin molecules immobilized on a slide glass in the presence of ATP [17, 18]. However, many myosin molecules are involved in this *in vitro* motility assay. Manipulation of single actin filaments allows single myosin molecules to interact with actin filaments.

10.3.1
Manipulation of Actin filaments

Actin filaments are manipulated by microneedle and laser trap and brought into contact with myosin molecules fixed on the surface of slide glasses. For laser trap experiments, both ends of actin filaments are attached to beads held by a laser trap (Figure 10.7A) [19]. Myosin or the head fragment of myosin (S1 of HMM), which contains a motor domain, is adsorbed sparsely onto a glass surface. Low density of myosin molecules on the glass surface ensures that the actin filaments interact only with single molecules of myosin. Direct interaction with the glass surface may interfere with the motile activity of myosin molecules. To eliminate the possibility that the direct interaction with the glass surface interferes with the motile activity, myosin molecules are incorporated into filaments [20]. The filaments are made by mixing myosin, which contains the motor domain, together with a large number of myosin rod molecules, a part of myosin that is responsible for formation of myosin filaments, so that there are only a few motor molecules on a single myosin filament. In this preparation, the motor domain does not interact with the glass surface directly. When the actin filament interacts with myosin molecules, the myosin molecule moves against the trap force by a laser. The displacement can be determined by monitoring the fluorescence from the beads.

For the microneedle method, the actin filaments were attached to a microneedle and brought into contact with single myosin molecules on the filaments [5]. The microneedle deflects when the actin filament moves. The tips of the microneedles are marked by nickel particles, which give the images of the tip of the microneedle better contrast.

The measurement of the displacement of beads and microneedle was performed as measurements of the displacement of actin filaments. For accurate measurements of the displacement, a pair of photodiodes is used. The images of the microneedle tips are projected onto a pair of photodiodes, and the changes in the difference of the light intensity give the displacement measurements in nanometer accuracy.

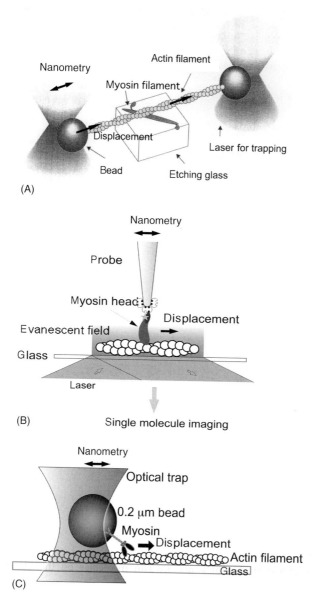

Fig. 10.7 Measurements of movement of single myosin molecule on an actin filament. Three systems have been used. (A) An actin filament is held by two beads at both ends by a laser trap and brought into contact with single myosin molecules immobilized. (B) Single myosin molecules are attached to the tip of a cantilever and brought into contact with actin filaments immobilized. (C) Single myosin molecules, which are attached to beads trapped by a laser, interact with actin filaments.

10.3.2
Manipulation of a Single Myosin Molecule

The manipulation of single myosin molecules instead of actin filaments has been challenging. Single fluorescent molecules were visualized in aqueous solution after refining fluorescence microscopes [3]. Single biomolecules can be visualized by attaching fluorescent probes to it. Single protein molecules were imaged as fluorescent spots. The size of the spots was several hundred nanometers, much larger than the real size, because of the diffraction limit. Then, single myosin S1 molecules were captured by a scanning probe (Figure 10.7B) [21]. Myosin S1 molecules placed sparsely on the actin filaments on the glass surface was scanned by a scanning probe. It was proved by monitoring the photobleaching behavior of the fluorescence from the tip of the scanning probe that the molecules captured were single molecules. Photobleaching from single molecules occur in a single step manner. Single myosin S1 molecules were then brought into contact with the bundle of actin filaments on the glass surface. Use of the actin bundle is advantageous for single molecule measurements. This is because for the actin bundle systems, (1) the target for single myosin S1 molecules to interact is larger, and (2) the system is tighter.

10.3.3
Unitary Steps of Myosin

Whether actin filaments or myosin molecules are manipulated, stepwise movement has been observed (Figure 10.8A). The step movement is associated with the binding and dissociation of myosin, which is monitored by the changes in the compliance. The beads or the tip of a scanning probe undergo Brownian motion. The amplitude of the fluctuation depends on the compliance of the system. The tight binding of myosin suppresses the Brownian motion, decreasing the compliance of the system. Utilizing this fact, we could monitor the dissociation and association events by measuring the variances. When myosin binds to actin filaments, myosin steps against the restoring force by the probes, and when myosin dissociates from actin filaments, myosin steps back to the original position owing to the restoring force. A single cycle of the stepwise movement corresponds to the hydrolysis of single ATP molecules. The time interval of the step movement increases as the concentration of ATP decreases, in agreement with Michaelis–Menten kinetic equation for simple enzymatic reactions. The relationship between the chemical reaction of the ATP hydrolysis and the mechanical events has been directly examined by simultaneously measuring both events.

The extent of the Brownian motion was dependent on the measurement system used. The fluctuation of the displacement was larger when actin filaments are manipulated than when myosin was manipulated with a scanning probe. The root mean square displacement due to thermal motion was 2.0 nm for the manipulation of single myosin molecules with a scanning probe compared with 4.5 nm for the manipulation of actin filaments with a laser trap system. This thermal fluctuation corresponds to the stiffness of the system of 1 pN nm^{-1} for the manipulation of myosin molecules compared with

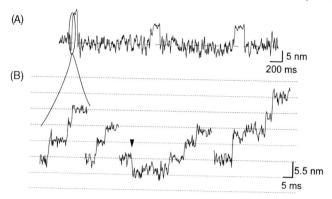
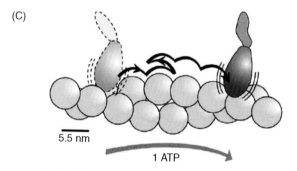

Fig. 10.8 Stepwise movement of muscle myosin on an actin filament. (A) Stepwise movement occurs when ATP is hydrolyzed. (B) A single step contains substeps with the size of 5.5 nm corresponding to the distance between actin monomers along single actin protomers. (C) The substeps are stochastic, suggesting that the movement of myosin is driven by a thermal Brownian ratchet mechanism.

<0.2 pN nm^{-1} for the manipulation of actin filaments. More compliant linkages exist between the myosin and the probes attached to actin filaments than the probes attached to myosin. The linkages between the probes and the actin filaments may be compliant, and also actin filaments themselves can be compliant. The large compliance makes the interpretation of the measurements difficult. Firstly, it was difficult to determine the step size for individual steps when the thermal motion was large, because the starting position of the step was not known. Only average step size was determined. Secondly, the space and time resolution are low for compliant systems. It was difficult to resolve fine processes in a step caused during the hydrolysis of single ATP molecules. Substeps within a step caused by the hydrolysis of single ATP molecules were resolved using probe microscopy, where the system is less compliant and thermal fluctuation is less (Figure 10.8B).

The step generated by a single ATP molecule contains several substeps of 5.5 nm, coinciding with the interval between adjacent actin monomers on one strand of the filament. These substeps are not simply coupled to specific processes in the ATP hydrolysis cycle. The substeps occur rather stochastically, and some of them (<10% of the total) occur in the backward direction. The interval time between the substeps was random and increased when the temperature was lowered. One to five steps are observed in a single displacement, giving rise to a total displacement of ∼5 to 30 nm. The average number of the substeps was three, and the average total displacement was 13 nm.

In the measurement system, the load applied to the myosin S1 molecules could be increased as the S1 molecules proceed and as the spring constant of the microneedle increases. At a high load, the number of steps in a single displacement was smaller, and the dwell time between the steps was larger, while the size of the step remained constant. The probability of the backward movement increased with an increase in the applied load. These data are not explained by a currently accepted model called a *lever-arm model*, in which the stepwise movement of myosin is derived by a structural change coupled with the chemical reaction of the ATP hydrolysis. This point is controversial, and a large body of research has been attempted to answer the question.

10.3.4
Step Size and Unconventional Myosin

The displacement caused by single myosin and ATP molecules or step size is a key to answer the question on the mechanism of the movement of myosin. Structural studies, including X-ray crystallography, on myosin have showed that the angle of the head part of myosin relative to the rest of the molecule depends on the chemical state of nucleotides on myosin molecules [22]. From these data, it is suggested that the rotation of the head of myosin causes the step movement and occurs at a step in the ATP hydrolysis reaction [23]. According to this model, the step size depends on the length of a lever of the rotation, which connects the head domain and the rod portion of myosin. The step size obtained for muscle myosin varies greatly between research groups. Our group has consistently reported a mean displacement of 15 to 20 nm, using both microneedle and laser trap methods [24], whereas other groups have reported smaller mean displacement of ∼4 to 6 nm using laser trap methods [19, 25, 26]. The latter numbers are consistent with the lever-arm model, and the former are not. One possible explanation for this difference in the step size is the method used to immobilize myosin on the glass surface or bead. We use a myosin filament to avoid direct attachment of the head portion of myosin, which contains the essential parts for its motile activity, to the glass surface. Other groups attach the fragment of the myosin head directly to the surface. Direct interaction of the myosin head to the glass surface might affect myosin activity. The other advantage of the filament is that the orientation of the head is known. The displacement generated by myosin has been demonstrated to be dependent on the orientation of the head relative to the actin filament. For experiments using myosin fragments, the orientation of the myosin heads is randomly distributed.

This model predicts that the step size would be less than 10 nm for conventional myosin and will depend on the length of the neck region, or lever arm. Recently, unconventional myosin V, which contains six repeats of the calmodulin or light chain binding site compared to two binding sites for conventional myosin, was shown to have a large step size (36 nm) in agreement with the lever-arm model [27]. However, a mutant of myosin V, which contains only one calmodulin binding site, also shows the same large step size [28], and additional studies are required to elucidate the mechanism for the large step size of myosin V. Myosin VI is a motor with a short neck region. However, the step size was the same as myosin V [29, 30]. These data were not explained by the lever-arm model.

Another theory is the Brownian ratchet model, in which the sliding movement is driven by directionally biased Brownian motion (Figure 10.8B) [31]. In this model, the step size can be larger than the size of the myosin head.

10.3.5
Manipulation of Kinesin

For the laser trap measurements, the size of beads is an important factor for the resolution of the measurements. Smaller beads have less friction coefficient and therefore high resolution. However, it is more difficult to visualize them clearly. Dark field illumination is convenient for small beads. Using the system with small beads (0.2 µm in radius as compared with 1-µm beads) and dark field illumination, accurate measurements of the movement of kinesin have been performed [31, 32].

In contrast to muscle myosin, kinesin moves processively like some types of unconventional myosins. In addition, stall force is ~7 pN greater than myosins. Because of these facts, kinesin is advantageous for single molecule measurements. For the mechanical measurements, kinesin is attached at low densities to beads trapped by a laser. Kinesin moves along microtubules immobilized on the glass surface against increasing trapping force until it reaches the stall force. Using a small bead 0.2 µm in radius and improved measurement systems, the stepwise movement of kinesin has been shown at all loads and ATP concentrations, and substeps have been suggested.

The time interval between the steps is dependent on ATP concentration, indicating that the mechanical event is tightly coupled. On the basis of this finding, a "hand-over-hand model" has been proposed, in which a kinesin walks on the tubulin heterodimers via its two heads in an alternating fashion. Thus, the movement of kinesin, in contrast to myosin, is tightly coupled to the ATP hydrolysis reaction.

Despite the tight coupling mechanism for double-headed kinesin movement, single-headed motors of the kinesin superfamily have exhibited thermal Brownian motion along the microneedle. Thermal movement might play a role in the sliding motion even for double-headed kinesin. Given that there is no lever arm in the structure of kinesin, another type of conformational change has been suggested. Substeps of kinesin have been suggested; however, the detailed mechanism underlying the sliding movement of kinesin will be explored using nanomanipulation techniques and protein engineering.

Dynein is responsible for the oscillating movement of flagella in eukaryotic cells, and the measurement using a laser trap showed that the force generated by very few dynein arms on an isolated doublet microtubule oscillated [33].

10.4
Different Types of Molecular Motors

Magnetic beads are trapped by a magnetic field rather than a laser.

The proton-transport ATP synthase, F_0F_1, is a rotary motor using the energy from the hydrolysis of ATP. The rotary motion of isolated F_1 and the F_0F_1 complex has been visualized by attaching a fluorescently labeled actin filament [34]. When the ATP concentration was lowered, a 120° step of the rotation was clearly identified. In this motor, the rotational motion is tightly coupled to the chemical reaction of the hydrolysis of ATP.

DNA-based molecular motors play an important role in regulating the transcription of DNA. A single DNA molecule attached to beads at both ends has been captured and suspended using a dual laser trap. The binding of fluorescently labeled RNA polymerase to the DNA followed by Brownian movement along the DNA molecule was visualized using total internal reflection microscopy [35]. After reaching the promotors, the RNA polymerase starts to transcribe the DNA to RNA. This process comprises a processive advancement of RNA polymerase along a DNA strand using the energy from the release of pyrophosphate from ribonucleoside triphosphates. The force and velocity produced by a single RNA polymerase during transcription has been measured using a laser trap [36]. RNA polymerase immobilized on the glass surface interacts with a DNA molecule attached at one end to a bead, which was captured and held in a laser trap, RNA polymerase moves at >10 nucleotides per second. The rotation of a DNA molecule during transcription by RNA polymerase has been directly observed.

10.5
Direct Measurements of the Interaction Forces

The interaction between biomolecules is described by a potential. The potential is reflected by the structure of the interaction surface and its physical properties such as charge distribution and hydrophobicity. The measurements of the potential therefore provide the information on the interface of the interaction. It is difficult to directly measure the profile of the surface without damaging the biomolecules.

One of the difficulties in constructing a microscope with which the protein–protein interaction can be measured is weak force. Protein–protein interaction is in the pN order, smaller by 3 orders than the forces that can be measured with atomic force microscopy (AFM). To measure the force of the order of pN in a noncontact mode, AFM microscopy has been improved. A flexible cantilever with the stiffness of 0.1 pN nm^{-1} was made, and the thermal bending motion of the cantilever was reduced using a feedback system with a laser radiation pressure. When the cantilever fluctuates beyond a certain amplitude, a laser turns on to exert the light pressure that keeps the position of the cantilever constant. The root mean square displacement of 6 nm was reduced to less than 1 nm.

10.5.1
Electrostatic Force Between Positively Charged Surfaces

Using this system, electrostatic repulsive forces were measured in solution (Figure 10.9A) [37]. The surfaces of glass and whiskers were modified with amino groups, to have positive charges. The force–distance relationship showed that electrostatic repulsive force is long-range, in agreement with theoretical electrostatic forces between charged conical stylus and infinite surface (Figure 10.9B). When a small amount of salt was added to the solution, the electrostatic forces reduced considerably owing to the shielding effect of ionic atmosphere. The effect of added salt was characterized by estimating Debye lengths in various ionic strengths. The dependencies of Debye length upon ionic strength were in agreement with the Debye–Hückel theory. Thus, the results confirmed experimentally that the theory is also valid in the region of macromolecular size.

10.5.2
Surface Force Property of Myosin Filaments

Surface electrostatic force images of proteins were obtained under noncontact condition [38]. Using a stylus tip modified with positive charges, myosin filaments were scanned in noncontact mode (Figure 10.9C). The images obtained are interpreted as electrostatic surface properties of myosin filaments. A repulsive force due to the interaction of like charges between the whisker tip and the glass surface contributed the background. Negatively charged myosin filaments neutralized the positive charge on

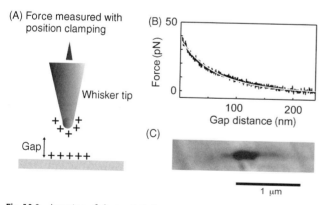

Fig. 10.9 Imaging of electrostatic interaction using noncontact force microscopy. (A) The electric interaction with charges on the glass surface was measured by scanning a whisker tip coated with positively charged groups. (B) The force–distance relationship of electrostatic interaction with charges coated on the glass surface was measured in agreement with model calculation. (C) Imaging of electrostatic potentials on the surface of myosin filaments.

the glass surface, resulting in dark images on a bright background. The detailed images of myosin filaments were consistent with the reported bipolar structure of myosin filament. Therefore, the electrostatic difference in the charge density between the center and both ends of filaments is attributable to the inhomogeneous distribution of the myosin head because the heads exist on the filament except in the central bare zone. Thus, the noncontact images provide the electrostatic features of the protein surface, rather than the surface topography.

References

1 Y. Ishii, T. Yanagida, *Single Mol.*, **1998**, *1*, 5–16.
2 M. Schliwa, G. Woehlke, *Nature*, **2003**, *422*, 759–765.
3 T. Funatsu, Y. Harada, M. Tokunaga, K. Saito, T. Yanagida, *Nature*, **1995**, *374*, 555–559.
4 R.D. Vale, T. Funatsu, D.W. Pierce, L. Ronberg, Y. Harada, T. Yanagida, *Nature*, **1996**, *380*, 451–453.
5 A. Ishijima, H. Kojima, H. Higuchi, T. Funatsu, T. Yanagida, *Nature*, **1991**, *352*, 301–306.
6 A. Ashkin, J.M. Dziedzic, T. Yamane, *Nature*, **1987**, *330*, 769–771.
7 A. Kishino, T. Yanagida, *Nature*, **1988**, *334*, 74–76.
8 H. Kojima, E. Muto, H. Higuchi, T. Yanagida, *Biophys. J.*, **1997**, *73*, 2012–2022.
9 Y. Tsuda, A. Yasutake, A. Ishijima, T. Yanagida, *Biophys. J.*, **1996**, *93*, 12931–12947.
10 Y. Harada, O. Ohara, A. Takatsuki, H. Itoh, N. Shimamoto, K. Kinosita, *Nature*, **2001**, *409*, 113–115.
11 M.S. Kellermayer, S.B. Smith, H.L. Granzier, C. Bustamante, *Science*, **1997**, *276*, 1112–1116.
12 L. Tskhovrebova, J. Trinick, J.A. Sleep, F.M. Simmons, *Nature*, **1997**, *387*, 308–312.
13 M. Rief, M. Gautel, F. Oesterheilt, J.M. Ferrandez, H.E. Gaub, *Science*, **1997**, *276*, 1109–1112.
14 D.J. Liphardt, B. Once, S.B. Smith, I.J. Tinoco, C. Bustamante, *Science*, **2001**, *292*, 733–737.
15 T. Nishizaka, H. Miyata, H. Yoshikawa, S. Ishhiwata, K. Kinosita, Jr., *Nature*, **1995**, *377*, 251–254.
16 K. Kawagushi, S. Ishiwata, *Science*, **2001**, *291*, 667–669.
17 S.J. Kron, J.A. Spudich, *Proc. Natl. Acad. Sci. U.S.A.*, **1986**, *83*, 6272–6276.
18 Y. Harada, K. Sakurada, T. Aoki, D.D. Thomas, T.J. Yanagida, *Mol. Biol.*, **1990**, *216*, 49–68.
19 J.T. Finer, R.M. Simmons, J.A. Spudich, *Nature*, **1994**, *368*, 113–119.
20 A. Ishijima, H. Kojima, H. Higuchi, Y. Harada, T. Funatsu, T. Yanagida, *Biophys. J.*, **1996**, *70*, 383–400.
21 K. Kitamura, M. Tokunaga, A.H. Iwane, T. Yanagida, *Nature*, **1999**, *397*, 129–134.
22 I. Rayment, H.M. Holden, M. Whittaker, C.B. Yohn, M. Lorenz, K.C. Holmes, R.A. Milligan, *Science*, **1993**, *261*, 58–65.
23 A. Houdusse, V.N. Katabokis, D. Himmel, A.G. Szent-Gyorgi, C. Cohen, *Cell*, **1999**, *97*, 459–470.
24 J.A. Spudich, *Nature*, **1994**, *372*, 515–518.
25 J.E. Molloy, J.E. Burns, J. Kendrick-Jones, R.T. Tregear, D.C. White, *Nature*, **1995**, *378*, 209–212.
26 M.J. Tyska, D.E. Dupuis, W.H. Guilford, J.B. Patlak, G.S. Waller, K.M. Trybus, D.M. Warshaw, S. Lowey, *Proc. Natl. Acad. Sci. U.S.A.*, **1999**, *96*, 4402–4407.
27 M. Rief, R.S. Rock, A.D. Mehta, M.S. Mooseker, M.S. Cheney, J.A. Spudich, *Proc. Natl. Acad. Sci. U.S.A.*, **2000**, *97*, 9482–9486.
28 H. Tanaka et al., *Biophys. J.*, **1998**, *75*, 1886–1894.
29 R.S. Rock, S.E. Rice, A.L. Wells, T.J. Purcell, J.A. Spudich, H.L. Sweeney, *Proc. Natl. Acad. Sci. U.S.A.*, **2001**, *98*, 13655–13659.

30 S. Nishikawa, K. Homma, Y. Komori, M. Iwaki, T. Wazawa, A.H. Iwane, J. Saito, R. Ikebe, E. Katayama, T. Yanagida, M. Ikebe, *Biochem. Biophys. Res. Commun.*, **2002**, *290*, 311–317.
31 M. Nishiyama, H. Higuchi, T. Yanagida, *Nat. Cell Biol.*, **2002**, *4*, 790–797.
32 M. Nishiyama, H. Higuchi, Y. Ishii, Y. Taniguchi, T. Yanagida, *Biosystems*, **2003**, *71*, 147–158.
33 C. Shingyoji, H. Higuchim, M. Yoshimura, E. Katayama, T. Yanagida, *Nature*, **1998**, *393*, 711–714.
34 H. Noji, K. Yasuda, M. Yoshida, K. Kinnosita, *Nature*, **1997**, *386*, 299–302.
35 Y. Harada, T. Funatsu, K. Murakami, Y. Nonomura, A. Iahihama, T. Yanagida, *Biophys. J.*, **1999**, *76*, 709–715.
36 H. Yin, D. Wang, K. Svoboda, R. Landick, S. Block, J. Gelles, *Science*, **1995**, *270*, 1653–1657.
37 T. Aoki, M. Hiroshima, K. Kitamura, M. Tokunaga, T. Yanagida, *Ultramicroscopy*, **1997**, *70*, 45–55.
38 T. Aoki, Y. Sowa, H. Yokota, M. Hiroshima, M. Tokunaga, Y. Ishii, T. Yanagida, *Single Mol.*, **2001**, *2*, 183–190.

11
Molecular Optobioelectronics

Eugenii Katz and Andrew N. Shipway

11.1
Introduction

Bioelectronics is a rapidly progressing interdisciplinary research field that aims to create systems with novel functions by integrating biomolecules with electronic elements (e.g., electrodes, field effect transistors or piezoelectric crystals) [1]. The contact engineering of biomaterials (such as enzymes, antigen–antibodies or DNA) on electronic supports controls the electrical properties of the biomaterial–transducer interface and can enable the electronic transduction of biocatalyzed transformations and biorecognition events. In this manner, bioelectronic sensing devices, biofuel elements and biotemplated circuitries have all been developed [2].

The reversible photostimulation of biological functions, such as biocatalysis and bioaffinity binding, offers a means for the transduction of an optical signal into a biochemical event [3, 4]. Various approaches to photostimulate biological functions have been developed, including the chemical modification of proteins by photoisomerizable units [5, 6], the application of photoisomerizable inhibitors [7], and the immobilization of enzymes in photoisomerizable polymers [8, 9]. Direct photostimulation of the redox functions of proteins provides a general route for the electrical transduction of optical signals. In this case, the biomaterial fulfils the dual purposes of light absorber and biochemical function.

Biomolecules can provide very specific recognition processes that result in the formation of functional complexes. The association of a biorecognition element such as an antigen (or antibody) or a DNA molecule with an electronic transducer allows electronic transduction of complex formation [10]. Bioaffinity complex formation in a thin film near a conductive or semiconductive surface yields a chemically modified film on the surface that alters its interfacial properties. Blocking an electrode surface with bulky biomolecules upon immunorecognition, or the formation of a charged interface upon hybridization of DNA molecules, increases the thickness of the double-charged layer, thus electrically insulating the electrode surface. This change allows the direct electronic readout of the bioaffinity binding events by examining the electrochemistry of redox probes that are in solution. The photochemical control of bioaffinity functions, or the association of the bioaffinity complexes with photocurrent-generating moieties, also yields routes to optobioelectronic systems based on biorecognition events.

Bioelectronics. Edited by Itamar Willner and Eugenii Katz
Copyright © 2005 WILEY-VCH Verlag GmbH & Co. KGaA, Weinheim
ISBN: 3-527-30690-0

Photocurrent generation on electrodes modified with organic dyes or semiconductor nanoparticles can be coupled to biocatalytic or biorecognition processes to yield an electronic readout signal reporting on the biochemical process occurring at the functionalized interface.

This chapter summarizes recent advances in molecular optobioelectronics. The electronic transduction of photochemically controlled biocatalytic or bioaffinity binding events is outlined, and the application of photocurrent-generating units as elements electronically reporting on the biocatalytic reactions or biorecognition events is discussed.

11.2
Electronically Transduced Photochemical Switching of Redox-enzyme Biocatalytic Reactions

A small optical signal that turns a biocatalytic process "on" or "off" can result in a very large response, since the biological system may have a very large turnover on the timescale of a measurement. The transduction of this biochemical response, usually a redox response, is therefore referred to as an "amplification" of the input optical signal. The photochemical switching of the biocatalytic functions of redox proteins can lead to the activation and deactivation of biocatalytic electron transfer cascades that translate the photonic signal into an electrical output [11–16]. These systems may provide the basis for future sensing and information-processing devices.

Three general methodologies for the application of photoisomerizable materials in bioelectrocatalytic systems may be envisaged (Figure 11.1). By the first method (Figure 11.1A), photoisomerizable units are tethered to a redox enzyme. In configuration **A**, the active-site environment of the enzyme is distorted, and the bioelectrocatalytic properties of the enzyme are blocked. Photoisomerization of the photoactive groups to state **B** restores the active-site structure, and the enzyme is activated for its bioelectrocatalytic process. Electrical contact between the biocatalyst and the electrode results in the transduction of a current to the macroscopic environment. By cyclic photoisomerization of the photoactive groups between the states **B** and **A**, the amperometric transduction can be switched between "on" and "off" states, respectively.

The second approach, which involves a photosensitive electrode, is shown in Figure 11.1B. The transducer element is functionalized with a photoisomerizable interface. When the photoisomerizable group is in state **A**, no affinity interactions exist between the biomaterial and the functionalized electrode. As a result, no electronical contacting occurs, and the system is in a mute state. Photoisomerization of the interface to state **B** results in the binding of the redox biomaterial to the surface by affinity interactions or intermolecular recognition properties. This results in the electronic coupling of the biomaterial and the electronic transducer, leading to the electronic transduction of the biocatalytic process. The photoisomerizable interface is said to act as a "photo-command" interface for controlling the electrical communication between the redox biomaterial and the electronic transducer.

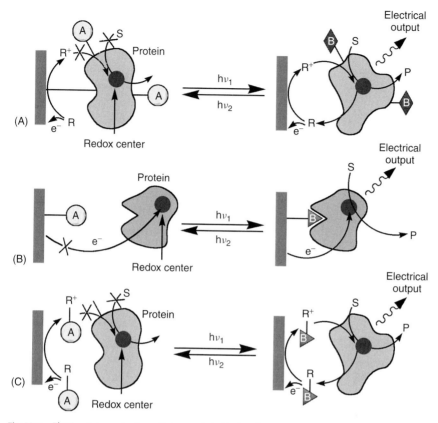

Fig. 11.1 Electronic transduction of photoswitchable bioelectrocatalytic functions of redox enzymes: (A) by the tethering of photoisomerizable units to the protein; (B) by application of a photoisomerizable command interface that controls the electrode contact between the redox protein and the electrode; (C) by application of photoisomerizable electron transfer mediators providing electron transport between the redox enzyme and the conductive support in one isomeric state, whereas being ineffective for the electron shuttling in another isomeric state. (S and P are the substrate and product of the enzymatic reaction, respectively; R and R^+ are redox relay units in the reduced and oxidized states, respectively).

The third approach is based on the application of photoswitchable electron transfer mediators. In one isomeric state, these structures are effective electron transfer mediators, while in the other state the redox units are ineffective (Figure 11.1C). Thus, photochemical switching between the two isomeric states of the units provides reversible switching of the bioelectrocatalytic process "on" and "off", even though the enzyme activity itself is unaffected.

In the forthcoming section, we will address various systems that were tailored along these concepts and lead to the electronic transduction of photoswitchable redox-enzyme biocatalytic functions.

11.2.1
Electronic Transduction of Biocatalytic Reactions Using Redox Enzymes Modified with Photoisomerizable Units

Enzymes modified with photoisomerizable groups can be used as light-switchable biocatalysts [17, 18]. The photoisomerizable groups can be randomly bound to the protein backbone or selectively coupled directly to the redox cofactor of the enzyme, thus providing the most efficient control over the biocatalytic process. Coupling of the biocatalytic process with the electrode support by use of diffusional electron transfer mediators results in the electronic transduction of the biocatalytic process, thus allowing the electronic readout of the photochemically switchable biocatalytic process.

11.2.1.1 Redox Enzymes Modified with Photoisomerizable Groups Randomly Tethered to the Protein Backbone

Glucose oxidase (GOx) has been employed as a redox enzyme to tailor a photoisomerizable enzyme electrode for the photoswitchable oxidation of glucose [19, 20]. Photoisomerizable nitrospiropyran (SP) units were randomly tethered to lysine residues of GOx, and the resulting photoisomerizable protein was assembled on a Au electrode as shown in Figure 11.2. A primary N-hydroxysuccinimide-functionalized thiol monolayer

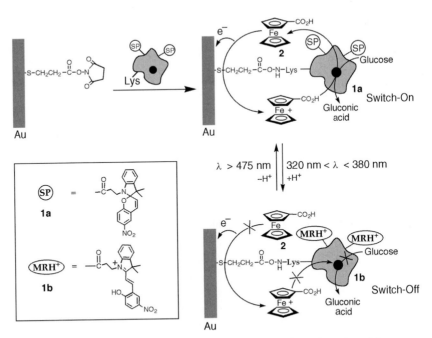

Fig. 11.2 Assembly of a photoisomerizable glucose oxidase monolayer electrode and the reversible photoswitchable activation/deactivation of the bioelectrocatalytic functions of the enzyme electrode.

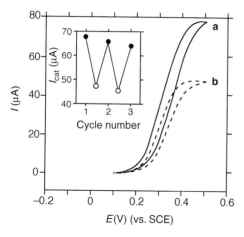

Fig. 11.3 Photostimulated bioelectrocatalyzed oxidation of glucose, 2.5×10^{-2} M, in the presence of ferrocene carboxylic acid **2** 5×10^{-3} M, as diffusional electron mediator in the presence of (**a**) the nitrospiropyran-tethered GOx **1a**; (**b**) the protonated nitromerocyanine-tethered GOx **1b**. Inset: Reversible photoswitchable amperometric transduction of the bioelectrocatalyzed oxidation of glucose by **1a** (●) and **1b** (○). (Adapted from Ref. [15], Figure 7.10, with permission).

was assembled on the conductive support, and the photoisomerizable enzyme was covalently coupled to the monolayer to yield the integrated photoactive enzyme electrode. The enzyme monolayer was found to undergo reversible photoisomerization, and irradiation of the nitrospiropyran-tethered GOx **1a** with UV light (320 nm $< \lambda <$ 380 nm) generated the protonated nitromerocyanine (MRH$^+$)-tethered GOx **1b**. Further irradiation of the **1b**-monolayer with visible light ($\lambda > $ 475 nm) restored the nitrospiropyran-tethered enzyme **1a**. The photoisomerizable enzyme monolayer electrode revealed photoswitchable bioelectrocatalytic activity (Figure 11.3). In the presence of ferrocene carboxylic acid **2** as a diffusional electron transfer mediator, the nitrospiropyran-tethered GOx **1a** revealed a high bioelectrocatalytic activity, reflected by a high electrocatalytic anodic current (Figure 11.3, curve a). The protonated nitromerocyanine-GOx **1b** exhibited a twofold lower activity, as reflected by the decreased bioelectrocatalytic current (Figure 11.3, curve b). By the reversible photoisomerization of the enzyme electrode between the **1a**- and **1b**-states, the current responses are cycled between high and low values, respectively (Figure 11.3, inset). Although the tethering of photoisomerizable units to the protein leads to photoswitchable bioelectrocatalytic properties, the "off"-state of the photoisomerizable GOx exhibits residual bioelectrocatalytic activity as the structural distortion of the protein in the **1b**-state is not optimized.

11.2.1.2 Redox Enzymes Reconstituted with FAD Cofactors Modified with Photoisomerizable Units

The point of action in a redox protein is its active site, and therefore the site-specific functionalization or mutation of an enzyme's active site is an ideal way to access its operation. Control over an enzyme by its site-specific modification with a photoisomerizable species has been accomplished by a semisynthetic approach involving the reconstitution of the flavoenzyme glucose oxidase with a synthetic photoisomerizable flavin adenine dinucleotide (FAD)-cofactor [21] (Figure 11.4). The photoisomerizable carboxylic acid–functionalized nitrospiropyran **3** was covalently coupled to N^6-(2-aminoethyl)-FAD **4** to yield the synthetic photoisomerizable nitrospiropyran-FAD cofactor **5a**

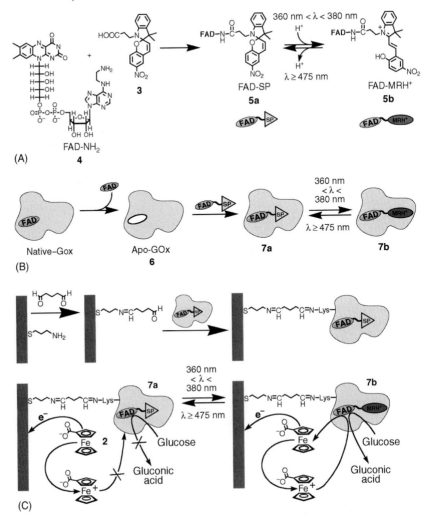

Fig. 11.4 (A) Synthesis of a synthetic photoisomerizable FAD cofactor. (B) Reconstitution of apo-glucose oxidase with the synthetic nitrospiropyran-FAD photoisomerizable cofactor to yield a photoisomerizable glucose oxidase. (C) Assembly of the nitrospiropyran-FAD-reconstituted GOx as a monolayer on the electrode and the reversible photoswitchable bioelectrocatalytic activation/deactivation of the enzyme electrode.

(Figure 11.4A). Upon illumination with the appropriate light, the nitrospiropyran-FAD cofactor **5a** could be reversibly photoisomerized to nitromerocyanine-FAD cofactor **5b**. The native FAD cofactor was extracted from glucose oxidase, and the synthetic photoisomerizable-FAD cofactor **5a** was reconstituted into the obtained apo-glucose oxidase (apo-GOx, **6**) to yield the photoisomerizable enzyme **7a** (Figure 11.4B). This reconstituted enzyme includes a photoisomerizable unit directly attached to the redox

11.2 Electronically Transduced Photochemical Switching of Redox-enzyme Biocatalytic Reactions

center of the protein, and hence the redox enzyme is anticipated to reveal optimized photoswitchable bioelectrocatalytic properties. The resulting enzyme was assembled on a Au electrode as outlined in Figure 11.4C to yield an integrated optobioelectronic assembly. The photoinduced bioelectrocatalytic oxidation of glucose was stimulated in the presence of ferrocene carboxylic acid **2** as a diffusional electron transfer mediator. The nitrospiropyran (SP) state of the reconstituted enzyme **7a** was inactive for the bioelectrocatalytic transformation, whereas photoisomerization of the enzyme electrode to the protonated nitromerocyanine (MRH$^+$) state **7b** activated the enzyme for the bioelectrocatalyzed oxidation of glucose (Figure 11.5). By the cyclic photoisomerization of the enzyme monolayer interface between the nitrospiropyran (SP) and the protonated nitromerocyanine (MRH$^+$) states, the biocatalyzed oxidation of glucose was cycled between "off-" and "on"-states, respectively (Figure 11.5, inset).

It was also found that the direction of the photobiocatalytic switch of the nitrospiropyran/nitromerocyanine-FAD-reconstituted enzyme is controlled by the electrical properties of the electron transfer mediator [22] (Figure 11.6). With ferrocene dicarboxylic acid **8** as a diffusional electron transfer mediator, the enzyme in the nitrospiropyran-FAD state **7a** was found to correspond to the "off"-state biocatalyst, while the protonated nitromerocyanine state of the enzyme **7b** exhibits "on" behavior. In the presence of the protonated 1-[1-(dimethylamino)ethyl]ferrocene **9**,

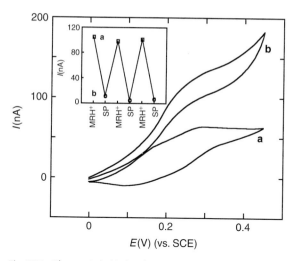

Fig. 11.5 Photoswitchable bioelectrocatalyzed oxidation of glucose, 5×10^{-2} M, by a photoisomerizable FAD-reconstituted GOx assembled as a monolayer on a Au electrode and using ferrocene carboxylic acid **2** 5×10^{-5} M as a diffusional electron mediator; (a) by the nitrospiropyran-FAD-reconstituted GOx **7a** monolayer; (b) by the protonated nitromerocyanine-FAD-reconstituted GOx **7b** monolayer. Inset: cyclic amperometric transduction of the bioelectrocatalyzed oxidation of glucose by the photoisomerizable reconstituted GOx monolayer electrode; (a) monolayer in the **7b**-state; (b) monolayer in the **7a**-state. (Adapted from Ref. [15], Figure 7.12, with permission).

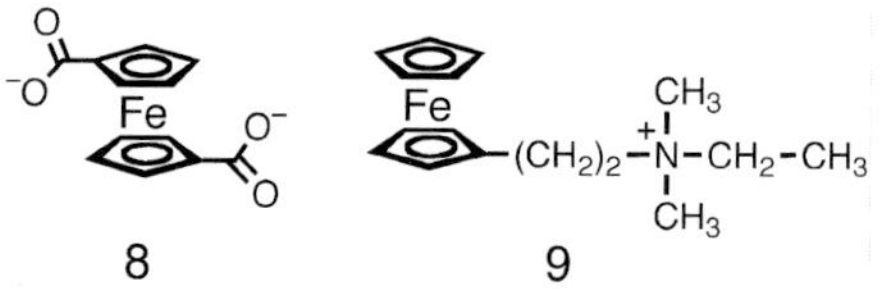

Fig. 11.6 Negatively and positively charged electron relays for the controlled switching of the bioelectrocatalytic activity of the reconstituted photoisomerizable Gox.

the direction of the photobioelectrocatalytic switch is reversed. The nitrospiropyran enzyme state **7a** is activated toward the electrocatalyzed oxidation of glucose, while the protonated nitromerocyanine enzyme state **7b** is switched "off", and is inactive for the electrochemical oxidation of glucose. This control of the direction of the photoisomerizable reconstituted enzyme photoswitch was attributed to electrostatic interactions between the charged diffusional electron mediator and the photoisomerizable unit linked to the FAD. The protonated nitromerocyanine photoisomer state (MRH$^+$) attracts the oxidized negatively charged electron mediator **8**, but repels the oxidized positively charged relay **9**. As a result, the photoisomer state of the enzyme **7b** is switched "on" in the presence of the negatively charged ferrocene derivative. With the positively charged electron transfer mediator, the opposite bioelectrocatalytic switch direction is observed. The **7b**-state of the enzyme repels the positively charged electron transfer mediator, and its biocatalytic functions are blocked, but the **7a**-state allows the mediator to approach and a bioelectrocatalytic current is observed.

11.2.2
Electronic Transduction of Biocatalytic Reactions Using Interactions of Redox Enzymes with Photoisomerizable "Command Interfaces"

The second approach to the organization of integrated optobioelectronic switches involves the organization of a photoisomerizable "command" interface [23–25]. A "command" interface controls the interfacial electrochemistry of a solution-state species by isomerization between two states. One of these states may actively block electron transfer (for instance, by electrostatically repelling the substrate), or alternatively one state may promote electrical communication (for instance, by binding the substrate through weak interactions). Thus, in one photoisomer state, electron transfer to a redox probe is significantly more favorable than in the other state. Various interactions, such as electrostatic interactions, host–guest, donor–acceptor or bioaffinity interactions, can contribute to this selectivity.

11.2.2.1   Electrochemical Processes of Proteins and Enzymes at "Command" Interfaces

Redox proteins usually lack direct electrical contact with electrodes since their active redox sites are necessarily embedded deeply in the protein [26]. This depth insulates the electrical communication between the redox site and the electrode by spatial separation and allows the enzyme to perform its functions efficiently in its highly reactive native environment. Suitable functionalization of the electrode is often able to stimulate electron transfer between the protein and the electrode, however, by aligning the enzyme in an optimized position or by offering electron mediation through a group that penetrates into the enzyme structure. A photoswitchable electrode that stimulates this electron transfer in one state can be used as a command surface for the enzyme's biocatalytic process. Figure 11.1B shows a surface-bound photoisomerizable unit that binds an enzyme in the **B** state (but not in the **A** state), stimulating its biocatalytic process.

For low molecular weight redox proteins, for example the hemoprotein cytochrome c (Cyt c), the chemical functionalization of electrode surfaces with molecular promoter units has been found to facilitate electron transfer communication between the protein redox center and the conductive supports [27–29]. The binding of the redox proteins to the promoter sites aligns the redox centers with respect to the electrode surface. This results in the shortening of the electron transfer distances and the electrical contacting of the redox centers. For example, pyridine units [27–29] or negatively charged promoter sites [30], assembled on electrode surfaces, have been reported to align Cyt c on electrodes by affinity or electrostatic interactions, and to facilitate the electrical communication between the heme-center and the electrode. Cytochrome c is a positively charged hemoprotein at neutral pH. This suggests that by the nanoengineering of an electrode with a composite layer consisting of Cyt c binding sites and photoisomerizable units that are transformed from neutral to a positively charged state, the electrostatic photoswitchable binding and dissociation of Cyt c to and from the electrode can occur [31]. In order to photoregulate the electrical communication between Cyt c and the electrode, a mixed monolayer consisting of pyridine sites and photoisomerizable nitrospiropyran units **10a** was assembled on a Au electrode (Figure 11.7A). This monolayer binds Cyt c to the surface and aligns the heme-center of the protein, and electrical contact with the electrode is thus stimulated (Figure 11.7B, curve a). Photoisomerization of the monolayer to the positively charged protonated nitromerocyanine state **10b** results in the electrostatic repulsion of Cyt c from the monolayer. As a result, the electrical communication between the hemoprotein and the electrode is blocked (Figure 11.7B, curve b). By cyclic photoisomerization of the monolayer between the nitrospiropyran **10a** and protonated nitromerocyanine **10b** states, the binding and dissociation of Cyt c to and from the monolayer occurs, and the electrical contact of the hemoprotein and the electrode is switched between "on" and "off" states, respectively.

Cytochrome c acts as an electron transfer mediator (cofactor) that activates many secondary biocatalyzed transformations by the formation of interprotein Cyt c-enzyme complexes [32–34]. Specifically, Cyt c transfers the electrons to cytochrome oxidase (COx), which mediates the four-electron reduction of oxygen to water. The photoswitchable electrical activation of Cyt c enables the photostimulated triggering of the COx-catalyzed reduction of O_2 (Figure 11.8A). In the presence of the pyridine-nitrospiropyran **10a** monolayer electrode, electron transfer to Cyt c activates the electron transfer cascade to COx, and the bioelectrocatalyzed reduction of O_2 to water proceeds [31]. The bioelectrocatalyzed reduction of O_2 is reflected by the transduced electrocatalytic cathodic current (Figure 11.8B, curve a). Photoisomerization of the

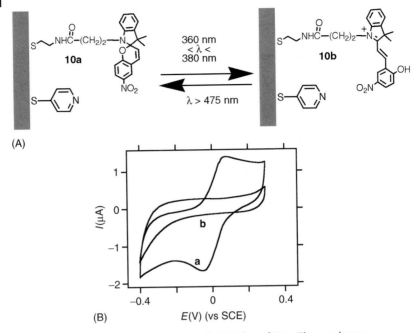

Fig. 11.7 (A) An interface for the light-controlled binding of Cyt c. The cytochrome electrochemistry is promoted by the pyridine component, but Cyt c is repelled from **10b**. (B) Cyclic voltammograms of Cyt c, 1×10^{-4} M, in the presence of (a) the pyridine-nitrospiropyran **10a** mixed monolayer electrode; (b) the pyridine-protonated-nitromerocyanine **10b** mixed monolayer electrode. Data recorded at a scan rate of 50 mV s^{-1}. (Adapted from Ref. [15], Figure 7.20, with permission).

monolayer to the protonated nitromerocyanine **10b** configuration results in the repulsion of Cyt c from the electrode interface. This blocks the interfacial electron transfer to Cyt c, and thus the COx-mediated reduction of O_2 is also inhibited (Figure 11.8B, curve b). Note that the bioelectrocatalytic cathodic current transduced by the Cyt c-COx protein assembly is ca 10-fold enhanced as compared to the amperometric current resulting from the Cyt c alone (measured at the same potential scan rate). This originates from the fact that COx induces a bioelectrocatalytic process, and the photonic

Fig. 11.8 (A) Reversible photoswitchable activation/deactivation of the electrical contact between Cyt c and the electrode and the secondary activation/deactivation of the COx-biocatalyzed reduction of oxygen using a photoisomerizable **10a/10b**-pyridine mixed thiolated monolayer as a command interface. (B) Cyclic voltammograms of the Cyt c/COx system corresponding to the photostimulated bioelectrocatalyzed reduction of O_2 in the presence of the photoisomerizable monolayer electrode: (a) bioelectrocatalyzed reduction of O_2 by Cyt c/COx in the presence of pyridine-nitrospiropyran **10a** mixed monolayer electrode; (b) cyclic voltammogram of the Cyt c/COx system under O_2 in the presence of the pyridine-protonated-merocyanine **10b** mixed monolayer electrode. Data recorded at scan rate 2 mV s^{-1}. (Adapted from Ref. [15], Figure 7.21, with permission).

11.2 Electronically Transduced Photochemical Switching of Redox-enzyme Biocatalytic Reactions | 319

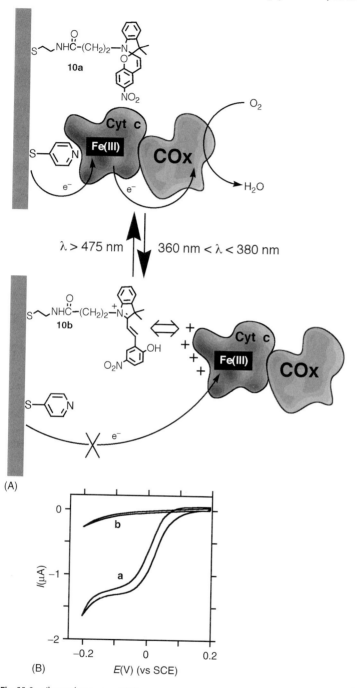

Fig. 11.8 (legend see page 318)

activation of the Cyt c-COx system drives the reduction of O_2 with a high turnover. Thus, the amperometric response of the Cyt c-COx layered electrode represents the amplified amperometric readout of the photonic information recorded in the monolayer. By the cyclic photoisomerization of the monolayer between the nitrospiropyran **10a** and protonated nitromerocyanine **10b** states, the cathodic amperometric responses from the system are reversibly cycled between "on" and "off" states, respectively. Similarly, the Cyt c-mediated oxidation of lactate in the presence of Cyt c-dependent lactate dehydrogenase (LDH) was photochemically switched "on" and "off" on the photoisomerizable interface, representing a biocatalytic system with the photochemical control of the anodic amperometric responses [20]. The photochemical control over the Cyt c-COx or Cyt c-LDH bioelectrocatalytic cascade represents a system analogous to the vision process. Photoisomerization of rhodopsin in the eye stimulates the binding of protein G to the light-active membrane, which activates a biocatalytic cascade that generates c-GMP and triggers the neural response.

The electrostatic control of the electrical contact between redox proteins and electrodes by means of photoisomerizable "command" interfaces was further demonstrated by the photochemical switching of the bioelectrocatalytic properties of glucose oxidase [35] (Figure 11.9). Ferrocene units were tethered to the protein backbone of glucose oxidase to yield an "electrically wired" enzyme that is activated for the bioelectrocatalyzed oxidation of glucose. The enzyme is negatively charged at neutral pH values ($pI_{GOx} = 4.2$) [36] and hence could be electrostatically attracted by positively charged surfaces. Accordingly, a thiolated nitrospiropyran **11a** monolayer was assembled on a Au electrode. In this state, inefficient electrical interactions between the ferrocene-modified GOx and the modified electrode exist, and only moderate bioelectrocatalyzed oxidation of glucose occurs. Photoisomerization of the monolayer to the protonated nitromerocyanine state **11b** results in the electrostatic attraction of the biocatalyst to the electrode support. The concentration of the enzyme at the electrode surface leads to effective electrical communication between the biocatalyst and the electrode, and yields the enhanced bioelectrocatalyzed oxidation of glucose. The transduced current represents an amplified signal resulting from the photonic activation of the monolayer. By the cyclic photoisomerization of the monolayer between the protonated nitromerocyanine state **11b** and the nitrospiropyran state **11a**, the enzyme can be switched between surface-associated and dissociated configurations, respectively, leading to reversible "on"–"off" control over the bioelectrocatalytic activity of the enzyme.

11.2.2.2 Surface-Reconstituted Enzymes at Photoisomerizable Interfaces

Another example of a "command" interface involves the photostimulated electrostatic control of the electrical contacting of a redox enzyme and an electrode in the presence of a diffusional electron mediator [37] (Figure 11.10). A mixed monolayer, consisting of photoisomerizable thiolated nitrospiropyran units **11a** and a synthetic N^6-(2-aminoethyl)-FAD cofactor **4** was assembled onto a Au electrode. Apo-glucose oxidase (apo-GOx) was reconstituted onto the surface FAD sites to yield an aligned enzyme electrode. Although it is well defined, this surface-reconstituted GOx (2×10^{-12} mol cm^{-2}) still lacks electrical communication with the electrode. In the

11.2 Electronically Transduced Photochemical Switching of Redox-enzyme Biocatalytic Reactions

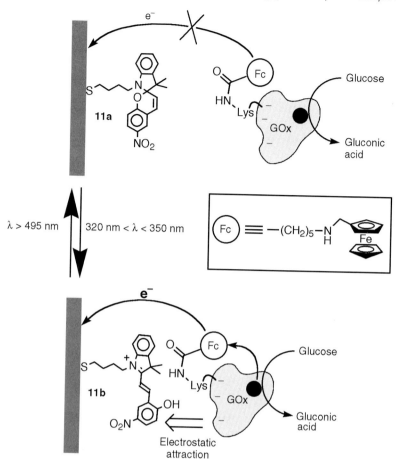

Fig. 11.9 Photochemical control of the electrical contact between a ferrocene-tethered glucose oxidase and the electrode using a thiolated nitrospiropyran as a command interface.

presence of the positively charged protonated 1-[1-(dimethylamino)ethyl]ferrocene **9**, as a diffusional electron mediator, the bioelectrocatalytic functions of the enzyme electrode are activated and controlled by the photoisomerizable component co-immobilized in the monolayer assembly (Figure 11.10). With the monolayer in the neutral nitrospiropyran state **11a**, the positively charged electron mediator **9** is oxidized at the electrode. This process allows it to act as an electron transfer mediator and to activate the bioelectrocatalyzed oxidation of glucose, as reflected by an electrocatalytic anodic current (Figure 11.11, curve a). Photoisomerization of the monolayer to the protonated nitromerocyanine state **11b** results in the electrostatic repulsion of **9** from the electrode, and the mediated electrical communication between the enzyme and the electrode is blocked. As a result, the bioelectrocatalytic functions of the GOx layer

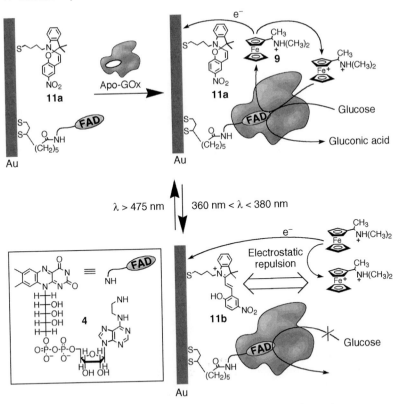

Fig. 11.10 Surface-reconstitution of apo-glucose oxidase on a mixed monolayer associated with an electrode consisting of an FAD cofactor and photoisomerizable nitrospiropyran units, and reversible photoswitching of the bioelectrocatalytic functions of the enzyme electrode in the presence of a positively charged diffusional relay.

are switched "off" (Figure 11.11, curve b). By the cyclic photoisomerization of the monolayer between the nitrospiropyran and protonated nitromerocyanine states, the modified bioelectrocatalytic interface is reversibly switched "on" and "off", respectively (Figure 11.11, inset).

11.2.3
Electronic Transduction of Biocatalytic Reactions of Redox Enzymes Using Electron Transfer Mediators with Covalently Bound Photoisomerizable Units

Photoswitchable electrical communication between enzymes and electrodes has also been achieved by the application of photoisomerizable electron transfer mediators [20, 38] (Figure 11.12A). Diffusional electron mediators (e.g. viologen **12** or ferrocene **13** derivatives) have been functionalized with photoisomerizable spiropyran/merocyanine units. These mediators can be reversibly photoisomerized from the spiropyran state

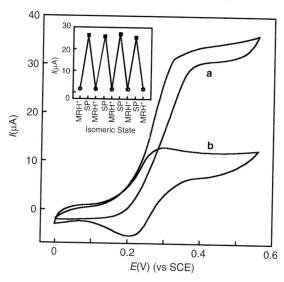

Fig. 11.11 Photoswitchable bioelectrocatalyzed oxidation of glucose, 8×10^{-2} M, by a composite monolayer consisting of GOx reconstituted onto FAD-units and nitrospiropyran photoisomerizable units in the presence of **9** as diffusional electron mediator: (**a**) in the presence of the nitrospiropyran state **11a**; (**b**) in the presence of the protonated nitromerocyanine state **11b**. Inset: Cyclic amperometric transduction of photonic signals recorded by the photoisomerizable monolayer electrode through the bioelectrocatalyzed oxidation of glucose. (Adapted from Ref. [15], Figure 7.24, with permission).

12a, **13a** to the merocyanine state **12b**, **13b** ($360 < \lambda < 380$ nm) and back ($\lambda > 475$ nm). An enzyme multilayer array composed of glutathione reductase (GR) or glucose oxidase was found to be electrically contacted only when the relay-linked chromophore was in the spiropyran state **12a**, **13a** (Figure 11.12A). Cyclic activation/deactivation of the enzyme arrays was achieved upon the photochemical isomerization of the electron transfer mediators between **12a**, **13a** and **12b**, **13b** states (Figure 11.12B). The lack of electrical interactions between the enzymes and the mediators in the merocyanine state could originate from steric constraints or from repulsive electrostatic interactions between the isomers and the enzymes.

11.3
Electronically Transduced Reversible Bioaffinity Interactions at Photoisomerizable Interfaces

The electronic transduction of antigen–antibody complex formation has been demonstrated as a basis for immunosensor devices [39, 40]. Several transduction means, including electrochemical (potentiometric [41, 42], amperometric [43–47] and

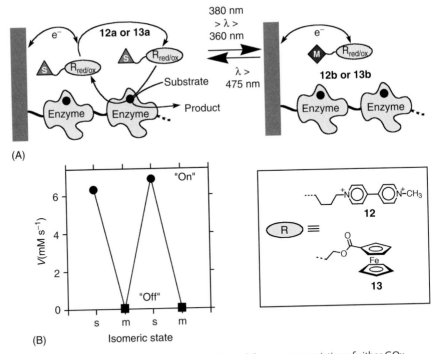

Fig. 11.12 (A) The electrochemical contacting of a multilayer array consisting of either GOx or GR in the presence of diffusional mediators (ferrocene and viologen, respectively) tethered to photoisomerizable units **12** and **13**. (B) Switching behavior of the bioelectrocatalytic function of a multilayer array of GR as a function of the state of a photoisomerizable group attached to a diffusional viologen mediator **12**: "S" and "M" represent the photoisomerizable units in the spiropyran **12a** and the merocyanine **12b** states, respectively.

impedimetric [10] signals) and piezoelectric (microgravimetric) [48–50], have been used to follow the formation of antigen–antibody complexes on surfaces. Figure 11.13 gives a schematic of amperometric, impedimetric and piezoelectric signals from the formation of antigen–antibody complexes on solid supports. The formation of an antigen–antibody complex on an electrode insulates the conductive support and introduces a barrier for electron transfer at an electrode interface, thus blocking the amperometric response of redox labels solubilized in the electrolyte solution (Figure 11.13A). The use of a large redox probe such as a redox-labeled enzyme ensures that the binding of the antibody to the electrode effectively blocks its bioelectrocatalytic function. Impedance spectroscopy, and specifically Faradaic impedance spectroscopy [10], is a useful method to probe the resulting changes in electron transfer resistance and capacitance at the interface. Upon the application of an alternating voltage close to the reduction potential of the redox label, the complex impedance (with real, $Z_{re}(\omega)$, and imaginary, $Z_{im}(\omega)$, components) is measured as a function of the applied frequency. The interfacial electron transfer resistance at the electrode support, R_{et}, is derived from a Nyquist plot (Z_{im} vs Z_{re}) in which the diameter of the semicircular region of the spectrum corresponds to the

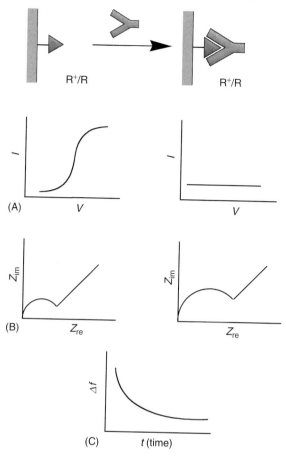

Fig. 11.13 Electronic transduction of the formation of antigen–antibody affinity complexes on transducers: (A) Amperometric transduction at an electrode (R^+/R is a redox label in the electrolyte solution). (B) Transduction by Faradaic impedance spectroscopy. (C) Piezoelectric Quartz Crystal Microbalance (QCM) transduction in the presence of a piezoelectric quartz crystal.

electron transfer resistance at the electrode surface. The formation of the hydrophobic antigen–antibody complex on the electrode surface insulates the electrode, as reflected by an increase in the interfacial electron transfer resistance and an enlarged semicircle diameter (Figure 11.13B). Changes in the mass associated with a piezoelectric crystal result in changes in its resonant frequency (Eq. 1). An increase in the quartz crystal mass as a result of the formation of the antigen–antibody complex is therefore accompanied by a decrease in the resonant frequency of the crystal, Figure 11.13C.

$$\Delta f = -C_f \Delta m \quad \text{where } C_f = 1.83 \times 10^8 \, (\text{Hz} \cdot \text{cm}^2 \cdot \text{g}^{-1}) \tag{1}$$

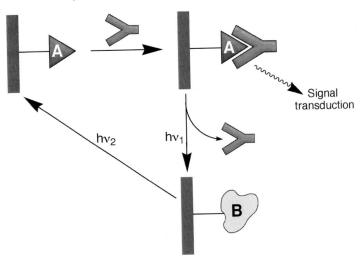

Fig. 11.14 Assembly of a reversible immunosensor using a photoisomerizable antigen-functionalized transducer.

11.3.1
Reversible Immunosensors Based on Photoisomerizable Antigens

Antigen–antibody affinity interactions usually exhibit very high binding constants ($K_a \approx 10^7 - 10^{10}$ M^{-1}). This turns most immunosensor devices into single-cycle bioelectronic systems, in which the regeneration of the sensor after interaction with the analyte is impossible. The concept of photoswitchable binding between a photoisomerizable substrate and a receptor has been used to address this problem [51, 52] (Figure 11.14). A photoisomerizable antigen is assembled on the sensing interface as a monolayer. In one photoisomer configuration (state **A**), the monolayer exhibits affinity for the antibody. The formation of the antigen–antibody complex on the surface is transduced to the environment, thus enabling the sensing of the antibody. After the sensing cycle, the monolayer is photoisomerized to state **B**, which lacks antigen affinity properties for the antibody. This enables the washing off of the antibody from the monolayer interface. The resulting monolayer is finally re-isomerized from state **B** back to state **A**, regenerating the original antigen interface. Thus, a two-step photochemical isomerization of the monolayer, with an intermediate rinsing process, enables the recycling of the sensing interface, and the functionalized transducer acts as a reusable immunosensor.

The reversible cyclic sensing of dinitrophenyl-antibody (DNP-Ab) was accomplished by this method, by the application of a photoisomerizable dinitrospiropyran monolayer on a Au support. A thiolated dinitrospiropyran photoisomerizable monolayer **14a** was assembled on a Au electrode or a Au-coated quartz crystal [51, 52] (Figure 11.15). The dinitrospiropyran monolayer acts as an antigen for the DNP-Ab, while the protonated dinitromerocyanine monolayer state **14b** lacks the antigen affinity for the DNP-Ab (Figure 11.15). The association of DNP-Ab to the dinitrospiropyran antigen-monolayer

11.3 Electronically Transduced Reversible Bioaffinity Interactions at Photoisomerizable Interfaces | 327

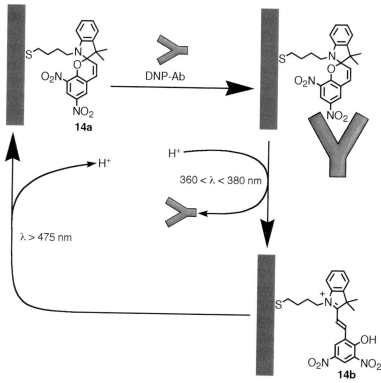

Fig. 11.15 Preparation of a photoisomerizable dinitrospiropyran monolayer on a transducer and the reversible sensing of DNP-Ab.

can be transduced electrochemically using amperometry [53] or Faradaic impedance spectroscopy [54], or by using piezoelectric (QCM) means [53].

The interfacial properties of the antigen-modified electrode were probed in the presence of an "electrically wired" redox enzyme and a respective substrate (i.e., ferrocene-tethered glucose oxidase and glucose, respectively) that provided an electronic signal in the form of an electrocatalytic current [53] (Figure 11.16A, curve a). The association of the DNP-Ab to the antigen-monolayer-functionalized electrode leads to the electrical insulation of the electrode support and to the introduction of an electron barrier at the electrode surface. Thus, the bioelectrocatalytic current is inhibited upon the formation of the DNP-Ab/dinitrospiropyran complex (Figure 11.16A, curve b). Similarly, the association of the DNP-Ab to the antigen-monolayer increases the interfacial electron transfer resistance, R_{et} [54]. In the presence of $[Fe(CN)_6]^{3-}/[Fe(CN)_6]^{4-}$ as a redox probe, the electron transfer resistance increases from 60 ± 2 kΩ to 80 ± 2 kΩ upon the formation of the DNP-Ab/(**14a**)-complex (Figure 11.16B, curves b and c, respectively). Photoisomerization of the DNP-Ab/dinitrospiropyran monolayer interface (360 nm $< \lambda <$ 380 nm) to the protonated dinitromerocyanine, followed by the rinsing-off of the DNP-Ab, regenerates the original

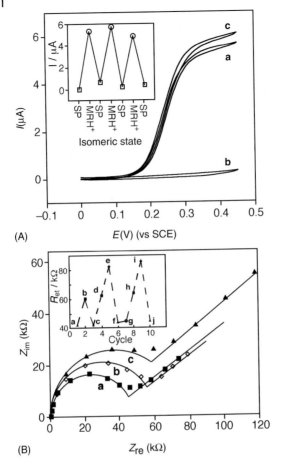

Fig. 11.16 (A) Cyclic voltammograms of: (a) the dinitrospiropyran **14a** monolayer electrode; (b) after addition of DNP-Ab to the dinitrospiropyran **14a** monolayer electrode; (c) after photoisomerization of the dinitrospiropyran/DNP-Ab to the protonated dinitromerocyanine **14b** monolayer and the washing off of the antibody. All data were recorded in the presence of ferrocene-tethered GOx as a redox biocatalyst and glucose, 5×10^{-2} M, scan rate 5 mV s^{-1}. Inset: Cyclic amperometric sensing of the DNP-Ab by the dinitrospiropyran photoisomerizable monolayer electrode. (B) Faradaic impedance spectra (Nyquist plots) of (a) the protonated dinitromerocyanine **14b** monolayer electrode; (b) the dinitrospiropyran **14a** monolayer electrode; (c) the dinitrospiropyran **14a** monolayer electrode upon addition of the DNP-Ab. Impedance spectra were recorded in the presence of [Fe(CN)$_6$]$^{3-/4-}$, 1×10^{-2} M, as a redox label. Inset: Interfacial electron transfer resistances, R_{et}, at the functionalized electrodes upon the cyclic photoisomerization of the monolayer and the reversible sensing of the DNP-Ab: (a) and (c) monolayer in the protonated dinitromerocyanine state **14b**; (b) (d) and (h) monolayer in the dinitrospiropyran state **14a**; (e) and (i) after binding of DNP-Ab to the dinitrospiropyran **14a** monolayer electrode; (f) and (j) after photoisomerization of the dinitrospiropyran **14a**/DNP-Ab monolayer electrode to the protonated dinitromerocyanine **14b** and the washing off of the DNP-Ab; (g) addition of the DNP-Ab to the protonated dinitromerocyanine monolayer electrode **14b**. (Adapted from Ref. [15], Figure 7.33, with permission).

amperometric response of the monolayer-functionalized-electrode (Figure 11.16A, curve c) and a low electron transfer resistance ($R_{et} = 47 \pm 2$ kΩ) in the Faradaic impedance spectrum, Figure 11.16B, curve a, indicating that the antibody was removed from the electrode support. The electron transfer resistance at the protonated dinitromerocyanine-functionalized electrode is lower than at the dinitrospiropyran-modified electrode using $[Fe(CN)_6]^{3-}/[Fe(CN)_6]^{4-}$ as redox probe because the positively charged protonated dinitromerocyanine monolayer interface electrostatically attracts the redox label. Further photochemical isomerization of the protonated dinitromerocyanine monolayer to the dinitrospiropyran interface ($\lambda > 475$ nm) regenerates the original sensing interface. The reuse of the interface is demonstrated in Figure 11.16(A and B), insets.

Piezoelectric (microgravimetric) transduction provides a further means to probe the binding interactions of the DNP-Ab with the photoisomerizable interface [53] (Figure 11.17A). The binding of DNP-Ab to the dinitrospiropyran results in a mass increase on the crystal, and a decrease in the crystal frequency of 120 Hz (Figure 11.17A,

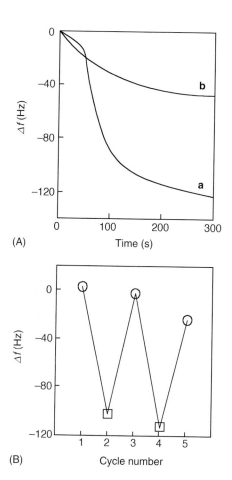

Fig. 11.17 (A) Time-dependent frequency changes of (a) the dinitrospiropyran **14a** monolayer associated with a Au-quartz crystal upon addition of DNP-Ab; (b) the protonated dinitromerocyanine **14b** monolayer on a Au-quartz crystal upon addition of the DNP-Ab. (B) Cyclic microgravimetric sensing of the DNP-Ab by the photoisomerizable dinitrospiropyran monolayer Au-quartz crystal. (□) Frequency of the crystal after addition of the DNP-Ab to the dinitrospiropyran **14a** functionalized crystal. (○) Frequency of the crystal after photoisomerization of the **14a**/DNP-Ab monolayer to the protonated dinitromerocyanine **14b** washing off of the DNP-Ab, and back isomerization to **14a**-monolayer state. (Adapted from Ref. [15], Figure 7.34, with permission).

curve a). From the extent of the frequency decrease, the surface coverage of the DNP-Ab on the dinitrospiropyran antigen layer was calculated to be ca 3.8×10^{-12} mol cm^{-2} (Eq. (1)). The frequency of the protonated dinitromerocyanine monolayer is only slightly affected upon interaction with the DNP-Ab ($\Delta f = -40$ Hz) (Figure 11.17A, curve b). By the cyclic photoisomerization of the monolayer between the protonated dinitromerocyanine and dinitrospiropyran states, the DNP-Ab can be washed-off from the transducer, and the sensing interface regenerated to yield a reusable immunosensor device (Figure 11.17B).

11.3.2
Biphasic Reversible Switch Based on Bioaffinity Recognition Events Coupled to a Biocatalytic Reaction

The ability to electrochemically transduce the photoswitchable electrocatalytic functions of photoisomerizable redox proteins associated with electrodes and to regulate

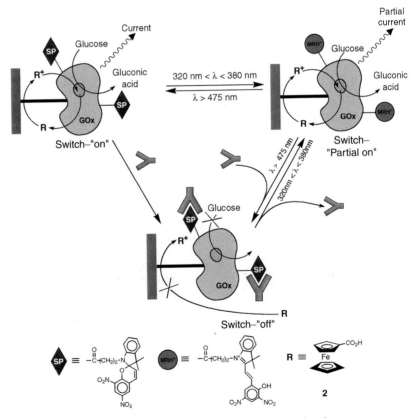

Fig. 11.18 Biphasic optobioelectronic switch by coupling of DNP-Ab with a dinitrospiropyran-functionalized GOx monolayer electrode.

the binding between a substrate and its receptor by light (e.g. the binding of a photoisomerizable antigen with the respective antibody) enables one to design complex photochemical bioswitches by the integration of several light-controlled biomaterials. A biphasic photochemical bioswitch exhibiting "on", "partial on" and "off" states was constructed by the assembly of a dinitrospiropyran-functionalized glucose oxidase (GOx) on a Au electrode [55] (Figure 11.18). The dinitrospiropyran-modified GOx monolayer electrode exhibits bioelectrocatalytic properties, and in the presence of ferrocene carboxylic acid **2** as a diffusional electron mediator, the biocatalytic interface electrocatalyzes the oxidation of glucose. This process is reflected by a high electrocatalytic current (Figure 11.19A, curve a). By the photoisomerization of the enzyme monolayer between the dinitrospiropyran state and the protonated dinitromerocyanine state, the system can be cycled between "on" and "partial on" states (Figure 11.19A, curves a and b, respectively). Since the dinitrospiropyran units tethered to the redox enzyme act as photoisomerizable antigen sites for DNP-Ab, the light-controlled association and dissociation of the antibody to and from the interface may control the bioelectrocatalytic functions of the enzyme-layered electrode. Interaction of DNP-Ab with the "on" state of the electrode results in the association of the antibody to the antigen sites, which blocks the electrical contacting of the enzyme

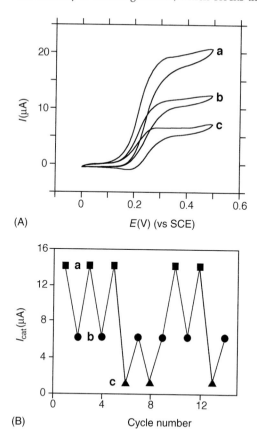

Fig. 11.19 (A) Cyclic voltammograms of (**a**) the dinitrospiropyran-tethered GOx monolayer electrode; (**b**) the protonated dinitromerocyanine-tethered GOx monolayer electrode; (**c**) the dinitrospiropyran-tethered GOx monolayer electrode in the presence of the DNP-Ab. Data were recorded in the presence of ferrocene carboxylic acid, 4×10^{-4} M, as diffusional electron mediator and glucose, 5×10^{-2} M, scan rate 2 mV s^{-1}. (B) Biphasic switchable amperometric transduction of photonic signals recorded by the photoisomerizable-GOx/DNP-Ab system: (**a**) the electrode in the dinitrospiropyran-GOx state; (**b**) the electrode in the protonated dinitromerocyanine state; (**c**) the electrode in the presence of the dinitrospiropyran-GOx monolayer with associated DNP-Ab.

and the electrode by the diffusional electron mediator (Figure 11.18). This prohibits the bioelectrocatalyzed oxidation of glucose by the enzyme-layered electrode, and the amperometric response of the system is fully switched "off" (Figure 11.19A, curve (c)). Photoisomerization of the GOx/DNP-Ab array to the protonated dinitromerocyanine state results in the dissociation of the antibody from the interface and "partial on" bioelectrocatalytic behavior. If the DNP-Ab is then washed from the system and the protonated dinitromerocyanine GOx-electrode is photoisomerized back to the dinitrospiropyran state, the "on" behavior is regenerated. Thus, the bioelectrocatalytic functions of the electrode may be cycled between three states: "on", "partial on" and "off" (Figure 11.19B).

11.4
Photocurrent Generation as a Transduction Means for Biocatalytic and Biorecognition Processes

Photocurrents generated on modified electrodes in the presence of organic dyes or semiconductor nanoparticles (e.g. CdS, CdSe) can be used as a transduction means to follow various biochemical processes, such as biocatalytic reactions and biorecognition events (particularly DNA hybridization).

11.4.1
Enzyme-Biocatalyzed Reactions Coupled to Photoinduced Electron Transfer Processes

Electron transfer reactions photoinduced by organic dyes or semiconductor nanoparticles, and resulting in the formation of reduced and oxidized organic species, can be coupled with enzyme-biocatalyzed processes [56]. Electron transfer mediators or cofactors can be photochemically activated into their oxidized or reduced forms, which then become involved in biocatalytic processes, thus driving the biochemical transformation by light. Alternatively, enzyme-catalyzed reactions can yield electron donors or acceptors that are subsequently consumed by the photochemical systems, thus supporting steady state electron flow through the photoactivated reaction centers. The coupling of these hybrid systems – composed of a photoactive reaction center (organic dyes or semiconductor nanoparticles) and an enzyme – to an electrode surface results in a photocurrent. Thus "true" signal transduction is achieved, that is, signal energy is applied as light, and is output as electricity.

This concept has recently been demonstrated by tailoring an acetylcholine esterase (AChE)-CdS nanoparticle hybrid monolayer on a Au electrode, and the activation of the photoelectrochemical functions of the nanoparticles by the biocatalytic process [57]. The CdS-AChE hybrid interface was assembled on the Au electrode by the stepwise coupling of cystamine-functionalized CdS to the electrode, and the secondary covalent linkage of the enzyme AChE to the particles (Figure 11.20A). In the presence of acetylthiocholine **15** as substrate, the enzyme catalyzes the hydrolysis of **15** to thiocholine **16** and acetate. Photoexcitation of the CdS semiconductor yields an electron-hole pair in the conduction

11.4 Photocurrent Generation as a Transduction Means for Biocatalytic and Biorecognition Processes

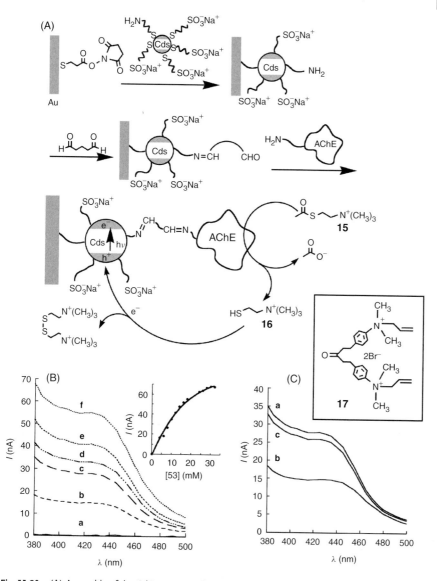

Fig. 11.20 (A) Assembly of the CdS nanoparticle/AChE hybrid system used for the photoelectrochemical detection of the enzyme activity. (B) Photocurrent action spectra observed in the presence of acetylthiocholine **15**: (a) 0 mM, (b) 6 mM, (c) 10 mM, (d) 12 mM, (e) 16 mM, (f) 30 mM. Inset: Calibration curve corresponding to the photocurrent at $\lambda = 380$ nm at variable concentrations of **15**. (C) Photocurrent spectra corresponding to the CdS/AChE system in the presence of **15**, 10 mM, (a) without the inhibitor, (b) upon addition of the inhibitor **17**, 1×10^{-6} M, (c) after rinsing the system and excluding of the inhibitor. (Adapted from Ref. [57], Figures 1 and 2(A), with permission).

band and the valence band, respectively. The enzyme-generated thiocholine **16** acts as an electron donor for valence-band holes. The scavenging of the valence-band holes results in the accumulation of the electrons in the conduction band and their transfer to the electrode with the generation of a photocurrent (Figure 11.20B). The addition of enzyme inhibitors, such as 1,5-*bis*(4-allyldimethylammoniumphenyl)pentane-3-one dibromide **17**, blocks the biocatalytic functions of the enzyme and as a result inhibits the photocurrent formation in the system (Figure 11.20C). Thus, the hybrid CdS/AChE system provides a functional interface for the sensing of the AChE inhibitors (e.g., chemical warfare agents) by means of photocurrent measurements. A similar system composed of a photoactivated CdS nanoparticle and co-immobilized formaldehyde dehydrogenase that utilizes formaldehyde as an electron donor has been reported [58]. In this hybrid system, direct electron transfer from the enzyme active center to the CdS photogenerated holes was achieved, and the steady state photocurrent signal in the system was reported to be directly related to the substrate concentration.

11.4.2
Biorecognition Events Coupled to Photoinduced Electron Transfer Processes

The photoelectrochemical transduction of DNA recognition processes has been demonstrated by the use of semiconductor (CdS) nanoparticles as the photochemically activating species [59]. Semiconductor CdS nanoparticles (2.6 ± 0.4 nm) were functionalized with one of the two thiolated nucleic acids **18** and **19**, which are complementary to the 5′ and 3′ ends of a target DNA sequence **20**. An array of CdS nanoparticle layers was then constructed on a Au electrode by a layer-by-layer hybridization process

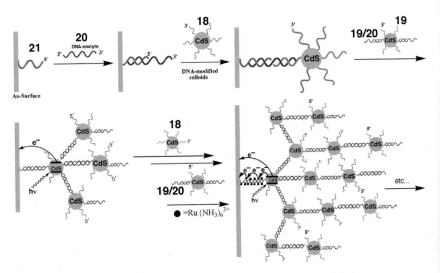

Fig. 11.21 The construction of CdS nanoparticle/DNA superstructures, and their use for the generation of photocurrents.

(Figure 11.21). A thiolated DNA monolayer of **21** was assembled on a Au electrode, and then the target DNA **20** acted as a cross-linking unit for the association of the **18**-modified CdS nanoparticles to the electrode by the hybridization of the ends of **20** to the **21**-modified surface and the **18**-functionalized CdS particles, respectively. The subsequent association of a second type of **19**-modified CdS particles prehybridized with the target DNA **20** to the first generation of the CdS particles resulted in a second layer of CdS particles. By the stepwise application of the two different kinds of nucleic acid–functionalized CdS nanoparticles hybridized with **20**, an array with a controlled number of nanoparticle layers could be assembled on the electrode. Illumination of the array resulted in the generation of a photocurrent. The photocurrent increased with the number of CdS nanoparticle generations associated with the electrode, and the photocurrent action spectra followed the absorbance features of the CdS nanoparticles, implying that the photocurrent originated from the photoexcitation of the CdS nanoparticles; that is, photoexcitation of the semiconductor induced the transfer of electrons to the conduction band and the formation of an electron-hole pair. Transfer of the conduction-band electrons to the bulk electrode, and the concomitant transfer of electrons from a sacrificial electron donor to the valence-band holes, yielded the steady state photocurrent in the system. The ejection of the conduction-band electrons into the electrode occurred from nanoparticles in intimate contact with the electrode support. This was supported by the fact that $Ru(NH_3)_6^{3+}$ units ($E°=-0.16$ V vs SCE), which bind electrostatically to the DNA, enhance the photocurrent from the DNA–CdS array; that is, the $Ru(NH_3)_6^{3+}$ units acted as electron wiring elements that facilitate electron hopping of conduction-band electrons from CdS particles that lack contact with the electrode through the DNA tether. The system is important not only because it demonstrates the use of photoelectrochemistry as a transduction method for DNA sensing but also since the system reveals the nanoengineering of organized DNA-tethered semiconductor nanoparticles on conductive supports.

11.5 Conclusions

This chapter has summarized recent advances in the development of optobioelectronic systems. Several different methods to photostimulate the functions of biomaterials by external light signals have been demonstrated. One method involves the chemical modification of redox enzymes (or the FAD cofactor of flavoenzymes) with photoisomerizable groups. Assembly of the resulting photoisomerizable enzymes as monolayers on electrodes leads to the reversible "amplified" amperometric transduction of the optical signal. Alternately, a nonfunctionalized enzyme attached to an electrode may be utilized in conjunction with a photoisomerizable electron mediator that only activates the bioelectrocatalytic current in one of its forms. A third approach involves the construction of a "photo-command" surface, which controls the binding of nonmodified enzymes at the electrode interface. This approach is reminiscent of the vision process in that a photochemical isomerization leads to changes in binding that activate a biocatalytic process. The integration of biological and electronic components in information storage and processing devices is of great interest to the scientific community. This approach is expected to lead to cheap *in vivo* biosensing devices and even to processors that take advantage of the unique strengths of biological functions. The systems described above may be considered to demonstrate information storage and transduction functions. The optical signal "writes" information to the assembly by defining the state of the photoisomerizable component. This state is then preserved (stored) until another signal changes the state again. The information is "read" by the detection of the biocatalytic current, and thus signal transduction from optical through biological to electronic forms is achieved.

Other aspects that have been addressed include the design of reversible amperometric immunosensor electrodes. This development is based on the application of photoisomerizable antigen monolayers that control the association and dissociation of antibodies to and from the electrode surface. Also, photocurrent generation on functionalized electrode surfaces has been coupled with enzyme-biocatalyzed reactions and DNA hybridization processes, thus providing an electronic means of reporting on biochemical events. Recent advances allow the application of semiconductor nanoparticles functionalized with enzymes or DNA as the photocurrent-generating units coupled with biochemical processes.

Interdisciplinary research is the hallmark of many recent technological advances. Many of the systems described herein are possible only by the combined application of molecular biology, classical synthetic chemistry, a little photochemistry and a lot of electrochemistry. This diversity can place a great stress on researchers who must be familiar with several fields at the highest level, but has the positive effect of encouraging interdepartmental collaborations. As systems of greater complexity are examined and applications become close, we can expect the boundaries between the sciences – and even between science, technology and engineering – to become even more blurred. Nanoparticle synthesis and physics, *in vitro* experimentation, electrical engineering and nanomachinery will surely enter the fray with great gusto, leading us shortly to new devices that redefine areas of medicine, analysis and information processing.

References

1 I. WILLNER, *Science*, 2002, 298, 2407–2408.
2 I. WILLNER, B. WILLNER, E. KATZ, in *Molecular Electronics: Bio-sensors and Bio-computers*, NATO Science Series, II, Mathematics, Physics and Chemistry – Vol. 96 (Eds.: L. BARSANTI, V. EVANGELISTA, P. GUALTIERI, V. PASSARELLI, S. VESTRI), Kluwer Academic Publishers, Dordrecht, 2003, 311–339.
3 I. WILLNER, B. WILLNER, in *Bioorganic Photochemistry, Biological Applications of Photochemical Switches* (Ed.: H. MORRISON), Vol. 2, Wiley, New York, 1993, 1–110.
4 I. WILLNER, S. RUBIN, *Angew. Chem. Int. Ed. Engl.*, 1996, 35, 367–385.
5 I. WILLNER, S. RUBIN, A. RIKLIN, *J. Am. Chem. Soc.*, 1991, 113, 3321–3325.
6 I. WILLNER, S. RUBIN, J. WONNER, E. EFFENHERGER, P. BÄUERLE, *J. Am. Chem. Soc.*, 1992, 114, 3150–3151.
7 P.R. WESTMARK, J.P. KELLY, B.D. SMITH, *J. Am. Chem. Soc.*, 1993, 115, 3416–3419.
8 I. WILLNER, S. RUBIN, T. ZOR, *J. Am. Chem. Soc.*, 1991, 113, 4013–4014.

9 I. WILLNER, S. RUBIN, *React. Polym.*, **1993**, *21*, 177–186.
10 E. KATZ, I. WILLNER, *Electroanalysis*, **2003**, *15*, 913–947.
11 E. KATZ, A.N. SHIPWAY, I. WILLNER, in *Encyclopedia of Electrochemistry, Bioelectrochemistry – Vol. 9* (Ed.: G.S. WILSON), (Editors-in-Chief A.J. BARD, M. STRATMANN), Wiley-VCH GmbH, Weinheim, **2002**, Chapter 17, 559–626.
12 I. WILLNER, B. WILLNER, in *Biological Applications of Photochemical Switches, Bioorganic Photochemistry* (Ed.: H. MORRISON), Vol. 2, Wiley, New York, **1993**, 1–110.
13 I. WILLNER, *Acc. Chem. Res.*, **1997**, *30*, 347–356.
14 I. WILLNER, E. KATZ, B. WILLNER, R. BLONDER, V. HELEG-SHABTAI, A.F. BÜCKMANN, *Biosens. Bioelectron.*, **1997**, *12*, 337–356.
15 E. KATZ, A.N. SHIPWAY, I. WILLNER, in *Photoactive Organic Thin Films* (Eds.: Z. SEKKAT, W. KNOLL), Academic Press, Amsterdam, **2002**, 219–268.
16 I. WILLNER, B. WILLNER, in *Molecular Switches* (Ed.: B.L. FERINGA), Wiley-VCH, Weinheim, **2001**, 165–218.
17 R. BLONDER, E. KATZ, I. WILLNER, V. WRAY, A.F. BÜCKMANN, *J. Am. Chem. Soc.*, **1997**, *119*, 11747–11757.
18 M. LION-DAGAN, I. WILLNER, *J. Photochem. Photobiol., A*, **1997**, *108*, 247–252.
19 M. LION-DAGAN, E. KATZ, I. WILLNER, *J. Am. Chem. Soc.*, **1994**, *116*, 7913–7914.
20 I. WILLNER, M. LION-DAGAN, S. MARX-TIBBON, E. KATZ, *J. Am. Chem. Soc.*, **1995**, *117*, 6581–6592.
21 I. WILLNER, R. BLONDER, E. KATZ, A. STOCKER, A.F. BÜCKMANN, *J. Am. Chem. Soc.*, **1996**, *118*, 5310–5311.
22 R. BLONDER, E. KATZ, I. WILLNER, V. WRAY, A.F. BÜCKMANN, *J. Am. Chem. Soc.*, **1997**, *119*, 11747–11757.
23 E. KATZ, B. WILLNER, I. WILLNER, *Biosens. Bioelectron.*, **1997**, *12*, 703–719.
24 I. WILLNER, B. WILLNER, *J. Mater. Chem.*, **1998**, *8*, 2543–2556.
25 I. WILLNER, B. WILLNER, *Bioelectrochem. Bioenerg.*, **1997**, *42*, 43–57.
26 A. HELLER, *Acc. Chem. Res.*, **1990**, *23*, 128–134.
27 F.A. ARMSTRONG, H.A.O. HILL, N.J. WALTON, *Q. Rev. Biophys.*, **1985**, *18*, 261–322.
28 P.M. ALLEN, H.A.O. HILL, N.J. WALTON, *J. Electroanal. Chem.*, **1984**, *178*, 69–86.
29 F.A. ARMSTRONG, H.A.O. HILL, N.J. WALTON, *Acc. Chem. Res.*, **1988**, *21*, 407–413.
30 M.J. TARLOV, E.F. BOWDEN, *J. Am. Chem. Soc.*, **1991**, *113*, 1847–1849.
31 M. LION-DAGAN, E. KATZ, I. WILLNER, *J. Chem. Soc., Chem. Commun.*, **1994**, 2741–2742.
32 W. JIN, U. WOLLENBERGER, F.F. BIER, A. MAKOWER, F.W. SCHELLER, *Bioelectrochem. Bioenerg.*, **1996**, *39*, 221–225.
33 A.E.G. CASS, G. DAVIS, H.A.O. HILL, D.J. NANCARROW, *Biochim. Biophys. Acta*, **1985**, *828*, 51–57.
34 D.A. POWIS, G.D. WATTUS, *FEBS Lett.*, **1981**, *126*, 282–284.
35 I. WILLNER, A. DORON, E. KATZ, S. LEVI, A.J. FRANK, *Langmuir*, **1996**, *12*, 946–954.
36 R. WILSON, A.P.F. TURNER, *Biosens. Bioelectron.*, **1992**, *7*, 165–423.
37 R. BLONDER, I. WILLNER, A.F. BÜCKMANN, *J. Am. Chem. Soc.*, **1998**, *120*, 9335–9341.
38 M. LION-DAGAN, S. MARX-TIBBON, E. KATZ, I. WILLNER, *Angew. Chem. Int. Ed. Engl.*, **1995**, *34*, 1604–1606.
39 P. SKLADAL, *Electroanalysis*, **1997**, *9*, 737–745.
40 A.L. GHINDILIS, P. ATANASOV, M. WILKINS, E. WILKINS, *Biosens. Bioelectron.*, **1998**, *13*, 113–131.
41 A.L. GHINDILIS, O.V. SKOROBOGAT'KO, V.P. GAVRILOVA, A.I. YAROPOLOV, *Biosens. Bioelectron.*, **1992**, *7*, 301–304.
42 U. PFEIFER, W. BAUMANN, *Fresenius' J. Anal. Chem.*, **1993**, *345*, 504–511.
43 W.O. HO, D. ATHEY, C.J. MCNEIL, *Biosens. Bioelectron.*, **1995**, *10*, 683–691.
44 L.X. TIEFENAUER, S. KOSSEK, C. PADESTE, P. THIÉBAUD, *Biosens. Bioelectron.*, **1997**, *12*, 213–223.
45 D. IVNITSKI, J. RISHPON, *Biosens. Bioelectron.*, **1996**, *11*, 409–417.
46 R. BLONDER, E. KATZ, Y. COHEN, N. ITZHAK, A. RIKLIN, I. WILLNER, *Anal. Chem.*, **1996**, *68*, 3151–3157.
47 E. KATZ, I. WILLNER, *J. Electroanal. Chem.*, **1996**, *418*, 67–512.
48 I. BEN-DOV, I. WILLNER, E. ZISMAN, *Anal. Chem.*, **1997**, *69*, 3506–3512.

49 A.A. Suleiman, G.G. Guilbault, *Analyst*, **1994**, *119*, 2279–2282.
50 B. König, M. Grätzel, *Anal. Chim. Acta*, **1993**, *276*, 323–333.
51 I. Willner, R. Blonder, A. Dagan, *J. Am. Chem. Soc.*, **1994**, *116*, 9365–9366.
52 I. Willner, B. Willner, *Biotechnol. Prog.*, **1999**, *15*, 991–1002.
53 R. Blonder, S. Levi, G. Tao, I. Ben-Dov, I. Willner, *J. Am. Chem. Soc.*, **1997**, *119*, 10467–10478.
54 F. Patolsky, B. Filanovsky, E. Katz, I. Willner, *J. Phys. Chem. B*, **1998**, *102*, 10359–10367.
55 I. Willner, M. Lion-Dagan, E. Katz, *Chem. Commun.*, **1996s**, 623–624.
56 E. Katz, A.N. Shipway, I. Willner, in *Electron Transfer in Chemistry* (Ed.: V. Balzani), Wiley-VCH, Weinheim, **2001**, 127–201.
57 V. Pardo-Yissar, E. Katz, J. Wasserman, I. Willner, *J. Am. Chem. Soc.*, **2003**, *125*, 622–623.
58 M.L. Curri, A. Agostiano, G. Leo, A. Mallardi, P. Cosma, M.D. Monica, *Mater. Sci. Eng., C*, **2002**, *22*, 449–452.
59 I. Willner, F. Patolsky, J. Wasserman, *Angew. Chem. Int. Ed.*, **2001**, *40*, 1861–1864.

12
The Neuron-semiconductor Interface

Peter Fromherz

12.1
Introduction

Both computers and brains work electrically, but their charge carriers are different – electrons in a solid ion lattice and ions in a polar fluid. Electrons in silicon have a mobility of about 10^3 cm^2/Vs, whereas the mobility of ions in water is around 10^{-3} cm^2/Vs. This enormous difference of mobility is at the root for the different architecture of the two information processors. It is an intellectual and technological challenge to join these different systems directly on the level of electronic and ionic signals as sketched in Figure 12.1.

In the eighteenth century, Luigi Galvani established the electrical coupling of inorganic solids and excitable living tissue. Now, after 50 years of dramatic developments in semiconductor microtechnology and cellular neurobiology, we may envisage such an integration by far more complex interactions, right on the level of individual nerve cells and microelectronic devices or even on the level of biomolecules and nanostructures. Today, however, we are not concerned whether brain–computer interfacing can be really implemented in the foreseeable future, with neuronal dynamics and digital computation fused to thinking-computing systems. The issue is an elucidation of the fundamental biophysical mechanisms on the level of nanometers, micrometers and millimeters, and also the development of a scientific and technological culture that combines the theoretical concepts and experimental methods of microelectronics, solid-state physics, electrochemistry, molecular biology and neurobiology. If we succeed in that endeavor, then we will be able to fabricate iono-electronic devices to solve problems in molecular biology, to develop neuroelectronic devices for an experimental physics of brain-like systems, and be able to contribute to medicine and information technology by creating microelectronic neuroprostheses and nerve-based ionic processors.

Having worked for some time with artificial biomembranes on semiconductor electrodes, I wrote in 1985 a note "Brain on line? The feasibility of a neuron–silicon junction" [1]. The idea of brain-computer interfacing was scaled down to the level of a real project: "The utopian question may be shaped into a proper scientific problem: How to design a neuron–silicon junction?" I outlined the mechanism of neuron–semiconductor interfacing in both directions. On that basis, the first experimental results were reported in 1991 and 1995 with nerve cells of the leech on

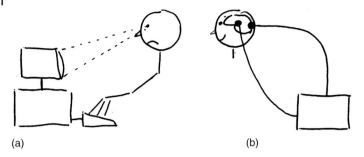

Fig. 12.1 Cartoon of brain–computer interfacing. (a) Communication through the macroscopic optical and mechanical pathways screen–eye and finger–keyboard. (b) Hypothetical microscopic interfacing of a computer with the visual and motor cortex [1].

open transistors and on capacitive stimulation spots of silicon chips [2, 3]. After those elementary steps, two directions were followed: (a) Downward, the microscopic nature of the cell–semiconductor contact was investigated with respect to its structure and electrical properties [4–25]. The goal is a physical rationalization of the junction in order to have a firm basis for a systematic optimization of neuron–silicon interfacing [26–31]. (b) Upward, hybrid systems were assembled with neuronal networks joined to microelectronic circuits [32–46]. Here, the goal is a supervision of numerous neurons in a network by noninvasive contacts to a semiconductor substrate as required for long-term studies of dynamical processes such as learning and memory. A transfer of the interfacing methods on simple silicon chips to chips fabricated by an industrial CMOS process is important [47, 48].

The present article relies on own publications [2–46] and reviews [49–54]. It discusses the physics of the cell–silicon junction, the electronic interfacing of individual neurons by transistor recording and capacitive stimulation, and first steps toward a connection of silicon chips with neuronal networks and with brain slices. Literature on the background of the field is found in the reference lists of the original publications.

12.2
Ionic–Electronic Interface

A neuron–silicon chip with an individual nerve cell from rat brain and a linear array of transistors is shown in Figure 12.2. A nerve cell (diameter about 20 μm) is surrounded by a membrane with an electrically insulating core of lipid. This lipid bilayer (thickness about 5 nm) separates the environment with about 150 mM (10^{20} cm^{-3}) sodium chloride from the intracellular electrolyte with about 150 mM potassium chloride. Ion currents through the membrane are mediated by specific protein molecules, ion channels with a conductance of 10 to 100 pS. Silicon is used as an electronically conductive substrate for three reasons: (a) Coated with a thin layer of thermally grown silicon dioxide (thickness 10–1000 nm), silicon is a perfect inert substrate for culturing nerve cells. (b) The

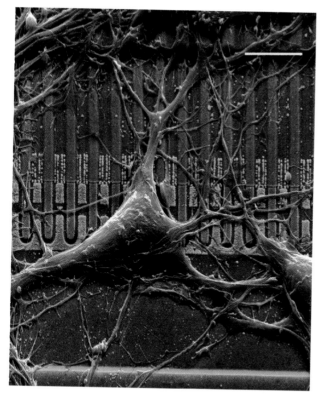

Fig. 12.2 Nerve cell from rat brain on silicon chip [22]. Colored electronmicrograph, scale bar 10 µm. The surface of the chip consists of thermally grown silicon dioxide (green). The metalfree gates of a linear array of field-effect transistors are visible as dark squares. The neuron (blue) is cultured on the chip for several days in an electrolyte.

thermally grown silicon dioxide suppresses the transfer of electrons and concomitant electrochemical processes that lead to a corrosion of silicon and to a damage of the cells. (c) A well-established semiconductor technology allows the fabrication of microscopic electronic devices that are in direct contact with the cells, shielded by the inert oxide layer.

In principle, a direct coupling of ionic signals in a neuron and electronic signals in the semiconductors can be attained by electrical polarization. If the insulating lipid layer of the neuron were in direct contact with the insulating silicon dioxide of the chip, a compact dielectric would be formed as sketched in Figure 12.3a and Figure 12.3b. An electrical field across the membrane – as created by neuronal activity – polarizes the silicon dioxide such that the electronic band structure of silicon and an integrated transistor is affected (Figure 12.3a). *Vice versa*, an electrical field across the silicon dioxide – as caused by a voltage applied to the chip – polarizes the membrane in a way that conformations of field-sensitive membrane proteins such as voltage-gated ion channels are affected (Figure 12.3b).

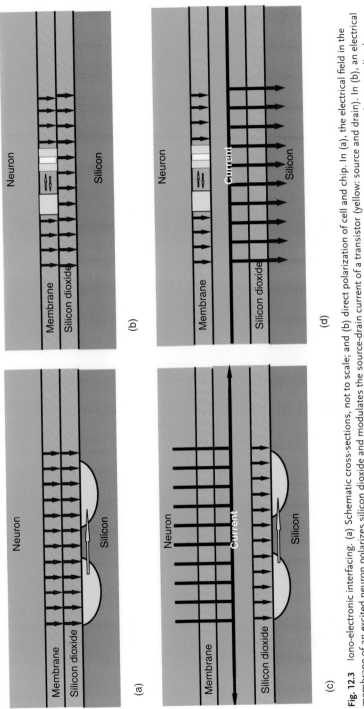

Fig. 12.3 Iono-electronic interfacing. (a) Schematic cross-sections, not to scale; and (b) direct polarization of cell and chip. In (a), the electrical field in the membrane of an excited neuron polarizes silicon dioxide and modulates the source-drain current of a transistor (yellow: source and drain). In (b), an electrical field in silicon dioxide polarizes the membrane and opens ion channels (yellow: closed and open conformations). (c) and (d) neuron–silicon coupling by electrical current. In (c), current through the membrane of an excited neuron leads to a transductive extracellular potential in the cleft between cell and chip, which polarizes the oxide and modulates the source-drain current. In (d), capacitive current through the oxide gives rise to a transductive extracellular potential which polarizes the membrane and opens ion channels.

However, when a nerve cell grows on a chip as illustrated in Figure 12.2, we cannot expect that the lipid layer of the cell and the oxide layer of silicon form a compact dielectric. Cell adhesion is mediated by protein molecules that protrude from the cell membrane (integrins, glycocalix) and that are deposited on the substrate (extracellular matrix proteins). These proteins keep the lipid core of the membrane at a certain distance from the substrate, stabilizing a cleft between cell and chip that is filled with electrolyte. The resulting conductive cleft shields electrical fields and suppresses a direct mutual polarization of silicon dioxide and membrane.

The cell–silicon junction forms a planar electrical core-coat conductor: the coats of silicon dioxide and membrane insulate the core of the conductive cleft from the conducting environments of silicon and cytoplasm. The first step of neuroelectronic interfacing is determined by the current flow in that core-coat conductor [11, 26]. (a) The activity of a neuron leads to ionic and displacement currents through the membrane (Figure 12.3c). The concomitant current along the core gives rise to a Transductive Extracellular Potential (TEP) between cell and chip. (b) A voltage transient applied to silicon leads to a displacement current through the oxide coat (Figure 12.3d). Again, a TEP appears between chip and cell because of the concomitant current along the cleft. In a second step of interfacing, the TEP in the core-coat conductor is detected by voltage-sensitive devices in the chip or in the cell: (a) The TEP induced by the neuron gives rise to an electrical field across the silicon dioxide that is probed by a field-effect transistor (Figure 12.3c). (b) The TEP induced by the chip gives rise to an electrical field across the membrane that is probed by voltage-gates ion channels (Figure 12.3d).

12.2.1
Planar Core-coat Conductor

The TEP in the cleft between cell and chip mediates the coupling of neurons and silicon. It is determined by the current balance in the core-coat conductor of the junction [11]. To describe current and voltage, we use the two-dimensional area-contact model or the zero-dimensional point-contact model (one-compartment model) as sketched in Figure 12.4.

Area-contact Model We describe the current in each area element of the junction by the area-contact model symbolized by the circuit of Figure 12.4a [11, 12]. The current along the cleft is balanced by the displacement current through silicon dioxide and by the ionic and displacement current through the attached membrane. The conservation of electrical charge per unit area of the junction is expressed by (1) where the left-hand side refers to the balance of current per unit length in the cleft, and the right-hand side to the current per unit area through membrane and oxide with the electrical potential V_M in the cell (membrane potential), the potential V_S in the substrate, the TEP V_J in the junction and the two-dimensional spatial derivative operator ∇. If the bath electrolyte is kept on ground potential ($V_E = 0$), V_M, V_S and V_J are the voltages between cell, silicon and junction and the bath.

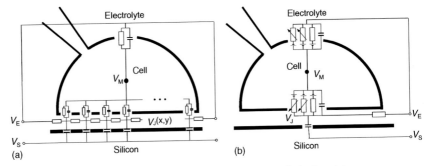

Fig. 12.4 Core-coat conductor of cell–semiconductor junction [11, 26]. The heavy lines indicate silicon dioxide, cell membrane and micropipette. The cross-sections are not to scale: the distance of membrane and chip is 10 to 100 nm, the diameter of a cell is 10 to 100 µm. (a) AC circuit of area-contact model. The infinitesimal elements of oxide, membrane and electrolyte film in the junction are represented as capacitors and ohmic resistances. (b) DC circuit of point-contact model with voltage-dependent ion conductances. Oxide, membrane and electrolyte film of the junction are represented by global capacitances and resistances. V_M is the electrical potential in the cell, V_J the transductive extracellular potential in the junction, V_S the potential of the substrate and V_E the potential of the bath.

$$-\nabla\left(\frac{1}{r_J}\nabla V_J\right) = c_S\left(\frac{\partial V_S}{\partial t} - \frac{\partial V_J}{\partial t}\right) + c_M\left(\frac{\partial V_M}{\partial t} - \frac{\partial V_J}{\partial t}\right) + g_{JM}(V_M - V_J) \quad (1)$$

Parameters are the sheet resistance r_J of the cleft, the area-specific capacitances c_M and c_S of membrane and substrate and an area-specific leak conductance g_{JM} of the attached membrane. Voltage-dependent ion conductances are not included in (1), for the sake of clarity. The specific capacitance c_M in the attached membrane is assumed to be the same as in the free membrane. The sheet resistance r_J can be expressed by the width d_J and the specific resistance ρ_J of the cleft with $r_J = \rho_J/d_J$.

Point-contact Model For many applications, it is convenient to describe the core-coat conductor by a one-compartment model with an equivalent circuit shown in Figure 12.4b [8, 11, 26, 29]. The conductive cleft is represented by a global Ohmic conductance G_J, attached membrane and silicon dioxide by the global capacitances C_{JM} and C_S. We take into account global ion specific conductances G_{JM}^i in the attached membrane. The reversal voltages V_0^i originate in the concentration differences of the ions between cell and environment. In a first approximation, they are assumed to be the same as in the free membrane. When we define area-specific parameters with respect to the area A_{JM} of attached membrane as $c_S = C_S/A_{JM}$, $c_M = C_{JM}/A_{JM}$, $g_{JM}^i = G_{JM}^i/A_{JM}$ and $g_J = G_J/A_{JM}$, Kirchhoff's law is expressed by (2) where V_J and V_E are the electrical potentials in junction and in bulk electrolyte.

$$g_J(V_J - V_E) = c_S\left(\frac{dV_S}{dt} - \frac{dV_J}{dt}\right) + c_M\left(\frac{dV_M}{dt} - \frac{dV_J}{dt}\right) + \sum_i g_{JM}^i(V_M - V_J - V_0^i) \quad (2)$$

12.2 Ionic–Electronic Interface

Area-contact versus Point-contact A comparison of (1) and (2) shows that the Laplace operator in a homogeneous area-contact model is replaced by a constant in the point-contact model with $-\nabla^2 \rightarrow r_J g_J$. To match the two models, we must express the area-specific conductance $g_J = G_J/A_{JM}$ by the sheet resistance r_J. Various averaging methods lead to a relation between global resistance and sheet resistance $G_J^{-1} = r_J/\eta\pi$ with a scaling factor $\eta = 4 - 6$ [11, 15, 20]. For a circular junction of radius a_J with $A_{JM} = a_J^2 \pi$, we obtain (3) with $r_J = \rho_J/d_J$.

$$g_J = \eta \frac{1}{r_J a_J^2} = \eta \frac{d_J}{\rho_J a_J^2} \tag{3}$$

Also, in the point-contact model, it is advantageous to use area-specific parameters (a) because the area-specific capacitances c_M and c_S of membrane and chip are usually known, (b) because area-specific membrane conductances g_{JM}^i are common in the neurophysiological literature and (c) because a single parameter g_J combines three properties of the junction that are often unknown, the specific resistance ρ_J, the width d_J and the radius a_J.

Intracellular Dynamics The TEP V_J is determined by the current in the junction alone according to (1) or (2), if the potentials V_M and V_S in cell and chip are under external control. Usually that condition holds for the chip, where V_S is held constant or is determined by a waveform $V_S(t)$ of stimulation. For the cell, V_M is held constant in voltage-clamp situations when the intracellular space is controlled by a micropipette (Figure 12.4). In situations of noninvasive extracellular recording and stimulation by a chip, the intracellular potential $V_M(t)$ obeys an autonomous dynamics, governed by the balance of ionic and displacement currents through the free and attached membrane as indicated in Figure 12.4. For the point-contact model, we obtain (4) using Kirchhoff's law.

$$c_M \left(\frac{dV_M}{dt} - \frac{dV_E}{dt} \right) + \sum_i g_{FM}^i (V_M - V_E - V_0^i) = -\beta_M \left[c_M \left(\frac{dV_M}{dt} - \frac{dV_J}{dt} \right) \right.$$
$$\left. + \sum_i g_{JM}^i (V_M - V_J - V_0^i) \right] \tag{4}$$

The left-hand side describes the outward current through the free membrane, and the right-hand side refers to the inward current through the attached membrane with the area-specific ion conductances g_{JM}^i and g_{FM}^i of attached and free membrane and with the ratio $\beta_M = A_{JM}/A_{FM}$ of attached and free membrane area. (2) and (4) together describe the coupled dynamics of the intracellular and extracellular potentials $V_M(t)$ and $V_J(t)$ for the point-contact model. In analogy, the area-contact model has to be amended by the intracellular dynamics. There, Kirchhoff's law for the cell is given by the outward current through free membrane according to the left-hand side of (4) and by an integral over all local inward currents through the attached membrane area [20].

Electrodiffusion The area-contact and the point-contact model as expressed by (1) and (2) imply that the ion concentrations in the narrow cleft between cell and chip are not changed with constant r_J, constant g_J and constant V_0^i. A change in the ion concentrations in the cleft is important if the ion channels in the junction are open for an extended time interval as in voltage-clamp experiments [23, 24]. An electrodiffusion version of the area-contact or the point-contact model accounts for these effects [24].

Conclusion The interfacing of neuron and semiconductors is mediated by a TEP. A large TEP results from high currents through membrane and silicon dioxide, and from a low conductance of the junction. Recording and stimulation of neuronal activity are promoted by a small distance d_J, a high specific resistance ρ_J and a large radius a_J of the cell-chip junction. Efficient recording requires high ion conductances g_{JM}^i in the attached membrane, efficient stimulation a high area-specific capacitance c_S of the chip.

12.2.2
Cleft of Cell-silicon Junction

The distance d_J between a cell membrane and a silicon chip is a fundamental parameter of cell–silicon junctions. The distance is measured by the method of fluorescence interference contrast (FLIC) microscopy, which relies on the formation of standing modes of light in front of the reflecting surface of silicon.

Fluorescence on Silicon We consider a lipid bilayer on oxidized silicon as sketched in Figure 12.5a. The membrane is labeled with amphiphilic dye molecules with transition dipoles in the membrane plane. Upon illumination, light is reflected at all interfaces, in

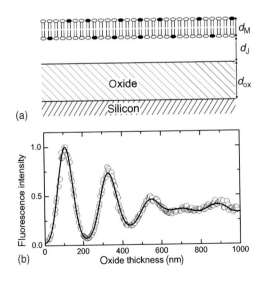

Fig. 12.5 Fluorescent lipid membrane on silicon [8]. (a) Schematic cross-section of lipid bilayer with incorporated dye molecules on oxidized silicon. The distance of membrane and chip is d_J, the thickness of the oxide is d_{ox}. (b) Experimental fluorescence intensity of a bilayer with the cyanine dye DiI versus oxide thickness. The data are fitted by the electromagnetic theory of dipole radiation with a single free parameter, the scaling factor of intensity.

particular, at the silicon-to-silicon dioxide interface. Also, the fluorescence light emitted by the dye molecules is reflected. Because of interference effects, the excitation and the fluorescence of the dye depend on the distance between membrane and silicon [4].

The electrical field of a light wave has a node in the plane of an ideal mirror. For normal incidence of light, the probability of excitation of a membrane-bound dye is described by the first factor of (5) with a thickness d_{ox} and a refractive index n_{ox} of silicon dioxide, with a width d_J and refractive index n_J of the cleft between membrane and chip and with a wavelength λ_{ex}. An analogous interference effect occurs for light that is emitted from the dye directly and with reflection. The probability of fluorescence at a wavelength λ_{em} in normal direction is described by the second factor of (5). The detected stationary fluorescence intensity $J_{fl}(d_J, d_{ox})$ is proportional to the product of excitation and emission probability according to (5) [4], which can be read as a function $J_{fl}(d_J)$ at constant d_{ox} or as a function $J_{fl}(d_{ox})$ at constant d_J.

$$J_{fl}(d_J, d_{ox}) \propto \sin^2\left[\frac{2\pi(n_{ox}d_{ox} + n_J d_J)}{\lambda_{ex}/2}\right] \cdot \sin^2\left[\frac{2\pi(n_{ox}d_{ox} + n_J d_J)}{\lambda_{em}/2}\right] \qquad (5)$$

For a cell on silicon, the complete electromagnetic theory of dipole radiation has to be applied. It leads to a more involved function $J_{fl}(d_{ox}, d_J)$, which takes into account the layered optical structure, the aperture of a microscope, the spectral bandwidth of illumination and detection and the nearfield interaction of dye and silicon [8].

For an experimental test of the modulated fluorescence on silicon, we attach a pure lipid bilayer with the cyanine dye DiI to a silicon chip with 256 terraces of silicon dioxide. The observed fluorescence intensity $J_{fl}(d_{ox}, d_J)$ is plotted in Figure 12.5b. We observe a damped periodic variation of the intensity, which levels out above 600 nm owing to the large aperture and the wide spectral bandwidth of detection. The experiment is perfectly fitted with the relation $J_{fl}(d_{ox}, d_J)$ of the complete electromagnetic theory using a single free parameter, the scaling factor of the intensity [8].

FLIC Microscopy The modulation of fluorescence on silicon is the basis of FLIC microscopy, which allows to determine the distance between a chip and a cell. A direct evaluation of d_J from the measured fluorescence intensity and the theoretical function $J_{fl}(d_{ox}, d_J)$ at a given value d_{ox} is not possible (a) because we cannot measure absolute intensities [4] and (b) because there is a background fluorescence from the upper membrane of the cell out of focus [5]. To overcome this problem, the intensity $J_{fl}(d_{ox}, d_J)$ of the membrane is measured on several oxide layers of different height d_{ox} at a certain unknown value of d_J. Usually 4 or 16 quadratic terraces are fabricated in a 10×10 µm unit cell of the silicon surface [5, 6]. The data are fitted by a function $\tilde{J}_{fl}$ according to (6) with three parameters, a scaling factor a, a background b and the optical width of the cleft $n_J d_J$.

$$\tilde{J}_{fl} = aJ_{fl}(d_{ox}, d_J) + b \qquad (6)$$

It is the main advantage of FLIC microscopy that the theoretical function $J_{fl}(d_{ox}, d_J)$ is dominated by the optics of the well-defined interface of silicon and silicon dioxide. Less known optical parameters of the cell – the thickness of the membrane including protein complexes and the refractive indices of membrane and cytoplasm – play almost no role. Prerequisite of FLIC microscopy is a similar geometry of cell adhesion on the different terraces and a homogeneous staining of the membrane.

Astrocyte on Laminin A fluorescence micrograph of glia cell from rat brain (astrocyte) on 16 different terraces is shown in Figure 12.6a. The chip is coated with a protein from the extracellular matrix (laminin) with a thickness of 3 nm in its dry state. The checkerboard pattern of fluorescence matches the oxide terraces [6]. Two features of the

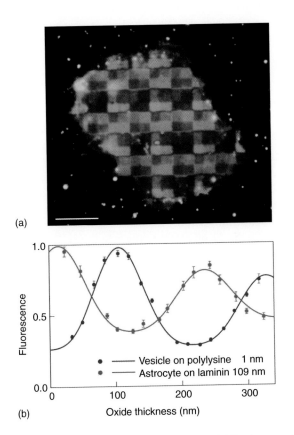

Fig. 12.6 Fluorescence interference contrast (FLIC) microscopy [6].
(a) Fluorescence micrograph of the adhesion region of a rat astrocyte on a silicon chip with quadratic 2.5 × 2.5 μm terraces of silicon dioxide. Scale bar 10 μm. The chip is coated with laminin. The membrane is stained with the dye DiI. (b) Fluorescence intensity versus height of the terraces for the astrocyte (red dots) and a lipid vesicle on polylysine (blue dots). The lines are computed by an electromagnetic theory assuming a water film between oxide and membrane of 109-nm thickness for the astrocyte and of 1-nm thickness for the lipid vesicle.

picture are important: (a) the intensity is rather homogeneous on each terrace; (b) the intensity is periodic with the unit cells of 4 × 4 terraces. These observations indicate that the membrane is stained homogeneously and that a well-defined distance of membrane and chip exists on all terraces.

The fluorescence intensity $J_{fl}^{exp}(d_{ox})$ on 16 terraces is plotted in Figure 12.6b versus the height of the terraces. It is highest on the thinnest oxide, drops and increases again on higher terraces. This result is quite in contrast to the model experiment of Figure 12.5b. For comparison, the result of a control experiment is plotted in Figure 12.6b where a stained vesicle made of a pure lipid bilayer is attached to the same microscopic terraces with polylysine and observed under the same optical conditions [14]. There the fluorescence starts with a minimum on the thinnest oxide as in the model experiment with a planar lipid bilayer (Figure 12.5). A fit of the data according to (6) leads to $d_J = 1$ nm assuming a refractive index of water. On the other hand, a fit of the data for the astrocyte membrane on laminin leads to $d_J = 109$ nm.

Focal Contact For comparison, we consider fibroblast cells on the extracellular matrix protein fibronectin. There, special cellular structures promote strong adhesive forces such that we may expect a particularly close distance of cell and chip. We visualize the focal contacts by fusing green fluorescent protein (GFP) to vinculin, one of their protein components, as shown in Figure 12.7a [7] Choosing a different illumination, we perform a FLIC experiment with the cyanine dye DiI on a chip with four different terraces as shown in Figure 12.7b. For a selected terrace depicted in Figure 12.7c, we compute a distance map $d_J(x, y)$. Figure 12.7d shows that even at focal contacts with their strong adhesion, the separation of the lipid core of the cell membrane and the chip is 50 nm within the lateral resolution of the microscope.

Conclusion The lipid core of a cell membrane and the silicon dioxide layer of a silicon chip are not in close contact. The large distance is caused by dangling polymer molecules that protrude from the membrane (glycocalix) and that are deposited on the chip (laminin) [9]. They give rise to repulsive entropic forces that balance the attractive forces of cell adhesion between the integrins in the membrane and laminin molecules. It will be an important task to lower the distance of cells and chips by special treatments of the chip surface and by genetical modifications of the membrane without impairing the viability of the cells.

12.2.3
Conductance of the Cleft

Given a cleft between cell and chip, we have to ask for the sheet resistance r_J of the junction in the area-contact model or for the area-specific conductance g_J in the point-contact model. Various approaches can be chosen to obtain r_J or g_J from measurements of the voltage transfer in the junction, considering the circuits of Figure 12.4. (a) We may apply a voltage $V_M - V_E$ between cell and bath or a voltage $V_S - V_E$ between chip

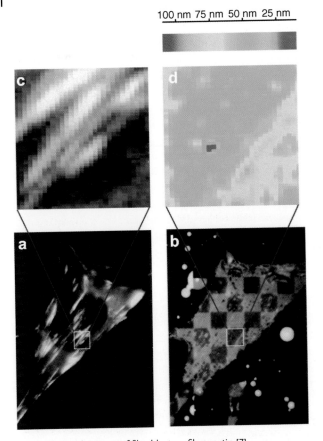

Fig. 12.7 Focal contact of fibroblast on fibronectin [7].
(a) Fluorescence micrograph in the light of GFP (green fluorescent protein) fused to vinculin showing elongated focal contacts.
(b) FLIC micrograph in the light of the cyanine dye DiI. The size of the four terraces of different height is 5 × 5 µm. (c) Blowup of a terrace of the vinculin picture. (d) Color-coded map of the distance between cell and chip obtained by FLIC microscopy on the selected terrace. Within the lateral resolution of about 400 nm, there is no close contact in the areas of focal adhesion.

and bath. (b) We may probe the voltage drop $V_J - V_S$ across the oxide with field-effect transistors or the voltage $V_M - V_J$ across the membrane with voltage-sensitive dye. (c) We may use AC voltages or voltage steps for stimulation. (d) We may use the point-contact model or the area-contact model to evaluate the data. Here we discuss three examples: intracellular AC stimulation/transistor recording/point-contact model, extracellular AC stimulation/transistor recording/area-contact model and extracellular stimulation with voltage step/optical recording with voltage-sensitive dye/point contact model. We start with a short summary of transistor recording.

Transistor Recording In a p-type metal oxide silicon field-effect transistor (MOSFET), the source-drain current I_D is controlled by the voltage V_{DS} between drain and source and the voltage V_{GS} between metal gate and source. Above the threshold $V_{GS} > V_T$ of strong inversion and below pinch-off, the current is described by (7) where the proportionally constant depend on the length and width of the channel, the mobility of the holes and the capacitance of the gate oxide.

$$I_D \propto [V_{DS}(V_{GS} - V_T) - V_{DS}^2/2] \tag{7}$$

An electrolyte replaces the metal gate in an electrolyte oxide silicon field-effect transistor (EOSFET). It is joined to an external metallic contact by a Ag/AgCl electrode that transforms ionic into electronic current. The source-drain current is controlled by the voltage $V_{ES} = V_E - V_S$ applied to the electrolyte. In (7), we substitute $V_{GS} \to V_{ES}$ and $V_T \to V_T^{(E)}$, where the threshold $V_T^{(E)}$ is determined by the work function of silicon, the redox potential of Ag/AgCl, the contact potential of the Ag/AgCl electrode and the electrical double layer at the interface electrolyte/silicon dioxide.

When we probe the electrical effect of a cell, the source-drain current of an EOSFET is modulated by the voltage $V_{JS} = V_J - V_S$ in the cell-silicon junction, of course. The cell affects the voltage drop $V_{JE} = V_J - V_E$ between junction and bulk electrolyte at a constant external voltage V_{ES}. With $V_{JS} = V_{JE} + V_{ES}$ we obtain for transistor recording (8).

$$I_D \propto [V_{DS}(V_{JE} + V_{ES} - V_T^{(E)}) - V_{DS}^2/2] \tag{8}$$

The change of the extracellular voltage occurs in the cleft between cell and chip, far beyond the electrical double layer, which has a thickness of 1 nm in 100 mM NaCl. Thus, we are dealing with a genuine modulation of the gate potential V_{JS}. Local voltage recording by an EOSFET has to be distinguished from the application of an EOSFET as an ion-sensitive transistor (ISFET). There, molecular interactions of protons and other ions in the electrical double layer modulate the threshold voltage $V_T^{(E)}$.

The characteristics $I_D(V_{DS}, V_{ES})$ of an EOSFET are measured in a calibration experiment by variation of the bath potential. The transconductance $(\partial I_D/\partial V_{ES})_{V_{DS}}$ is determined at a working point defined by the potentials V_E, V_D and V_S. When we assume that the transconductance of the calibration experiment is valid for the local recording of a potential, we obtain from the experimental ΔI_D the extracellular potential V_J with $V_E = 0$ according to (9).

$$\Delta I_D = \left(\frac{\partial I_D}{\partial V_{ES}}\right)_{V_{DS}} V_J \tag{9}$$

We use p-type transistors where all parts of the silicon chip are held at a positive voltage with respect to the bath with $V_{ES} = V_E - V_S < 0$, $V_{DS} > V_{ES}$, and with bulk silicon on source potential $V_B = V_S$. The bias voltages at the working point prevent cathodic corrosion of the chip and an invasion of sodium ions into the transistors. The thickness of the gate oxide is around 10 nm. In arrays with close spacing, the transistors

Cell Stimulation with Transistor Recording We apply an intracellular AC voltage with an amplitude $\underline{V}_M(\omega)$ at an angular frequency ω to a cell using a patch-pipette in whole-cell configuration. We record the complex response $\underline{V}_J(\omega)$ in the junction with a transistor [10, 11]. A leech neuron on a transistor is shown in Figure 12.8a.

Amplitude and phase of the transfer spectrum $\underline{V}_J/\underline{V}_M$ are plotted in Figure 12.8b and Figure 12.8c versus the frequency $f = \omega/2\pi$. We find two types of spectra: (a) The A-type spectrum has a small amplitude at low frequencies, an increase of the phase around 10 Hz and an increase of the amplitude above 1000 Hz. (b) The B-type spectrum has a high amplitude at low frequency and a further increase at 1000 Hz. There is only a minor change of the phase around 1000 Hz. Similar measurements can be made with an array of transistors beneath a single leech neuron as illustrated in Figure 12.9. In this case, the voltage transfer as a function of frequency and space coordinate is evaluated with the area-contact model [12].

We evaluate the experiment of Figure 12.8 with the point-contact model. We insert in (2) an intracellular stimulation $V_M = \underline{V}_M \exp(i\omega t)$ with a complex response $V_J = \underline{V}_J \exp(i\omega t)$. When we take into account a leak conductance g_{JM} in the attached membrane, we obtain (10) at $dV_S/dt = 0$ and $V_E = 0$ with the time constants τ_J and τ_{JM} of the junction and the attached membrane.

$$\frac{\underline{V}_J}{\underline{V}_M} = \frac{c_M}{c_M + c_S} \cdot \frac{\tau_J/\tau_{JM} + i\omega\tau_J}{1 + i\omega\tau_J}, \quad \tau_J = \frac{c_S + c_M}{g_J + g_{JM}}, \quad \tau_{JM} = \frac{c_M}{g_{JM}} \tag{10}$$

The high frequency limit of the amplitude $|\underline{V}_J/\underline{V}_M|_\infty = c_M/(c_M + c_S)$ is determined by the capacitances, the low frequency limit $|\underline{V}_J/\underline{V}_M|_0 = g_{JM}/(g_{JM} + g_J)$ by the conductances. There is no phase shift in the limits of low and high frequency. If an intermediate frequency range exists with $\omega\tau_{JM} \gg 1$ and $\omega\tau_J \ll 1$, a phase shift of $\pi/2$ appears where the current is determined by the membrane capacitance and the junction conductance in series. We interpret the spectra of Figure 12.8 in terms of (10) using a membrane capacitance $c_M = 5\,\mu F\,cm^{-2}$ of leech neurons and a stray capacitance $c_S = 0.3\,\mu F\,cm^{-2}$ of the chip.

In the A-type spectrum, the small amplitude at low frequencies indicates a low membrane conductance g_{JM}. Concomitantly, the increase of the phase at a rather low frequency reflects a large time constant τ_{JM} of the membrane. The increase of the amplitude at a high frequency indicates a small time constant τ_J and a large conductance g_J. When we fit the data, we obtain and $\tau_J = 25\,\mu s$ and the conductances $g_{JM} = 0.36\,mS\,cm^{-2}$ and $g_J = 217\,mS\,cm^{-2}$. In the B-type spectrum, the enhanced amplitude at low frequencies indicates a large membrane conductance g_{JM}, the further increase at a high frequency is due to a large conductance g_J. The minor change of phase suggests that a range with $\omega\tau_{JM} \gg 1$ and $\omega\tau_J \ll 1$ does not exist, that is, that the two time constants are similar. When we fit the data, we obtain $\tau_{JM} = 130\,\mu s$ and $\tau_J = 66\,\mu s$ and the conductances $g_{JM} = 38.5\,mS\,cm^{-2}$ and $g_J = 40.8\,mS\,cm^{-2}$.

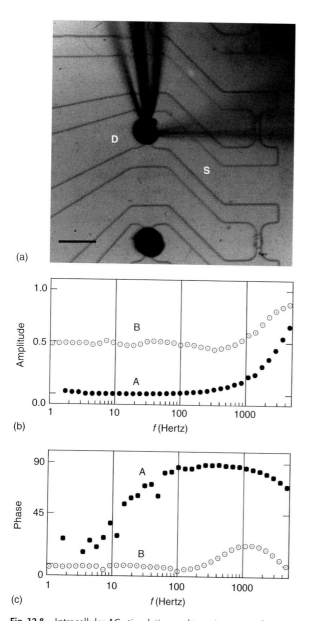

Fig. 12.8 Intracellular AC stimulation and transistor recording [10, 11]. (a) Micrograph of leech neuron on EOSFET with source S and drain D. The cell is contacted with a patch-pipette. From the right, a second pipette is impaled to measure the actual voltage $\underline{V}_M$ in the cell. (b) Amplitude of voltage transfer $\underline{V}_J/\underline{V}_M$ from cell to junction versus frequency f. (c) Phase of the voltage transfer. The dots mark the A-type spectrum, the circles the B-type spectrum.

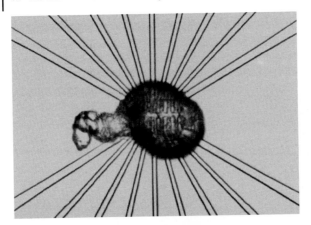

Fig. 12.9 Leech neuron on array of field-effect transistors (diameter of the cell about 60 µm) [12]. The array consists of two rows with eight transistors shining through the cell body. The drain contacts are radially directed upward and downward. The contact of the common source is at the left and right of the array. The transistors and contacts are separated by local field oxide (LOCOS process.) The cell is connected with a patch-pipette, and the transistor array is used to measure the profile of the extracellular voltage in response to applied intracellular AC voltage.

The crucial difference of A-type and B-type junctions is the leak conductance of the attached membrane [10, 11]. Whereas in an A-type contact the membrane conductance is normal, it is enhanced in B-type junction by 2 orders of magnitude. From the specific junction conductances $g_J = 217$ mS cm^{-2} and $g_J = 40.8$ mS cm^{-2}, we obtain with (3) with $\eta = 5$ and with an estimated contact area $A_{JM} = 1000$ µm^2 the sheet resistances $r_J = 7.7$ MΩ and $r_J = 41$ MΩ. If the cleft is filled with bulk electrolyte ($\rho_J = 100$ Ωcm), its width would be $d_J = 130$ nm and $d_J = 24.4$ nm.

Bath Stimulation with Transistor Recording In a second type of experiment, extracellular AC stimulation is applied, and the response of the junction is mapped with a transistor array [13–15]. We describe an experiment with a pure lipid membrane [14]. A giant vesicle is sedimented onto the chip and attached by polylysine as shown in Figure 12.10a. An AC voltage $\underline{V}_E$ is applied to the electrolyte with respect to ground potential, and the modulation of the extracellular voltage $\underline{V}_J$ with respect to ground is observed in amplitude and phase. The amplitude of voltage transfer $\underline{V}_J/\underline{V}_E$ is plotted in Figure 12.10b versus the position of the transistors and the frequency f. At low frequencies, the cleft perfectly follows the voltage in the bath. This coupling is mediated by the conductance of the cleft considering Figure 12.4a. Already, around $f = 2$ Hz the voltage transfer drops in the center of the junction, and a hammock-like profile appears. There the capacitive current through membrane and oxide begins to contribute. At high frequencies where the capacitive current dominates, a plateau is observed again.

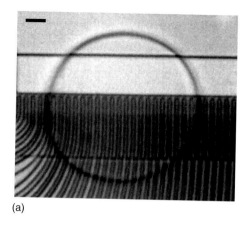

(a)

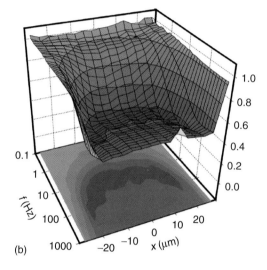

(b)

Fig. 12.10 Membrane–silicon junction probed by extracellular stimulation and transistor recording [14]. (a) Micrograph of giant lipid vesicle on a linear transistor array. Scale bar 10 µm. The gates are between the ends of the dark lanes of local field oxide. (b) Amplitude of voltage transfer $\underline{V}_J/\underline{V}_E$ (ratio of voltage amplitude in the cleft and voltage amplitude in the bath with respect to ground) versus position x and frequency f.

We use the area-contact model of (1) to evaluate the profile of the transfer function $\underline{V}_J/\underline{V}_E$ [14]. We do not consider explicitly the current balance of the intracellular space for the area-contact model but assign the area elements of the free membrane serially to the area elements of the attached membrane. Assuming that the properties of the free and attached membrane are identical, we define an effective area-specific capacitance and conductance $\tilde{c}_M = c_M/(1+\beta_M)$ and $\tilde{g}_{JM} = g_{JM}/(1+\beta_M)$. This "local approximation" [14, 15] avoids an integration over the attached membrane [20, 21]. The voltage transfer from the electrolyte to the junction in a circular junction with a radius a_J is given by (11) as a function of the radial coordinate a and the angular frequency ω with the modified Bessel function I_0 and the time constants τ_{JM} and $\tilde{\tau}_{JM}$ and the complex reciprocal length constant $\tilde{\gamma}_J$ of the core-coat conductor.

$$\frac{V_{\rm J}(a,\omega)}{V_{\rm E}(\omega)} = \frac{I_0(\tilde{\gamma}_{\rm J}a)}{I_0(\tilde{\gamma}_{\rm J}a_{\rm J})} + \frac{i\omega\tau_{\rm JM}}{1+i\omega\tilde{\tau}_{\rm JM}}\left[1 - \frac{I_0(\tilde{\gamma}_{\rm J}a)}{I_0(\tilde{\gamma}_{\rm J}a_{\rm J})}\right]$$

$$\tau_{\rm JM} = \frac{c_{\rm M}}{g_{\rm JM}}, \quad \tilde{\tau}_{\rm JM} = \frac{\tilde{c}_{\rm M}+c_{\rm S}}{\tilde{g}_{\rm JM}}, \quad \tilde{\gamma}_{\rm J}^2 = r_{\rm J}\tilde{g}_{\rm JM}(1+i\omega\tilde{\tau}_{\rm JM}) \tag{11}$$

For a radius $a_{\rm J} = 25$ µm and an area ratio $\beta_{\rm M} = 0.7$ estimated from the shape of the vesicle, with the capacitance $c_{\rm M} = 0.6$ µF cm^{-2} for solvent-free lipid bilayers, we obtain a perfect agreement of theory and experiment, when we assume a sheet resistance $r_{\rm J} = 130$ GΩ and a membrane conductance $g_{\rm JM} < 1$µS cm^{-2}. The low conductance reveals the perfect quality of the lipid bilayer. The sheet resistance is surprisingly high. With $d_{\rm J} = 1$ nm measured by FLIC microscopy, we obtain from $r_{\rm J} = \rho_{\rm J}/d_{\rm J}$ a specific resistance $\rho_{\rm J} = 13\,000$ Ωcm, which is far higher than the specific resistance $\rho_{\rm E} = 250$ Ωcm of the bulk electrolyte. The discrepancy can be assigned to a lowered concentration of ions in the narrow cleft, caused by the image force near the oxide and the membrane with their low dielectric constants.

Chip Stimulation with Optical Recording In a third type of analysis, a voltage $V_{\rm S} - V_{\rm E}$ is applied between chip and electrolyte, and the response of the voltage $V_{\rm M} - V_{\rm J}$ across the attached membrane is observed with a voltage-sensitive dye [19–21]. We consider here an experiment with cells of the line HEK 293 (human embryonic kidney cells) on a chip coated with fibronectin, a protein from the extracellular matrix. The outer surface of the cell membrane is stained with the voltage-sensitive dye diButyl-Naphthylamine-Butylsulfonato-IsoQuinolinium (BNBIQ) [16]. At selected wavelengths of excitation and emission, the dye responds with a decrease of fluorescence when a positive voltage is applied to the cytoplasm [17]. Voltage pulses with a height of $V_{\rm SE}^0$ are applied to a highly p-doped silicon chip, and the fluorescence change is recorded by signal averaging. A rather thick oxide ($d_{\rm ox} = 50$ nm) is chosen to get high fluorescence intensity in front of the reflecting silicon. Optical transients in the attached and free membrane are depicted in Figure 12.11. After a negative voltage step, the fluorescence transient is negative in the adhesion region, indicating a positive change of the membrane voltage $V_{\rm M} - V_{\rm J}$. For a positive voltage step, the change of the membrane voltage $V_{\rm M} - V_{\rm J}$ is negative. The data are fitted with exponentials. For the attached membrane, the time constant is 2.9 µs.

We evaluate the experiment with the point-contact model [19]. A step stimulation with an amplitude $V_{\rm SE}^0$ applied to the chip with respect to the bath is inserted into (2) with $c_{\rm S} dV_{\rm S}/dt = c_{\rm S} V_{\rm SE}^0 \delta(t)$. When we neglect all ionic currents, we obtain from (4) an exponential response of the voltage across the attached membrane according to (12) with a time constant $\tilde{\tau}_{\rm J}$ where the effective capacitance per unit area is $\tilde{c}_{\rm M} = c_{\rm M}/(1+\beta_{\rm M})$.

$$\frac{V_{\rm M} - V_{\rm J}}{V_{\rm SE}^0} = -\frac{1}{1+\beta_{\rm M}}\frac{c_{\rm S}}{\tilde{c}_{\rm M}+c_{\rm S}}\exp\left(-\frac{t}{\tilde{\tau}_{\rm J}}\right), \quad \tilde{\tau}_{\rm J} = \frac{\tilde{c}_{\rm M}+c_{\rm S}}{g_{\rm J}} \tag{12}$$

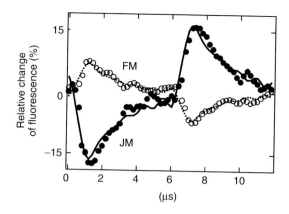

Fig. 12.11 Membrane–silicon junction probed with voltage-sensitive dye BNBIQ [19]. A positive voltage of +6 V pulse is applied to the chip from 0 to 3 ms, a negative pulse of −6 V from 3 to 6 ms. The transient change of fluorescence is plotted for the attached (JM) and free membrane (FM) of an HEK293 cell. The data are fitted with exponentials convoluted with the transfer function of the chip and the response function of the photomultiplier.

From the experimental time constant $\tilde{\tau}_J = 2.9\,\mu s$ with $c_M = 1\,\mu F\,cm^{-2}$, $c_S = 0.07\,\mu F\,cm^{-2}$ and $\beta_M = 0.4$ we obtain a specific conductance $g_J = 270\,mS\,cm^{-2}$ of the junction. Using (3) with $\eta = 5$ and a contact area $A_{JM} = 725\,\mu m^2$, a sheet resistance of $r_J = 8\,M\Omega$ is evaluated. That result is similar to the A-type junction of leech neuron. For the HEK293 cells, however, we are able to measure the width of the cleft by FLIC microscopy. We find $d_J = 50\,nm$. From $r_J = \rho_J / d_J$ we obtain a specific resistance $\rho_J = 40\,\Omega cm$ in the cleft. This value is quite similar to the surrounding bath with a specific resistance $74\,\Omega cm$. Improved experiments with AC stimulation and phase mapping of fluorescence show that the difference is not significant [20]. The specific resistance in the cleft is indistinguishable from the specific resistance in the bath for HEK293 cells and nerve cells on various substrates in various electrolytes [21].

Conclusion The cleft between neuronal cells and chips has an electrical resistance that corresponds to a thin film of bulk electrolyte. The sheet resistance is in the order of $r_J \approx 10\,M\Omega$ with a global resistance around $G_J^{-1} \approx 1\,M\Omega$. There is no gigaohm seal between neuronal cells cultured on a chip. It should be noted that the width of the cleft is far larger than the thickness of the diffuse electrical double layer at the silicon dioxide and at the membrane with a Debye length around $\kappa_D^{-1} \approx 1\,nm$ in 100 mM NaCl and is also far larger than the Bjerrum length $l_B \approx 0.7\,nm$ of Coulombic interactions that govern the interaction with image charges in membrane and silicon dioxide. It will be a difficult task to enhance the sheet resistance by lowering the width or by enhancing the specific resistance of the cleft.

12.2.4
Ion Channels in Cell-silicon Junction

During neuronal excitation, the TEP $V_J(t)$ depends on the current through ion conductances in the attached membrane. During capacitive stimulation of neurons, the primary target of the TEP $V_J(t)$ are the ion conductances in the attached membrane. Thus, we have to ask: (a) Are there functional ion channels in the contact region at all? (b) Is the density of ion channels in the contact the same as in the free membrane? (c) Can we record open channels with a transistor? (d) Can we activate channels from the chip by capacitive stimulation? Here, we consider the recording of two systems: intrinsic potassium channels in rat neurons and recombinant potassium channels in HEK293 cells.

Rat Neurons Neurons from rat hippocampus are cultured on a chip with transistors as shown in Figure 12.2. The intracellular voltage of a cell is varied by the whole-cell patch-clamp technique. Simultaneously, we measure the current I_M through the total membrane with the micropipette and the extracellular voltage in the contact area V_J with a transistor, holding the bath at ground potential [22]. The sodium current is inhibited by tetrodotoxin. The intracellular voltage V_M, the outward current I_M and transistor record V_J are plotted in Figure 12.12.

At a depolarization $V_M = 20$ mV there is a stationary current $I_M = 0.25$ nA and a superposed transient current. These two current components are due to two different potassium conductances, a K-type conductance and an inactivating A-type conductance. The extracellular voltage V_J detected by the transistor shows a stationary response that matches the stationary K-type current, but no component corresponding to the A-type

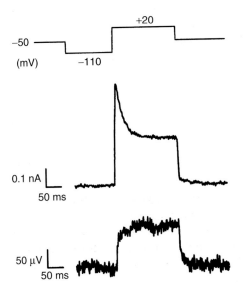

Fig. 12.12 Rat neuron on a transistor under voltage-clamp [22]. The protocol of the intracellular voltage V_M is shown at the top. In the center, the total membrane current I_M is plotted, at the bottom the extracellular voltage V_J recorded by a transistor obtained by averaging 30 records.

current [22]. (The slow relaxation of the transistor signal after the depolarizing and hyperpolarizing step is due to electrodiffusion effects.)

We discuss the result in terms of the point-contact model using (2) and (4) [22, 23]. At a constant voltage V_M, the membrane current I_M for a single ion conductance with an average area-specific conductance g_M^i in the whole cell membrane is given by (13) with the total membrane area A_M, assuming that the extracellular voltage is small with $V_J \ll V_M - V_0^i$. The extracellular voltage V_J is described by (14) with the specific conductance g_{JM}^i in the attached membrane.

$$I_M = A_M g_M^i (V_M - V_0^i) \tag{13}$$

$$V_J = \frac{1}{g_J} g_{JM}^i (V_M - V_0^i) \tag{14}$$

If the channels in the attached and free membrane have the same functionality, the relative conductances $g_M^i/\bar{g}_M^i$ and $g_{JM}^i/\bar{g}_{JM}^i$ follow the same voltage-dependence where $\bar{g}_M^i$ and $\bar{g}_{JM}^i$ are the maximum conductances with open channels. Considering (13) and (14), the transistor record V_J and the pipette record I_M are proportional to each other for all voltages V_M according to (15).

$$\frac{V_J}{I_M} = \frac{1}{g_J A_M} \frac{\bar{g}_{JM}^i}{\bar{g}_M^i} \tag{15}$$

Considering Figure 12.12 with (15), we conclude: (a) the absence of transient in the transistor response indicates that there is no A-type potassium conductance in the attached membrane with $\bar{g}_{JM}^A \ll \bar{g}_M^A$; (b) the visible response of the transistor shows that functional K-type channels exist in the junction. To evaluate the ratio $\bar{g}_{JM}^K/\bar{g}_M^K$ from the experimental $V_J/I_M = 480\ k\Omega$ with (15), we need a value of the scaling factor $(g_J A_M)^{-1}$. $(g_J A_M)^{-1}$ and A_M are obtained by an AC measurement when the channels are closed. From (2) and (4) we obtain the complex response of the current $\underline{I}_M$ and of the extracellular voltage $\underline{V}_J$ (16) and (17) at $g_{JM}^i = 0$ and $\omega \tau_J \ll 1$. The scaling factor is given by the ratio $\underline{V}_J/\underline{V}_M$ according to (18).

$$\underline{I}_M = i\omega c_M A_M \underline{V}_M \tag{16}$$

$$\underline{V}_J = \frac{i\omega c_M}{g_J} \underline{V}_M \tag{17}$$

$$\frac{\underline{V}_J}{\underline{I}_M} = \frac{1}{g_J A_M} \tag{18}$$

From the responses by AC stimulation at $\omega = 200$ Hz, we obtain with (16) and (17) an area $A_M = 1100\ \mu m^2$ and a junction conductance $g_J = 1000\ mS\ cm^{-2}$ at $c_M = 1\ \mu F\ cm^{-2}$. With the resulting scaling factor $(g_J A_M)^{-1} = 91\ k\Omega$ and with $V_J/I_M = 480\ k\Omega$ the ratio of the maximum conductances in the attached and total membrane is $\bar{g}_{JM}^K/\bar{g}_M^K = 5.3$ using (15). We conclude: the K-type potassium channels are significantly accumulated in the junction.

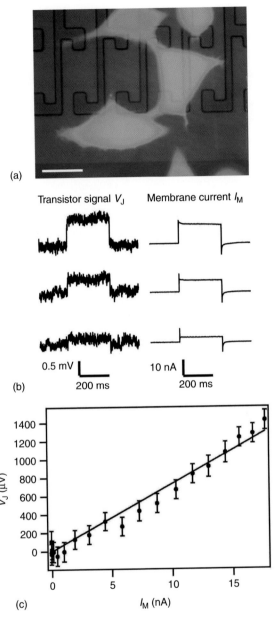

Fig. 12.13 Recombinant hSlo potassium channels on a transistor [23]. (a) HEK293 cells on linear transistor array. With a blue illumination, the transfected cells appear in the color of the fluorescence of GFP (green fluorescent protein) used as a marker. (b) Iono-electronic coupling at three intracellular voltages $V_M = 30, 45, 58$ mV. Left: extracellular voltage V_J in the cell-silicon junction. Right: current I_M through the total cell membrane. Intracellular potassium concentration 152 mM, extracellular concentration 100 mM. (c) Extracellular voltage V_J versus membrane current I_M. The regression line has a slope $V_J/I_M = 73$ kΩ.

Recombinant Channels A more detailed investigation is possible with recombinant hSlo potassium channels that are overexpressed in HEK293 cells. No signal averaging is required, and the voltage-dependent gating can be analyzed in detail. The cells are cultured on a transistor array as depicted in Figure 12.13a. The voltage V_M in the cell is changed step-by-step with a patch-pipette, and the total membrane current I_M and the extracellular voltage V_J are simultaneously recorded [23]. Examples are shown in Figure 12.13b for a high extracellular potassium concentration. A depolarization leads to an enhancement of the current and a correlated enhancement of the transistor signal. Thus, functional hSlo potassium channels must exist in the junction.

In that experiment, we were able to check whether the functionality of the channels in the attached and free membrane is the same [23]. We plot V_J versus I_M in Figure 12.13c for all voltages V_M and obtain a strict linear relation over the whole range of gating. The result shows that (15) is valid with a constant ratio $V_J/I_M = 73\ k\Omega$. We measure the scaling factor $(g_J A_M)^{-1}$ with AC stimulation at $\omega = 6\text{--}20$ Hz. From the current and the transistor signal, we obtain with (16) and (17) a membrane area $A_M = 2360\ \mu m^2$ and a junction conductance $g_J = 1960\ mS\ cm^{-2}$ at $c_M = 1\ \mu F\ cm^{-2}$.

The high junction conductance indicates a small effective radius a_J of the junction according to (3). It is due to a peripheral location of the transistor in the area of cell adhesion. With the scaling factor $(g_J A_M)^{-1} = 22\ k\Omega$ and with $V_J/I_M = 73\ k\Omega$, the ratio of the maximum area-specific conductance in the attached and total membrane is $\bar{g}_{JM}^{hSlo}/\bar{g}_{M}^{hSlo} = 3.3$ using (15). Thus, the recombinant hSlo potassium channels are accumulated in the attached membrane.

For low physiological potassium concentrations, the fast response of the transistor to open potassium channels is followed by a slow signal. It is due to an enhanced potassium concentration in the cleft and an interaction of potassium ions with the negatively charged gate oxide of the transistor. The effect can be quantitatively rationalized by an electrodiffusion version of the point-contact model, a Stern–Graham model for the electrical double layer [24].

Sensorics An overexpression of ligand-gated channels and transistor recording is a promising approach to develop cell-based biosensors. In such systems, the intracellular voltage is not controlled. The TEP is obtained by combining (2), (3) and (4) for small signals with $V_J \ll V_M - V_0^i$ and $dV_J \ll dV_M$ at $V_E = 0$ according to (20) [52]: A cellular electronic sensor relies on an accumulation or depletion of channels with $g_{JM}^i - g_{FM}^i \neq 0$, a high driving voltage $V_M - V_0^i$ and a junction with a large radius a_J, a high specific resistance ρ_J and a small distance d_J.

$$V_J = \frac{\rho_J a_J^2}{5 d_J (1 + \beta_M)} (g_{JM}^i - g_{FM}^i)(V_M - V_0^i) \tag{19}$$

Conclusion The combination of transistor recording with whole-cell patch-clamp shows that functional ion channels exist in the area of cell adhesion. Important for neuronal interfacing and cellular biosensorics is the observation that ion channels are selectively accumulated and depleted. A control of the expression and sorting of ion channels is an important task to optimize cell-chip contacts. An activation of channels is achieved by capacitive stimulation from chips with high capacitance in an electrolyte with high resistance [25].

12.3
Neuron–Silicon Circuits

Cell–silicon junctions are the basis for an integration of neuronal dynamics and digital electronics. The first step is an interfacing of individual nerve cells and silicon microstructures (Figure 12.14a) with (a) eliciting neuronal activity by capacitive stimulation from the chip and (b) recording of neuronal activity by a transistor. In the next level, pairs of nerve cells are coupled to a chip with two fundamental pathways: (a) stimulation of a neuron, signal transfer through a neuronal network with synapses to a second neuron and recording of neuronal activity there by a transistor (Figure 12.14b). (b) Recording activity of one neuron by a transistor, signal transfer through the microelectronics of the chip and capacitive stimulation of a second neuron (Figure 12.14c). In a further step, defined neuronal networks are created on the chip such that an intimate communication of network dynamics and computation can be envisaged (Figure 12.14d).

Identified neurons from invertebrates are preferred in these experiments because they are large and easy to handle, because they form strong neuroelectronic junctions, and last but not least, because small neuronal networks have a distinct biological function in invertebrates and may be reconstituted and studied on a chip.

12.3.1
Transistor Recording of Neuronal Activity

The activity of a nerve cell – an action potential – consists in a fast opening of sodium channels with a concomitant current into the cell and a delayed opening of potassium channels with a compensating outward current. The ionic currents are coupled by the membrane potential $V_M(t)$, which controls the opening and closing of the channels. The voltage transient $V_M(t)$ is confined by the reversal voltages of sodium and potassium channels. The neuronal excitation drives ionic and capacitive current through the membrane attached to a chip. This current is squeezed through the cleft between cell and chip and gives rise to a TEP $V_J(t)$ that is recorded by a transistor. First, we derive $V_J(t)$ expected for neuronal excitation using the point-contact model. Then, we consider transistor recordings of leech and rat neurons.

Small Signal Approximation The TEP is determined by the coupled dynamics of the extracellular and intracellular potentials V_M and V_J according to (2) and (4). We assume that the extracellular potential is small with $V_J \ll V_M - V_0^i$ and $dV_J \ll dV_M$ and that

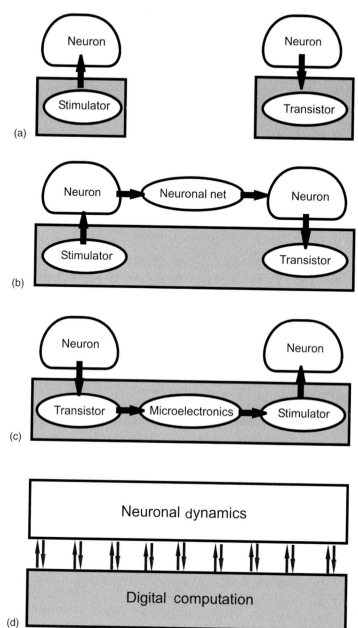

Fig. 12.14 Neuroelectronic hybrids. (a) Capacitive stimulation and transistor recording of individual neurons from semiconductor. (b) Two-neuron pathway with capacitive stimulation, signal transmission through neuronal network and transistor recording at a second neuron. (c) Two-neuron pathway with transistor recording, signal processing by microelectronics on the chip and capacitive stimulation of a second neuron. (d) Integration of neuronal dynamics and digital electronics by bidirectional signaling on a microscopic level.

the capacitive current to the chip is negligible. With $V_E = 0$ we obtain (20) and (21) with the current j_{INJ} injected by a pipette per unit area of the cell membrane [26]: The extracellular potential $V_J(t)$ is determined by the capacitive and ionic current through the attached membrane. The intracellular potential $V_M(t)$ is governed by the currents through the attached and the free membrane.

$$g_J V_J = \sum_i g_{JM}^i (V_M - V_0^i) + c_M \frac{dV_M}{dt} \tag{20}$$

$$(1 + \beta_M) c_M \frac{dV_M}{dt} = -\sum_i (g_{FM}^i + \beta_M g_{JM}^i)(V_M - V_0^i) + (1 + \beta_M) j_{INJ} \tag{21}$$

A-, B- and C-type Response When the attached membrane contains no voltage-gated conductances, we obtain (22) with a leak conductance g_{JM} in the attached membrane using (20).

$$g_J V_J = g_{JM} V_M + c_M \frac{dV_M}{dt} \tag{22}$$

For negligible leak conductance, the capacitive current through the attached membrane dominates. Then the TEP is proportional to the first derivative of the intracellular voltage with $V_J \propto dV_M/dt$ [2]. This situation corresponds to the A-type junction observed in AC measurements. We call it an A-type response. For a dominating ohmic leak conductance, the TEP reflects the intracellular waveform itself with $V_J \propto V_M$ [10, 11, 27]. We call it a B-type response in analogy to the B-type junction found in AC experiments.

When we insert (21) into (20), the capacitive current is expressed by the ionic current through the free membrane and we obtain (23). The TEP is determined by the differences $g_{JM}^i - g_{FM}^i$ of the area-specific ion conductances in the attached and free membrane [26].

$$g_J V_J = \frac{1}{1 + \beta_M} \sum_i (g_{JM}^i - g_{FM}^i)(V_M - V_0^i) + j_{INJ} \tag{23}$$

This striking relation shows that the TEP of an action potential relies on an inhomogeneity of the membrane. In particular, (23) reveals that a selective accumulation or depletion of voltage-gated channels can give rise to a wide spectrum of waveforms $V_J(t)$. We call them *C-type responses* [26]. Details must be treated by numerical simulation. A special signal is expected when all conductances are accumulated by the same accumulation factor $\mu_J^i = g_{JM}^i/g_{FM}^i$. For $\mu_J > 1$, the response is proportional to the negative first derivative of the intracellular voltage according to (24).

$$V_J = \frac{1 - \mu_J}{1 + \mu_J \beta_M} \frac{c_M}{g_J} \frac{dV_M}{dt} \tag{24}$$

Leech Neuron Transistor records of leech neurons are shown in Figure 12.15. Two positions of the cells are illustrated in Figure 12.15a and Figure 12.15b, with the cell body right on a transistor [27] and with the axon stump placed on a transistor array [28]. The cells are impaled with a micropipette and action potentials are elicited by current injection. The intracellular potential $V_M(t)$ is measured with the pipette. The response of the transistors is calibrated in terms of the extracellular potential $V_J(t)$ on the gate. Three types of records are depicted in Figure 12.15c: (a) With a cell body on a transistor, $V_J(t)$ resembles the first derivative of the intracellular voltage $V_M(t)$. (b) With a cell body on a transistor, $V_J(t)$ resembles the intracellular voltage $V_M(t)$ itself. (c) With an axon stump on a transistor, $V_J(t)$ resembles the inverted first derivative of $V_M(t)$.

The results perfectly match the A-type, the B-type and the special C-type response considered above. The C-type record is observed only when the axon stump is placed on the transistor (Figure 12.15b). That region is known for its enhanced density of ion channels. The small amplitude is due to the small size of the axon stump with a high junction conductance g_J according to (3). The density of channels is depleted in the cell body where we observe A-type and B-type records.

Rat Neuron Neurons from rat hippocampus are cultured for seven days in neurobasal medium on a chip coated with polylysine [29, 30]. Selected cells, such as depicted in Figure 12.2, are contacted with a patch-pipette and action potentials are elicited by current pulses. Transistor records are obtained by signal averaging, locking the transistor signal to the maximum of the intracellular transient. A result obtained with 63 sweeps is shown in Figure 12.16.

The amplitude of the extracellular potential $V_J(t)$ is around 0.15 mV. The action potential $V_M(t)$ gives rise to two positive transients in $V_J(t)$, one in the rising phase and one in the falling phase. The upward and downward jumps in the record match the upward and downward steps of the injection current according to (23). The small amplitude of the transistor record is a consequence of a high junction conductance g_J, which is expected for the small size of rat neurons considering (3). From an AC measurement, we estimate $g_J = 600$–700 mS cm^{-2}. The shape of the transistor response can be interpreted in terms of (23): (a) the positive peak in the rising phase is related to the sodium current. Considering (23) with $V_M - V_0^{Na} < 0$, it must be connected to $g_{JM}^{Na} - g_{FM}^{Na} < 0$, that is, a depletion of sodium channels in the junction. In other words, the sodium inward current through the free membrane gives rise to a capacitive outward current through the attached membrane. (b) The positive peak in the falling phase can be assigned to a potassium outward current through the attached membrane, considering an accumulation of potassium channels with $g_{JM}^K - g_{FM}^K > 0$ and $V_M - V_0^K > 0$.

Conclusion Neuronal activity is detected by field-effect transistors. The response is rationalized by a TEP in the cell-chip contact that plays the role of a gate voltage. There is no unique response to action potentials; the shape of the extracellular record depends on the cell type and the cell area attached to the chip. The amplitude of the extracellular records is small, because the junction conductance is high compared

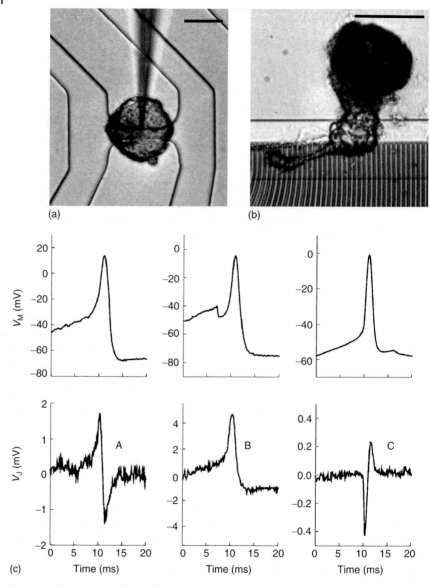

Fig. 12.15 Transistor recording of neuronal excitation in leech neurons [27, 28]. (a) Cell body of a neuron on the open gate oxide of a field-effect transistor. Scale bar 50 μm. The cell is impaled with a micropipette. (b) Axon stump of a neuron on a linear array of field-effect transistors. Scale bar 50 μm. (c) A-, B- and C-type coupling. The upper row shows the intracellular voltage $V_M(t)$, the lower row the extracellular voltage $V_J(t)$ on the gate oxide. A- and B-type couplings are observed for arrangement (a), C-type couplings for arrangement (b).

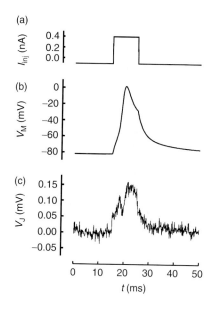

Fig. 12.16 Transistor record of neuronal excitation in rat neuron [30]. (a) Injection current applied by a patch-pipette; (b) intracellular potential $V_M(t)$; (c) transistor record scaled as an extracellular potential $V_J(t)$ on the gate (63 averaged signals).

to the effective ion conductances in the contact with $g_J \gg g_{JM}^i - g_{FM}^i$. The signals are particularly weak for mammalian neurons because of their small size. We attempt to optimize recording by (a) lowering the noise of the transistors with improved design and fabrication [31], (b) enhancing the sheet resistance between cell and chip or (c) inducing a large inhomogeneity of channel distribution.

12.3.2
Capacitive Stimulation of Neuronal Activity

A changing voltage $V_S(t)$ applied to a stimulation spot beneath a neuron leads to a capacitive current through the insulating oxide. The concomitant current along the cleft between cell and chip gives rise to a TEP $V_J(t)$ beneath the neuron. As a result, voltage-gated ion channels may open in the membrane, and an action potential $V_M(t)$ may arise. We consider first the extracellular and intracellular voltage of the point-contact model after stimulation with a voltage step. Then we discuss experiments with a voltage step and with a burst of voltage pulses.

A-type Stimulation Stimulation starts from the resting state of a neuron with low ion conductances. For the initial phase, we completely neglect the ion conductances in the attached and free membrane. Considering the coupling of the intracellular and extracellular potentials $V_M(t)$ and $V_J(t)$ with (2) and (4), we obtain (25) and (26) for $V_E = 0$ with $\tilde{c}_M = c_M/(1 + \beta_M)$ [3, 19].

$$(c_S + \tilde{c}_M)\frac{dV_J}{dt} + g_J V_J = c_S \frac{dV_S}{dt} \tag{25}$$

$$\frac{dV_M}{dt} = \frac{\beta_M}{1+\beta_M}\frac{dV_J}{dt} \tag{26}$$

For a voltage step of height V_S^0 at time $t = 0$, the perturbation of the junction is $c_S dV_S/dt = c_S V_S^0 \delta(t)$. The voltages across the attached and free membrane $V_M - V_J$ and V_M respond with exponentials for $t > 0$ according to (27) with the time constant $\tilde{\tau}_J$.

$$V_M - V_J = -\frac{1}{1+\beta_M}\frac{c_S}{\tilde{c}_M + c_S} V_S^0 \exp\left(-\frac{t}{\tilde{\tau}_J}\right)$$

$$\tilde{\tau}_J = \frac{c_S + \tilde{c}_M}{g_J} \tag{27}$$

$$V_M = \frac{\beta_M}{1+\beta_M}\frac{c_S}{\tilde{c}_M + c_S} V_S^0 \exp\left(-\frac{t}{\tilde{\tau}_J}\right)$$

For a positive voltage step $V_S^0 > 0$, the voltage drop is negative (hyperpolarizing) across the attached membrane and positive (depolarizing) across the free membrane. For $V_S^0 = 5$ V, $c_S = 0.35$ µF cm^{-2}, $\tilde{c}_M = 5$ µF cm^{-2} and $\beta_M = 1/6$ as estimated for leech neurons, the voltage amplitude is $|V_M - V_J|_0 \approx 300$ mV across the attached membrane and $|V_M|_0 \approx 50$ mV across the free membrane. With $g_J \approx 200$ mS cm^{-2} of an A-type junction of a leech neuron, we expect a time constant $\tilde{\tau}_J \approx 25$ µs. For mammalian neurons, the time constants are even shorter due to the larger g_J and smaller c_M as verified by optical recording (Figure 12.11). The crucial question is how such short transients can affect the ion conductance of a membrane.

B-type Stimulation We consider the role of a leak conductance g_{JM} in the attached membrane. To avoid complicated equations, we assume that the intracellular voltage is small compared to the extracellular voltage with $V_M \ll V_J$ and $dV_M \ll dV_J$. From (2) we obtain (28) for the extracellular voltage at $V_E = 0$ with the exponential solution of (29) after a voltage step of height V_S^0.

$$(c_M + c_S)\frac{dV_J}{dt} + (g_{JM} + g_J)V_J = c_S \frac{dV_S}{dt} \tag{28}$$

$$V_J = \frac{c_S}{c_M + c_S} V_S^0 \exp\left(-\frac{t}{\tau_J}\right) \quad \tau_J = \frac{c_M + c_S}{g_{JM} + g_J} \tag{29}$$

The upward jump of the extracellular transient $V_J(t)$ injects a charge pulse into the cell by capacitive polarization. This charge is withdrawn during the decaying exponential. Additional charge is injected during the exponential transient through a leak conductance in the attached membrane. Using (2) we obtain the injected current per unit area according to (30).

$$j_{JM} = V_S^0 c_S \left\{ \frac{c_M}{c_M + c_S} \left[\delta(t) - \frac{\exp(-t/\tau_J)}{\tau_J} \right] + \frac{g_{JM}}{g_J + g_{JM}} \frac{\exp(-t/\tau_J)}{\tau_J} \right\} \quad (30)$$

Considering (2), the perturbation of the intracellular potential $V_M(t)$ is obtained from (31).

$$c_M \frac{dV_M}{dt} + \sum_i g_{FM}^i (V_M - V_0^i) = \beta_M j_{JM} \quad (31)$$

When we neglect the voltage-gated conductances g_{FM}^i in the free membrane in the stimulation phase, the intracellular potential $V_M(t)$ is given by (32) with the resting potential V_M^0.

$$V_M = V_M^0 + V_S^0 \beta_M \frac{c_S}{c_M} \left[\left(\frac{c_M}{c_M + c_S} - \frac{g_{JM}}{g_{JM} + g_J} \right) \exp\left(-\frac{t}{\tau_J}\right) + \frac{g_{JM}}{g_{JM} + g_J} \right] \quad (32)$$

There is a jump of the intracellular voltage at $t = 0$ due to the capacitive effect. The subsequent relaxation within the time τ_J levels out at a potential determined by the leak conductance. There is a stationary change of the intracellular potential (33)

$$V_M - V_M^0 = V_S^0 \beta_M \frac{c_S}{c_M} \frac{g_{JM}}{g_{JM} + g_J} \quad (33)$$

If that potential change is above a threshold, an action potential is elicited, if it is below the threshold, it relaxes with the time constant of the total cell. For a positive voltage step $V_S^0 = 5$ V applied to the B-type junction of a leech neuron with $\beta_M = 1/6$, $c_S = 0.35$ μF cm^{-2}, $c_M = 5$ μF cm^{-2}, $g_{JM} \approx 40$ mS cm^{-2} and $g_J \approx 40$ mS cm^{-2} we expect a depolarization of $\Delta V_M = 25$ mV.

C-type Stimulation When we take into account voltage-gated conductances in the attached and free membrane, different neuronal responses are expected to a positive or negative voltage step, depending on channel sorting. Such junctions must be treated as numerical stimulation on the basis of an assumed dynamics of voltage-gated channels.

Step Stimulation of Leech Neuron Neurons from the leech are stimulated by a single positive voltage step applied to a capacitive stimulation spot on a silicon chip as illustrated in Figure 12.17a [3]. The height of the steps is $V_S^0 = 4.8, 4.9, 5.0$ V. The intracellular response $V_M(t)$ is shown in Figure 12.17b. When the height exceeds a threshold, an action potential is elicited.

A positive step in an A-type contact leads to a hyperpolarizing effect on the attached membrane. No sodium channels can open there to induce an action potential. The free membrane is affected by a depolarizing transient with a small amplitude $|V_M|_0 \approx 50$ mV and a time constant $\tau_J \approx 25$ μs. It is difficult to imagine how sodium channels with a time constant in the millisecond range are able to respond to such short transients. Thus, it is likely that a B-type junction exists. The TEP $V_J(t)$ injects ohmic current into

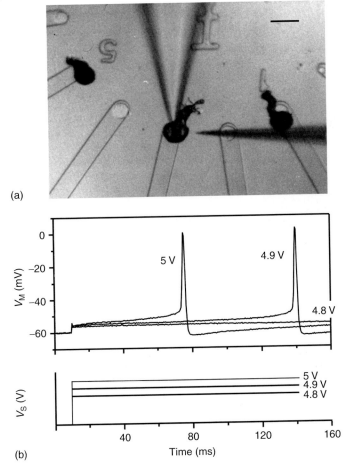

Fig. 12.17 Capacitive stimulation by silicon chip [3]. (a) Leech neurons are attached to circular stimulation spots (20–50 μm diameter) covered by 10-nm silicon dioxide. The rest of the chip is insulated by a 1-μm thick field oxide. One neuron is impaled with a micropipette electrode. An additional electrode (right) measures the local bath potential. Scale bar 100 μm.
(b) Top: intracellular voltages $V_M(t)$. Bottom: voltage steps $V_S(t)$ applied to the chip (not on scale).

the cell through a leak conductance g_{JM} according to (30). A resulting quasi-stationary depolarization of about $\Delta V_M = 25$ mV is sufficient to elicit an action potential.

Burst Stimulation of Snail Neuron In many junctions of neurons from leech and snail, excitation is achieved only when a burst of voltage pulses is applied to a chip [32, 41, 42]. An example is shown in Figure 12.18. A snail neuron is attached to a two-way junction made of a stimulation area and a transistor (Figure 12.18a). When a burst of voltage

12.3 Neuron–Silicon Circuits | 371

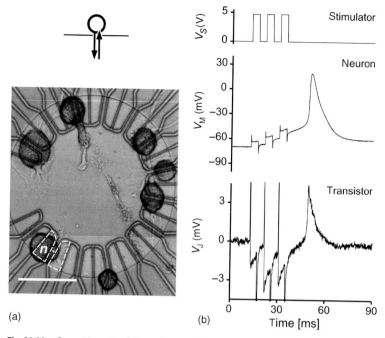

Fig. 12.18 Capacitive stimulation of neuron by a burst of voltage pulses [41].
(a) Micrograph of snail neurons on a chip with a circular arrangement of two-way contacts. The stimulation area with two wings under neuron n is marked with a dashed line, the transistor is located between the two wings. Scale bar 100 μm.
(b) Top: Voltage $V_S(t)$ applied to the stimulation area. Center: Intracellular voltage $V_M(t)$ measured with an impaled pipette. Bottom: Extracellular voltage $V_J(t)$ measured with the transistor.

pulses is applied to the stimulation area, the intracellular voltage $V_M(t)$ responds with short capacitive transients at the rising and falling edge of each pulse and a stationary depolarization during the pulses as illustrated in Figure 12.18b. After the third pulse, the intracellular potential rises such that an action potential is elicited [41]. The transistor allows us a look into the junction. The rising and falling edges of each pulse lead to capacitive transients $V_J(t)$. At the rising edge, an additional negative transient $V_J(t)$ is initiated that slowly decays during the pulse and during the subsequent pulse interval. We conclude that the positive capacitive transients in the cleft at the onset of the pulses induce an ionic inward current through the attached membrane. This conductance decays slowly and is not affected by the negative capacitive transient. It is responsible for the intracellular depolarization $V_M(t)$.

However, a positive extracellular transient has a hyperpolarizing effect on the attached membrane and cannot open voltage-gated channels there. We have to consider that a short transmembrane transient with an amplitude $|V_M - V_J|_0 \approx 300$ mV is sufficient to induce a transient, reversible electroporation of the membrane. Possibly, the negative capacitive transient is not sufficient to induce electroporation, because its electrical field is opposite to the electrical field in the resting state. Further studies are required to reveal the contribution of electroporation and to eliminate it by proper engineering of the cell-chip contact with voltage-gated sodium channels.

Conclusion The concept of a core-coat conductor can guide a rationalization of capacitive stimulation of neuronal activity on silicon chips. But the situation is less clear than with transistor recording. Optical recording directly reveals that fast voltage transients actually exist in the attached and free membrane. But how those transients affect the cell is uncertain. Current injection through a leaky membrane, capacitive gating of ion channels and transient electroporation are difficult to distinguish. Further studies on neuronal excitation are required with a recording of the local voltage by transistors or voltage-sensitive dyes, comparing voltage-clamp and current-clamp. Detailed studies on the capacitive gating of ion channels and on electroporation will be most helpful. An optimized stimulation may be achieved (a) by lowering the junction conductance, (b) by inserting recombinant ion channels into the junction and (c) by fabricating stimulation spots with an enhanced specific capacitance.

12.3.3
Two Neurons on Silicon Chip

We consider two hybrid circuits with two neurons on a silicon chip, the signal transmission from a neuron through a chip to another neuron and the signal transmission from the chip through two synaptically connected neurons back to the chip (Figure 12.14).

Neuron–Chip–Neuron The equivalent circuit of recording and stimulation of two neurons on a chip is shown in Figure 12.19: A field-effect transistor probes the extracellular voltage V_J in the junction between the first cell and the chip as caused by the membrane currents of an action potential. Capacitive stimulation induces a voltage V_J in the junction of the second cell to elicit neuronal excitation by activating membrane conductances. The processing unit accomplishes five tasks (Figure 12.19): (a) the source-drain current of the transistor is transformed to a voltage and amplified; (b) the response to an individual action potential is identified by a threshold device (Schmitt trigger); (c) a delay line is started; (d) a train of voltage pulses is generated and applied to a capacitive stimulator; (e) the cross talk from stimulator to transistor in the chip is eliminated by a refractory circuit: the delay line is not started directly by the output of the Schmitt trigger, but by the onset of a flip-flop as triggered by an action potential. The flip-flop is reset after stimulation.

In a first implementation, the chip consists of two parts: (a) an interface unit with transistors and stimulation spots and (b) an interneuron unit implemented as a conventional integrated circuit for discrimination, delay line and pulse shaping [33], [34]. Figure 12.20a shows two snail neurons attached to two-way contacts of the interface

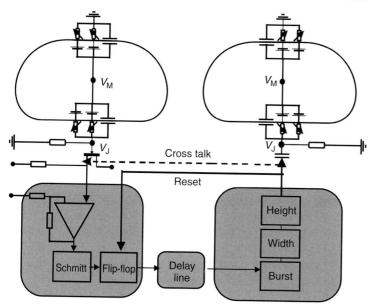

Fig. 12.19 Electronic coupling of two disconnected neurons [33, 34]. In the upper part, the interfacing between the neurons and the semiconductor is represented as an equivalent circuit with the point-contact model. In the lower part, the function of the interneuron unit is sketched as a block circuit with three stages: (1) amplification and signal recognition, (2) delay line and (3) pulse generator. The cross talk from stimulation to recording is marked by a dashed line, the reset of the flip-flop after stimulation by a drawn line.

unit. The two silicon chips of the interfacing and the interneuron unit are bonded side by side with the interface unit exposed to the culture medium in a chamber as illustrated in Figure 12.20b.

The connection from a spontaneously firing neuron A along the chip to a separated neuron B is shown in Figure 12.21 [34]. Both neurons are impaled by micropipette electrodes to observe their intracellular voltage V_M. On the left, we see four action potentials of neuron A and the response of the transistor after amplification. The negative transient of the source-drain current is caused by an outward current through the cell membrane. On the right, we see four delayed bursts of voltage pulses (17 pulses, height 2 V, width 0.3 ms, interval 0.3 ms) that are applied to the stimulator beneath neuron B. The firing of neuron B is in strict correlation to the firing of neuron A.

The first step of processing on the chip is the assignment of a digital signal to each action potential of neuron A (Figure 12.21). However, in addition, the transistor records strong, short perturbations in coincidence with the stimulation pulses at neuron B. They are caused by capacitive cross talk on the chip. A digital signal is assigned to these artifacts, too, as shown in Figure 12.21. A second processing step on the chip prevents these artifacts from feeding back into the pulse generator: the response to each action potential activates a flip-flop (Figure 12.21). The onset of the flip-flop starts

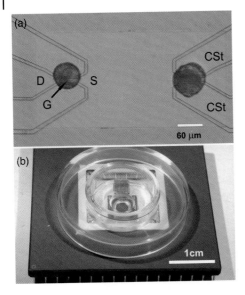

Fig. 12.20 Neurochip [33, 34]. (a) Micrograph of a silicon chip with an all-silica surface. Two snail neurons are attached to two two-way interface contacts. The source (S), drain (D) and gate (G) of a transistor, and the wings of a capacitive stimulator (CSt) are marked. The surface of the chip is made of silicon dioxide. The bright rectangle is the area of thin oxide. (b) Perspex chamber with chip. The quadratic interface unit forms the bottom of the circular Perspex chamber in contact with the culture medium. The interneuron unit is bonded to it side by side, shielded from the electrolyte.

the delay line that triggers the voltage generator. The flip-flop is reset after completed stimulation. Thus, the stimulation artifact meets an activated flip-flop and cannot interfere.

The pathway neuron–silicon–neuron demonstrates that single-action potentials from individual nerve cells can be reliably fed into a digital electronic processor and that after computations a single-action potential in an individual nerve cell can be reliably elicited, all on a microscopic level. The study relies on the established physiology of neurons, on the known physics of interfacing and on standard electronics. It is a fundamental exercise in neuroelectronic engineering. The cross talk may be avoided (a) by an enhanced capacitance of the stimulators and a lower amplitude of the stimulation pulses, (b) by transistors with lower noise to provide better-defined waveforms of the records such that they can be distinguished from artifacts and (c) by blocking devices on the chip to reduce the interferences.

Chemical Synapse on Chip To achieve the coupling of a chemical synapse to a silicon chip, we used two different identified neurons from the snail [35]. *In vivo* the neuron VD4 from *Lymnaea stagnalis* forms a cholinergic synapse with the neuron LPeD1. This synapse can be reconstituted *in vitro* in a soma–soma configuration. By using a soma–soma contact, we avoid a displacement of neurons by outgrowth. VD4 and LPeD1 neurons are paired on a linear array of microelectronic contacts for stimulation and recording as illustrated in Figure 12.22. The presynaptic neuron is stimulated from a capacitor, and postsynaptic activity is recorded by a transistor.

The experiment proceeds in three steps [35]: (a) it is tested whether action potentials in VD4 are elicited by capacitor stimulation, while keeping the cell impaled with a

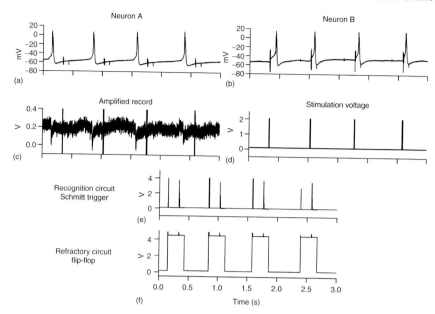

Fig. 12.21 Pathway neuron–silicon chip–neuron [34]. (a) Intracellular voltage of neuron A. (b) Intracellular voltage of neuron B, which fires with a constant delay after neuron A. (c) Response of the transistor after current–voltage conversion and amplification. Weak downward signals of the action potentials and strong perturbations due to cross talk of the stimulation voltages at neuron B. (d) Bursts of voltage pulses at the stimulator beneath neuron B. Each burst of 10 ms duration consists of 17 voltage pulses (width and separation 0.3 ms). (e) Output of the Schmitt trigger. Signals are assigned to the action potentials and to the cross talk perturbations. (f) State of the refractory flip-flop. The flip-flop is set by an action potential. This transition triggers the delay line that elicits the stimulation voltages. It is reset after the end of the stimulation burst.

microelectrode. For this purpose, two pulses with +3 V amplitude and 0.5 ms duration are applied to a capacitor (Figure 12.23). A sequence of three paired pulses gives rise to short responses of the intracellular voltage that induce a sustained intracellular depolarization. Action potentials generally occur after the second or third stimulus. (b) Excitatory postsynaptic potentials (EPSP) in LPeD1 are induced by the capacitively elicited activity in VD4, as observed again by intracellular recording. These EPSPs are indistinguishable from EPSPs induced by intracellular stimulation of VD4. Yet, transistor recording of the EPSPs is not possible because of noise. But postsynaptic action potentials are observed after two to three presynaptic spikes that are elicited by capacitor stimulation (Figure 12.23). (c) To complete electronic interfacing of synaptic transmission, we tested whether this postsynaptic activity could be observed with a transistor. In fact, the action potential elicited in the LPeD1 neuron by synaptic transmission is recorded by a transistor as a sharp peak of about 3 mV in its rising phase (Figure 12.23).

A particularly interesting aspect of the VD4-LPeD1 synapse is its capability to exhibit short-term potentiation that is thought to form the basis of working memory in

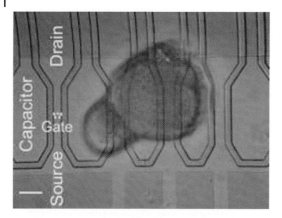

Fig. 12.22 Silicon chip with chemical synapse [35]. Micrograph of a silicon chip with a linear array of capacitors and transistors. A presynaptic VD4 neuron (left) and a postsynaptic LPeD1 neuron (right) from *Lymnaea stagnalis* are attached in soma–soma configuration and form a cholinergic chemical synapse. Scale bar 20 μm.

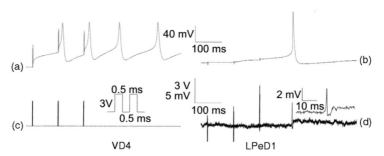

Fig. 12.23 Synaptic transmission by chemical synapse on silicon chip [35].
(a) Intracellular voltage of VD4 with four action potentials (holding voltage −60 mV).
(b) Intracellular voltage of LPeD1 with one postsynaptic action potential.
(c) Voltage at the capacitor beneath VD4 with three double-pulse stimuli (blowup).
(d) Transistor record of LPeD1 with the response to the postsynaptic action potential (blow up). The short transients in the transistor records are due to extracellular voltages beneath the neuron pair and to electrical cross talk on the chip.

animals. Specifically, a presynaptic tetanus in VD4 of 5 to 10 action potentials is able to enhance the amplitude of subsequent EPSPs, which generate postsynaptic spikes in LPeD1. In the neuron–chip system, potentiation is probed by the appearance of postsynaptic action potentials. First, as a control, a single-action potential is elicited in VD4 by a pair of voltage pulses applied to the capacitor. In this case, postsynaptic depolarization was not sufficient to elicit an action potential in LPeD1. Then a capacitive tetanus of six single voltage pulses is applied to VD4. To test for successful potentiation, again an action potential is elicited in VD4 a few seconds after the tetanus. That post-tetanic action potential in the presynaptic cell reproducibly causes a postsynaptic spike in LPeD1 that was recorded by the transistor [35].

12.3.4
Toward Defined Neuronal Nets

The function of neuronal networks is generally based on two features: (a) a mapping of a set of neurons onto another set or a mapping of a set of neurons onto itself as in the symmetrical Hopfield net illustrated in Figure 12.24. (b) Hebbian learning rules with an enhancement of synaptic strength as a consequence of correlated presynaptic and postsynaptic activity. Systematic experiments on network dynamics require (a) a noninvasive supervision of all neurons with respect to stimulation and recording to induce learning and to observe the performance of the net on a long time scale, and (b) a fabrication of neuronal maps with a defined topology of the synaptic connections. To achieve the second goal, we have to control the position of the neuronal cell bodies, the direction and bifurcation of grown neurites and the formation of synapses.

Chemical Guidance The motion of neuronal growth cones is guided by chemical patterns. Defined arborizations of leech neurons are achieved by chemical guidance with lanes of extracellular matrix protein [36–38]. They are fabricated by UV photolithography of a homogeneous film of extracellular matrix protein using metal masks. When a cell body is placed on the root of a treelike pattern with orthogonal branchings, the growth cone perfectly follows the lanes and is perfectly split at the branch points into daughter cones as illustrated in Figure 12.25a [37].

Using linear chemical patterns, we are able to guide the outgrowth of two neurons such that their growth cones are forced to collide and to form a synapse. This experiment

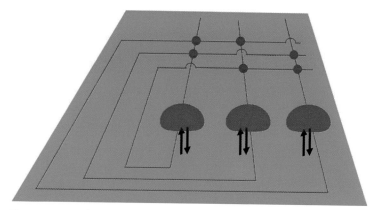

Fig. 12.24 Hypothetical defined neuronal network (red) with symmetrical connections of axons, synapses and dendrites supervised from a semiconductor chip (blue) by two-way interfacing (black).

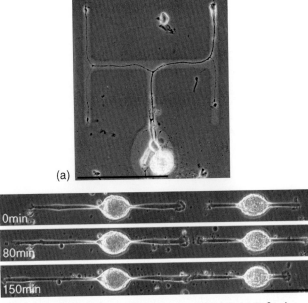

Fig. 12.25 Guided outgrowth by chemical patterns [37, 38]. (a) Defined bifurcations of a leech neuron. The pattern with orthogonal branchings is made by UV photolithography of extracellular matrix protein. Scale bar 100 µm. (b) Controlled formation of a synapse between two snail neurons. The lanes are made by UV photolithography of brain-derived protein adsorbed to the substrate. Scale bar 100 µm.

is performed with snail neurons [38]. The chip is coated with polylysine and incubated with dissected snail brains. Secreted proteins are adsorbed on the chip and patterned by UV photolithography. Three stages are illustrated in Figure 12.25b. After the encounter of the growth cones, synapse formation is checked as discussed above.

Though the process of chemical guidance is rather perfect, there are three problems: (a) the positions of cell bodies are not stable. Sprouting neurites exert strong forces on the cell bodies and may draw them along the chip surface such that interfacing to localized capacitors and transistors is broken. (b) The stability of neuritic trees is limited. Neurites have a tendency to shorten. When they are guided around a corner, they dissociate from the guiding lane and cross the nonguiding environment. (c) A patterned substrate provides a restricted area of growth, but it does not guide a neurite in a certain direction on a branched pattern. With several neurons on the patterns, it does not guide a certain neuron into a certain direction.

The first problem is illustrated in Figure 12.26. Several snail neurons are attached to two-way contacts in a circular arrangement. Within two days, a network of neurites is formed in the central area of the chip where the neurons are connected by electrical synapses. But the cell bodies are removed from their two-way contacts [41]. The mechanical instability of the arrangement dramatically lowers the yield of a

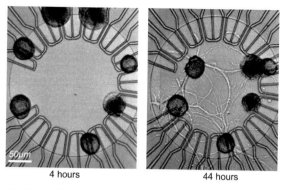

Fig. 12.26 Displacement of neuronal cell bodies by neuronal outgrowth [41]. Snail neurons are placed on a chip with a circular array of two-way contacts. The chip is coated with polylysine and the culture medium is conditioned with snail brains. (a) 4 hours after mounting; (b) 44 hours after mounting.

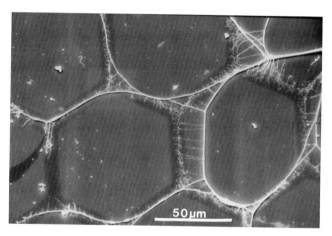

Fig. 12.27 Incomplete control by chemical guidance [36]. Electronmicrograph of the neurites of leech neurons. Scale bar 50 μm. A hexagonal pattern of extracellular matrix protein is made by UV photolithography. Grown neurites dissociate from the guiding lanes. Neurites grow without control on all given lanes of the pattern.

simultaneous interfacing of several neurons. The second and third problems are illustrated in Figure 12.27 with leech neurons on a hexagonal pattern [36]. Apparently, neuronal growth cones are perfectly guided by the lanes of extracellular matrix protein. But, grown neurites dissociate from the bent lanes such that the defined shape of the neuritic tree is lost. Different neurites grow on the same lane such that the neuritic tree is not defined by the pattern.

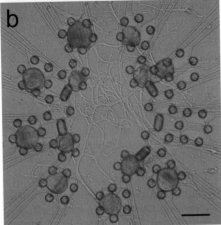

Fig. 12.28 Mechanically stabilized network of neurons on silicon chip [42]. (a) Electronmicrograph of snail neuron immobilized by a picket fence on a two-way contact after three days in culture. Scale bar 20 µm. (b) Micrograph of neuronal net with cell bodies (dark blobs) on a double circle of two-way contacts with neurites grown in the central area (bright threads) after two days in culture. Scale bar 100 µm. Pairs of pickets in the inner circle are fused to barlike structures.

Immobilization of Neurons The displacement of cell bodies is overcome by mechanical fixation. Picket fences are fabricated around each two-way contact by photolithography of a polyimide [42]. Neuronal cell bodies are inserted into the cages as illustrated in Figure 12.28a. They are immobilized even after extensive outgrowth. A chip with a network on a circular array is shown in Figure 12.28b. Mechanical immobilization allows an interfacing of more than one neuron in a net with sufficient probability. Two aspects of the technology are crucial: (a) the fabrication of the cages is a low-temperature process and does not interfere with the semiconductor devices; (b) the neurites grow on the same surface where the cell bodies are attached. No forces arise that lift the cell bodies from the two-way contacts.

Signaling Chip–Neuron–Neuron–Chip An experiment with a signaling pathway silicon–neuron–neuron–silicon (Figure 12.14) is shown in Figure 12.29 using a network of immobilized snail neurons [42]. Two neurons are selected, which are connected by an electrical synapse and which are placed on two-way junctions. A burst of seven pulses is applied to excite neuron 1 as checked with an impaled micropipette.

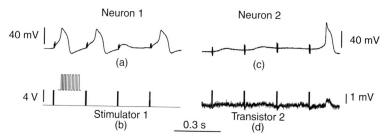

Fig. 12.29 Electronic interfacing of two synaptically connected snail neurons [42]. (a) Intracellular voltage of neuron 1. (b) Bursts of voltage pulses applied to the stimulator (seven pulses, amplitude 5 V, duration 0.5 ms). (c) Intracellular voltage of neuron 2. (d) Transistor record of neuron 2.

In neuron 2, we observe a subthreshold postsynaptic depolarization. A second burst of voltage pulses elicits another action potential in neuron 1 and leads to a further depolarization of neuron 2. After the third burst, which fails to stimulate neuron 1, the fourth burst gives rise to an action potential in neuron 1, which finally leads to a postsynaptic excitation in neuron 2. This postsynaptic action potential is recorded by the transistor underneath neuron 2, completing an electronically interfaced monosynaptic loop.

Correlated with the burst of stimulation pulses, we observe perturbations of the transistor records and of both microelectrode signals. These perturbations do not reflect actual changes of the voltage on the gate or of the intracellular voltage, respectively. Control experiments without neurons reveal a direct pathway of capacitive coupling through the chip from stimulators to transistors. Control experiments with an open stimulator and a neuron far away reveal a capacitive coupling to the micropipette through the bath. A subsequent depolarization of the cell is not observed in that case. Shape, delay and temporal summation of the postsynaptic signals correspond to the experiments with intracellular presynaptic stimulation. We conclude that the depolarization of the postsynaptic neuron in Figure 12.29 is induced by synaptic transmission and not by direct chip stimulation.

Topographical Guidance The instability of grown neuritic trees is overcome by topographical guidance. There, the grown neurites are immobilized by microscopic grooves that are used as cues for the guidance of the growth cones. Microscopic grooves are fabricated from a polyester photoresist on the chip [39]. An example of a topographical structure obtained by photolithography with five pits and connecting grooves is depicted in Figure 12.30a. The resin is compatible with cell culture, and the low temperature process does not damage microelectronic devices of the chip. The insert in Figure 12.30a shows the perfectly vertical walls of the pits and grooves. The chip is coated with polylysine. Secreted proteins from excised snail brains in the culture medium are adsorbed and render the surface of silicon dioxide on the bottom of grooves and pits as well as the surface of the polyester equally suitable for outgrowth. Cell bodies

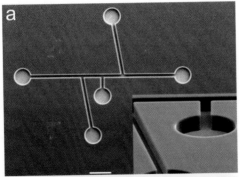

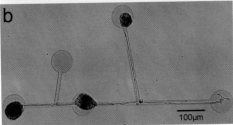

Fig. 12.30 Topographical guidance [39]. (a) Polyester structure on silicon chip with pits and grooves. Electronmicrograph, scale bar 100 μm. (b) Micrograph of a network of three snail neurons formed by topographical guidance. The neurons are connected by electrical synapses.

of snail neurons are placed into the pits of the polymer structure. Neurites grow along the grooves and are split at bifurcations of the groove to form neuritic trees as shown in Figure 12.30b. As a micrograph does not allow an assignment of the neurites to the different neurons, we sequentially stain individual neurons by injection with the dye Lucifer Yellow, which cannot pass the electrical synapses of snail neurons. The synaptic connections are checked by impaling with micropipettes as discussed above.

Electrical Guidance Topographical guidance does not solve the problem that a neuritic tree is not uniquely defined by the guiding pattern. We must combine it with control signals that are only effective at certain positions and certain times to promote or inhibit the growth at a crossing or at branch points and that induce or prevent synapse formation at certain places. Electrical manipulation of growth cones may be a tool to accomplish that task [40].

Conclusion In the next step, we envisage joining small networks with defined geometry made by mechanical guidance to silicon chips with two-way contacts [43]. It remains unclear, however, whether large neuronal nets with hundreds and thousands of neurons can be joined in a defined way by various methods of guidance. An alternative strategy must be kept in mind: a disordered neuronal net is grown on a chip with thousands of closely packed interface structures. Most neurons are on an interfacing contact at any time and the rearranging network is continuously supervised for its structure and dynamics.

12.4
Brain–Silicon Chips

Instead of culturing defined neuronal nets by artificially controlled outgrowth, we may use naturally occurring neuronal nets in mammalian brains. Considering the planar nature of semiconductor chips, planar networks are preferred in order to attain an adequate interfacing with the chip. Organotypic brain slices are particularly promising as they are only a few cell layers thick and conserve major neuronal connections. When we succeed in coupling a brain-grown net to numerous closely packed transistors and stimulation spots, we are able to study the distributed dynamics of the neuronal network.

An interfacing of a transistor or a stimulation spot with an individual neuron in neuronal tissue can hardly be achieved. Thus, with brain slices, we have to consider the stimulation and recording of local populations of neurons. The concept of a core-coat conductor of individual cells is no longer adequate. To guide the development of appropriate chips and to evaluate experimental data, we discuss a simple concept [44] that relates (a) neuronal currents in a slice to the extracellular potential that is recorded by transistors and (b) capacitive stimulation currents from a chip to the extracellular potential that elicits neuronal excitation from the tissue. On that basis, we discuss experiments with transistor recording and with capacitor stimulation in brain slices [45, 46].

12.4.1
Tissue-sheet Conductor

In an organotypic brain slice, the neurons are embedded in a tissue of about 70 µm thickness between the insulating silicon dioxide of a chip and an electrolytic bath on ground potential as illustrated in Figure 12.31a. Excited neurons are local sources or sinks of current that flows to adjacent regions of the tissue layer and to the bath. As a consequence, an extracellular potential appears in the tissue that may be recorded by transistors in the substrate. On the other hand, capacitive contacts in the substrate may locally inject current that flows to adjacent regions of the tissue and to the bath. The resulting extracellular potential may elicit neuronal excitation.

Volume Conductor In a tissue of densely packed neurons, we cannot consider the true extracellular voltage on a submicrometer level between the cells, but an average field potential V_{field} that arises from currents per unit volume j_{source} of cellular sources or j_{stim} of stimulation electrodes. The continuity relation of current in three dimension leads to (34), with the three-dimensional spatial derivative operator ∇ assuming an isotropic average specific resistance ρ of the tissue.

$$-\nabla \left(\frac{1}{\rho} \nabla V_{field} \right) = j_{stim} + j_{source} \qquad (34)$$

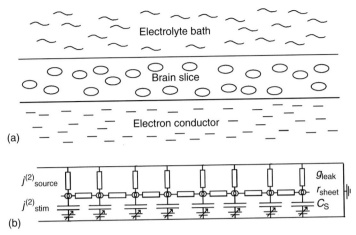

Fig. 12.31 Brain slice on silicon substrate [44]. (a) Geometry of tissue layer between electron conductor and electrolyte bath. (b) Sheet-conductor model. The neurons in the slice give rise to a current per unit area. The current flows along the slice (sheet resistance r_{sheet}) and to the bath (leak conductance g_{leak}). The slice is stimulated by current from the chip by capacitive contacts (specific capacitance c_S).

We consider two boundary conditions [44]: (a) the slice is on ground potential at its surface. With the z-direction normal to the layer plane and a height h of the slice, the constraint is $V_{field}(z = h) = 0$, when we neglect voltage drops in the bath. (b) At the substrate, a capacitive current per unit area $j_{stim}^{(2)}$ determines the gradient of the potential with $-(dV_{field}/dz)_{z=0} = \rho j_{stim}^{(2)}$. In principle, the field potential can be computed from (34) with the boundary conditions for an arbitrary pattern of stimulation electrodes and an arbitrary distribution of neuronal excitation. However, to get a simple picture of the electrical properties of a slice, we will use a two-dimensional approximation based on volume conductor theory.

Sheet-conductor Model A brain slice between chip and bath is a planar conductor with a capacitively coupled bottom and a leaky cover. Neglecting the z-dimension normal to the plane, we describe the thin tissue by a sheet resistance $r_{sheet}(x, y)$, the shunting effect of the bath by an ohmic conductance per unit area $g_{leak}(x, y)$, and the substrate by a capacitance per unit area $c_S(x, y)$ as illustrated by the circuit of Figure 12.31b. The neuronal current sources per unit area $j_{source}^{(2)}(x, y)$ and the stimulation current per unit area due to a changing voltage V_S at the capacitive contacts are balanced by the current along the sheet and by ohmic and capacitive shunting to the bath and the substrate according to (35) with the extracellular potential $V_{field}(x, y)$ and the two-dimensional spatial derivative operator ∇.

$$-\nabla \left(\frac{1}{r_{sheet}} \nabla V_{field} \right) + g_{leak} V_{field} + c_S \frac{\partial V_{field}}{\partial t} = j_{source}^{(2)} + c_S \frac{\partial V_S}{\partial t} \qquad (35)$$

The sheet-conductor model formulated by (35) describes (a) the extracellular field potential that arises from neuronal activity with $\partial V_S/\partial t = 0$, as it may be recorded with transistors, and (b) the extracellular field potential that is caused by capacitive stimulation with $j^{(2)}_{source}(x, y) = 0$, as it may elicit neuronal excitation.

Note the similarity and the difference between Figure 12.31b with (35) and Figure 12.4a with (1): in both cases, the circuits and the differential equations describe the continuity relation of electrical current in a two-dimensional system. However, Figure 12.4a and (1) refer to the junction of an individual cell and a chip, whereas Figure 12.31b and (35) describe a tissue sheet with numerous neurons on a chip.

12.4.2
Transistor Recording of Brain Slice

We culture a slice from rat hippocampus on a silicon chip with an all-oxide surface and a linear array of transistors as shown in Figure 12.32. The slice has a thickness $h = 70$ μm. It is stimulated with a tungsten electrode in the gyrus dentatus. A profile of evoked field potentials is recorded across the CA1 region [45].

Field Potential Two transistor records from stratum radiatum (layer of dendrites) and from stratum pyramidale (layer of cell bodies) are shown in Figure 12.33a. Excitatory postsynaptic potentials of neuronal populations are observed. There is a negative amplitude in the region of the dendrites where current flows into the cells and a positive amplitude in the region of the cell bodies with a compensating outward current. For comparison, the records of micropipette electrodes are plotted in Figure 12.33a. The transistor records have a similar shape, but a higher amplitude than the micropipette records. The amplitudes are proportional to each other as shown in Figure 12.33b. The similar shape and the similar and proportional amplitude validate the approach of transistor recording. Larger amplitudes are expected for a measurement near the insulating substrate where the shunting effect of the bath is smaller as compared to a measurement near the surface where the microelectrodes are positioned.

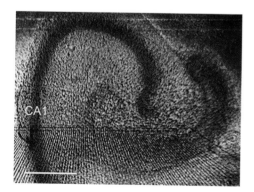

Fig. 12.32 Organotypic slice from rat hippocampus on silicon chip [45]. Nissl staining of a slice cultured for 14 days. Scale bar 400 μm. The dots are neuronal cell bodies. A linear array of field-effect transistors is aligned perpendicular to the CA1 region through stratum pyramidale and stratum radiatum to gyrus dentatus.

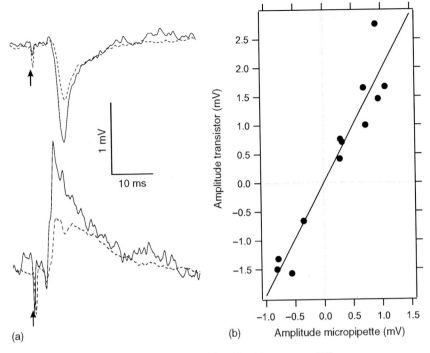

Fig. 12.33 Transistor recording in organotypic slice of rat hippocampus [45]. (a) Evoked field potentials in the stratum radiatum (top) and stratum pyramidale (bottom) of the CA1 region. The arrows mark the stimulation artifact. The dashed lines are the simultaneous records by micropipette electrodes. (b) Amplitude of transistor records versus amplitude of microelectrodes. The regression line has a slope of 1.96.

Potential Profile Across CA1 The amplitudes of the evoked field potentials across the CA1 region are plotted versus the position of the transistors in Figure 12.34 [45]. The region of negative potentials matches stratum radiatum; the region of positive potentials, stratum pyramidale.

We compare the experimental field potential with the sheet-conductor model. Along the CA1 layer of the hippocampus, the electrical activity is usually assumed to be constant. In that case, (35) can be reduced to a one-dimensional relation across the CA1 region along the soma-dendrite direction x. Without stimulation we obtain (36) with the length constant $\lambda_{\text{sheet}} = 1/\sqrt{g_{\text{leak}} r_{\text{sheet}}}$ when we neglect the capacitive current with $c_S dV_{\text{field}}/dt \ll g_{\text{leak}} V_{\text{field}}$.

$$-\lambda_{\text{sheet}}^2 \frac{d^2 V_{\text{field}}}{dx^2} + V_{\text{field}} = \frac{j_{\text{source}}^{(2)}}{g_{\text{leak}}} \tag{36}$$

We express the sheet resistance as $r_{\text{sheet}} = \rho/h$ by the specific resistance ρ of the slice and its thickness h and the leaks as $g_{\text{leak}} = 2/\rho h$ by the conductance from the center of the slice to the bath. At a specific resistance $\rho = 300\,\Omega\text{cm}$, the sheet resistance is

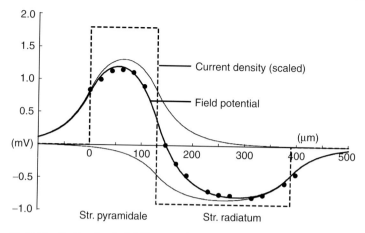

Fig. 12.34 Profile of evoked field potentials across the CA1 region [45]. The amplitude of voltage transients plotted versus position of the transistors (black dots). The data are fitted with the field potential V_{field} computed by the sheet-conductor model with a constant negative current density in the stratum radiatum and a balancing constant positive current density in the stratum pyramidale. The contributions from the two strata are drawn as thin lines. The scaled profile of current density $j^{(2)}_{source}/g_{leak}$ is indicated (dashed line).

$r_{sheet} = 43$ kΩ and the leak conductance is $g_{sheet} = 1000$ mS cm^{-2}. The length constant is $\lambda_{sheet} = h/\sqrt{2}$. For $h = 70$ μm we obtain $\lambda_{sheet} = 50$ μm.

The simplest current profile that is physiologically meaningful is a constant density of synaptic inward current in the stratum radiatum that is balanced by a constant outward current density in the stratum pyramidale. For a constant current-source density $j^{(2)}_{source}$ in a range $-x_0 < x < x_0$, the analytical solution of (36) is given by (37).

$$V_{field}(x) = \frac{j^{(2)}_{source}}{g_{leak}} \begin{cases} 1 - \exp\left(-\frac{x_0}{\lambda_{sheet}}\right) \cosh\left(\frac{|x|}{\lambda_{sheet}}\right) & |x| < x_0 \\ \sinh\left(\frac{x_0}{\lambda_{sheet}}\right) \exp\left(-\frac{|x|}{\lambda_{sheet}}\right) & |x| > x_0 \end{cases} \quad (37)$$

The width of stratum radiatum is about 260 μm, of stratum pyramidale about 130 μm. Using (37) we compute the potential profiles of the two regions and superpose them. We obtain a perfect fit of the experimental data as shown in Figure 12.34, when we choose scaled current densities $j^{(2)}_{source}/g_{leak} = -0.9$ mV in stratum radiatum and $j^{(2)}_{source}/g_{leak} = 1.8$ mV in stratum pyramidale. In the center of stratum radiatum, the curvature of the field potential $V_{field}(x)$ is small. Considering (36), there the potential reflects the synaptic current density with $V_{field} \propto j^{(2)}_{source}$. This result is in contrast to the volume conductor theory for bulk brain where the curvature is proportional to the local current-source density according to (35).

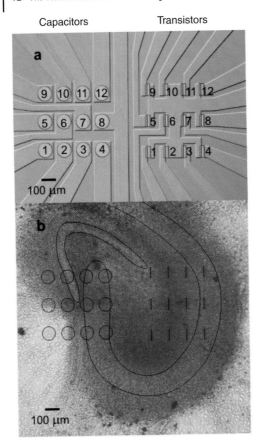

Fig. 12.35 Silicon chip for capacitive stimulation of cultured hippocampus slice [46]. (a) Micrograph of silicon chip with array of capacitors (diameter 100 μm) and array of field-effect transistors (gate 78 × 10 μm). Scale bar 100 μm. The reflection colors mark the different areas of the chip (blue: undoped chip, red: doped areas, green: thin oxide). (b) Micrograph of silicon chip with unstained hippocampus slice used for the representative measurements. The approximate position of CA1, CA3 and gyrus dentatus is indicated by thin lines.

12.4.3
Capacitive Stimulation of Brain Slices

We culture a slice from rat hippocampus on a silicon chip with two arrays of relatively large capacitors and of relatively large transistors as shown in Figure 12.35a [46]. The capacitors are insulated with a thin layer of titanium dioxide in order to achieve a high area-specific capacitance of $c_S = 1.8\,\mu\text{F cm}^{-2}$. The slice is arranged such that CA3 is stimulated by capacitors and the response of CA1 is recorded with transistors as illustrated in Figure 12.35b.

Capacitive Stimulation and Synaptic Transmission We apply triangular voltage pulses with a slow positive ramp and a fast downward step to capacitor No.7 in stratum pyramidale of CA3 (Figure 12.36a). We record with transistor No.2 in stratum radiatum of CA1 (Figure 12.36b). There, an evoked field potential appears with typical delay, shape and amplitude [46]. The amplitude of the records increases with the amplitude of the stimulus. The experiment shows that capacitive stimulation using a silicon chip

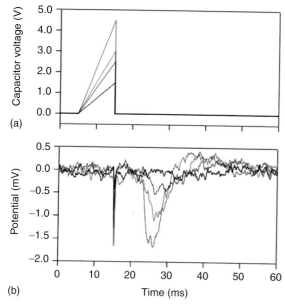

Fig. 12.36 Capacitor stimulation in CA3 and transistor recording in CA1 of organotypic hippocampus slice [46]. (a) Stimulation in CA3 by triangular voltage pulses of different amplitude applied to capacitor No.7. (b) Transistor records in CA1 with transistor No.2 in stratum radiatum.

is possible and that a complete pathway chip–neuron–synapse–neuron–chip can be implemented with brain tissue.

Extracellular Potential above Capacitor For a circular capacitor with radius a_S and area-specific capacitance c_S, the field potential $V_{\text{field}}(a)$ as a function of radius coordinate a is given by (38) when we disregard neuronal activity and assume $dV_{\text{field}}/dt \ll dV_S/dt$ [44].

$$-\lambda_{\text{sheet}}^2 \left(\frac{\partial^2 V_{\text{field}}}{\partial a^2} + \frac{1}{a} \frac{\partial V_{\text{field}}}{\partial a} \right) + V_{\text{field}} = \frac{c_S}{g_{\text{leak}}} \frac{\partial V_S}{\partial t} \tag{38}$$

For a ramp-shaped stimulus voltage $dV_S/dt = \text{const.}$ with amplitude ΔV_S and duration Δt_S applied to a capacitor of radius a_S, we obtain (39) with modified Besselfunctions I_0, I_1, K_0 and K_1.

$$V_{\text{field}}(a) = \frac{c_S}{g_{\text{leak}}} \frac{\Delta V_S}{\Delta t_S} \begin{cases} 1 - \frac{a_S}{\lambda_{\text{sheet}}} K_1\left(\frac{a_S}{\lambda_{\text{sheet}}}\right) I_0\left(\frac{a}{\lambda_{\text{sheet}}}\right) & a < a_S \\ \frac{a_S}{\lambda_{\text{sheet}}} I_1\left(\frac{a_S}{\lambda_{\text{sheet}}}\right) K_0\left(\frac{a}{\lambda_{\text{sheet}}}\right) & a > a_S \end{cases} \tag{39}$$

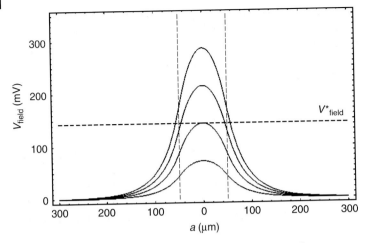

Fig. 12.37 Theory of capacitive stimulation [44], [46]. Potential profiles as a function of radius are plotted for voltage ramps with amplitudes $|\Delta V_S| = 1$ V, 2 V, 3 V, 4 V at a duration $\Delta t_S = 10$ μs applied to a circular capacitor of radius $a_S = 50$ μm with a specific capacitance $c_S = 1.8$ μF cm^{-2}. A threshold $|V^*_{\text{field}}| = 143$ mV is marked with a dashed line. The threshold of stimulation is $|\Delta V_S| = 2$ V. The area of neuronal excitation expands with increasing amplitude of the stimulus.

Some computed potential profiles $V_{\text{field}}(a)$ are plotted in Figure 12.37 for voltage ramps with amplitudes $|\Delta V_S| = 1$ V, 2 V, 3 V, 4 V and a duration $\Delta t_S = 10$ μs applied to a contact of radius $a_S = 50$ μm with specific capacitance $c_S = 1.8$ μF cm^{-2}. The slice is modeled by a thickness $h = 70$ μm, a specific resistance $\rho = 300$ Ωcm ($\lambda_{\text{sheet}} = 50$ μm, $r_{\text{sheet}} = 43$ kΩ, $g_{\text{leak}} = 1000$ mS cm^{-2}). The field potential V_{field} spreads beyond the capacitor. For a certain threshold V^*_{field} of extracellular stimulation, the area of excitation expands with increasing amplitude ΔV_S. When we assume that brain tissue is excited beyond a local field potential $|V^*_{\text{field}}| = 143$ mV, a critical stimulus voltage in the center of the capacitor $|\Delta V_S| = 2$ V can be calculated for the given system. Excitation is restricted to an area within the capacitor for $|\Delta V_S| < 3.4$ V and spreads beyond for larger stimuli.

Stimulus-response Relation The amplitude of the evoked field potential in CA1 depends on the amplitude of the voltage step applied to CA3 as shown in Figure 12.36. Beyond a threshold of 2 V for the stimulation voltage in CA3, the amplitude of the recorded field potential in CA1 increases [46]. Considering Figure 12.37, we explain this input–output relation as follows: When a downward voltage step is applied to the capacitor in CA3, the induced current across and along the slice gives rise to a negative extracellular potential profile $V_{\text{field}}(a)$. Nerve cells can be excited if the extracellular potential reaches a negative threshold V^*_{field} in the center of the capacitor. By further enhancement, the area where the potential is beyond the threshold expands. As a result, an increasing number of CA3 neurons are excited. Concomitantly, we may expect an increasing number of activated synapses in CA1. The increasing synaptic inward current in stratum radiatum gives rise to a larger field potential V_{field} in CA1 that is recorded by a transistor. Thus, Figure 12.36 reflects an increasing synaptic activation in CA1 that is caused by an increasing area of excitation in CA3. Neuronal excitation starts when the stimulus ΔV_S in the center $a = 0$ of the capacitor gives rise to an extracellular threshold potential V^*_{field} according to (40).

$$V^*_{field} = \frac{c_S}{g_{leak}} \frac{\Delta V^*_S}{\Delta t_S} \left[1 - \frac{a_S}{\lambda_{sheet}} K_1 \left(\frac{a_S}{\lambda_{sheet}} \right) \right] \qquad (40)$$

We fit (40) to the experimental onset of stimulation $|\Delta V^*_S| = 2$ V. At given parameters $a_S = 50$ μm and $\lambda_{sheet} = 50$ μm we obtain a threshold $|V^*_{field}| = 143$ mV, as used to draw Figure 12.37. It is in a proper order of magnitude. The area of excitation as a function of the stimulus $|\Delta V_S|$ is obtained by inverting (39). The result is plotted in Figure 12.38a. Above threshold, excitation is well restricted within the capacitor. For example, at

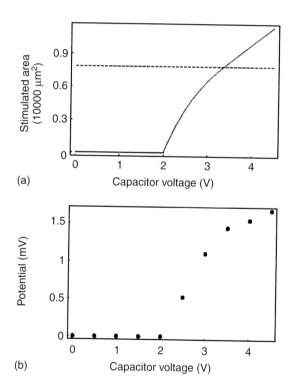

Fig. 12.38 Transistor record in CA1 and capacitor stimulus in CA3 [46]. (a) Theory (sheet-conductor model). Area of neuronal excitation in CA3 versus amplitude of voltage step applied to a capacitor. The area of the capacitor is marked by a dashed line. (b) Experiment. Amplitude of evoked field potentials in CA1 recorded by transistor versus amplitude of downward voltage step applied to capacitor in CA3.

$\Delta V_S = -2.5$ V the radius of excitation is $a = 35$ µm. Excitation spreads beyond the capacitor for higher stimuli.

The recorded field potential V_{field} in CA1 is proportional to the synaptic current $j_{\text{source}}^{(2)}$, that is, it depends on the number of excited neurons in CA3. We may expect that the field potential in CA1 is proportional to the area of excitation in CA3. In fact, the amplitude of the transistor records in CA1 as a function of presynaptic capacitor stimulation as plotted in Figure 12.38b resembles the computed function of the excitable area in CA3 (Figure 12.38a) [46]. The relative scaling of the two plots depends on the strength of synaptic coupling through the Schaffer collaterals and the properties of CA1 tissue. For $2x_0 = 130$ µm, $\lambda_{\text{sheet}} = 50$ µm and $g_{\text{leak}} = 1000$ mS cm^{-2}, we derive from an experimental field potential $V_{\text{field}} = -1$ mV at a stimulus $\Delta V_S = -2.5$ V a synaptic current density $j_{\text{source}}^{(2)} = -14$ µA mm^{-2}. It is in a proper order of magnitude.

12.5
Summary and Outlook

The present chapter shows that the basic questions on the electrical interfacing of individual nerve cells and semiconductor chips are fairly well answered, the nature of the core-coat conductor, the properties of the cleft, the role of accumulated ion channels, the mechanism of transistor recording and capacitive stimulation. With respect to the latter issue, however, studies on the capacitive gating of ion channels and on electroporation on planar stimulation contacts are required. In the near future, we are faced with important steps of optimization: (a) For the semiconductor, the capacitance of the stimulation contacts must be enhanced, and the noise of the transistors has to be lowered. (b) For the neurons, the structural and electrical properties of the cell membrane (glycocalix, ion channels) in the neuron–semiconductor junction must be studied and optimized by recombinant methods.

With respect to systems of neuronal networks and digital microelectronics, we are in a rather elementary stage. Two directions may be envisaged in the future: (a) small defined networks of neurons from invertebrates and mammals must be created with learning chemical synapses and with a defined topology of synaptic connections. (b) Large neuronal nets may be grown on closely packed arrays of two-way interface contacts such that the rearranging structure and dynamics of the net is under continuous control of the chip. For that direction of research, an adaptation of the industrial standard of CMOS technology is crucial [47, 48].

With respect to the interfacing of brain slices, two directions will be followed: (a) Arrays of two-way contacts will lead to a complete spatiotemporal mapping of brain dynamics. (b) Learning networks on a chip will be implemented and will allow systematic studies of memory formation. Here, probing with large arrays of transistors and capacitors of CMOS chips will also be important.

The availability of involved integrated neuroelectronic systems will help unravel the nature of information processing in neuronal networks and will give rise to new and fascinating physical–biological–computational questions. Of course, visionary dreams of bioelectronic neurocomputers and microelectronics neuroprostheses are unavoidable and exciting, but they should not obscure the numerous practical problems.

Acknowledgments

The work reported in this chapter was possible only with the cooperation of numerous students who contributed with their skill and enthusiasm. Most valuable was also the help of Bernt Müller (Institute for Microelectronics, Technical University Berlin), who fabricated several chips and gave his advice for our own chip technology. The neurochip project was generously supported by the University Ulm, the Fonds der Chemischen Industrie, the Deutsche Forschungsgemeinschaft, the Max-Planck-Gesellschaft and the Bundesministerium für Bildung und Forschung.

References

1 P. Fromherz, 20th Winter Seminar "Molecules, Memory and Information", Klosters, **1985** (available at http://www.biochem.mpg.de/mnphys/).
2 P. Fromherz, A. Offenhäusser, T. Vetter, J. Weis, Science, **1991**, 252, 1290–1293.
3 P. Fromherz, A. Stett, Phys. Rev. Lett., **1995**, 75, 1670–1673.
4 A. Lambacher, P. Fromherz, Appl. Phys. A, **1996**, 63, 207–216.
5 D. Braun, P. Fromherz, Appl. Phys. A, **1997**, 65, 341–348.
6 D. Braun, P. Fromherz, Phys. Rev. Lett., **1998**, 81, 5241–5244.
7 Y. Iwanaga, D. Braun, P. Fromherz, Eur. Biophys. J., **2001**, 30, 17–26.
8 A. Lambacher, P. Fromherz, J. Opt. Soc. Am. B, **2002**, 19, 1435–1453.
9 G. Zeck, P. Fromherz, Langmuir, **2003**, 19, 1580–1585.
10 P. Fromherz, C.O. Müller, R. Weis, Phys. Rev. Lett., **1993**, 71, 4079–4082.
11 R. Weis, P. Fromherz, Phys. Rev. E, **1997**, 55, 877–889.
12 R. Weis, B. Müller, P. Fromherz, Phys. Rev. Lett., **1996**, 76, 327–330.
13 V. Kiessling, B. Müller, P. Fromherz, Langmuir, **2000**, 16, 3517–3521.
14 P. Fromherz, V. Kiessling, K. Kottig, G. Zeck, Appl. Phys. A, **1999**, 69, 571–576.
15 V. Kiessling, P. Fromherz, in preparation.
16 H. Ephardt, P. Fromherz, J. Phys. Chem., **1993**, 97, 4540–4547.
17 B. Kuhn, P. Fromherz, J. Phys. Chem., **2003**, 107, 7903–7913.
18 A. Lambacher, P. Fromherz, J. Phys. Chem., **2001**, 105, 343–346.
19 D. Braun, P. Fromherz, Phys. Rev. Lett., **2001**, 86, 2905–2908.
20 D. Braun, P. Fromherz, Biophys. J., **2004**, 87, 1351–1359.
21 R. Gleixner, P. Fromherz, in preparation.
22 S. Vassanelli, P. Fromherz, J. Neurosci., **1999**, 19, 6767–6773.
23 B. Straub, E. Meyer, P. Fromherz, Nat. Biotechnol., **2001**, 19, 121–124.
24 M. Brittinger, P. Fromherz, J. Phys. Chem., submitted.
25 M. Ulbrich, P. Fromherz, Phys. Rev. Letters, submitted.
26 P. Fromherz, Eur. Biophys. J., **1999**, 28, 254–258.
27 M. Jenkner, P. Fromherz, Phys. Rev. Lett., **1997**, 79, 4705–4708.
28 R. Schätzthauer, P. Fromherz, Eur. J. Neurosci., **1998**, 10, 1956–1962.
29 S. Vassanelli, P. Fromherz, Appl. Phys. A, **1997**, 65, 85–88.
30 S. Vassanelli, P. Fromherz, Appl. Phys. A, **1998**, 66, 459–464.
31 M. Völker, P. Fromherz, Small, submitted.

32 A. Stett, B. Müller, P. Fromherz, *Phys. Rev. E*, **1997**, *55*, 1779–1782.
33 M. Ulbrich, P. Fromherz, *Adv. Mater.*, **2001**, *13*, 344–347.
34 P. Bonifazi, P. Fromherz. *Adv. Mater.*, **2002**, *14*, 1190–1192.
35 R.A. Kaul, N.I. Syed, P. Fromherz, *Phys. Rev. Lett.*, **2004**, *92*, 038102/1–4.
36 P. Fromherz, H. Schaden, T. Vetter, *Neurosci. Lett.*, **1991**, *129*, 77–80.
37 P. Fromherz, H. Schaden, *Eur. J. Neurosci.*, **1994**, *6*, 1500–1504.
38 A.A. Prinz, P. Fromherz, *Biol. Cybern.*, **2000**, *82*, L1–L5.
39 M. Merz. P. Fromherz, *Adv. Mater.*, **2002**, *14*, 141–144.
40 S. Dertinger, P. Fromherz, in preparation.
41 M. Jenkner, B. Müller, P. Fromherz, *Biol. Cybern.*, **2001**, *84*, 239–249.
42 G. Zeck, P. Fromherz, *Proc. Natl. Acad. Sci. U.S.A.*, **2001**, *98*, 10457–10462.
43 M. Merz, P. Fromherz, *Adv Funct Material*, submitted.
44 P. Fromherz, *Eur. Biophys. J.*, **2002**, *31*, 228–231.
45 B. Besl, P. Fromherz, *Eur. J. Neurosci.*, **2002**, *15*, 999–1005.
46 M. Hutzler, P. Fromherz, *Eur. J. Neurosci.*, **2004**, *19*, 2231–2238.
47 B. Eversmann, M. Jenkner, C. Paulus, F. Hofmann, R. Brederlow, B. Holzapfl, P. Fromherz, M. Brenner, M. Schreiter, R. Gabl, K. Plehnert, M. Steinhauser, G. Eckstein, D. Schmitt-Landsiedel, R. Thewes, *IEEE J. Solid State Circ.*, **2003**, *38*, 2306–2317.
48 A. Lambacher, M. Merz, R.A. Kaul, M. Jenkner, B. Eversmann, F. Hoffmann, R. Thewes, P. Fromherz, *Appl. Phys. B*, **2004**, *79*, 607–611.
49 P. Fromherz, *Ber. Bunsen-Ges. Phys. Chem.*, **1996**, *100*, 1093–1102.
50 P. Fromherz, *Phys. Blätter*, **2001**, *57*, 43–48.
51 P. Fromherz, *Chem. Phys. Chem.*, **2002**, *3*, 276–284.
52 P. Fromherz, in *Electrochemical Microsystems Technology* (Eds.: J.W. Schultze, T. Osaka, M. Datta), Taylor & Francis, Andover, **2003**, 541–559.
53 P. Fromherz, *Physica*, **2003**, *16*, 24–34.
54 P. Fromherz, in *Nanoelectronics and Information Technology* (Ed.: R Waser), Wiley-VCH, Weinheim, **2003**, 783–805.

13
S-Layer Proteins in Bioelectronic Applications

Stefan H. Bossmann

13.1
Introduction

According to the generl definition by Willner and Katz, "the basic feature of a bioelectronic device is the immobilization of a biomaterial onto a conductive support, and the electronic transduction of the biological functions associated with the biological matrices." [1]. In the meantime, the classic definition of biomaterials includes, among other systems, the immobilization and application of various redox enzymes, receptors, antibodies or antigens, DNA – single and double strands of various lengths [1]. All that these systems have in common is that their biological functions were triggered or transduced by various electronic or electrochemical signals. The application of S-layer proteins in bioelectronic applications must be regarded as an extension of this definition because S-layer proteins possess no functions that can be triggered electronically or electrochemically. However, recent advances in genetic engineering are very promising with regard to the future integration of specific functions, such as supramolecular hosts and building blocks of bioelectronic devices into S-layer lattices [2–5]. Various enzymes, such as glucose oxidase [6], have already been incorporated into S-layer protein matrices.

One of the most striking advantages of employing S-layer proteins is their ability to form highly ordered lattices at various surfaces, thus permitting the construction of extended nanopatterned surfaces. Depending on the biological nature of the S-layer protein used and eventual tailored chemical modifications, some of their lattices form a regular pattern of nanopores at electrode surfaces. These pores can be used as insulating templates [7] for the synthesis of metal nanoparticles and nanorods (nanowires). This approach offers a straightforward pathway toward the design of "hybrid bioelectronic devices", combing the advantages of typical supramolecular nanosystems [8] and bioelectronic systems [9]. In these "hybrid systems", protein-embedded metal nanoparticles can be used to perform electrochemical functions, thus mimicking the functions of enzymes. This approach is commonly referred to as *biomimicry* [10]. However, its application to S-layer protein lattices is only in the beginning stages.

Bioelectronics. Edited by Itamar Willner and Eugenii Katz
Copyright © 2005 WILEY-VCH Verlag GmbH & Co. KGaA, Weinheim
ISBN: 3-527-30690-0

13.1.1
Upcoming Nanotechnology Applications

The domain of advanced nanotechnology is unquestionably the most promising upcoming technology [11]. Its recent developments offer future applications in very different areas, such as electron tunneling (superconduction) at feasible temperatures [12], (photo)electrocatalysis [13, 14] or the development of quantum computers [15]. To date, the most developed nanosystems for quantum applications consisted of doped carbon nanotubes [16]. However, this particular template material, which contains a high degree of sp^2 carbon centers, leads inevitably to at least some degree of electrical connection between the embedded nanoparticles. Therefore, possible quantum effects could be cloaked by unwanted electrical connection of the individual quantum dots.

The use of stable protein (S-layer) lattices as nanotemplates at electrode surfaces for the stabilization of nanoparticles and nanowires would provide an electrically insulating matrix for the electrosynthesis [17] of metal nanoparticles or nanowires. Furthermore, protein matrices possess the ability of self-organization under *in vitro* conditions. The aptitude of self-organization of S-layer proteins is very close to ideal [18, 19].

In this article, an approach is presented to describe the most important applications of S-layer proteins toward the field of bioelectronics, as defined in a very broad sense. In particular, the following topics will be dealt with:

2. S-Layer Proteins and Porins.
3. Experimental Methods Developed for Hybrid Bioelectronic Systems.
4. Applications of S-Layer Proteins at Surfaces.
5. Molecular Nanotechnology Using S-Layers.
6. Immobilization and Electrochemical Conducting of Enzymes in S-Layer Lattices (Sensor-Applications).

13.2
S-layer Proteins and Porins

This review is concerned with the development of possible applications of bacterial proteins. S-layer proteins form two-dimensional crystalline surface arrays and have been identified in hundreds of different species from nearly every taxonomic group of walled Eubacteria and are also present in many archeabacteria [20]. S-layer proteins from numerous species were isolated and characterized and are principally available for technical applications. Porins are channel proteins in bacterial outer membranes and allow the diffusion of small and hydrophilic compounds [21–24]. In contrast to the S-layer proteins, various porins have been intensively studied with respect to their biophysical properties [25–27]. Especially the mechanisms and the triggering factors of the channel opening are of great interest because of possible applications of ion- and diffusive channels [28–32]. Furthermore, the crystal structures of some porins of gram-negative bacteria have been solved, and the diffusion of ions through porin channels have been computer-simulated [33]. Channel porins from gram-positive bacteria and especially from mycobacteria [34] are of special interest with respect to their application in nanotechnology because of their unique channel lengths [35]. However, the generation of well-defined, highly ordered lattices of porins has failed up to this time. The incorporation of porins and other proteins of interests (such as enzymes [6], or S-layer streptavidin fusion proteins [36]) into S-layer lattices offers a promising pathway to combine both advantages and to generate well-ordered and functional bioelectronic nanostructures. Since the structures (and especially also the biological functions) of S-layer proteins have been optimized in evolution, we are able to use these versatile biological tools directly instead of mimicking their structures and working principles in artificial membranes and supramolecular systems.

Nevertheless, it must be noted that the evolution of S-layer proteins was – obviously – not directed toward any applications in nanotechnology but rather headed, among other purposes, for the optimized stabilization and protection of the bacterial cell wall. It is apparent that the chemical stability of bacterial proteins (S-layer subunits and porins) against thermal and photochemical decomposition reactions must be sufficient in order to permit successful application in the field of nanotechnology [37].

13.2.1
The Building Principles of Tailored S-layer Proteins Layers

The surfaces of many bacteria are covered by two-dimensional crystalline arrays of proteins or glycoprotein subunits, which are commonly referred to as "S-layers" [20, 38]. Consequently, S-layer proteins are among the most prominent proteins in prokaryotes comprising up to 15% of the total cellular protein mass [20]. Substantial knowledge has been accumulated in the last decade on the structure, assembly principles, biochemistry and genetics of S-layers. Slowly, their natural function principles are also being revealed, permitting the use of S-layer proteins in many future applications.

S-layers are thought to function as molecular sieves, protective or adhesive coatings or ion traps. Furthermore, in all bacteria that possess S-layers as exclusive cell wall components, the cell shape is determined by the (glyco)proteins [39]. Whereas the S-layer proteins of gram-negative Archaebacteria and of gram-negative Eubacteria are not chemically linked to their cytoplasmic or outer membranes, respectively, the S-layer proteins of at least some gram-positive eu- and Archaebacteria are indeed chemically associated to macrobiomolecules situated underneath the outer membrane as shown in Figure 13.1 [6]. It must be noted that considerable variations in structure, complexity and lattice symmetry of S-layer proteins exist, allowing very different kinds of applications. However, many S-layers are similar with respect to their chemical composition, formation principles, biophysical properties and molecular size distribution.

The thickness of S-Layers ranges from 5 to 30 nm [6, 39]. The outer surface often appears to be smooth, whereas the inner surface is more corrugated [40, 41]. Structural investigations of S-layers *in vivo* and after recrystallization on various substrates have been performed using a combination of very different techniques, such as metal-shadowing, negative staining, ultrathin-sectioning, freeze-etching, scanning probe microscopy experiments and underwater atomic force microscopy [20]. The

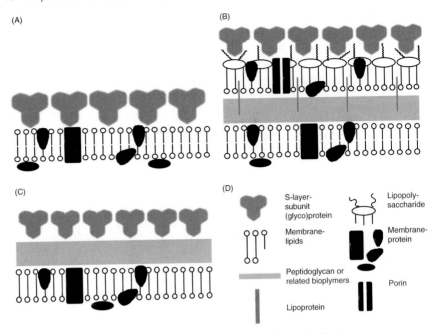

Fig. 13.1 The structures of bacterial cell walls. (A) gram-negative Archaebacteria; (B) gram-negative Eubacteria; (C) gram-positive Archae- and Eubacteria (according to [20]); (D) Legend (with permission from [40]).

structural information available reveals information down to the subnanometer range. Consequently, the formation principles of the two-dimensional S-layer protein arrays and their self-repeating structures are principally known and can be applied on various substrates (see below) [42]. S-layer lattices, which form as arrays of protein or glycoprotein subunits, possess either oblique (p1, p2), square (p4) or hexagonal (p3, p6) symmetry as shown in Figure 13.2. The symmetry class, in which the S-layer proteins crystallize, is strongly dependent on the bacterial species; however, the self-repeating two-dimensional crystals of S-layers at artificial surfaces do not always follow the structure of the naturally observed pattern. Obviously, the experimental conditions determine whether the formation of well-ordered lattices at surfaces occurs spontaneously or not. One of the results from the structure elucidation of S-layers is that they form highly porous mesh works with porosity in the range between 30 and 70%. The assemblies of identical protein or glycoprotein subunits possess masses in the range of 35 000 to 230 000 Dalton and exhibit pores of identical size and morphology. Many S-layers feature two or even more distinct classes of pores in the range of 2 to 8 nm. The subunits of most S-layers interact with each other during the process of S-layer formation through noncovalent forces, including hydrophobic interactions, hydrogen bonds, ionic bonds and the interaction of polar groups [39, 41]. Most striking is the fact that very large surface regions ($>10\,000\,\mu m^2$) can be easily patterned.

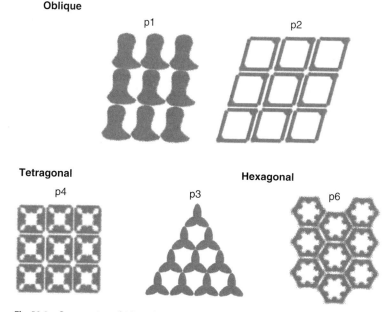

Fig. 13.2 Symmetries of S-layer lattices (schematic representations) with permission from [40].

The overall chemical composition of S-layer (glyco)proteins from all phylogenetic branches was found to be surprisingly similar [39, 41]. The typical S-layer is composed of an acidic protein or glycoprotein, possessing an isoelectric point in the region from pH 3 to 5. Since the isoelectric point of a protein results from the superposition of the isoelectric points of those amino acids that are exposed to the aqueous electrolyte, it is very interesting to take a look at the chemical composition. Typical S-layer proteins have high amounts of glutamic and aspartic acid. Together, they resemble up to 15 mol%. The lysine content of S-layer proteins is in the range of approximately 10 mol%. Therefore, many S-layer proteins are very suitable biotemplates for the defined binding and subsequent electroreduction of more or less noble metal nanoparticles (for instance, *Cu, Ni, Au, Pd, Pt*). Note that ionic amino acids make up about a quarter of the total composition of S-layer proteins, leading to a relatively high ionic strength in the direct vicinity of the S-layer (nano)structures at surfaces. This finding indicates clearly the importance of ionic interactions during the formation of the S-layer lattice structure. Furthermore, it facilitates the electrodeposition of nanoparticles within the S-layer protein structures. It should be noted that the content of sulfur-containing amino acids in the S-layer proteins is close to zero. The main fraction of amino acids of S-layer proteins is hydrophobic in nature. It is even more interesting that hydrophobic and hydrophilic amino acids do not form extended clusters but alternating segments with a more hydrophilic region at the terminal N-end. Information regarding the secondary structures of S-layer proteins was obtained by using circular dichroism measurements,

FT-IR-spectroscopy and especially from secondary structure prediction based on the available protein sequence data [43]. In a typical S-layer, up to 70% of the proteins form α-helices and 20% form β-sheets. Aperiodic foldings and β-turn contents account for most of the missing fraction. In many of the S-layers, the N-terminal regions, which are among the parts of the protein sequence self-organized as α-helices, are able to recognize distinct structures in the underlying cell envelope [44]. Because of the noncovalent nature of the bonds between the subunits, which are hydrophobic as well as ionic in nature, the growth of the S-layer, which surrounds the whole bacterium in most of the investigated cases, proceeds very rapidly. One can approximate that a coherent S-layer at the surface region of an average cell is formed by the supramolecular interaction of approximately 5×10^5 protein or glycoprotein subunits. For S-layered bacteria with a generation time of 20 min, approximately 500 copies of the S-layer subunits have to be synthesized inside the cell, translocated, and integrated into the S-layer per second [20]. Because of their unusual building mechanisms and dynamics, S-layers can be described as "dynamic closed surface crystals", with the intrinsic capability of assuming a structure of low free energy during cell growth and also during their growth and assembling processes at membranes and artificial surfaces.

13.2.2
Chemical Modification of S-layers [45]

The cross-linking procedure using glutaraldehyde produces a net negative charge. Carbodiimide activation offers the opportunity of linking carboxylic acids to a manifold of chemical reagents and groups. Thus, either a strong net negative charge or a strong positive net charge can be achieved. Also, the hydrophobicity of SUMs can be modulated following the same experimental approach. It is of special importance that in the quoted work, the distinct geometry of the self-repeating S-layer lattice did not change noticeably when the free carboxylic acids are chemically modified [40]. S-layer glycoproteins offer the same opportunities via the activation of their glycan chains either by periodate oxidation or by cyanogens-bromide activation. A recent example of this versatile strategy was the application of S-layer-coated liposomes for entrapping and binding of functional molecules:

The S-layer protein from *Bacillus stearothermophilus* PV727P2 was recrystallized on positively charged unilamellar liposomes of an average diameter of 180 nm formed by dipalmitoyl phosphatidylcholine (DPPC), cholesterol, and hexadecylamine (HDA) (molar ratio: 10/5/4) [46]. The S-layer subunits possess the molecular weight of 97 000 and form a two-dimensional lattice of oblique symmetry (see Figure 13.3). Cross-linking was achieved using glutaraldehyde. Then, the free carboxylic groups were activated with 1-ethyl-3-(dimethylaminopropyl)-carbodiimide (EDC). The advantage this reagent has is that it can be successfully employed in aqueous solutions also. The cross-linking process could be observed employing S-layer-coated carbonic anhydrase containing liposomes (SCALs). Whereas the cross-linking reaction proceeded exclusively between adjacent S-layer subunits within the two-dimensional lattice at the outer surface of the liposomes, the enzyme carbonic anhydrase, located inside the liposomes, was

Fig. 13.3 Oblique S-layer lattice of *Bacillus stearothermophilus* (adapted from [20] and [40], with permission from [40]).

not affected by the chemical treatment as it could be demonstrated using SDS-polyacrylamide gel electrophoresis (SDS-PAGE). When the cross-linking of the S-layer subunits was performed using Bis(sulfosuccinimidyl)suberate (BS), evidence was found for heterologous cross-linking between the S-layer protein and membrane-incorporated HDA.

13.2.3
Interaction by Noncovalent Forces

The supramolecular as well as the highly specific binding of molecules and polymers to S-layer lattices can be also achieved by using noncovalent forces. Important factors determining the nature and the selectivity of the binding are the S-layer lattice type, the geometric sizes (or the van-der-Waals radii, respectively) of the morphological units, and the chemical nature, the physicochemical properties and the distribution of binding sites on the array. An example of this binding type is the binding of polycationic ferritin (PCF) on the S-layer lattice of *Bacillus coagulans* E38-66 [47].

The hexagonally ordered S-layer lattices from *Bacillus stearothermophilus* PV72 and *Thermoanaerobacter thermohydrosulfuricus* L111-69 [48] and the square S-layer lattice from *Bacillus sphaericus* CCM 2120 can bind anionic, cationic, zwitterionic and noncharged macromolecules. Note that only the surface of *T. thermohydrosulfuricus* is glycosylated [20].

13.3
Experimental Methods Developed for Hybrid Bioelectronic Systems

In this chapter, selected emerging techniques for the characterization and tailoring of protein-layer interfaces are described. They all have in common that they permit the elucidation of particular structure elements or functions in the nanoscale. These techniques should generally be supported by more conventional techniques, such as FT-IR [49], cyclic voltammetry [50, 51], impedance spectroscopy (IS) [52], surface plasmon resonance (SPR)[53], and contact angle measurements [54]. The intelligent application of these more conventional techniques is vital for the construction and operation of bioelectronic devices.

13.3.1
Electron Microscopy

The analysis of the two- and three-dimensional nanostructures can be performed using REM (raster electron microscopy) [55] and TEM (transmission electron micrography) [56]. Whereas AFM permits a highly resolved spatial resolution (see below), TEM is the method of choice for the analysis of extended nanostructured surfaces. The analysis of biological structures by means of electron microscopy can be enhanced using mathematical tools. The software package IMAGE, generously provided by the National Institute of Health (USA), provides detailed insight in combination with a calibration procedure employing commercially available PAMAM-starburst dendrimers [56].

13.3.2
Combined X-Ray and Neutron Reflectometry [57]

The joint refinement of X-ray and neutron reflectometry data [56] offered an unprecedented insight into the structural details of the interaction of an S-layer structure coupled to a phospholipid monolayer. The reconstitution of the S-layer protein isolated from B. sphaericus CCM2177 at dipalmitoylphosphatidylethanolamine (DPPE) was studied in detail. Particular emphasis has been made to elucidate the protein–lipid interface: A strong interaction of the S-layer protein with the lipid headgroups was observed. However, the interaction of the protein is essentially confined to the region of the lipid headgroup. The peptide interpenetrates the phospholipid headgroups almost in its entire depth but does not affect the hydrophobic lipid chains of DPPE. Coupling to S-layers does not reduce the chain order in the neighboring lipid (mono)layer. Interaction with the S-layer protein reduced the headgroup hydration by approximately 40%. The imaging of the spatially resolved electron densities, the interconnection principles of the S-layer proteins and the adjacent lipids and the local mobility of the functional groups, which form functional S-layer-assemblies, is of profound importance in understanding the working principles and their further development.

13.3.3
Atomic Force Microscopy Using Protein-functionalized AFM-cantilever Tips

Atomic force microscopy (AFM) has been successfully used for studying numerous nanostructured systems at high resolution [58]. Furthermore, AFM was successfully employed for the measurement of interfacial forces as shown in Figure 13.4 [59], intermolecular [60] and intramolecular interactions [61]. The design of tailored S-layer devices will crucially depend on the spatially resolved elucidation of adhesive forces and especially ligand-receptor interactions [62]. An especially promising approach consists of the functionalization of an AFM-cantilever with suitable receptor molecules or protein. The work that is quoted here employed the well-known biotin–avidin system [58], but principally, any other host–guest system of interest could be applied here. This approach will offer the experimental possibility of probing different sites of nanostructured S-layer lattices. The understanding of the particular physical properties of these reactive sites or pore structures – in contrast to the surrounding S-layer matrices featuring no particular chemical or electrochemical functions – will lead to a detailed mechanistic understanding of the interaction between hydrophobic and polar forces, hydrogen bonding effects and tailored host–guest interactions.

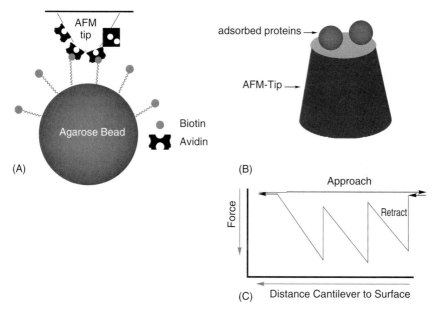

Fig. 13.4 (A) Schematic representation of an avidin-functionalized atomic force microscope (AFM) tip and the biotinylated agarose bead according to reference [58]. Biotin is covalently linked to the elastic agarose filaments of the bead. (B) Schematic representation of an atomic force microscope with adsorbed large proteins or protein aggregates on its surface. (C) Typical AFM force-extension curve.

13.3.4
Scanning Electrochemical Microscopy

The working principle of the scanning electrochemical microscope (SECM) combines the advantages of AFM-spectroscopy with the spatially resolved information about the electrochemical performance of building components of bioelectronic devices, such as redox enzymes and metal nanoparticles and nanowires. In addition to the elucidation of already built bioelectronic devices, SECM will further allow the construction of nanodevices, such as the on-site reduction of nanoparticles within the pore structures of S-layer protein lattices or the regular nanopattern of protein-containing surfaces [63].

Most SECM measurements involve steady state current measurements. However, transient current measurements have also been performed and have provided vital information about the kinetics of homogeneous reactions and time-dependent systems [64]. The SECM transient response has previously been simulated and experimentally observed for planar electrodes, microdisks, and thin-layer cells over a wide time range [65]. When the tip radius is known, transient current measurements principally permit the determination of diffusion coefficients without knowledge of solution concentration and the number of electrons transferred [66]. The measurement of transient currents allows the investigation of a diversity of membrane structures, which differs in their diameters as well as their biological/chemical structures. An important contribution, with respect to elucidation of nanopore systems using SECM, is the investigation of ion-transport (K^+ and Tl^+) through gramicidin D channels in a horizontal bilayer lipid membrane [64, 67, 68]. Although chemically different, ion channels and nanopores formed by S-layer lattices have several important features in common: (a) they possess similar pore diameters and length; (b) binding/transport-selectivities for selected cations are observed; and (c) site-directed mutation allows – principally – the tailoring of the pores for a variety of organic, metal-organic or inorganic host structures.

13.4
Applications of S-layer Proteins at Surfaces

The physicochemical properties and especially the permeability of S-layer cell envelopes are of profound interest because they offer (a) an insight into the function principles of the bacterial transport mechanisms and (b) the possibility of designing permeability filters for nano- and biotechnological applications [20, 69].

13.4.1
S-layer Proteins as Permeability Barriers

S-layer proteins [70] from mesophilic and thermophilic *Bacillaceae* were studied in detail. Contrary to the underlying peptidoglycan containing "wall skeleton", no net negative surface charge was determined. The charge density found in the native S-layer from *B. sphaericus* was 1.6 carboxylic acid functions per nanometer squared.

However, the free carboxylic acid functions are neutralized by free amino groups of lysine. The presence of an approximately equimolar amount of amino groups and carboxylic acid groups in the pore areas of many *Bacillaceae* prevents the adsorption of charged macromolecules inside the pores. This offers the interesting application of S-layer proteins as ion-exchanger membranes in bioelectrochemical devices and offers a pathway toward the development of biofuel cells and biobatteries.

Furthermore, permeability studies performed on the same S-layers demonstrates that – in spite of the considerable differences in the lattice types – the molecular weight cutoff was found to be 30 000 to 40 000. This data is in agreement with a pore diameter of approximately 4 to 5 nm. On the basis of this finding, the pores formed by S-layer lattices have diameters comparable to the porin structures isolated from mycobacteria (for instance, MspA, isolated from *Mycobacterium smegmatis*) [71].

13.4.2
S-layer Proteins at Lipid Interfaces

13.4.2.1 Semifluid Membranes

A simple and straightforward method for generating coherent S-layers on lipid films is the injection of isolated S-layer protein subunits in the subphase of the Langmuir-Blodgett (LB)-trough. The crystallization of the two-dimensional S-layer lattices begins at several nucleation points and proceeds underneath the preformed phospholipid or tetraetherlipid films until coalescence of the single crystal domains occurs. Generally, the orientation of S-layers, which can be crystallized on any kind of surface, is determined by the anisotropy of their physicochemical surface characteristics. The more hydrophilic surface, which possesses net negative charges, is usually directed toward the negatively charged or zwitterionic head groups of phospholipid or tetraetherlipid films. Since the direction is determined by the charged head groups embedded in the S-layer structure, the more hydrophobic "outer" face (with respect to its natural orientation adhering to the bacterial cell) is directed toward the water phase during its formation. However, ordered phases with adhering S-layers can be either turned or transferred onto organic, ceramic or metal supports. It is noteworthy that the fluidity of the lipid films, which were previously used as self-assembled mono- or bilayers, is a very critical parameter for generating coherent S-layer lattices (see Figure 13.5). When the formation of S-layers adhering to lipid films is achieved, cross-linking can be performed using glutaraldehyde. The latter procedure remarkably increases the mechanical stability of the S-layers on lipid film.

In a very similar manner, so-called semifluid membranes can be generated [6]. In these cases, where free carboxylic or amine groups are left after the cross-linking procedure using glutaraldehyde, lipid molecules of the LB film can be chemically linked to the S-layer protein lattice. The linking of the S-layer protein subunits to individual lipid molecules, which interact nevertheless with all lipid molecules forming the lipid monolayer or bilayer, decreases the lateral diffusion. Consequently, the fluidity of the whole membrane assembly decreases. Using standard procedures, a manifold of functional biomolecules (for instance, ion channels, molecular receptors, proton

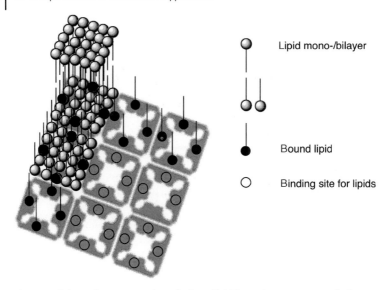

Fig. 13.5 Schematic representation of a "semifluid" membrane, composed of an S-layer lattice possessing binding sites for lipid molecules and a lipid mono-/bilayer.

pumps or porins) can be integrated in semifluid membranes. Depending on the structure and hydrophobicity of the biomolecules, the integration into the lipid film can be achieved either before the chemical fixation of the adhering S-layer lattice or after this procedure. Because of the flexibility of this experimental approach, this technology has a big potential to initiate a broad spectrum of developments in the fields of optical or electronic nanodevices, diagnostics in medicine, and life-science engineering.

13.4.2.2 Site-selective S-layer Reconstitution at Phospholipid Monolayers and Transfer onto Silicon Wafers

The S-layer protein from *B. coagulans* E38-66 shows a site-directed recrystallization at negatively and positively charged phospholipid monolayers [72]. The S-layers from *B. coagulans* are formed by the supramolecular interaction of identical nonglycosylated protein subunits (M_r 97 000), which form an oblique lattice structure ($a = 9.4$ nm, $b = 7.4$ nm, $\gamma = 80°$) possessing nanopores of identical size and symmetry (pore diameter: 4.5 ± 1 nm). Furthermore, the protein subunits possess an "inner" and "outer" surface (the latter being directed away from the cell *in vivo*) as well as two different types of cation binding sites (see Figure 13.6). The "natural inner surface" is more corrugated than the "outer" surface and is featuring a net negative charge at pH > 4.3. The "outer" surface is more hydrophobic than the "inner" surface. Whereas the more elevated surface areas appear to be uncharged, negative charge residues are concentrated within the pore regions.

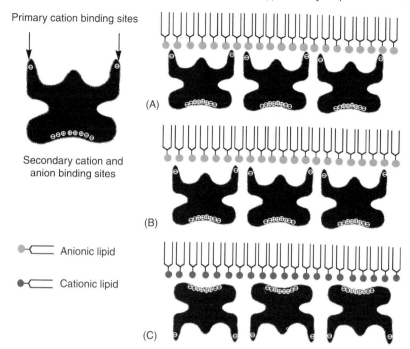

Fig. 13.6 Two types of binding sites for cations and anions on the S-layer protein subunits from B. coagulans. Distinct anionic sites for the binding of cations are found on the "inner surface" (directed toward the cell body of the bacterium, "up" in the schematic drawing). Distributed and therefore far less specific anionic and cationic binding sites can be found at the "outer surface" (naturally directed toward the aqueous environment of the bacterium). The direction of the S-layer protein subunits and, therefore, the structure of the formed crystalline protein lattice at the hydrophobic surface, can be tailored by choosing the appropriate experimental conditions. (A) Condensed phosphatidylcholine (PC)-monolayer, pH > 5, no alkaline and alkaline earth cations added; (B) Phosphatidylethanolamine- or phosphatidylcholine-monolayer, Ca(II) added, pH > 5; (C) Cationic monolayer, pH > 5, the same result was obtained with and without adding Ca(II).

Under optimized conditions [73], carpets of crystalline S-layer domains possessing a maximal diameter of 10 000 nm (10 µm) can be obtained. The pH of the buffer-solutions during the crystallization of S-layer protein lattices and the net charge of the lipid bilayers, which serve as crystallization template, define the orientation of the S-layer with respect to the lipid monolayer. At pH = 4.0, the S-layer proteins isolated from *B. sphaericus* CCM 2177 reconstituted attached to the lipid monolayer in an inside–outside orientation. However, when the pH was ranging between 5.7 and 9.0, the "natural" orientation of the S-layer subunits was observed. It must be noted that the binding of S-layer proteins in their "reverse" orientation (with their "outer" surfaces directed toward the lipid interface) occurred much less specifically and ordered than the binding in their "native" ("inner") orientation. These findings indicate that hydrophobic interactions, which were dominant only very close to the natural p*I*-value of 4.3 of this S-layer protein, cause the inside–outside orientation, whereas ionic interactions, which were superior at higher pH, trigger the natural orientation.

After crystallization at lipid interfaces, S-layer lattices from *B. coagulans* E38-66 could be transferred onto silicon wafers. Both orientations of the protein subunits could be transferred without a significant loss of their crystalline geometry. This procedure offers a straightforward and relatively simple procedure for the lateral nanopatterning of semiconductor electrodes. The nanopores within the S-layer lattice appear to be insulated by the protein matrix, opening nano-pathways toward the silicon surface. These pathways could serve as cation-conducting nanopores because they contain negatively charged residues in their interior.

13.4.2.3 Lateral Patterning and Self-assembly of S-layer Lattices at Silicon Supports

The successful patterning and self-assembly of 2D S-layer lattices at silicon supports [74] were achieved using the SbpA protein isolated from *B. sphaericus* CCM 2177. Micromolding in capillaries (MIMIC) [75] was used as the state-of-the-art soft lithography technique. It could be demonstrated that lateral structuring is possible across various scales ranging from tens of microns down to submicron dimensions. It is of special interest that the natural chemical functionality was preserved, as indicated by the covalent attachment of human IgG antibodies and subsequent recognition by FITC (fluoresceinisothiocyanate) labeled antihuman IgG antigen. The applied experimental techniques are summarized in Figure 13.7.

The structure and also the differences in height were investigated by using atomic force microscopy employing air-dried samples. Similar results were obtained by contact-mode AFM under aqueous buffers. Furthermore, the expected crystalline pattern of SbpA, possessing a lattice constant of approximately 13 nm, was also detected here. This can be regarded as another proof of the functional crystallization of this particular S-layer protein under the more technical conditions employed here. The height of the S-layer structures was determined to be 15 nm. This value could be regarded as an indication of the occurrence of S-layer bilayers because it exceeds the naturally found height of a monolayer by 6 nm. Note that O_2-plasma treatment of the silicon supports increased the velocity of mold microchannel filling by the S-layer protein and also the observed ordering of the lattice structures at the surface, most likely due to an increase in the hydrophilic effect. The results from AFM-imaging are reported in Figure 13.8.

13.4.2.4 Incorporation of Enzymes PLA$_2$ and α-hemolysin Inside S-layer Protein Lattices of *Bacillus sphaericus* at Hydrophobic DMPE Interfaces: Bulk hydrolysis and Single-channel Conductance

An important paradigm in membrane biology implies that the total sum of the physical properties of lipid membrane components can modulate the activity of enzymes, which are reconstituted within the membrane structure. The model enzymes *porcine*

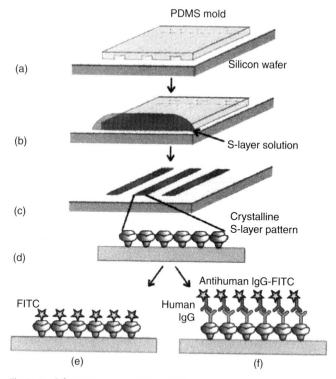

Fig. 13.7 Schematic representation of S-layer protein patterning and assembly by MIMIC. (a) Channels are formed when aPDMS (poly(dimethylsiloxane) mold contacts a silicon wafer support. (b) Channels are filled with a protein solution by capillary forces. (c–d) Following mold removal, crystalline protein patterns are observed on the support surface. (e) S-layer patterns may be labeled with a fluorescence marker or (f) used as substrates for an antibody–antigen immunoassay (with permission from [75]).

pancreatic phospholipase A_2 (PLA$_2$) hydrolyzes the hydrophobic "surfactant" dimyristoylphosphatidylethanolamine (DMPE), which can form hydrophobic monolayers when in contact with aqueous buffer solution. This particular enzyme possesses an approximately 10 000-fold increase in its hydrolytic activity, when aggregated with a hydrophobic membrane, compared to its activity in aqueous buffers [76]. The hydrolysis leads to the formation of the "free" fatty acid and 1-acyllysophospholipids.

Bulk Hydrolysis of DMPS by PLA$_2$ The S-layer protein isolated from *Bacillus sphaericus* CCM 2177 (Czech Collection of Microorganisms) was shown to crystallize spontaneously on DMPE-monolayers generating a highly structured S-layer protein lattice. PLA$_2$ was able to fit into the pores of the S-layer lattice and to partially reconstitute within the DMPE-monolayer. The global kinetic constant of ester hydrolysis, Q_m, was

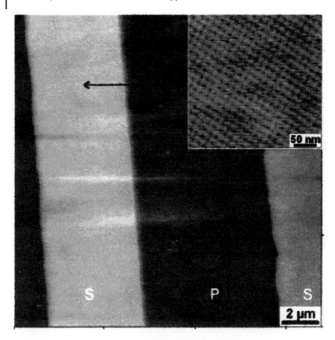

Fig. 13.8 Contact-mode AFM image of 6 μm-wide S-layer protein tracks patterned and recrystallized on a hydrophilic silicon wafer support (Z-range 50 nm). Inset shows the square lattice symmetry of the SbpA S-layer protein (Z-range 1.2 nm). The images were acquired under 100 mM NaCl. The regions denoted S and P represent S-layer tracks and PDMS residue, respectively (with permission from [75]).

estimated in the presence and absence of crystalline S-layer. A significant influence of the initial surface pressure Π (mN m^{-1}) on Q_m was found in DMPE-monolayers. Note that the isotherms of DMPE showed the well-known transition from the liquid-expanded to the liquid-condensed phase at an area of approximately 0.67 nm^2 per DMPE-molecule. Incorporation of PLA$_2$ from underneath the monolayer led to a decrease in Π after an initiation period (see Figure 13.9). In contrast, crystallization of SbpA from the subphase, on the DMPE-monolayer led to a significant increase of Π (from 10.5 to 22.7 ± 0.6 0.4 mN m^{-1}). Remarkably, the global kinetic constant Q_m did not drop to zero after the S-layer lattice was completely crystallized underneath the DMPE-monolayer. It was concluded from the experiments that 20% of the enzymatic activity is retained. It is of special importance in the design of future bioelectronic systems on the basis of crystalline protein lattices as stabilizing support that the highly ordered S-layer does not prevent the diffusion of the substrate to the enzyme PLA$_2$. Since the substrate DMPE is stabilized by the DMPE-monolayer, an even higher conversion of other substrates, for example, by redox enzymes, can be expected. Furthermore, it must be noted that many proteins, such as the enzymes glucose oxidase, alcohol dehydrogenase, and urease, denature rapidly at the air–water interface to form

13.4 Applications of S-layer Proteins at Surfaces

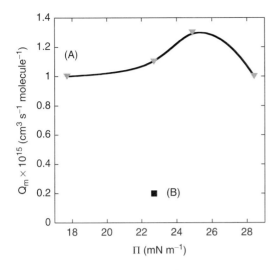

Fig. 13.9 Plot of the global kinetic constant Q_m versus the surface pressure Π. (A) PLA_2 activity when incorporated into DMPE-monolayers; (B) PLA_2 activity when incorporated in a DMPE/crystalline SbpA lattice superstructure (calculated according to data taken from [74]).

peptide sheets of various thickness [77]. This denaturing process can be – almost completely – prevented by using the combination of hydrophobic monomembrane and stabilizing S-layer lattice.

13.4.2.5 Introduction of Gramicidin into Tetraether Lipid-supported S-layer Lattices

The reconstitution of channel proteins into lipid membranes, which are stabilized by adherent S-layer protein lattices [78], can be regarded as an ideal method to combine a highly insulating membrane system that possesses conducting channel systems. Ideally, these ion, water and/or electrically conducting channel systems should possess similar dimensions to those of the membranes. In great similarity to their natural environment, most porins reconstitute within the lipid membranes because they provide naturally pathways through various bacterial cell membranes. This strategy offers two distinct advantages when compared to more simple systems: (a) the number of conducting channels can be chosen independently from the structure of the stabilized lipid membrane and (b) from the multitude of known porins, channel forming systems with tailored structural and biophysical properties can be selected (ion- vs cation selectivity, size of the pore diameter, lengths of the porin channel, the magnitude of the gating voltage) [79, 80, 81]. However, reconstitution within hydrophobic pores, formed by S-layer protein lattices, could also occur. In these cases "dead-end-channels" can be formed, which are unable to provide pathways through the hydrophobic membranes.

An excellent example of this experimental strategy can be found in Ref. [78]. The main polar lipid (MPL) from *Thermoplasma acidophilum*, an archea originally isolated from self-heating coal refuse piles, which has adapted to very acidic conditions (pH = 1 − 2) and high temperature ($T > 55\,°C$), possesses a bipolar structure. It consists of an *sn*-3-glycerophosphate and a β-L-gulose unit as the two headgroups, which are ether-linked by two C40 isoprenoidic chains each possessing one cyclopentane ring. Tetraether lipids are known to provide excellent "surfactants" for the generation of liposomes and planar artificial membranes with remarkable long-term stability, especially at elevated temperatures [82]. The construction of the gramicidin-modified, stabilized hydrophobic membrane was performed in four consecutive steps:

1. An S-layer ultrafiltration membrane (SUM) was generated by depositing the S-layer protein subunits isolated from *Bacillus sphaericus* CCM 2120 on microfiltration membranes, followed by chemical cross-linking.
2. The second step consisted in the preparation of lipid membranes, formed MPL surfactants and DPhPC (diphytanoyl-sn-glycero-3-phosphatidylcholine)/MPL-surfactant mixtures, on the SUM.
3. Further stabilization of the SUM-supported lipid membranes was achieved by recrystallization of a second S-layer lattice, isolated from *B. sphaericus* CCM 2177, on the nonstabilized side of the lipid membrane. Both S-layer membranes used in this study possess similar lattice dimensions and physical properties, because they originate from related bacteria.
4. Finally, gramicidin was injected into the aqueous buffer and reconstituted spontaneously within the S-layer stabilized liquid membrane.

The construction of gramicidin-modified, SUM-supported monolayers yields highly insulating membranes with lifetimes in the range of one day. All SUM-supported membranes permitted the spontaneous reconstitution of various amounts of gramicidin, and single-pore measurements could be performed (see Figure 13.10). Therefore, these pore-containing membrane structures fulfill the demands for future biotechnological applications with respect to membrane structure, fluidity and long-term stability. Further applications could be concerned with membrane-based biosensors and could eventually result in the accomplishment of the "lab-on-a-biochip" technology.

13.4.3
Introduction of Supramolecular Binding Sites into S-layer Lattices

The application of S-layer–coated liposomes for entrapping and binding of functional molecules via the avidin- or streptavidin-biotin bridge demonstrated this potential [6]. Two biotin residues accessible to avidin binding were bound to every S-layer subunit within the cross-linked S-layer lattice. Using biotinylated ferritin, which serves as an excellent Electron Paramagnetic Resonance (EPR) marker, it could be demonstrated that a well-ordered layer of streptavidin formed on the accessible surface of the S-layer-coated liposomes (see Figure 13.11).

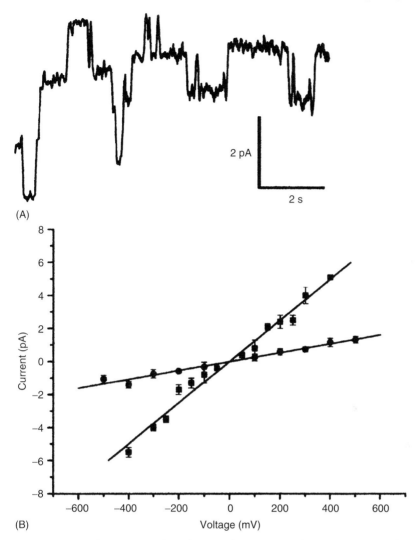

Fig. 13.10 (A) Single-pore current through gramicidin reconstituted in an SUM-supported MPL membrane bathed in aqueous KCl buffer (1 M KCl) at $V_m = -150$ mV.
(B) Current–voltage curves for gramicidin reconstituted in SUM-supported MPL membranes at room temperature. Squares: 1 M KCl, pH = 5.8; circles, 1 M NaCl, pH = 5.8. In both cases, good linear regressions with $r = 0.990$ (KCl) and 0.987 (NaCl) were obtained (with permission from [79]).

The same strategy also proved valuable when biotinylated antihuman IgG was attached via streptavidin to the biotinylated S-layer-coated liposomes. The biological activity of the S-layerbound antihuman IgG was confirmed using the enzyme-linked immunosorbent assay technique (ELISA).

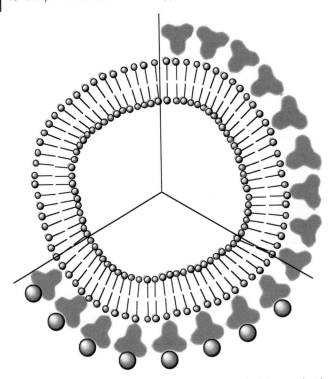

Fig. 13.11 Schematic representation of a liposome, which is coated with an S-layer lattice. The S-layer subunits are chemically linked to functional macromolecules, such as enzymes (with permission from [40]).

13.5
Molecular Nanotechnology Using S-layers

13.5.1
Patterning of S-layer Lattices by Deep Ultraviolet Irradiation (DUV)

One important and very successful approach toward the molecular nanotechnology [83] of the future consists in the generation of so-called "self-assembled-monolayers" (SAMs) at metal (especially gold), silicon or highly oriented pyrolithic graphite (HOPG)-surfaces. SAMs are formed by chemically attached lipids, which are bound to these surfaces in such a density that cooperative effects, such as two-dimensional condensation, occur. Therefore, SAMs are highly ordered and can serve as substrates for the binding and/or incorporation of hydrophobic substrates of all kinds. A very interesting alternative to the formation of SAMs is provided by the recrystallization of S-layer subunits of surfaces, which are suitable for nanofabrication. Silicon or gallium arsenide wafers are two examples of the successful creation of an ordered S-layer on a semiconductor material. S-layer lattices have great potential for the development of molecular manufacturing procedures and biological nanoresists. The first successful use of recrystallized S-layer lattices on semiconductor materials was achieved by the method of deep ultraviolet irradiation (DUV) as shown in Figure 13.12 [84]. This method is principally related to the conventional negative photoresists, which will not be able to satisfy the hardware needs of the information age. However, it can be regarded as an example of the modern nanotechnology yet to be developed, because it uses the self-replication properties of S-layer lattices for lateral nanopatterning of wafer surfaces.

The method of deep ultraviolet irradiation employs the argon–fluorine (ArF) excimer (or more correctly, exciplex) laser system as a coherent light source. The current restriction in the availability of optimal mask systems only permits the generation of submicron structures. Exposure with deep ultraviolet photons, followed by treatment with an aqueous buffer solution, leads to the complete removal of the S-layer lattice located on top of the wafer material. It is very likely that the absorption of high-energy photons causes chemical transformations of the amino acids that are forming the S-layer subunit proteins. Possible photoreactions are Norrish I and II reactions [85] as well as photoinitiated oxidation reactions, which produce water-soluble groups such as carboxylic acids and alcohols. Because of this deep UV-induced chemical transformation, the ability of the S-layer subunits to form ordered lattices is greatly diminished. If the exposure time is optimized, the exposed S-layer surfaces can be completely removed by this procedure. It is noteworthy that the unexposed S-layer surface areas retain their structural and, therefore, also their functional integrity.

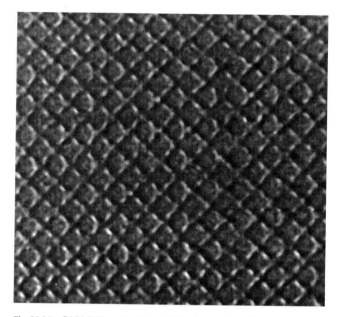

Fig. 13.12 DUV (193 nm)-patterned S-layer-coated surface (550 × 550 nm); (for the original images, see [84], with permission from [41]).

13.5.2
Synthesis of Semiconductor and Metal Nanoparticles Using S-layer Templates Design of Gold and Platinum Superlattices Using the Crystalline Surfaces Formed by the S-layer Protein of *Bacillus sphaericus* as a Biotemplate

As pointed out earlier, the use of organic and inorganic templates for the synthesis of semiconductor and metal nanoparticles is a very promising strategy. S-layer lattices represent ideal templates because they represent highly ordered pore structures. By choosing the right S-layer type and patterning procedure, nanopores can be obtained. These nanopores can support either the crystallization of inorganic nanoparticles or their chemical or electrochemical reduction [6]. Principally, any combination of S-layer lattice symmetries (oblique, square or hexagonal), lattice dimensions and pore sizes with a suitable nanomaterial (metal, metal oxide or semiconductor particle, for instance, $H[AlCl_4]$, K_2PtCl_6, $PdCl_2$, $NiSO_4$, $Cu(SO_4)_2$, $Pb(NO_3)_2$ or $K_3[Fe(CN)_6]$) can be achieved.

The thermal- and electrosynthesis of noble metal nanoparticles using the pores of the crystalline lattice formed by the S-layer protein isolated from *B. sphaericus* on hydrophobic supports represents a very important step toward the design of bioelectronic devices using S-layer proteins as building elements.

13.5.2.1 Synthesis of S-layer Stabilized Gold nanoparticles
The crystallization of the S-layer protein from *B. sphaericus* was performed on electron microscope grids (HOPG) as stabilizing and conductive support [86]. The crystalline S-layer shows square symmetry (p4) with a lattice constant of 12.8 nm. The morphological units consist of four identical nonglycosylated protein subunits ($M_r = 120\,000$). At pH = 9.0, the S-layer crystallized with its "natural outer face" on the supporting TEM grid, exposing its inner corrugated surface. The crystalline S-layers have been stabilized employing a calcium containing borate buffer of pH = 9.0 first and then a cross-linking procedure with glutaraldehyde. Subsequently, the S-layer lattices on electron microscope grids reacted with iminothiolane in a triethanolamine/HCl buffer of pH = 8.4. Subsequently, the disulphide bonds, which were introduced by reaction with iminothiolane, have been reductively cleaved by reaction with dithiothreitol in sodium bicarbonate solution [87]. More than 50% of the free amino groups exposed at S-layer lattices have been converted into free thiol functions by this procedure.

13.5.2.2 Reductive Precipitation of Au(0)
The slow and defined reductive precipitation of gold nanoparticles in the interior of the pores embedded by the S-layer protein lattice were allowed to react with 10 mM $HAuCl_4 \times 3H_2O$. Transmission electron micrography and energy dispersive X-ray analysis of the precipitated materials proved unambiguously that Au(0) nanoparticles are formed under irreversible oxidation of the S-layer protein. It is noteworthy that the thermal nucleation into distinct gold nanoparticles occurred only after chemical modification of the S-layer lattice. Otherwise, no formation of Au(0) was observed.

Therefore, it was concluded that an optimal density of free amino and thiol groups is mandatory for the formation of gold nanoparticles, which proceeds in three distinct steps: (a) binding of Au(III) cations by the thiol groups. This process may be facilitated by the presence of aliphatic amino groups; (b) slow thermal reduction of Au(0) to Au(I) and then Au(0); (c) the newly formed Au(0) centers catalyze the further reduction of Au(III) to Au(0) and act therefore as nucleation centers; (d) reduction of the gold nanoparticles was completed by electron bombardment under TEM conditions as well as the energy input from this process, which promoted the lateral diffusion of the very fine (d < 1 nm) initially formed Au(0) particles. Since their concentration appears to be highest within the nanopores embedded by the S-layer lattice and because diffusion is no problem within the time frame of four days required for the completion of this reaction, gold nanoparticles possessing a diameter of 4.5 ± 0.5 nm were formed. Their unusual shape resembled the square morphology of the pore region within the S-layer lattice. A comparison of the shape of the S-layer lattice, as obtained by using cross correlation averaging (and further interpreted using the program IMAGE) and the resulting Au(0) nanoparticles, is shown in Figure 13.13. Note that a preferential orientation was present in the short range order, where Au(0) nanoparticles were found to be rotated by 45° with respect to the base vectors of the square S-layer lattice.

This procedure offered the first experimental evidence that even highly corrugated S-layer lattices, which most often are turned "upside down", can successfully host metal nanoparticles, forming the so-called *hybrid bioelectronic devices*. In this case, the nanoparticles were formed, electrically insulated by the embedding S-layer lattice but depending on the selected type of S-layer, an electrical connection to the underlying electrode surface also appears to be possible. Therefore, S-layer lattices seem to be an ideal biotemplating material for the generation of highly ordered hybrid nanocrystal superlattices.

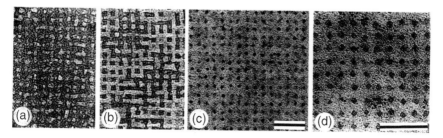

Fig. 13.13 Transmission electron micrographs of the chemically modified and Au(III)chloride treated S-layer lattice of *Bacillus sphaericus* CCM 2177 under increasing electron doses (with permission from Ref. [87]). (a) A coherent film of fine grainy gold precipitates is found under low electron dose conditions. (b and c) Upon increase of the electron dose, regularly arranged monodisperse gold clusters are formed in the pore region of the S-layer. (d) Frequently, the square-shaped gold particles were rotated by 45° with respect to the base vectors of the S-layer lattice. (Bar size for a–d: 50 nm).

13.5.3
Generation of S-layer Lattice-supported Platinum Nanoclusters

The chemical reduction of hexachloroplatinate(II) ($K_4[PtCl_4]$) by sodium azide NaN_3 under physiological reaction conditions (37 °C) in the presence of the S-layer protein isolated from *Sporosarcina ureae* provided the opportunity of exploring the advantages of a parallel manipulating approach of a very complex hybrid nanostructure using the periodicity of the biomolecular self-recognition of S-layer proteins. The parallel approach consists of the simultaneous reduction of Pt(II) chloro/aquo complexes obtained from hexachloroplatinate(II), preaged in an aqueous buffer solution, during the self-assembling process of the S-layer protein subunits from *S. ureae* shown in Figure 13.14 [88, 89]. For the fabrication of large nanoclustered arrays, as they are mandatory for their possible future use as fast recording and memory materials in future computer devices, parallel nanofabrication possesses a considerable advantage compared to classic serial methods such as the manifold of lithographic methods (for instance, electron beam lithography and soft lithography) and the nanomanipulation techniques [90]. After the self-assembling/chemical reduction and deposition process, the nanocrystal superlattices that were obtained were transferred onto graphite-covered copper grids for transmission electromicrography from aqueous buffer solution and then dehydrated in vacuum.

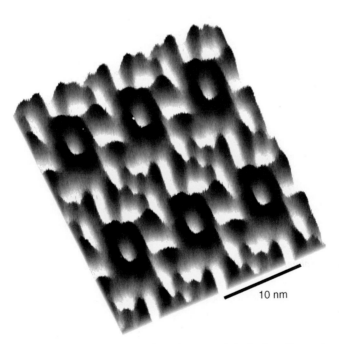

Fig. 13.14 Surface of the crystalline S-layer protein lattice from *Sp. ureae* calculated using IMAGE (according to data provided in [89]).

Similar to *B. sphaeroides* (see above), this S-layer protein also crystallized in p4-symmetry (unit cell: 13.2 × 13.2 nm). Also in this case, the more corrugated "inner surface" of the S-layer lattice served as a biotemplate for the reduction of Pt(II). Each unit cell covered a surface area of 4×10^{12} cm^{-2} and contained seven possible "nanocontainers" for the deposition of platinum nanoparticles. In contrast to the S-layer protein from *B. sphaeroides*, this particular protein subunit contained a sufficient amount of "docking sites" for Pt(II). Generally, cysteine and lysine especially can serve this function. Thus, a chemical derivatization of the S-layer prior to metal deposition was not necessary. Also the interlinking of individual S-layer subunits by a particular cross-linking reaction could be saved because the chemically generated Pt(0) nanoparticles served as interconnectors.

As it becomes clear from the comparison of the self-repetitive lattice structure formed by S-layer subunits from *S. ureae* with and without the presence of protein nanoparticles, the resulting *hybrid system* closely resembles the structure of the "native" lattice. The only major difference is a slightly enlarged size of the unit cell (12.8 × 12.8 nm). This difference, which exceeds the experimental error, can be attributed to the presence of the interlinking Pt(0) clusters, which slightly enlarged the distance, but not the assembling principles of the protein subunits.

Besides their spatial and geometric distribution, their diameter distribution of the obtained Pt(0) particles is of great interest. It could be measured using a series of high-resolution TEM micrographs and a computerized image analysis system (QUANTIMET). The main diameter of the Pt(0) particles has been calculated to be 1.9 nm. The full width at half maximum of the obtained statistical distribution function is 1.2 nm. This result indicates the profound influence of the protein matrix not only in the position and the shape of the Pt(0) nanoparticles but also in the growth of the noble metal clusters. Furthermore, the Pt(0) clusters appear to be a purely metallic phase, as indicated by the results from high-resolution TEM. The lattice fringes of the clusters (lattice constant: 0.39 nm) are well resolved, indicating that the separation of the metallic phase and the surrounding (and insulating) protein is close to perfect. Compared to the synthesis of gold nanoparticles described in Section 13.5.2.1, the approach described here features three distinct advantages: (a) a parallel approach has been successfully utilized in the fabrication of a hybrid system, (b) the platinum particles are distinctly smaller (diameter 1.9 nm, compared to 4–5 nm) and (c) the size distribution is narrower.

13.5.4
Formation and Selective Metallization of Protein Tubes Formed by the S-layer Protein of *Bacillus sphaericus* NCTC 9602

Using a different crystallization method (treatment with 7.5 M guanidine hydrochloride in 50 mM TRIS hydrochloride buffer in order to create S-layer protein monomers, followed by dialysis in 10 mM CaCl$_2$ to reassemble the monomers), S-layer tubes featuring a diameter up to 1 µm and a length of approximately 10 µm have been formed. Adsorption of these protein tubes on the graphite surfaces of the TEM grids leads to the formation of flat-lying, double-layered, band-shaped protein structures. The more corrugated cytoplastic ("inner") surfaces have been directed toward the inside of the protein tubes during the initial adsorption at the surface. However, folding along the [1,1]-axis of the protein crystal occurs after deposition. Therefore, the larger corrugations come "in touch" and then form small cavities located between the alternately oriented layers of the tubes.

These cavities can then be used as templates for either the electron beam–induced reduction of K_2PtCl_4 and K_2PdCl_4 or their chemical reduction employing sodium azide as reductive agent.

1. Chemical reduction: Analogous to the experiments described in Section 13.5.2.1, the long-term treatment of the S-layer structures of *B. sphaericus* with previously hydrolyzed Pt(II) and Pd(II) complexes leads to the formation of highly ordered metal(0) cluster arrays, embedded by the biotemplate. The well-separated clusters are somewhat smaller than the Au(0) nanoparticles synthesized using a very similar strategy (see: [86]), (d = 2–3 nm for Pt(0) and Pd(0), compared to 4–5 nm for Au(0)). However, comparison of the symmetry and ordering principles of the hybrid S-layer lattices reveals a very high degree of similarity.
2. A completely different result is obtained when the S-layer lattice is only treated shortly (less than 2 h) with K_2PtCl_4 or K_2PdCl_4 in the presence of NaN_3 (see Figure 13.15). Nucleation and metal(0) particle growth did not occur, because there was not sufficient time. However, accumulation of rather high concentrations of Pt(II) and Pd(II) salts within the nanopores of the S-layer structure, which served again as binding sites, have been observed. Typically, the same binding sites were discovered as during the process of "negative staining". It can be expected that charge attraction plays an important role in this process, because uranyl acetate (UO_2^{2+}, typically used as 2% (per weight) solution) is also positively charged! After impregnation, the S-layer structures have been dehydrated in high vacuum. Exposure to electrons, emitted by the TEM, leads to electron beam–induced reduction of Pt(II) and Pd(II) to the metallic state. This procedure has the advantage

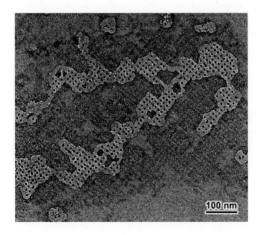

Fig. 13.15 TEM micrograph of highly ordered platinum particles formed by electron beam–induced nucleation in a recrystallized S-layer tube of *B. sphaericus* NCTC 9602 after 80 min. of treatment with K_2PtCl_4 (with permission from [89]).

of being site-specific, but it is also far more time consuming because of its serial approach. The Pt(0)- and Pd(0) particles created by impregnation and subsequent electron beam exposure exhibit a square-like shape with a lateral size of 5 to 7 nm. Again, the squares are somewhat tilted with respect to the major symmetry axis of the protein array. Note that long-term exposure leads to the inevitable destruction of the S-layer matrix and then agglomeration of the metal(0) nanoparticles.

13.5.5
S-layer/Cadmium Sulfide Superlattices

Inorganic superlattices of Cadmium Sulfide (CdS) with either oblique or square lattice symmetries were fabricated by exposing self-assembled S-layer lattices to Cd^{2+}-ion solutions. After cation exchange, a slow reaction with hydrogen sulfide was carried out. Precipitation of CdS was confined to the nanopores only and therefore the resulting CdS-doped S-layer superlattices possess the same symmetry as the S-layer lattices used as templates. Using this method, extended arrays of a very defined nanostructure, extending to 5×10^4 nm^2, could be fabricated [91].

13.6
Immobilization and Electrochemical Conducting of Enzymes in S-layer Lattices

Amperometric and optical bioanalytical sensors have been developed using suitable S-layer lattices as nanoscale immobilization matrix. The concepts of enzyme immobilization and of the manufacture of S-layer ultrafiltration membranes (SUMs) have already been discussed. For the fabrication of a single enzyme sensor, for instance, a glucose sensor, glucose oxidase was bound to the surface of an S-layer ultrafiltration membrane. The electrical connection to the sensing layer was achieved by sputtering a nanoscopic gold or platinum layer onto the S-layer bound enzyme layer. The whole sensor assembly is then usually stabilized using a conventional gold or platinum electrode by which the electrical contact to the nanolayer is achieved. In this configuration, the analyte reaches the immobilized enzyme through the pores of the S-layer.

13.6.1
S-layer and Glucose Oxidase-based Amperometric Biosensors

Glucose oxidase (M_r 150,000) was chemically immobilized at the S-layers proteins of *B. stearothermophilus* PV72 and *T. thermohydrosulfuricus* L111-69 (see Figure 13.16) [6]. The labeling density corresponded to 2 to 3 enzymes per hexametric unit cell of the S-layer lattice. This finding corresponds to the formation of a monomolecular enzyme layer on the surfaces of both S-layer lattices. Glucose oxidase retains approximately 35% of its original enzymatic activity. When spacer molecules are used (4-amino butyric

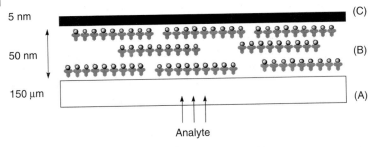

Fig. 13.16 Schematic representation of an amperometric biosensor. (A) microfiltration membrane, (B) multiplayer, consisting of enzyme loaded S-layer fragments, (C) sputtered metal layer (with permission from [40]).

acid or 6-amino caprotic acid), the measured activity of glucose oxidase increases to about 60%. The use of spacers obviously prevents the binding of whole enzymes or partial enzyme structures within the S-layer pores and is especially advantageous for the generation of enzyme assays. However, the measurable current using only one monolayer of enzyme remains, principally, rather low and, although the function principles of an amperometric glucose oxidase/S-layer sensor can be elucidated by employing electrochemical techniques, a commercial application is rather difficult [91].

13.6.2
S-layer and Glucose Oxidase–based Optical Biosensors

A very similar sensing principle, however, combined with an optical detection system (optode), provides a state-of-the-art glucose sensor (see Figure 13.17). Its construction principles are closely related to the amperometric glucose biosensor, as depicted in

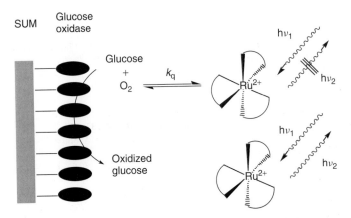

Fig. 13.17 Working principles of an optode for the detection of glucose.

Section 13.6.1. Also for the development of this fiber-optic glucose biosensor, a monomolecular layer of glucose oxidase was covalently immobilized on the surface of an S-layer ultrafiltration membrane [92, 93]. Glucose is measured employing the following measuring principle: If glucose is present, the SUM-stabilized glucose oxidase monolayer oxidizes the sugar under consumption of oxygen. As optical transducer, a Ru(II) complex was chosen. Under irradiation with UVA and blue light, Ru(II) polypyridyl complexes show an intense luminescence in the red region of the visible spectrum. If the irradiation of the Ru(II) complex is pulsed, the fluorescence response can also be measured time-resolved. If no glucose is present, the luminescence of the Ru(II) complex is dynamically quenched by dissolved molecular oxygen. However, if glucose is present, the activity of glucose oxidase diminishes the concentration of O_2. As a consequence, the luminescence intensity, as well as the luminescence lifetime increase and can be detected separately. The performance of this biosensor, in terms of response time, linear range and stability, is comparable to existing optodes. Given its minute size, the system presented is considered to hold great promise in the development of microintegrated optical biosensors.

13.7
Conclusions

Although the use of S-layer components as structural building components in bioelectronic applications is in the early stages of development, it nevertheless offers striking advantages. Straightforward and versatile, 2D-patterning and, with a certain extension into the third dimension, 3D-patterning also can be achieved at various interfaces. Crystalline S-layers can serve as almost ideal host systems in complex bioelectronic systems. In these systems, S-layer lattices will provide "well-like" geometries in which electrically connected redox enzymes and well-confined nanoparticles will achieve the desired biomimetic functions.

References

1 I. Willner, E. Katz, *Angew. Chem., Int. Ed.*, **2000**, *39*, 1180–1218; *Angew. Chem.*, **2000**, *112*, 1230–1269.
2 E. Couture-Tosi, H. Delacroix, T. Mignot, S. Mensage, M. Chami, A. Fouet, G. Mosser, *J. Bacteriol.*, **2002**, *184*, 6448–6456.
3 M. Pleschberger, A. Neubauer, E.M. Egelseer, S. Weigert, B. Lindner, U.B. Sleytr, S. Muyldermans, M. Sára, *Bioconjugate Chem.*, **2003**, *14*, 440–448.
4 E.M. Riedmann, J.M. Kyd, A.M. Smith, S. Gomez-Gallego, K. Jalava, A.W. Cripps, W. Lubitz, *FEMA Immunol. Medial Microbiol.*, **2003**, *37*, 185–192.
5 J. Antikainen, L. Anton, J. Sillanpää, T.K. Korhonen, *Mol. Microbiol.*, **2002**, *2*, 381–394.
6 B. Schuster, U.B. Sleytr, *Rev. Mol. Biotechnol.*, **2000**, *74*, 233–254.
7 A. Huczko, *Appl. Phys. A: Mater. Sci. Process.*, **2000**, *70*, 365–376.
8 A.N. Shipway, I. Willner, *Chem. Commun.*, **2001**, *2001*, 2035–2045.
9 Y. Xiao, F. Patolsky, E. Katz, J.F. Hainfeld, I. Willner, *Science*, **2003**, *299*, 1877–1881.

10 R. Esfand, D.A. Tomalia, *Drug Discovery Today*, **2001**, *6*, 427–436.
11 Z.L. Wang, *Charact. Nanophase Mater.*, **2000**, 1–11.
12 M. Tinkham, D. Davidovic, D.C. Ralph, C.T. Black, *J. Low Temp. Phys.*, **2000**, *118*, 271–285.
13 L.D. Burke, P.F. Nugent, *Gold Bull.*, **1998**, *31*, 39–50.
14 S. Link, M.A. El-Sayed, *J. Phys. Chem. B*, **1999**, *103*, 8410–8426.
15 G.Y. Tseng, J.C. Ellenbogen, *Science*, **2001**, *294*, 1293–1294.
16 D. Srivastava, M. Menon, K. Cho, *Comput. Sci. Eng.*, **2001**, *3*, 42–55.
17 C.A.J. Foss, *Metal Nanoparticles*, **2002**, 119–139.
18 D. Pum, A. Neubauer, E. Gyorvary, M. Sara, U.B. Sleytr, *Nanotechnology*, **2000**, *11*, 100–107.
19 U.B. Sleytr, M. Sara, D. Pum, *Supramol. Polymers*, **2000**, 177–213.
20 M. Sara, U.B. Sleytr, *Prog. Biophys. Mol. Biol.*, **1996**, *65*, 83–111.
21 H. Nikaido, S.H. Kim, E.Y. Rosenberg, *Mol. Microbiol.*, **1993**, *8*, 1025–1030.
22 G.E. Schulz, *Curr. Opin. Struct. Biol.*, **1996**, *6*, 485–490.
23 T. Schirmer, *J. Struct. Biol.*, **1998**, *121*, 101–109.
24 R. Koebnik, K.P. Locher, P.V. Gelder, *Mol. Microbiol.*, **2000**, *37*, 239–253.
25 T. Lichtinger, A. Burkovski, M. Niederweis, R. Krämer, R. Benz, *Biochemistry*, **1998**, *37*, 15024–15032.
26 F.G. Riess, T. Lichtinger, R. Cseh, A.F. Yassin, K.P. Schaal, R. Benz, *Mol. Microbiol.*, **1998**, *29*, 139–150.
27 E. Maier, G. Polleichtner, B. Boeck, R. Schinzel, R. Benz, *J. Bacteriol.*, **2001**, *183*, 800–803.
28 X. Jiang, M.A. Payne, Z. Cao, S.B. Foster, J.B. Feix, S.M.C. Newton, P.E. Klebba, *Science*, **1997**, *276*, 1261–1264.
29 W. Welte, K. Diederichs, M. Przybylski, M.O. Glocker, R. Benz, J. Breed, *NATO ASI Ser. C Math. Phys. Sci. – Adv. Study Inst.*, **1998**, *510*, 239–276.
30 A. Mathes, H. Engelhardt, *Biophys. J.*, **1998**, *75*, 1255–1262.
31 D.J. Müller, A. Engel, *J. Mol. Biol.*, **1999**, *285*, 1347–1352.
32 T. Schirmer, T.A. Keller, Y.-F. Wang, J.P. Rosenbusch, *Science*, **1995**, *267*, 512–514.
33 W. Im, B. Roux, *J. Mol. Biol.*, **2002**, *319*, 1177–1198.
34 M. Niederweis, S. Ehrt, C. Heinz, U. Klocker, S. Karosi, K.M. Swiderek, L.W. Riley, R. Benz, *Mol. Microbiol.*, **1999**, *33*, 933–945.
35 H. Engelhardt, C. Heinz, M. Niederweis, *J. Biol. Chem.*, **2002**, *227*, 37567–37572.
36 D. Moll, C. Huber, B. Schlegel, D. Pum, U.B. Sleytr, M. Sára, *Proc. Natl. Acad. Sci.*, **2002**, *99*, 14646–14651.
37 C. Weiss-Wichert, M. Smetazko, M. Valina-Saba, T. Schalkhammer, *J. Biomol. Screen.*, **1997**, *2*, 11–18.
38 U.B. Sleytr, M. Sara, D. Pum, *Microelectron. Eng.*, **1989**, *9*, 13–20.
39 H. Bahl, H. Scholz, N. Bayan, M. Chami, G. Leblon, T. Guli-Krzywicki, E. Shechter, A. Fouet, S. Mesnage, E. Tosi-Couture, P. Gounon, M. Mock, E.C.d. Macario, A.J.L. Macario, L.A. Fernandez-Herrero, G. Olabarria, J. Berenguer, M.J. Blaser, W. Lubitz, B. Kuen, M. Sara, P.H. Pouwels, C.P.A.M. Kolen, H.J. Boot, A. Palva, M. Truppe, S. Howorka, G. Schroll, S. Lechleitner, S. Resch, *FEMS Microbiol. Rev.*, **1997**, *20*, 47–98.
40 M. Niederweis, S.H. Bossmann, *Encyclopedia of Nanotechnology*, **2004**, *7*, 851–867.
41 M. Sara, U.B. Sleytr, *J. Bacteriol.*, **2000**, *182*, 859–868.
42 B. Kuen, W. Lubitz, *Cryst. Bacterial Cell Surf. Prot.*, **1996**, 77–102.
43 M. Sara, E.M. Egelseer, *Cryst. Bacterial Cell Surf. Prot.*, **1996**, 103–131.
44 P. Messner, *Cryst. Bacterial Cell Surf. Prot.*, **1996**, 35–76.
45 S. Kupcue, M. Sara, U.B. Sleytr, *J. Membr. Sci.*, **1991**, *61*, 167–175.
46 S. Howorka, M. Sara, Y. Wang, B. Kuen, U.B. Sleytr, W. Lubitz, H. Byley, *J. Biol. Chem.*, **2000**, *275*, 37876–37886.
47 C. Mader, S. Kupcue, U.B. Sleytr, M. Sara, *Biochim. Biophys. Acta*, **2000**, *1463*, 142–150.
48 S. Kupcue, C. Mader, M. Sara, *Biotechnol. Appl. Biotechnol.*, **1995**, *21*, 275–286.

49 K. Taga, R. Kellner, U. Kainz, U.B. Sleytr, *Anal. Chem.*, **1994**, *66*, 35–39.
50 I. Willner, V. Heleg-Shabtai, E. Katz, H.K. Rau, W. Haehnel, *J. Am. Chem. Soc.*, **1999**, *121*, 6455–6468.
51 B. Lundholm-Sethson, J. Nyström, P. Geladi, R. Koeppe, A. Nelson, C. Whitehouse, *Anal. Bioanal. Chem.*, **2003**, *377*, 478–485.
52 A.B. Kharitonov, J. Wasserman, E. Katz, I. Willner, *J. Phys. Chem. B*, **2001**, *105*, 4205–4213.
53 T. Liebermann, W. Knoll, *Colloids Surf., A: Physicochem. Eng. Aspects*, **2000**, *171*, 115–130.
54 X. Wang, Z. Gershman, A.B. Kharitonov, E. Katz, I. Willner, *Langmuir*, **2003**, *19*, 5413–5420.
55 M. Niederweis, C. Heinz, K. Janik, S.H. Bossmann, *Nano Lett.*, **2002**, *2*, 1206–1210.
56 M. Niederweis, C. Heinz, K. Janik, S.H. Bossmann, *Nano Lett.*, **2001**, *1*, 169–174.
57 M. Weygand, B. Wetzer, D. Pum, U.B. Sleytr, N. Cuviller, K. Kjaer, P.B. Howes, M. Lösche, *Biophys. J.*, **1999**, *76*, 458–468.
58 M. Micic, A. Chen, R.M. Leblanc, V.T. Moy, *Scanning*, **1999**, *21*, 394–397.
59 W.A. Ducker, T.J. Senden, *Nature*, **1991**, *353*, 239–241.
60 G.U. Lee, D.A. Kidwell, R.J. Colton, *Langmuir*, **1994**, *10*, 354–361.
61 M. Rief, M. Gautel, F. Oesterhelt, J.M. Fernandez, H.E. Gaub, *Science*, **1997**, *276*, 1109–1112.
62 E.L. Florin, V.T. Moy, H.E. Gaub, *Science*, **1994**, *264*, 415–417.
63 C. Cannes, F. Kanoufi, A.J. Bard, *Langmuir*, **2002**, *18*, 8132–8141.
64 J. Mauzeroll, M. Buda, A.J. Bard, *Langmuir*, **2002**, *18*, 9453–9461.
65 A.J. Bard, G. Denuault, R.A. Friesner, B.C. Dornblaser, L.S. Tuckerman, *Anal. Chem.*, **1991**, *63*, 1282–1288.
66 R.D. Martin, P.R. Unwin, *Anal. Chem.*, **1998**, *70*, 276–284.
67 S. Amemiya, A.J. Bard, *Anal. Chem.*, **2000**, *72*, 4940–4948.
68 S. Amemiya, Z. Ding, J. Zhou, A.J. Bard, *J. Electron. Chem.*, **2000**, *483*, 7–17.
69 U.B. Sleytr, H. Bayley, M. Sara, A. Breitwieser, S. Kuepcue, C. Mader, S. Weigert, F.M. Unger, P. Messner, B. Jahn-Schmid, B. Schuster, D. Pum, K. Douglas, N.A. Clark, J.T. Moore, T.A. Winningham, S. Levy, I. Frithsen, J. Pankov, P. Beale, H.P. Gillis, D. Choutov, K.P. Martin, *FEMS Microbiol. Rev.*, **1997**, *20*, 151–175.
70 S. Kupcue, M. Sara, S. Weigert, U.B. Sleytr, *Adv. Filtrat. Separat. Technol.*, **1993**, *7*, 416–419.
71 S.H. Bossmann, K. Janik, M.R. Pokhrel, C. Heinz, M. Niederweis, *Surf. Interface Anal.*, **2004**, *36*, 127–134.
72 B. Wetzer, A. Pfandler, E. Györvary, D. Pum, M. Lösche, U.B. Sleytr, *Langmuir*, **1998**, *14*, 6899–6906.
73 B. Schuster, P.C. Gufler, D. Pum, U.B. Sleytr, *Langmuir*, **2003**, *19*, 3393–3397.
74 E.S. Györvary, A. O'Riordan, A.J. Quinn, G. Redmond, D. Pum, U.B. Sleytr, *Nano Lett.*, **2003**, *3*, 315–319.
75 Y. Xia, G.M. Whitesides, *Gen. Rev.*, **1998**, *28*, 153–184.
76 R. Verger, M.C.E. Mieras, G.H.D. Haas, *J. Biol. Chem.*, **1973**, *248*, 4023–4034.
77 D. Gidalevitz, Z. Huang, S.A. Rice, *Proc. Natl. Acad. Sci.*, **1999**, *96*, 2608–2611.
78 B. Schuster, S. Weigert, D. Pum, M. Sára, U.B. Sleytr, *Langmuir*, **2003**, *19*, 2392–2397.
79 Y. Cheng, R.J. Bushby, S.D. Evans, P.F. Knowles, R.E. Miles, S.D. Ogier, *Langmuir*, **2001**, *17*, 1240–1242.
80 M. Rentschler, P. Fromherz, *Langmuir*, **1998**, *14*, 547–551.
81 P.V. Gelder, N. Saint, P. Phale, E.F. Eppens, A. Prilipov, R.V. Boxtel, J.P. Rosenbusch, J. Tommassen, *J. Mol. Biol.*, **1997**, *269*, 468–472.
82 Q. Fan, A. Relini, D. Cassinadri, A. Gambacorta, A. Gliozzi, *Biochem. Biophys. Acta*, **1995**, *1240*, 83–88.
83 D. Pum, A. Neubauer, E. Gyorvary, M. Sara, U.B. Sleytr, *Nanotechnology*, **2000**, *11*, 100–107.
84 D. Pum, G. Stangl, C. Sponer, K. Riedling, P. Hudek, W. Fallmann, U.B. Sleytr, *Microelectron. Eng.*, **1997**, *35*, 297–300.
85 N.J. Turro, *Modern Molecular Photochemistry*, University Science Books, Mill Valley, **1991**, 45–48.
86 S. Dieluweit, D. Pum, U.B. Sleytr, *Supramol. Sci.*, **1998**, *5*, 15–19.

87 P. Messner, M. Wellan, W. Kubelka, U.B. Sleytr, *Appl. Micobiol. Biotechnol.*, **1993**, *40*, 7–11.

88 M. Mertig, R. Kirsch, W. Pompe, H. Engelhardt, *Eur. Phys. J. D.*, **1999**, *9*, 15–48.

89 M. Mertig, R. Wahl, M. Lehmann, P. Simon, W. Pompe, *Eur. Phys. J. D.*, **2001**, *16*, 317–320.

90 K. Douglas, N.A. Clark, K.J. Rothschild, *Appl. Phys. Lett.*, **1986**, *48*, 676–678.

91 D. Pum, A. Neubauer, U.B. Sleytr, S. Pentzien, S. Reetz, W. Kautek, *Ber. Bunsen-Ges.*, **1997**, *101*, 1686–1689.

92 A. Neubauer, D. Pum, U.B. Sleytr, I. Klimant, O.S. Wolfbeis, *Biosens. Bioelectron.*, **1996**, *11*, 317–325.

93 A. Neubauer, S. Pentzien, S. Reetz, W. Kautek, D. Pum, U.B. Sleytr, *Sens. Actuators, B: Chem.*, **1997**, *B40*, 231–236.

14
Computing with Nucleic Acids

Milan N. Stojanovic, Darko Stefanovic, Thomas LaBean, and Hao Yan

14.1
Introduction

Exactly one decade has passed since Adleman's seminal demonstration that biochemical operations can be performed on DNA in order to solve certain combinatorial problems [1]. So, the time seems ripe to critically analyze where the field stands, what has been accomplished, what needs to be accomplished, and perhaps most importantly, what are the realistic expectations in this field. A rational assessment is also pertinent because the whole field of DNA computing has come to be considered by many as being "more hype than substance" [2], and the practitioners in the field are partly to blame for this. Namely, initial opening in the field was so impressive and inspirational that it garnered understandable yet, in retrospect, completely unjustified exuberance about the potential applications of "DNA computers". As a result, it became standard in the field to put forward incredible predictions in the abstracts and conclusions of papers. This led to a situation where many very important advances in the field of DNA computing are viewed with suspicion by computer scientists and are misunderstood by biochemists. Misinterpretations in the popular press did not help either (not to mention the Schön affair in the somewhat related field of molecular electronics). In this respect, it is sobering to read the latest *Science* paper from Adleman [3], in which he scales the "Mount Everest" of DNA parallel computation (solving a 20-variable instance of 3-SAT), yet he puts the field in a realistic perspective.

At the very beginning, we have to define what DNA computation is. Many examples in the literature do not compute anything tangible (in layman's terms, at least), and in many cases, actual computation is not performed by DNA, but by a human operator. So we decided to discuss those approaches in which authors claim to be able to perform computation by using some type of DNA hybridization procedure. We will specifically exclude methods in which DNA is only incidentally used and in which, for example, Boolean behavior is actually displayed by proteins [4] or by cells. In particular, for reasons of space, we decided to focus only on those approaches that have appeared in mainstream peer-reviewed literature more than once (i.e., that established themselves), and we could not, for the most part, cover the abundant literature that appeared in various DNA Computation conferences.

This review is split into four main sections according to the major thrusts in what could loosely be defined as "the field of DNA computation": (a) Adleman's experiment

Bioelectronics. Edited by Itamar Willner and Eugenii Katz
Copyright © 2005 WILEY-VCH Verlag GmbH & Co. KGaA, Weinheim
ISBN: 3-527-30690-0

and massively parallel approaches, (b) The Seeman–Winfree self-assembly paradigm, (c) Rothemund–Shapiro automata, and (d) nucleic acid catalysts in computation. The section on the Seeman–Winfree approach was written by Yan and LaBean, while the remaining text was prepared by Stojanovic and Stefanovic, who also take full responsibility for any omissions and errors. We tried to make this review particularly friendly to an audience without a computer science background, thinking of it as both primer and critical retrospective.

14.2
Massively Parallel Approaches

14.2.1 Adleman's First Experiment

The Hamiltonian path problem is a prominent illustration of a class of problems studied in computational complexity theory (which will not be discussed in detail here, but interested readers could consult, e.g., [5] for a standard introduction), known as *NP-complete problems*: given a number N of cities and a set of M one-way roads connecting pairs of cities, is there a way to start in city 0 and end in city N, while passing through each city exactly once? The most direct approach would be to try all possible orders of visiting the cities and see which one, if any, complies with the roadmap. Given that the number of such possible orders increases enormously with N, this solution becomes rapidly impractical, even for the fastest silicon-based computers. It is exactly this approach that Adleman demonstrated with DNA molecules for one instance of the problem with 7 cities and 14 roads connecting them [1]. Adleman began by synthesizing seven 20-mer oligonucleotides representing each city, and fourteen 20-mer oligonucleotides representing the 14 roads between cities (Figure 14.1). In the oligonucleotide corresponding to a road that goes from X to Y, the first ten bases were complementary to the last ten bases of the oligonucleotide corresponding to city X, and the second ten bases were complementary to the first ten bases of the oligonucleotide corresponding to city Y. To give an example, if city 3 is represented by the sequence 5'-CCTAGTCAGAACGTTCGAAA and city 4 by 5'-CCCATTAAAGATTACCCGTC, the path connecting them is represented by 3'-TGCAAGCTTTGGGTAATTTC. Thus, the DNA computation procedure solving this problem consists of mixing all roads and cities together in the presence of T4 DNA ligase. The cities hybridize roads, and ligase sticks the cities (+ strand) and roads (− strand) together. Among the 10^{13} double-stranded molecules formed in this experiment, at least one (probabilistically) contains a solution to the problem, that is, all seven cities ligated in the proper order, and this solution is formed almost instantaneously.

However, the deconvolution of this mixture to isolate the solution and prove it correct took an additional seven days and consisted of the following biochemical steps (for a popular treatise, see [6]): The solution had to start in city 0 and end in city 6; thus, its double-stranded molecule had to start and end with the corresponding two oligonucleotides. Therefore, double-stranded molecules formed in the mixture were PCR amplified in the presence of primers corresponding to these cities (one primer was biotinylated), eliminating all other oligonucleotides into the background. The solution

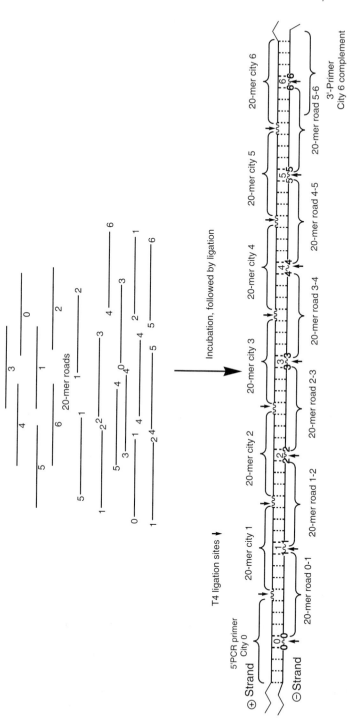

Fig. 14.1 Schematic representation of the Adleman's experiment. Single-number lines represent cities, while two-number lines represent one-way roads connecting cities. Upon mixing, and ligation to form a set of possible solutions, the PCR reaction is used to amplify oligonucleotides with 0 and 6 city-based primers. The amplicons with proper size (i.e., containing 7 cities) are gel purified, followed by the sequential affinity capture of those solutions which contain all cities.

also had to contain seven cities; thus, after the PCR amplification, it had to be 140 bp long. Gel electrophoresis was used to isolate all products of length 140. Finally, each city had to be passed through once, that is, each of the original seven oligonucleotides had to be present in the 140-mer. A five-step purification scheme was designed to assure this. A biotinylated strand was eliminated on the streptavidine, and single-stranded solution was subjected to five successive purification steps on five affinity materials containing complementary strands to each of the cities that had to be visited in between cities 0 and 6. A correct solution should be preferentially enriched through this procedure. Finally, a series of six PCRs was performed; in each PCR, one primer corresponded to city 0, and the other primer corresponded to one of the remaining six cities in turn. This procedure generated a ladder of PCR products that could be used to read out the order of cities in the solution.

Adleman's approach had three main characteristics: First, it was massively parallel, that is, all solutions (or many solutions) were generated at the same time. Importantly, this massive parallelism was achieved in a very small volume. Second, the desired solution was isolated from the mixture by positive selection. Both computation and selection steps were essentially performed through hybridization. Third, the approach was essentially nondeterministic, unlike with electronic computers.

Adleman's experiment started the field of DNA computation and generated huge enthusiasm, on the basis of the following arguments: Adleman performed 10^{14} operations per second in a test tube (counting each ligation as a single operation), while the fastest computers at that time operated at 10^{12} operations per second. This was performed with an energy efficiency of 2×10^{19} operations per joule, much higher than modern computers and closer to the maximal efficacy allowed by the second law of thermodynamics. Finally, the density of information storage in DNA is much higher than in any commercial storage medium. All these arguments supported the *impression* that DNA could be used in a massively parallel approach to crack some of the most difficult problems in computation, and these very same arguments are still occasionally used to justify the intrinsic computing ability of DNA.

However, it was immediately clear that there were some difficulties with this approach. Mainly, critics pointed toward the physical limitations of the approach, arguing that solving harder problems (to be precise, larger problem instances) would require unrealistic amounts of DNA [7, 8]; for example, if the number of cities were to be increased to 70, the amount of DNA required would be 1025 kg. Despite these objections, which still remain in place, Adleman's result was so intellectually impressive that it quickly generated large interest and follow-up work.

A flurry of activity was triggered by Lipton's suggestion [9] that there is another set of problems for which a massively parallel approach is particularly suitable, so-called satisfiability problems, or SAT. In these problems, there is a Boolean formula (logic formula), with a certain number of clauses and a certain number of variables. Each variable can be assigned either the value *true* (T) or the value *false* (F), and the problem is to determine if there exists an assignment of values to the variables that makes the overall formula true. For example, the logic formula $\Phi = (x_1 \text{ OR } \sim x_2) \text{ AND } (\sim x_1 \text{ OR } \sim x_2)$, where $\sim$ stands for negation, is a simple two-clause, two-variable (x_1 and x_2) SAT problem. The formula is in so-called conjunctive normal form: the clauses connect variables in an OR arrangement, meaning that *at least one of the variables* mentioned in a clause has to have a matching value for that clause to evaluate to T; when a variable appears negated, such as ~x_1 in the second clause, the matching value is F, otherwise the matching value is T. Because a formula consists of clauses connected by conjunctions (AND), *all* clauses have to evaluate to T for the formula as a whole to evaluate to T. The difficulty of the problem lies in the fact that setting a variable to a matching value in one clause also sets its value for all other clauses that mention it. Clearly, as it was the case with the Hamiltonian path problem, one way to address this question frontally is simply to try the values of T and F for all variables, until some combination proves good. For our simple example, the solutions that make $\Phi = $ T are two; one is $x_1 = $ T and $x_2 = $ F, and the other is $x_1 = $ F and $x_2 = $ F. However, the number of combinations to be tested rises exponentially in this brute-force approach with the number of variables, and even the fastest computers cannot (and do not) take this approach for exceedingly large formulae; to make matters worse, as long as clauses with more than two variables are permitted, no known approach exists that does not take exponential time in the worst case [10]. Yet, the massively parallel approach with DNA computing is exactly brute-force. Similar to Adleman's first experiment, researchers assign certain variable values to arbitrary oligonucleotides, construct a library of all possible solutions, and find an experimental procedure to eliminate any wrong answers, or select those answers in which the individual clauses are all true. However, note that if the solution is not found by this algorithm, we cannot say whether that is because none exists or our algorithm was not good (there was some shortcoming in the laboratory procedure) or perhaps we did not start with enough DNA (on the other hand, if the algorithm finds a solution, we cannot say whether there exist additional solutions or how many; this is not a shortcoming given the problem statement as above, since we only care if a solution exists.)

Before we discuss this procedure on the latest example by Adleman, solving a 20-variable instance of the 3-SAT problem [3], we will just point the reader to Figure 14.2, which describes the procedure for the simple formula we used above. The particular instance Adleman solved had 24 clauses (Figure 14.3). He chose to demonstrate his procedure on the problem with a unique solution, that is, for this instance, exactly one assignment of true (T) or false (F) values to each of the variables x_1 to x_{20} satisfies the formula. Each clause had three variables or their negations, thus 3-SAT. Each variable was encoded by two oligonucleotides, one for the value of T and the other for the value of F, 40 15-mers in total. Then a library of all possible solutions was designed. With 20 variables, each library had to encode T or F values in 20 different positions. This can be accomplished with 300-mers, consisting of 20 15-mers, each defining T or F values of variables x_1 to x_{20}. There are a total of 2^{20} (1 048 576) combinations that the library had to cover. The library was assembled in convergent combinatorial synthesis, using mix-and-split synthesis on beads (cf. [11]). For example, half of the beads were initially exposed to a series of synthetic steps on the oligonucleotide synthesizer, defining the x_1(T) oligonucleotide, while the other half was derivatized with a series of monomers, defining the x_1(F) oligonucleotide. Then, the beads were combined, split again, and one half was derivatized to assemble x_2(T), while the other half was used to derivatize x_2(F). Two separate libraries were synthesized, encoding all combinations of x_1 to x_{10}, and x_{11}

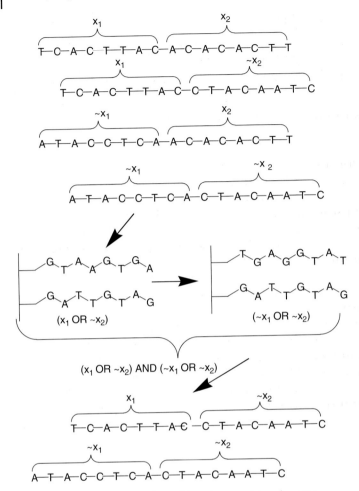

Fig. 14.2 Simplified example of a DNA-based approach to solve a two clause-two variable SAT problem $\Phi = (x_1 \text{ OR } \sim x_2) \text{ AND } (\sim x_1 \text{ OR } \sim x_2)$. All possible values of x_1 and x_2 are represented with four oligonucleotide stretches (x_1, $\sim x_1$, x_2 and $\sim x_2$), and all possible (four) combinations of these values representing solutions are defined by four oligonucleotides combining two of these stretches. These four combinations are culled at two successive affinity columns, each representing one of the two clauses. Two oligonucleotides, selected at both columns represent possible solutions of this SAT problem.

to x_{20}, and these two halves were combined in the final library using the polymerase extension method.

The next step was selection for those library members that satisfy individual clauses. There were 24 clauses, thus, 24 individual selection steps had to be performed. First, for each selection step, an affinity matrix was synthesized. The affinity matrix was polyacrylamide gel, with complementary oligonucleotides attached during polymerization

$\Phi = (\sim x_3 \text{ or } \sim x_{16} \text{ or } x_{18})$ and $(x_5 \text{ or } x_{12} \text{ or } \sim x_9)$ and $(\sim x_{13} \text{ or } \sim x_2 \text{ or } x_{20})$ and $(x_{12} \text{ or } \sim x_9 \text{ or } \sim x_5)$ and $(x_{19} \text{ or } \sim x_4 \text{ or } x_6)$ and $(x_9 \text{ or } x_{12} \text{ or } \sim x_5)$ and $(\sim x_1 \text{ or } x_4 \text{ or } \sim x_{11})$ and $(x_{13} \text{ or } \sim x_2 \text{ or } \sim x_{19})$ and $(x_5 \text{ or } x_{17} \text{ or } x_9)$ and $(x_{15} \text{ or } x_9 \text{ or } \sim x_{17})$ and $(\sim x_5 \text{ or } \sim x_9 \text{ or } \sim x_{12})$ and $(x_6 \text{ or } x_{11} \text{ or } x_4)$ and $(\sim x_{15} \text{ or } \sim x_{17} \text{ or } x_7)$ and $(\sim x_6 \text{ or } x_{19} \text{ or } x_{13})$ and $(\sim x_{12} \text{ or } \sim x_9 \text{ or } x_5)$ and $(x_{12} \text{ or } x_1 \text{ or } x_{14})$ and $(x_{20} \text{ or } x_3 \text{ or } x_2)$ and $(x_{10} \text{ or } \sim x_7 \text{ or } \sim x_8)$ and $(\sim x_5 \text{ or } x_9 \text{ or } x_{12})$ and $(x_{18} \text{ or } \sim x_{20} \text{ or } x_3)$ and $(\sim x_{10} \text{ or } \sim x_{18} \text{ or } \sim x_{16})$ and $(x_1 \text{ or } \sim x_{11} \text{ or } \sim x_{14})$ and $(x_8 \text{ or } \sim x_7 \text{ or } \sim x_{15})$ and $(\sim x_8 \text{ or } x_{16} \text{ or } \sim x_{10})$.

Fig. 14.3 The 3-SAT problem with 24 clauses solved by Adleman.

through 5′-acrydite modifications. For example, for the clause ($\sim x_3$ OR $\sim x_{16}$ OR x_{18}), the affinity material contained three oligonucleotides, complementary to $x_3(F)$, $x_{16}(F)$, and $x_{18}(T)$. When the library passes through this affinity material, only those members for which this clause is true, that is, that contain complements to one of these three oligonucleotides, are captured. These members are then released, and the procedure is repeated for each clause independently, eventually leading to the preferential isolation of a true solution (and many false positives, which are present in significantly smaller number of copies). The correct solution in the end, in Adleman's case at least, was sufficiently enriched after 24 rounds of amplification that it could be cleanly read out in the PCR procedure.

Before we discuss the readout, let us describe the most interesting experimental set-up, or the DNA "computer", that is, its hardware (Figure 14.4). The hardware consists of an electrophoresis box with two chambers, a hot one kept at 65 °C by a circulatory bath and a cold one kept at 15 °C. The two chambers are connected by a glass tube filled with polyacrylamide gel in three layers. In the hot chamber, this glass tube contains a probe-releasing layer, whereas the cold chamber contains a capture (affinity) layer, synthesized as described above, and there is an ordinary gel between them. The capture

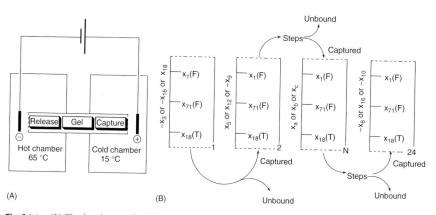

Fig. 14.4 (A) The hardware of a DNA computer. Two chambers (hot and cold) are connected through a tube filled with three gels (release, connecting and capture gel. (B) Schematic presentation of the computation process. Sequences bound to the first capture gel are released into the second capture gel, and this process is continued.

layer is clause-specific, as described above. The capture layer from one selection cycle is used as a release layer for the next cycle, and a total of 24 cycles, one for each clause, is performed. In the first cycle, the release layer contains the double-stranded library, with one strand covalently bonded to polyacrylamide. In this set-up, in each cycle, the release layer in the hot chamber releases the pool, and the affinity selection is performed in the cold chamber by hybridization for one of the three oligonucleotides attached to the polyacrilamide. Any of the three complements present in the library member would be sufficient to bind that particular oligonucleotide to the affinity matrix in that cycle of selection. Thus, the logic function performed in each affinity selection is OR (true if any of them is present). The library member has to survive all 24 selections (one for each clause), in order to be read out at the end. The logic function performed in successive affinity selections is AND (true if all are true). Hence, the library member that was selected at the end will contain T or F values for each of the variables encoded in its structures and will satisfy the given formula, that is, it will be a solution of this SAT problem.

Finally, after 24 rounds of selection, the readout was performed by an elegant series of PCR reactions. With some redundancy, the readout for a variable x_n was done by four PCR reactions containing four sets of primers corresponding to each combination of $x_n(F)$ or $x_n(T)$ variable and $x_1(F)$ or $x_1(T)$ variable. For each variable n, only one out of these four combinations gave an amplicon band of the expected length ($15 \times n$), and this set of primers represent values for variables x_1 and x_n.

We should note that by discussing only the first and the last experiment by Adleman, we are doing an injustice to numerous important efforts by other scientists. However, these two experiments are extremely characteristic of the massively parallel computation approach, and in this respect, further discussion would be somewhat redundant. In the intervening period, there was a gradual increase in complexity of the problems that were addressed in this field. Some of the other instances of the SAT problems solved were as follows: (a) Groups led by Smith and Corn, and Yoshida and Suyama solved four-variable (2^4 or 16 possible combinations) examples [12, 13], (b) Sakamoto and colleagues solved a six-variable (2^6 or 64 possible combinations) instance using an interesting hairpin DNA system [14], (c) Landweber and Lipton solved a nine-variable (2^9 or 512 possible solutions) example [11]. However, the 20-variable SAT problem solved by Adleman is the first and only example (in the peer-reviewed literature, at least) of a problem instance beyond what a human could reasonably solve without computational aids. Smith–Corn's approach is interesting because it includes computation on surfaces and a destructive algorithm (viz., nonsolutions are eliminated); many important technical improvements, applicable in other fields (e.g., gene chips) were made along the way, including important advances in fluorogenic detection [15]. The Landweber–Lipton approach is the first to demonstrate that RNA can be used for computation, and again uses a destructive algorithm; nonsolutions are removed by RNAse H. This group was also among the first to report the combinatorial split-and-pool synthesis of libraries for computation [11], and this method was used by Adleman in his final report. Landweber's team is also developing alternative hardware designs (microreactors) that could significantly improve the performance of DNA computers [16].

14.3
The Seeman–Winfree Paradigm: Molecular Self-assembly

Since soon after the excitement following Adleman's first experiment, limits have been noted on the size of combinatorial search problems that can be implemented in DNA, owing especially to the exponential growth of search spaces and the volume constraints on wet computing techniques [17]. In addition to volume constraints, approaches involving biochemical manipulation for massively parallel computation (as discussed in the previous section) suffered from rather inefficient and tedious hands-on laboratory steps, the total number of which increased at least linearly with problem size. These concerns have been sidestepped by more recent theoretical and experimental advances, including the development of various autonomous biochemical systems (see Section 14.15.4 below) and by the strategy of computation by algorithmic self-assembly described in this section.

Self-assembly, in general, can be defined as the spontaneous formation of ordered structure by components that not only act as structural building blocks but also encode information, specifying how the building blocks should be organized into the final construction. In other words, a set of structural units carrying smart address labels that specify neighbor relations are able to self-assemble into a desired structure. In DNA computing by self-assembly, a diverse library of address labels is available via the programmable molecular recognition afforded by Watson–Crick complementary sequence matching. As we shall see, DNA can be used not only as the information carrying "smart glue" but also as the structural material from which the building blocks are formed.

Following Adleman's first experiment, biochemical reactions such as *anneal, ligate,* and *separate* were mapped to computational primitives, and sets of primitives were evaluated theoretically to determine their inherent computational power. The fundamental insight that spawned DNA computing by self-assembly was made by Winfree when he noted that DNA annealing by itself was theoretically capable of performing universal computation [18]. He went further and recognized that certain stable DNA structures being developed by Seeman for nanoengineering and crystallography could serve as physical incarnations of a mathematical model known as *Wang tiles*, which had already been shown to be capable of Turing universal computation. In Wang tiling, unit tiles are labeled with symbols on each edge such that tiles are allowed to associate only if their edge symbols match [19]. Tile sets have been designed that successfully simulate computing devices known as *Turing Machines* and are therefore capable of universal computation [20]. The recognition that DNA tiles, exemplified by DX and TX complexes (see Figure 14.5), could represent Wang tiles in a physical system, where edge symbols are encoded in the base sequence of sticky ends, led to proofs that DNA tilings are capable of universal computation [21]. Computation by self-assembly of DNA tiles is a significant advance over biochemical manipulation computing schemes because self-assembly involves a single-step procedure in which the computation occurs during the annealing of carefully designed oligonucleotides. Contrast this with Adleman's experiment in which the annealing step generated all possible solutions and where a

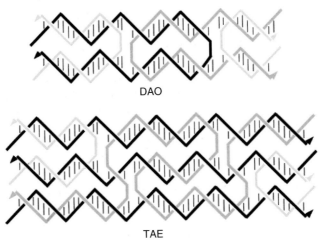

Fig. 14.5 Example DX and TX tiles drawn as an idealized projection of 3D helices onto the plane of the page with helix axes lying horizontal on the page. Strands are shaded for ease of tracing individual oligonucleotides through the complexes. Each straight strand segment represents a half-turn around the helix. Vertical segments of strands indicate strand exchange (branch junction) sites where strands cross over from one helix to another. Note that two strands are exchanged at each crossover point. Arrowheads indicate 3′ ends of strands. Thin vertical hashes indicate base pairing between strands. Unpaired segments on 5′ ends represent sticky ends. The top complex is a DAO double-crossover, so called because of its Double (two) ds-helices, Anti-parallel strand exchange points, and Odd number of helical half-turns between junctions. The bottom complex is a TAE (Triple, Anti-parallel, Even number of half-turns between crossovers). Anti-parallel crossovers cause strands to reverse their direction of propagation through the complex upon exchanging helices. For example, the lightest gray strand in the DAO begins on the right-hand side of the top helix; it propagates left until it crosses over to the bottom helix, then it continues back to the right until it reaches the right-hand end of the tile. The effect of spacing between crossover points can be seen by comparing the strand trace of the DAO with that of the TAE. The TAE contains three strands (black) that span the entire width of the tile; they are the nonexchanging strands at each of the crossover points. With an odd number of half-turns between crossovers (see DAO), no strands span the width of the tile. Many other strand topologies are possible; these shown and several others have been experimentally tested. Note that the Figure also shows how the minor groove of one helix is designed to pack into the major groove of neighboring helices.

long series of laboratory steps is required to winnow the set by discarding incorrect answers. Self-assembly (without errors) will theoretically only allow formation of valid solutions during the annealing step, thereby eliminating the laborious phase involving a large number of laboratory steps. An interesting theoretical extension of this approach is a one-step approach to SAT problems using graph self-assembly, which also takes advantage of Seeman's early work on synthesizing geometric figures [22].

Let us come back to DNA algorithmic assembly for computation in a moment after reviewing some of Seeman's earlier work on DNA-based construction. Seeman had been working since the early 1980s on the design and construction of objects and periodic matter composed of synthetic DNA oligonucleotides (reviewed in [23]). He noted that simple double-stranded DNA (dsDNA), being linear, could only be used for the construction of linear assemblies and that more complex building blocks would be required for two- and three-dimensional constructions. He also noted the use of branched junction structures in biological DNA, most importantly the Holliday junction found in homologous recombination complexes. A Holliday junction is formed by four strands of DNA (two identical pairs of complementary strands) where double-helical domains meet at a branch point and exchange base-pairing partners. The branch junctions in recombination complexes are free to diffuse up and down the paired homologous dsDNA domains since the partners share sequence identity along the vast majority of their lengths. Seeman showed that by specifically designing sequences that were able to exchange strands at a single specified point and by breaking the sequence symmetry that allowed the branch junction to migrate, immobile junctions could be constructed and used in the formation of stable and rigid DNA building blocks [24, 25]. These building blocks, especially double-crossover (DX) complexes, became the physical incarnation of Wang tiles for the construction of periodic assemblies and the formation of two-dimensional crystals of DX DNA tiles [26].

The first published report of a successful computation by DNA self-assembly demonstrated example cumulative-XOR calculations on fixed input strings [27]. XOR (a function that takes two input bits and returns a zero if the inputs are equal and returns a one if the inputs are not equal) was performed using binary-valued tiles designed to assemble an input layer that then acted as a foundation upon which output tiles specifically assembled on the basis of the values encoded on the input tiles. The system made use of triple-crossover (TX) tiles composed of three double-helical domains linked by strand exchange at four crossover points [28]. The implementation required eight tile types – two input tiles, four output tiles, and two corner tiles that connected the input layer to the output. The binding slot for the first output tile was formed from a sticky end from the first input and a sticky end from a corner tile. Subsequent output binding slots were formed from one sticky end from the next input tile and one sticky end from the previous output tile. In this way, the cumulative-XOR was calculated over the encoded input string. An example computational assembly is shown in Figure 14.6. The described system also demonstrated the use of readout from a reporter strand that was formed by ligation of strands carrying a single bit-value from each tile in the superstructure. Technically, the computation occurred during the self-assembly step, but the readout stage required ligation of the reporter strand segments, PCR amplification, restriction cleavage, and separation of the fragments by polyacrylamide gel electrophoresis. Algorithmic aperiodic self-assembly requires greater fidelity than periodic self-assembly, because correct tiles must compete with partially correct tiles. For example, if the output binding slot displays sticky ends for inputs of 1 and 0, two output tiles would match the first input, and two would match the second input, but only one output tile type would match both input sticky ends. An error rate of between 2 and 5% was observed on the readout gels, presumably because of a low probability of incorrect output tiles being trapped in the final complex. The tiling system presented is capable of performing parallel computations on multiple,

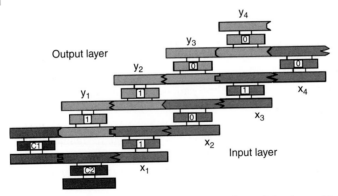

Fig. 14.6 A cartoon of an example cumulative-XOR computation is shown. The input layer (x_i) tiles are medium grey; corner (C_i) tiles are dark grey; and output layer (y_i) tiles are light grey. All DNA tiles follow the TAO strand trace [28]. The operation is designed to proceed from lower left to upper right, because the x_i and C_i tiles have longer sticky ends than the y_i tiles and thus assemble first. After the first y tile has been added, $x_1 = 1$ and $y_1 = 1$, because according to design, $y_1 = x_1$. In the next step, the array contains $x_2 = 0$: and since $x_2 \neq y_1$, then y_2 should be 1, and only the y tile with value 1 and inputs $y_{i-1} = 1$ and $x_i = 0$ fits properly into the binding slot between y_1 and x_2, as shown. Further steps proceed and use input string 1010 to compute output string 1100, as expected.

randomly assembled input layers, however, the examples implemented were prototype calculations on fixed inputs.

The first parallel molecular computation using DNA tiling self-assembly in which a large number of distinct inputs were simultaneously processed has now been described [29]. The system followed the concept of "string tile" assembly, which derives from the observation that by allowing neighboring tiles in an assembly to associate by sticky ends on each side, one could increase the computational complexity of languages generated by linear self-assemblies [30]. Surprisingly sophisticated calculations can be performed with single-layer linear assemblies when contiguous strings of DNA trace through individual tiles and the entire assembly multiple times [31]. In essence, a "string tile" is the collapse of a multi-layer assembly into a simpler superstructure by allowing individual tiles to carry multiple bits of information on multiple segments of the reporter strands, thereby allowing an entire row of a truth table (both input and output bits) to be encoded within each individual tile. "String tile" arithmetic implementations allow input and output strings to assemble simultaneously, with each pair-wise operation being directly encoded in the structure of a tile, while control of information flow in multidigit calculations is handled by the tile-to-tile associations in the tile assembly superstructure. The experimentally implemented "string tile" arithmetic used linear self-assembly of DNA double-crossover tiles (DX) to perform pair-wise XOR calculations [29]. Experimental execution proceeded as follows: first, the set of purified oligonucleotides for each individual tile type (four tile types – one for each combination of possible inputs) were slowly annealed in separate tubes from 95 °C to 65 °C to ensure the formation of valid tiles before parallel self-assembly of multi-tile superstructures. Separately annealed computational and corner tiles were then mixed together at 65 °C and further annealed to 16 °C. Then the reporter strand segments were ligated to one another to produce reporter strands that contain the inputs and outputs of the calculations. Readout of the calculations was accomplished by first selectively purifying ligated reporter strands corresponding to example (4-bit) computations by extracting the band from a denaturing polyacrylamide gel of the ligation products. The purified reporter strands (containing in this case, 24 possible calculations) were then amplified by PCR. Purified PCR products of the proper length were then ligated into a cloning vector. Finally, dideoxy sequencing was performed on mini-prep DNA from five randomly selected clones. The sequencing results obtained from the five randomly selected clones matched the designed sequence words and encoded valid computations.

Theoretical assessments of DNA-based computation by self-assembly indicate computation power far beyond that which has yet been experimentally implemented. The great advance over Adleman's approach to 20-variable SAT is that computation is performed without human intervention, but the problem of scaling (i.e., of the amount of DNA needed) for computationally difficult problems remains. One realm in which algorithmic DNA self-assembly provides great promise is in the nanofabrication of specific aperiodic structures for templating of nanoelectronics devices. Such nanopatterned materials could be used not only for communications and computational devices but also for sensors, biosensors, medical diagnostics, and molecular robotics applications [32].

14.4
The Rothemund–Shapiro Paradigm: Simulating State Machines

In 1995, Rothemund proposed using a cascade of cleavages by a restriction enzyme as a way of mimicking a finite automaton [33]. These theoretical considerations were recently put into practice and improved upon by Shapiro and colleagues [34]. We will now discuss the principles behind this important contribution in greater detail, focusing in particular, on the biochemical details of their experiment. Shapiro's group published two papers on this subject [34, 35], and we will discuss only the second paper [35] for reasons of space and simplicity.

The FokI restriction enzyme recognizes the double-stranded sequence GGATG and cleaves at positions 9 and 13 bases away, leaving a 5′ overhang shown in Figure 14.7. The exact sequence at the restriction site and the intervening sequence are not relevant for this enzyme. Shapiro et al. discovered that the enzyme can actually recognize nicked duplexes and still perform restriction. Thus, one sequence (a guide sequence) containing the GGATG double-helical region that is precomplexed with Fok1 (dubbed "hardware" according to Shapiro) can cleave another sequence that has a complementary overhang.

Let us now assume that there is a specific input molecule (**I**, Figure 14.8) with an overhang complementary to a T4 complex (called the *software/hardware complex* in Shapiro's terminology) and that both are present in solution. There is also a

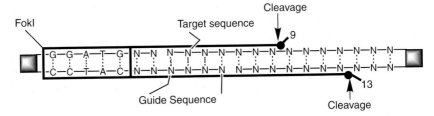

Fig. 14.7 The complex between FokI (box structure) and guide sequence recognizes the target sequence and cleaves at positions 9 and 13, regardless of the target structure, if their overhangs are complementary.

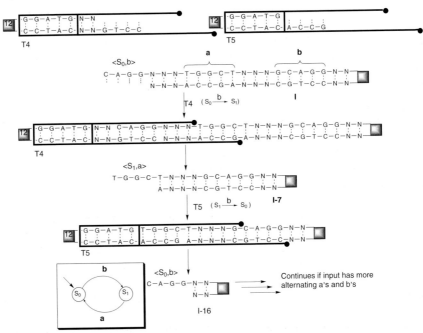

Fig. 14.8 The input molecule is recognized by the **T4** complex and cleaved. The newly revealed overhang is recognized by the **T5** complex and cleaved again. The process can be repeated if the remaining sequence in the input contains properly spaced regions **a** and **b**, which are cleaved by FokI to yield overhangs recognized by **T4** and **T5**, respectively. The boxed region represents schematically an automaton capable of transitioning between two states according to rules **T4** and **T5**, in other words, capable of accepting an input tape with alternating letters **a** and **b** (i.e. **abababababab**).

second complex in solution, T5, whose role is similar to T4. The complexes T4 and T5 and the input together constitute a finite-state automaton; we now describe the chemical/catalytic/enzymatic behavior of the system and how it is interpreted as an automaton. The overhang at **I** (CAGG) defines an initial state (which we can arbitrarily call S_0) and a current input symbol, which we can call **b**. Thus, the automaton (mixture of input and software/hardware) is at the beginning of its "calculation" (actually an enzymatic cascade) in the state S_0 and will "read" the next symbol on the tape (input), namely, **b**. T4 complexes the input and cleaves at the position 9/13 bases away within the region defining symbol **a** (five bases long, TGGCT, and defining two possible states upon cleavage, S_1 through TGGC and S_0 through GGCT, but both still reading as **a**). Upon cleavage, the complex disintegrates, leaving the shortened input (**I-7**) with a new overhang in place, TGGC. This overhang represents the next state S_1 and the next current symbol **a**. Thus, we say that through the binding between the complex T4 and the input, the automaton performed an elementary step wherein it read (consumed) the first input symbol, **b**, and transitioned from state S_0 into state S_1 ($S_0 \rightarrow S_1$), moving along the input (tape) to the second symbol, **a**. Thus, the first 12 bases of the input could be interpreted as a tape, reading two symbols **ba**. Now, the T5 complex is also present in solution, and it will recognize the **I-7** overhang. Then, T5 will cleave **I-7** at the 9/13 position, within the region defined as **b** (GCAGG, defining two states S_0 as CAGG and S_1 as GCAG, both still representing **b**), leaving the new CAGG overhang. Thus, the automaton now transitioned, by the action of the T5 complex, from state S_1 into state S_0, having read the input symbol **a**. It also moved to the following input symbol **b**, on the (ever shrinking) input, now at **I-16**.

As for **I**, its first 20 bases read **bab**, because these sequences are present at precisely spaced distances. We can now easily imagine that, with a judicious choice of input, and with regions **a** and **b** properly (regularly) spaced in the input, and in the presence of various transition molecules (software) in solution, we could continue this cascade (in our example, started by actions of T4 and T5). The T's would move continuously along the input and cleave it, until no more **a**'s and **b**'s could be found. In effect, what we just described is an automaton capable of consuming input strings over a two-letter alphabet {**a**, **b**} and transitioning between states, depending on the input consumed and in accordance with a preprogrammed control (a transition function), defined through the presence of particular transition complexes. For example, the transition function consisting of transition rules T4 and T5 and no others can move the automaton back and forth between states S_0 and S_1 endlessly, moving along an input made of symbols **a** and **b** in alternation (**baba**...). An input with a different arrangement of symbols (e.g., **abbabababa**) would stall this automaton; on the other hand, some other automaton could read this other input – for instance, the automaton whose transition function includes all transition rules T1-T8 from [35].

In conclusion, through the choice of a set of transition rules, or software/hardware complexes (transition rules of the form: if you find the system in state S_M and read a symbol $\sigma \in \{a, b\}$ from the input, go to state S_N and move to the next input symbol), one could write a molecular "program" able to walk through and degrade an input DNA molecule, if this input molecule encodes a defined set of symbols. Only those input molecules for which transitions always succeed will be completely degraded (consumed); if during the reading of an input molecule at some moment there is no transition complex T present in the system that matches the current configuration (i.e., the current state and the current input symbol), the automaton stalls, and the input molecule is not completely degraded. Thus, it can be said that this system indeed works as a finite-state automaton and thus recognizes the *regular language* abstractly specified by its set of transition rules.

We can now see why Shapiro's accomplishment is an impressive one, and why it received justified attention. For the first time, behavior akin to a finite automaton's was demonstrated on the molecular scale. Moreover, several different automata were modularly constructed ("programmed") through different choices of the complexes present in solution. From the synthetic biology point of view (which, unfortunately, the authors do not stress in their papers), this is an equally significant accomplishment: a reaction cascade that can be tailored, and that can oscillate between two states! It is easy to envisage various sensory inputs, inhibitory feedbacks, oscillators, and loops as future accomplishment. But what about computation, is this automaton ever going to compute something practical and compete with silicon-based computers? On the basis of this work, various claims as to computation power have been made in the literature, scientific and otherwise, including a patent on molecular Turing machines for universal computation! However, in these reviewers' opinion, these automata are more significant as a truly beautiful intellectual/experimental exercise, and as a system through which one could study new and emerging behaviors in complex enzymatic networks.

14.5
Nucleic Acid Catalysts in Computation

Modern electronic (silicon) computing is based on very different principles from the DNA-based computation methodologies described above. The basic electronic devices that perform Boolean calculations are called *logic gates*, and in a bottom-up approach elementary logic gates are organized into more complex electronic circuits, which are used (along with other devices) to build computers. Logic gates (listed in Figure 14.9) and electronic circuits take one or more electrical inputs and provide electrical outputs on the basis of a certain rule. For example, a NOT gate, or inverter, is connected to one input and one output, and provides the output (the output voltage is high) if and only if the input is absent (the input voltage is low), while an AND gate, with two inputs and one output, provides the output if and only if both its inputs are present. These are very simple rules, in which the output depends only on the instantaneous values of the inputs; with more complicated rules, the output may also depend on past values of the inputs. The similarity between circuits in electrical engineering and various solution-phase processes in living organisms, such as allosteric regulation and metabolic and genetic control circuitry, was noted early (see [36] for a survey). The extension of this analogy resulted over the last decade in an impressive array of demonstrations of "digital" behavior at the molecular scale and in cells.

Before proceeding with biochemical details, we have to define some terminology: when we say that molecules behave like logic gates, what we really mean is that their behavior can be influenced by the presence of other molecules (or ions) in solution. This is a significantly looser definition than in electrical engineering, and is actually

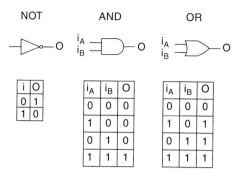

Fig. 14.9 Schematic representation of logic gates, including their corresponding truth tables. i represents an input signal, while O represents an output signal. 1 represents presence of an input or an output, while 0 represents absence of either an input or an output. NOT gate inverts an input, AND gate integrates two inputs to produce an output when both inputs are present, while OR gate integrates two inputs to produce an output when either of the two, or both of the inputs are present.

somewhat too general to be meaningful. For example, any enzyme undergoing allosteric regulation can be redefined as a logic gate (sensor, detector, or, in chemists' terminology, YES gate), with an allosteric effector as the input, and a catalytic reaction as the output. A small fluorescence molecule interacting with a quencher can be interpreted as a NOT gate, with the quencher as the input, and fluorescence as the output. While the latter concept indisputably led to the very interesting and novel advances in the field of "intelligent" molecular sensors, particularly through the amazing work of de Silva [37], and similar principles were recently extended to nucleic acid-based recognition [38], its significance for the actual ability to compute something useful has been challenged. For example, Williams has pointed out that systems with inputs and outputs of different phases cannot be organized into circuits, and that applications of these systems will be severely limited without the clear-cut ability to interconnect gates [39]. This statement, in our opinion, is generally correct, and this limitation is very serious. However, cells arrange molecules, the behavior of which can be reinterpreted as logic-gate-like behavior, into complex metabolic circuits. Metabolic circuits and gene expression control circuits can be rearranged to a limited extent, and our ability to do so in the laboratory is continuously growing. These approaches may not be suited to supplant silicon in the near future, if ever, but may be most useful in engineering cellular behavior, for example.

Nucleic acid catalysts, which were discovered in the last 25 years, are oligonucleotides that catalyze reactions, primarily of other oligonucleotides. Ribozymes are RNA-based catalysts, while the corresponding DNA-based catalysts are called *deoxyribozymes*. Of the variety of deoxyribozymes available, we will focus here on phosphodiesterases (Figure 14.10), which cleave the essential bond in other oligonucleotide substrates (and

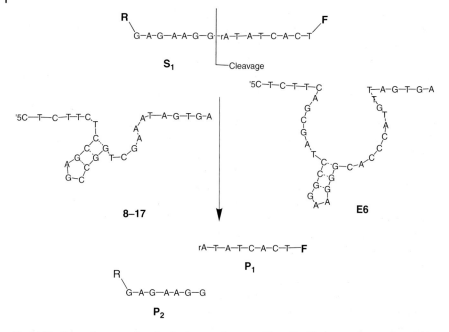

Fig. 14.10 Deoxyribozyme phosphodiesterases. Enzymes **E6** and **8–17** cleave substrate **S** to yield two products, **P1** and **P2**. The increase in fluorescence is caused by the separation of a donor fluorophore (in this case fluorescein **F**) and a quencher (in this case energy acceptor tetramethylrhodamine **R**).

are therefore sometimes called *cleavases*), although we will also briefly mention ligases (which combine two smaller oligonucleotides into a larger product) as well.

Some time ago, two of us (Milan Stojanovic and Darko Stefanovic) set out to develop a computation system based on DNA enzymes, with the following ideas on our mind: (a) the basic units should behave as molecular-scale logic gates; (b) it should be possible to arrange these logic gates in parallel and in series in more complex arrays (in hindsight, similar to metabolic or genetic circuitry control networks); (c) the system should be biocompatible, solution-phase, and fully autonomous – that is, once computation is triggered by the presence of inputs, no human intervention should be necessary; and, lastly, (d) instead of trying to achieve complex computation and compete with silicon, we would focus on what we initially defined as "decision-making in solution". Decision-making is defined as a Boolean calculation that takes into account several inputs and produces various outputs on the basis of the presence or absence of these inputs. Thus, from a computational perspective, these goals were quite modest, and no outstanding problems from computer science or electrical engineering would be solved. However, we looked at this plan from a different perspective, that is, if successful, we hoped to establish the unprecedented control and flexibility of behavior in solution on the molecular scale, and we hoped that we would eventually be able to connect these networks to drug release or molecular movement.

The decision to use nucleic acid catalysts (phosphodiesterases) was based on the realization that they would be well suited to yield molecular-scale logic gates producing molecules (oligonucleotides) as outputs and that these outputs could be used as inputs to other gates. In order to be able to handle gates easily, we settled on DNA catalysts, or deoxyribozymes, because these are stable (unlike ribozymes) and can be handled without any special precautions and, with the advances in custom oligonucleotide synthesis, could be viewed as off-the-shelf reagents. Furthermore, particularly well timed was the development of the general approach to allosteric control of nucleic acids, that is, rational modular design, which we were able to anticipate on the basis of our reading of two influential back-to-back reviews that appeared in 1997 in the special issue of Chemical Reviews on Combinatorial Chemistry [40, 41]. The first review, by Ellington and Osborne, covered the methods for selection of oligonucleotide-based recognition domains, or aptamers, while the second one, by Breaker, described approaches for construction of nucleic acid catalysts. Soon afterward, the first report of a combination of these methods was published by Breaker's group [42]; it described a ribozyme that was negatively regulated by ATP (Figure 14.11). These authors constructed an allosteric ribozyme by judiciously combining two separate modules (domains): the first module consisted of an aptamer binding ATP, while the second consisted of a hammerhead ribozyme [cf. [43]]. This approach is now known as *rational modular design* of allosteric nucleic acid catalysts (sometimes called *allozymes* or *aptazymes*). In a series of follow-up papers, the same group perfected this concept by combining it with an *in vitro* selection in order to generate a series of molecular sensors gates (YES gates) and described the first array of RNA switches. Using this technology, the first nucleic acid catalyst with two recognition modules for small molecules (FMN and theophilline) was constructed as well [44]. The principles of modular design were also applied to ligases by Ellington's group, which reported simple molecular switches for small molecules, nucleoproteins sensing protein analytes and an enzymatic AND ligase gate, with a special case oligonucleotide and a small molecule as input [45]. Ellington must also be credited for the first published suggestions that nucleic acid enzymes could be used in a simple feed-forward computation to release a drug. Specifically, he suggested using an AND gate sensor enzyme to sense glucose and glucagon and release insulin.

The construction of detector gates (switches) for small molecules and proteins is particularly important from the perspective of analytical chemistry. Namely, with certain limitations, for the first time a seemingly general and straightforward approach from the analyte to the catalytic enzyme was developed. However, these advances were unlikely to translate to any new computational approaches, because, from this perspective, allosterically regulated nucleic acid catalysts still suffer from the discordant inputs and outputs. In order to address this issue, we decided to focus on oligonucleotides as inputs, that is, the same type of molecule that nucleic acid catalysts usually produce as outputs. Most importantly, with the choice of oligonucleotides as inputs, an additional flexibility, that of having a large number of easily exchangeable inputs, would be gained. At the onset of our work in this field, there were scattered reports of oligonucleotides influencing the activity of nucleic acid enzymes; these reports, however, could not be generalized to arbitrary input oligonucleotides or were

(A)

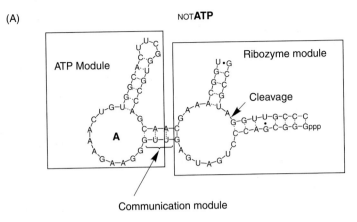

(B)

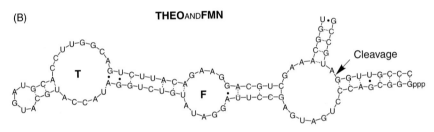

Fig. 14.11 (A) NOT logic demonstrated by a ribozyme allosterically regulated by ATP. (B) AND gate logic demonstrated by a ribozyme allosterically regulated by FMN (F module) and theophilline (T module).

geared toward intracellular applications and could not be readily used to construct logic gates. In a manner inspired by Tyagi and Krammer's sensors for oligonucleotides (molecular beacons) [47], we designed a general approach to allosteric control by oligonucleotides – catalytic molecular beacons, or YES gates. This approach was further expanded to construct a first complete set of deoxyribozyme-based logic gates (NOT, ANDNOT, ANDANDNOT), which could be arranged in solution to accomplish complex Boolean calculations [48].

Stoichiometric molecular beacons exist in solution as stable stem-loop structures in which the fluorescence of a reported dye attached to the 5′ end is quenched by a proximate quencher attached to the 3′ end. In the presence of a complementary nucleic acid, the stem opens, and this event is coupled to a loss of quenching and an increase in fluorescence. We combined in the YESi$_1$ gate a catalytic module from the core deoxyribozyme (E6 in Figure 14.10) [49], with a stem-loop module complementary to i$_1$. The stem-loop inhibits the catalytic module through the overlap of the stem with the 5′ (or 3′, if so desired) substrate recognition region. Hybridization of i$_1$ to the complementary loop opens the stem, reversing intramolecular competitive inhibition to allow substrate binding to proceed. The YESi$_1$ gate behaves as a two-state switch, with the active state in the presence of input. In order to improve visualization of the output (cleaved oligonucleotide) in this system, we developed a fluorogenic cleavage technique method. We placed a fluorescein donor at the 5′ terminus of **S**, and its fluorescence emission was partially quenched by the tetramethyl rhodamine acceptor or Black-Hole 1 quencher positioned at the 3′ terminus. Cleavage of this double end-labeled substrate to products results in at least tenfold increase in fluorescein emission.

The single-input $NOTi_6$ gate (Figure 14.12) is inhibited by a specific oligonucleotide, and is constructed by replacing the nonconserved loop of the E6 catalytic core with a stem-loop sequence complementary to the input oligonucleotide. Hybridization of input with the loop opens the required stem structure of the core, distorts its shape and inhibits the catalytic function. Thus, the presence of the input oligonucleotide will hinder the increase in fluorescence. The $NOTi_6$ gate behaves as a two-state switch, with the active state in the absence of input.

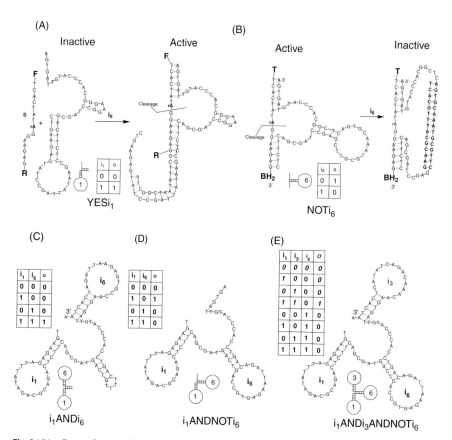

Fig. 14.12 Deoxyribozyme-based gates: (A) $YESi_1$ gate, active in the presence of input i_1; (B) $NOTi_6$ gate, inactive in the presence of i_6; (C) i_1ANDi_6 gate, active only in the presence of both inputs; (D) two-input $i_1ANDNOTi_6$ gate, active in the presence of one, but not the other input; (E) three-input $i_1ANDi_3ANDNOTi_6$ gate, active in the presence of two inputs and in the absence of the third. Truth tables and schematic representations for individual gates are also shown.

To construct the AND gate, we needed a deoxyribozyme allosterically regulated by two different input oligonucleotides. We achieved this by attaching controlling elements to each end of a single catalyst. In the absence of its proper input, either of the attached stem-loop structures independently inhibits output formation. Only upon hybridization of both loops to complements do both stems open, allowing recognition of S and its catalytic cleavage. The i_1ANDi_6 gate behaves as a four-state switch, with one state, in the presence of both inputs, active.

We also combined YES and NOT gates in a single molecule to construct an $i_1ANDNOTi_6$ gate. We attached a stem-loop recognizing i_1 to a position at the 5' end, in which it inhibits the catalysis, as in a YES gate, and a stem-loop recognizing i_6 to the internal position of the E6 catalytic motif, where it does not influence the catalytic reaction, until it recognizes the input oligonucleotide. The $i_1ANDNOTi_6$ is active only in the presence of i_1 and the absence of i_6. Finally, we combined AND gates with NOT gates in a single molecule to construct an eight-state switch with one active state, $i_1ANDi_3ANDNOTi_6$.

The YES, NOT, AND, ANDNOT, and ANDANDNOT deoxyribozyme-based logic gates we constructed represented a basic set of molecular-scale gates (AND and NOT are a basic set in electronics). Importantly, these gates are generic and modular, in the sense that other deoxyribozymes (or ribozymes) could be combined in similar constructs, with the expectation of similar behavior (though some limitations based on secondary structures have to be taken into account). That means that we can now, in principle at least, construct enzymatic networks that perform Boolean calculations of any complexity, that is, that we can have an arbitrary number of inputs and outputs. Monomolecular systems of more than three inputs could be envisioned, but would not be general, and we were not interested in pursuing them. Instead, we opted to start developing parallel and serial arrangements of deoxyribozyme-based gates in order to achieve more complex Boolean calculations.

One of the first systems we developed was a half-adder [50], and the key to this success was fully modular behavior of deoxyribozyme-based logic gates. The logic circuits that perform addition within central processing units of computers are called *adders*, and a half-adder, which adds two single binary digits (bits), is a building block for adders. While it is highly unlikely that DNA will ever be called upon to compete with electronic computers in performing additions, the construction of adders is one of the first tests for any new computational medium, and we wanted to assess our ability to construct enzymatic systems that make more complex, multi-input, multi-output decisions. The system of three enzymes, two ANDNOT gates (based on deoxyribozyme E6) and one AND gate (based on deoxyribozyme 8–17), two inputs, two substrates, red fluorogenic cleaved by E6 and green-fluorogenic cleaved by 8–17, which behave as a half-adder, is shown in Figure 14.13. This molecular half-adder system analyzes the presence of two input molecules and comes up with two different outputs (red or green fluorescence) in accordance with the following set of rules: (a) the absence of both inputs leaves the system as is, that is, without any output, (b) the presence of either single input leads to the cleavage of the red fluorogenic substrate, while (c) the presence of both inputs leads to the cleavage of only the green fluorogenic substrate. The significance of this accomplishment is that it is the first fully artificial, solution-phase molecular-scale

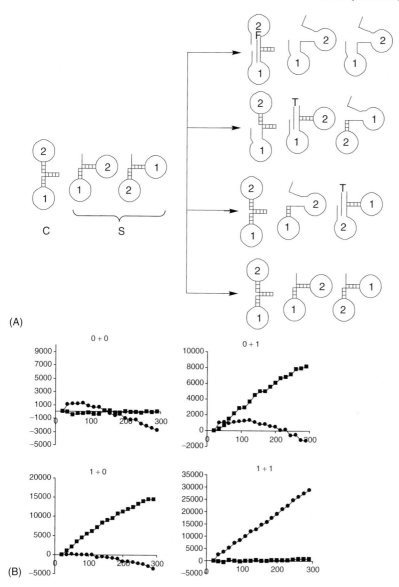

Fig. 14.13 (A) Array of deoxyribozymes behaving as a half-adder. The array consists of two ANDNOT gates cleaving TAMRA-containing substrate and one AND gate cleaving fluoroescein-containing substrate. (B) Fluorescence changes under four different conditions, in the presence of no inputs (0 + 0) the presence of either one of the inputs (0 + 1, 1 + 0), and in the presence of both inputs (1 + 1). Squares represent changes in TAMRA fluorescence over times (sum digit) and diamonds changes in fluorescein emission over time (carry digit). This behavior mimics the half-adder in engineering, if we presume that inputs represent the digits of two binary numbers being added together.

450 | 14 Computing with Nucleic Acids

system in which an enzymatic reaction can be triggered or inhibited under such a precise set of conditions. Interestingly, first half-adder behavior of nucleic acids was reported soon after Adleman's initial report, in a paper suggesting general addition by nucleotides, using human-guided procedures [51]. A similar procedure was published by Yurke and colleagues [52].

The next application of this system was in the construction of the first DNA-based Boolean automaton capable of autonomously responding to human inputs in a meaningful fashion [53]. We opted to construct an automaton to play tic-tac-toe

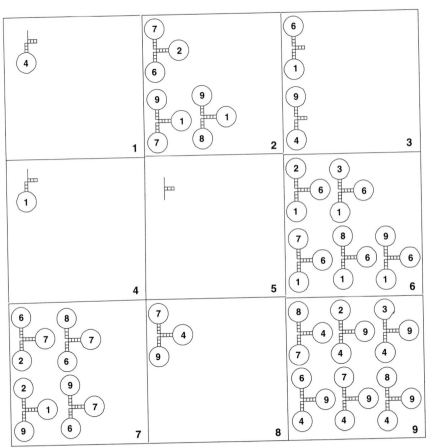

Fig. 14.14 Schematic representation of the gate distribution in the tic-tac-toe playing automaton MAYA. The automaton consists of 24 enzymes, 1 constitutively active in the central well, two YES gates, two AND gates and 19 ANDANDNOT gates. Bold numbers in lower corners of each well represent the well number; addition of the input i_1 to all wells means that human played in square 1. Numbers in the gates represent inputs, keyed to particular moves. For example, 1 in the YES gate in well 4 (center-left) means that a YESi_1 gate is in this well; it will be the only fully active gate after the addition of input i_1 to all wells. In well 8 (bottom-center), 7, 9, and 4 mean that this is the i_7ANDi$_9$ANDNOTi$_4$ gate, active when inputs i_7 and i_9, but not i_4, are present.

against a human player (see Figure 14.14). This choice was made because tic-tac-toe is a traditional challenge for any new computational system. Also, we could not envisage any other solution-phase system capable of playing this simple game autonomously. The game is played in nine wells of a well plate (which could be arranged in a 3 × 3 board). Human moves are represented by an oligonucleotide keyed to a particular move and added to all wells, while the automaton's moves are presented by a large fluorescence increase in a particular well (readout in the fluorescence plate reader). The automaton plays a simplified game, always claiming the center first, and the first human move is symmetry-restricted to one corner or one side move. We also settled on a single, perfect strategy, that is, we "hard-wired" the automaton's game to give the human no chance to win. These simplifications led to a representation of the game as a series of Boolean formulae that compute the automaton's output in each well, on the basis of the human inputs present in all wells, and these formulae were "technology-mapped" to 23 deoxyribozyme-based logic gates (and one constitutively active enzyme in the central well) by arranging gates in the individual wells around a common substrate. For example, the output in well 1 (upper left corner) was calculated by a single $YESi_4$ gate, which cleaved the fluorogenic substrate only in the presence of the input keyed to the human move to well 4 (but added to all wells). The most complex calculation was performed in well 9 (lower right corner) with six three-input gates operating in parallel around the same fluorogenic substrate. Each of these gates was activated under a different set of conditions dynamically arising in various games, depending on the human opponent's moves. This technology is currently being expanded to construct an even larger automaton, with over 120 gates, to play the fully general game.

We will now go through the details of one game against the automaton, in order to clarify how such automata work. The enzymes of the automaton are organized hierarchically. One enzyme is constitutively active, that is, it is active without added inputs; two enzymes are allosterically regulated by single inputs (responding to the first human move), two enzymes are regulated by two inputs (responding to the human second move), and the remaining 19 enzymes are three-input gates (responding to the human third and fourth moves). They cover all possible combinations of inputs that the human player may add to the wells in the course of a legal game, and are distributed to wells in order to activate in precisely one well a fluorogenic cleavage corresponding to the correct response (according to the predetermined winning strategy) to a particular human move. The game starts with the addition of a necessary cofactor metal ion (Mg^{2+} or Zn^{2+}) to all wells, which "turns on" the automaton, and the fluorogenic cleavage of substrate is observed in the center well. Then, for example, the human move into well 4 is communicated to the automaton by adding the input oligonucleotide i_4 to all wells. Well 1 contains the $YESi_4$ gate, which is activated and cleaves the substrate. In order to avoid losing immediately, the human must block a three-in-a-row by the automaton, and must add i_9 to all wells, and this in turn switches on the automaton's response in well 3. Note that all wells at this moment contain inputs i_4 and i_9, but only well 3 contains the gate activated by these two inputs (i_4ANDi_9). The human is now in trouble, because the automaton formed a "fork", and any move of the human will now trigger a three-in-a-row for the automaton, either vertically or diagonally. For example, addition of input i_2 to block three-in-a-row in well 2 will trigger the triple input gate

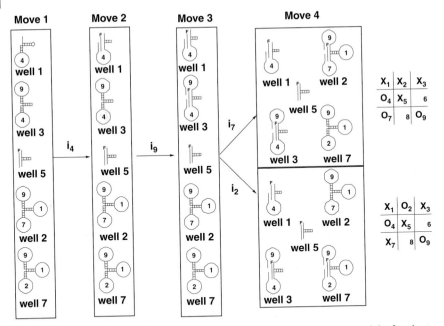

Fig. 14.15 Two representative games (differing in the third move by the human and the fourth move by the automaton) representing the winning strategy implemented by automaton. We show only those gates that participate in these two games; all other gates remain silent. In the first move by the automaton, a constitutively active gate in well 5 gives a fluorescent signal. After the human's first move, conveyed to the automaton by the addition of i_4, the gate in well 1 is activated. The second human move (i_8) triggers the fluorogenic cleavage in well 3. The human has no choice now, as the automaton has formed a fork: any move by the human will trigger a three-in-a-row. We show two human moves, an attempt to block the diagonal three-in-a-row by adding i_7 to all wells, which leads to a vertical three-in-a-row by activating $i_7\text{AND}i_9\text{ANDNOT}i_1$, and addition of i_2, which leads to a diagonal three-in-a-row by activating $i_2\text{AND}i_9\text{ANDNOT}i_1$.

$i_9\text{AND}i_2\text{ANDNOT}i_1$ (ANDNOTi_1 was substituted for ANDi_4 in the technology-mapping process). Addition of the input i_7 to block a three-in-a-row in well 7 will activate the corresponding gate in well 2 (Figure 14.15).

The tic-tac-toe automaton is an example of engineering circuits of allosterically controlled enzymes and, together with similar automata now under construction in our laboratory, it represents the heyday of parallel arrangements of solution-phase molecular-scale logic gates. Their main limitation is that only up to three inputs can be considered within a single molecule, and to go around this limitation we have to start arranging gates serially. Two types of serial communication between gates were developed; in one, an upstream deoxyribozyme-based logic gate cleaves the inhibitor of a downstream logic gate, while in the other an upstream ligase logic gate forms the allosteric regulator of a downstream gate. It can easily be demonstrated that these two types of "inter-gate communication" eventually can be generalized to achieve Boolean calculations of any complexity.

So, what are the applications of deoxyribozyme-based networks? Obviously, this type of computation cannot compete with modern computers in performing complex calculations. The enzymes are too slow (the most active cleavage enzyme has a turnover rate below 10 per minute!). Also, in many cases we noticed that although an isolated gate at 200-nM concentration exhibits perfect digital behavior (i.e., negligible fluorescence increase in the absence of proper inputs), already three gates at these concentrations show significant amount of nondigital behavior. This indicates that the more gates we put together in solution, the more unpredicted behavior and cross talk between gates will occur, even if we start reducing concentrations. Some of this behavior may actually be interesting and worth studying on its own, owing to the possibility that we can evolve the systems toward new behaviors. Despite these unresolved issues, we see three areas of potential applications for deoxyribozyme-based and related networks in the future. First, the ability of molecules to perform, for the first time, arbitrary Boolean calculations with great precision and flexibility can be used to control molecular devices (we are reluctant to use the much abused term "nanorobots" here). For example, a series of inputs could be tied to the Boolean calculation, which would result in triggering a therapeutic effect. Also along these lines, we are constructing mechanical devices that could be controlled by logic gates. Second, ribozyme-based logic gates could form circuits capable of controlling cell behavior; the ease with which a deoxyribozyme-based automaton was constructed argues that such networks may have greater flexibility than corresponding protein analogs [2]. Finally, we should stress the significance of studying these networks for the theories of early life, and for mimicking early metabolic networks. Somewhat beyond the scope of this review, but definitely very intriguing, are the possibilities of interconnecting these enzymatic networks with various self-evolving systems that are used by RNA chemists. In particular, in lieu of a conclusion to this section, we are referring the reader to Joyce's self-evolving ribozymes [54] and, more recent, Ellington's oscillatory network of deoxyribozymes [55]. Could such networks evolve to perform some computational function? In a historic example (Donald Michie's MENACE), a set of matchboxes and glass beads was trained to play a decent game of tic-tac-toe, and training a set of molecules to perform this task should be possible as well. After all, the analogy between training neural networks and evolutionary selection of cell-signaling circuitry has entered even the basic biochemistry textbooks [56].

14.6
Conclusion

After ten years of intensive efforts, and large investment, we have to admit that DNA computation is unlikely to make modern silicon computers obsolete, or, indeed, ever to solve any useful computational problem much faster than the average human can. However, DNA computation is taking new and unexpected forms over time. While many of these approaches may not even be called DNA computation any more, the field will in various forms survive and will have clear implications in nanotechnology, diagnostics, synthetic biology, intelligent drug delivery, and studies of prebiotic life. Along these lines, let us end with a citation from Adleman: "Despite our successes, and those of others, in the absence of technical breakthroughs, optimism regarding the creation of a molecular computer capable of competing with electronic computers on classical computational problems is not warranted. However, molecular computers can be considered in a broader context. ... They may provide a much-needed means for controlling chemical/biological systems in the same way that electronic computers have provided a means for controlling electrical/mechanical systems." [3].

References

1 L.M. Adleman, *Science*, **1994**, *266*(5187), 1021–1024.
2 J.J. Tabor, A.D. Ellington, *Nat. Biotechnol.*, **2003**, *21*(9), 1013–1015.
3 R.S. Braich, N. Chelyapov, C. Johnson, P.W.K. Rothemund, L. Adleman, *Science*, **2002**, *296*, 499–502.
4 J.E. Dueber, B.J. Yeh, K. Chak, W.A. Lim, *Science*, **2003**, *301*, 1904–1908.
5 H.R. Lewis, C.H. Papadimitriou, *Elements of the Theory of Computation*, Prentice Hall, Englewood cliffs, New Jersey, **1981**.
6 L.M. Adleman, *Sci. Am.*, **1998**, 34–41.
7 J. Hartmanis, *Bull. EATCS*, **1995**, *55*, 136–138.
8 W.D. Smith, DNA computers in vitro and vivo, in *DNA Based Computers, Proceedings of a DIMACS Workshop*, Princeton, NY, **1995**, 121–185.
9 R.J. Lipton, *Science*, **1995**, *268*, 542–545.
10 D.S. Johnson, A catalog of complexity classes, in *Handbook of Theoretical Computer Science* (Ed.: J. van Leeuwen), Vol. A, Elsevier/MIT Press, Amsterdam, NY & Cambridge, MA, **1990**.
11 D. Faulhammer, A.R. Cukras, R.J. Lipton, L.F. Landweber, *Proc. Natl. Acad. Sci. U.S.A.*, **2000**, *97*(4), 1385–1389.
12 Q. Liu, L. Wang, G.A. Frutos, A.E. Condon, R.C. Corn, L.M. Smith, *Nature*, **2000**, *403*, 175–179.
13 H. Yoshida, A. Suyama, Solutions to 3-SAT by breadth first search, in *Preliminary Proceedings of the Fifth International Meeting on DNA Based Computers*, Cambridge, MA, **1999**, 9–20.
14 K. Sakamoto, H. Gouzu, K. Komiya, D. Kiga, S. Yokoyama, T. Yokomori, M. Hagiya, *Science*, **2000**, *288*, 1223–1226.

15 L. Wang, J.G. Hall, M. Lu, Q. Liu, L.M. Smith, *Nat. Biotechnol.* **2001**, *19*, 1053.
16 M.S. Livstone, L. Landweber, Mathematical considerations in the design of microreactor-based DNA computers, in *DNA 9, Ninth International Meeting on DNA Based Computers*, Madison, Wisconsin, **2003**.
17 J.H. Reif, Paradigms for biomolecular computation, in *First International Conference on Unconventional Models of Computation*, Auckland, New Zealand; *Unconventional Models of Computation* (Eds.: C.S. Calude, J. Casti, M.J. Dinneen), Springer, Singapore, New York, **1998**, pp. 72–93.
18 E. Winfree, On the computational power of DNA annealing and ligation, in *DIMACS Series in Discrete Mathematics and Theoretical Computer Science* (Eds.: R. Lipton, E.B. Baum); *Proceedings of the 1st DIMACS Workshop on DNA Based Computers*, Princeton, NJ, 199–221.
19 H. Wang, *Bell Syst. Tech. J.*, **1961**, *40*, 1–141.
20 H. Wang, *Fundam. Math.*, **1975**, *82*, 295–305.
21 E. Winfree, X. Yang, N.C. Seeman, Universal computation via self-assembly of DNA: some theory and experiments, in *DNA Based Computers II: DIMACS Workshop* (Eds.: L.F. Landweber, E.B. Baum), Princeton, NJ.
22 P. Sa-Ardyen, N. Jonoska, N. Seeman, The assembly of graphs whose edges are DNA helix axes, in *DNA 9, Ninth International Meeting on DNA Based Computers*, Madison, Wisconsin, **2003**.

23 N.C. Seeman, *Nature*, **2003**, *421*, 427–431.
24 N.C. Seeman, *J. Theor. Biol.*, **1982**, *99*, 237–247.
25 N.C. Seeman, *Chem. Intell.*, **1995**, *1*, 38–47.
26 E. Winfree, F. Liu, L.A. Wenzler, N.C. Seeman, *Nature*, **1998**, *394*, 539–544.
27 C. Mao, T.H. LaBean, J.H. Reif, N.C. Seeman, *Nature*, **2000**, *407*, 493–496.
28 T.H. LaBean, H. Yan, J. Kopatsch, F. Liu, E. Winfree, J.H. Reif, N.C. Seeman, *J. Am. Chem. Soc.*, **2000**, *122*, 1848–1860.
29 H. Yan, L. Feng, T.H. LaBean, J.H. Reif, *J. Am. Chem. Soc.*, **2003**, *125*, 14246–14247.
30 T. Eng, in *DNA Based Computers III: DIMACS Workshop* (Eds.: H. Rubin, D.H. Wood), American Mathematical Society, Philadelphia, PA, **1999**.
31 E. Winfree, T. Eng, G. Rozenberg, String tile models for DNA computing by self-assembly, in *DNA Computing: Proceedings of the 6th International Workshop on DNA-based Computers*, Leiden, June **2000** (Eds.: A. Condon, G. Rozenberg), Springer, **2001**, 63–88.
32 H. Yan, S.H. Park, G. Finkelstein, J.H. Reif, T.H. LaBean, *Science*, **2003**, *301*, 1882–1884.
33 P.W.K. Rothemund, A DNA and restriction enzyme implementation of turing machines, in *DNA Based Computers, Proceedings of a DIMACS Workshop*, **1995**, 75–119.
34 Y. Benenson, T. Paz-Elizur, R. Adar, E. Keinan, Z. Livneh, E. Shapiro, *Nature* **2001**, *414*, 430–434.
35 Y. Benenson, R. Adar, T. Paz-Elizur, Z. Livneh, E. Shapiro, *Proc. Natl. Acad. Sci. U.S.A.*, **2003**, *100*(5), 2191–2196.
36 A. Goldbeter, *Biochemical Oscillations and Cellular Rhythms: The Molecular Bases of Periodic and Chaotic Behaviour*, Cambridge University Press, Cambridge, **1996**.
37 A.P. de Silva, I.M. Dixon, H.Q.N. Gunaratne, T. Gunnlaugsson, P.R.S. Maxwell, T.E. Rice, *J. Am. Chem. Soc.*, **1999**, *121*(6), 1393–1394.
38 A. Saghatelian, N.H. Voelcker, K.M. Guckina, V.S.-Y. Lin, M.R. Ghadiri, *J. Am. Chem. Soc.*, **2000**, *125*, 346–347.
39 P. Ball, *Nature*, **2000**, *406*, 118–120.
40 S.E. Osborne, A.D. Ellington, *Chem. Rev.*, **1997**, *97*, 349–370.
41 R.R. Breaker, *Chem. Rev.*, **1997**, *97*, 371–390.
42 J. Tang, R.R. Breaker, *Chem. Biol.*, **1997**, *4*, 453–459.
43 M. Araki, Y. Okuno, Y. Sugiura, *Nucleic Acids Res.*, **1998**, *26*, 3379–3384.
44 R.R. Breaker, *Curr. Opin. Biotechnol.*, **2002**, *13*, 31–39.
45 M.P. Robertson, A.D. Ellington, *Nat. Biotechnol.*, **2001**, *19*, 650–655.
46 J.C. Cox, A.D. Ellington, *Curr. Biol.*, **2001**, *11*(9), R336.
47 S Tyagi, F.R. Krammer, *Nat. Biotechnol.*, **1996**, *14*, 303–309.
48 M.N. Stojanovic, T.E. Mitchell, D. Stefanovic, *J. Am. Chem. Soc.*, **2002**, *124*(14), 3555–3561.
49 R.R. Breaker, G.F. Joyce, *Chem. Biol.*, **1995**, *2*, 655–660.
50 M.N. Stojanovic, D. Stefanovic, *J. Am. Chem. Soc.*, **2003**, *125*(22), 6673–6676.
51 F. Guarnieri, M. Fliss, C. Bancroft, *Science*, **1996**, *273*, 220–223.
52 B. Yurke, A.P. Mills, Jr., S.L. Cheng, *BioSystems*, **1999**, *52*(1–3), 165–174.
53 M.N. Stojanovic, D.A Stefanovic, *Nat. Biotechnol.*, **2003**, *21*, 1069–1074.
54 M.C. Wright, G.F. Joyce, *Science*, **1997**, *276*, 614.
55 M. Levy, A.D. Ellington, *Proc. Nat. Acad. Sci. U.S.A.*, **2003**, *100*, 6416–6421.
56 B. Alberts, D. Bray, J. Lewis, M. Raff, K. Roberts, J.D. Watson, in *Molecular Biology of the Cell*, 3rd ed., Garland Sciences, New York, London, **1994**, Chapter 15.

15
Conclusions and Perspectives

Itamar Willner and Eugenii Katz

The different topics reviewed in the book highlight the tremendous progress in the area of bioelectronics. While primary efforts in the field challenged the application of simple electronic units such as electrodes with biological materials for sensor applications (e.g., oxygen electrode for the potentiometric analysis of glucose), later ingenious methods to fabricate integrated biomolecule-functionalized electronic elements (e.g., electrodes, field effect transistors, piezoelectric crystals) led to an explosion of potential applications and technologies based on these systems that include the tailoring of biosensors, biofuel cells, biomaterial-based machines, computers and other intelligent devices. For almost two decades, substantial research activities were directed toward development of chemical methods to integrate the bioactive materials with the electronic units in biologically active functional configurations. In parallel, the fundamental understanding of charge transport phenomena (electron/hole or proton transfer) through biomolecules, which are the basic processes activating bioelectronic systems, attracted the interest of theoreticians. The theoretical paradigms that were formulated for charge transport through proteins or nucleic acids, the ion/proton transport through channels, and the understanding of the formation of molecular or macromolecular charge transport tunneling barrier had a tremendous impact on the development of the area of bioelectronics. The understanding of the fundamental difficulties involved in the integration of biomolecules with "foreign" electronic units sparked the imagination of chemists, biologists and material scientists to overcome these limitations by developing new chemical concepts, new functional materials and modified, tailored-made biomolecules to compose the biomolecule–electronic hybrid systems. The encapsulation of enzymes in polymer matrices composed of conductive polymers, redox-active polymers or hydrogels immobilized on conductive supports proved as an effective means to establish electrical contact between redox proteins and electrodes. The incorporation of ion channels in membranes, or artificial membrane mimetic systems, associated with electrodes, generated perforated, porous matrices for specific and selective electronic biosensing. The advances in the functionalization of surfaces with monolayers or thin films introduced new dimensions to the field of bioelectronics. Ordered layered arrays could be deposited on surfaces using Langmuir–Blodget methods, supramolecular architectures or covalent bindings. Aligned protein structures

Bioelectronics. Edited by Itamar Willner and Eugenii Katz
Copyright © 2005 WILEY-VCH Verlag GmbH & Co. KGaA, Weinheim
ISBN: 3-527-30690-0

on surfaces were assembled by the use of surface-reconstitution techniques or by the application of genetically engineered biomaterials. Similarly, the deposition of other biomolecules such as antigens/antibodies or nucleic acids on surfaces paved the way for the fabrication of electronic arrays for high-throughput parallel analysis of many analytes. This progress and the scientific advances led bioelectronics to be a technologically ripe discipline. Numerous small and large enterprises commercializing bioelectronic systems, mainly in the field of electronic biosensing, were established. The practical applications of biosensors are broad and span from clinical diagnostics (e.g., detection of genetic disorders and infectious diseases, detection of cancers or the analysis of enzyme disorders), environmental control (air/aquatic/earth pollutants), food quality control, forensic applications (e.g., DNA matching or drugs analysis) to defense and homeland security applications (e.g., the detection of nerve gases, pathogens or toxins). All of these advances promise a bright and prosperous future for the field.

Did the field reach the level of maturation where only technological applications may be envisaged? The answer to this question is definitely "NO". New concepts and methods are constantly introduced to the field, suggesting the development of new routes in the ground of bioelectronics. The development of intelligent composite bioelectronic systems operating by feedback loop mechanisms is one important direction to follow. This may include integrated sensor systems that activate flow devices (e.g. pumps or permeable membranes) that could diagnose a medical disorder and automatically release the appropriate dose of a drug (e.g., detect elevated glucose concentration and release insulin), implanted biofuel cells that use body fluids as the fuel for generating electricity for activating prosthetic devices or hearing aids, or photoresponsive transistor arrays that activate neural networks that may lead to artificial vision devices.

The miniaturization of electronic circuits by lithographic methods had a revolutionary impact on microelectronics and computer engineering. While in the 1970s a density of a few thousands of transistors per chip was feasible, thirty years later ultradense arrays with densities corresponding to 10^8 transistors per chip are viable. Not surprisingly, these miniaturization concepts were adapted by numerous scientists active in the area of bioelectronics. Ingenious methods to pattern surfaces with biomolecules and to fabricate interfaces with addressable biomolecules were developed. Studies on the assembly of active interfaces that analyze in parallel multiple analytes and read out electronically the combinatorial assays were already proven successfully and the extension of these systems to operating technologies are promising. The photolithographic patterning of surfaces with nucleic acids or proteins to form arrays consisting of 100 000 different genes or proteins is already a commercialized technology. DNA arrays and protein arrays became key tools in gene analysis and genetic engineering. At present, the bioaffinity complexes occurring on these arrays are read by optical signals. It is, however, easy to anticipate that these advances will be soon adapted by the bioelectronic community. We may envisage that in the near future we will be able to construct arrays of electronic elements (electrodes, capacitors or field-effect transistors) on which numerous sensing matrices are deposited on the different units. The simultaneous electronic readout of the many sensors will then enable the high-throughput analysis of numerous genetic disorders, or the parallel analysis of many infectious pathogens and their variants. Once these biomolecular electronic chips are established, their *in vivo* implanting seems to be the next generation challenge. Implanting of such biomolecular chips may lead to the continuous parallel sensing of many analytes in body fluids, for example blood, and instantaneously alert on clinical disorders, infections, and cancerous cell formation. Once such "bioelectronic chips" are implanted in the organizms, their integration with appropriate feedback devices seems feasible. Self-regulated release of the appropriate therapeutic drugs, the use of photoresponsive biomolecular chips to activate neural networks, or the use of implanted biofuel cells as implanted batteries that use body fluids as fuels to activate artificial organs seem reachable challenging topics. Figure 15.1 depicts in a schematic view our vision on the future perspectives of bioelectronics. The past 50 years have revolutionized civilization with the developments of microelectronics and computer science, Figure 15.1(A). The ability to store ultradense information in small volumes, and process rapidly this information, paved the way to complex computations and high-tech vehicles and devices such as space shuttles or aircrafts. Similarly, Figure 15.1(B) depicts the cartoon of the future bioelectronic-assisted human being. Implanted bioelectronic chips will transduce medical failures and will transduce instantaneous diagnosis. External electronic devices will then actuate implanted arrays for controlled drug release. Are these perspectives science fiction or reality? On the basis of the progress and advances in the field of bioelectronics, we may feel optimistic that the seeds of bioelectronics that have sprouted and developed impressively in the past two decades will bloom and flourish in the near future.

Nanotechnology added new dimensions to the area of bioelectronics. Nanobiotechnology and nanobioelectronics became leading "buzz" words in the scientific community. The development of different scanning probe microscopies, such as atomic force

Fig. 15.1 From electronics to bioelectronics:
(A) A cartoon describing the developments in electronics in the last three decades.
(B) A cartoon representing the perspectives of future bioelectronics.

microscopy (AFM), scanning tunneling microscopy (STM), near-field scanning optical microscopy (NSOM), and others, provide means not only to image individual biomolecules on surfaces, such as enzymes, DNA or protein-DNA complexes but also to manipulate and pattern biomolecules on surfaces. These scanning microscopes allow the translocation of biomolecules on surfaces and their specific positioning. Recently, the use of AFM tips as "writing tools" employing "chemical inks" to generate nanoscale structures on surfaces became a popular method to pattern solid supports. This dip-pen lithography method is certainly of high value for the future nanoscale patterning of surfaces with biomolecules. The technological infrastructure developed with the various scanning probe microscopy techniques (e.g., precise atomic-scale positioning of cantilevers, optical detection of minute lever deflections, fabrication of cantilever arrays, etc.) together with the theoretical understanding of individual molecular interactions or tunneling through single molecules provide invaluable tools for future nanobiotechnology. The imaging of the bond interactions of individual antigen–antibody complexes or base-pairing in DNA, the monitoring of DNA hybridization on cantilever arrays by following surface stress on these nanoobjects, or the modification of AFM tips with enzymes that act as biocatalytic tools to pattern chemically modified surfaces represent several recent applications of these microscopic tools. It is certain that these microscopy tools will continue to develop and will play a central role in the assembly and imaging of nanoscale bioelectronic systems and, particularly, will provide new functional operating units for biosensing and biomolecule-driven nanomechanics and nanomachinery.

Most importantly, nanotechnology provides new materials for bioelectronic applications. The unique electronic, optoelectronic and catalytic properties of metal or semiconductor nanoparticles/nanorods or different nanotubes/nanowires (e.g., carbon nanotubes) provide new opportunities in the area of nanobioelectronics. New quantum-controlled phenomena such as electron confinement or coulomb blockade single-electron charging in these nanostructures, or the encapsulation of biomolecules in these nanostructures open new possibilities to yield functional bioelectronic systems. The similar dimensions of biomolecules and inorganic/organic nanoparticle or nanotube structures open the way for fabrication of hybrid systems of novel functions. Recent advances in the field demonstrated the electrical contacting of redox proteins with electrodes by means of metal nanoparticles or carbon nanotubes, the use of DNA templates as a charge-transporting matrix for photoexcited electrons formed in semiconductor nanoparticles and the generation of photocurrents, or the use of enzyme-functionalized carbon nanotubes as a gate material controlling the activity of a field-effect transistor. These primary studies highlight the potential contributions of nanoscale materials to the future development of bioelectronics.

A further important aspect related to nanobiotechnology is the use of biomolecules as templates to synthesize metallic, semiconducting or insulating nanowires or nanotubes. Mother nature provides unique biological structures consisting of DNA or self-assembled proteins that might act as attractive templates for the fabrication of wires or solid nanotube containers. Furthermore, the ability to modify DNA by synthetic means or the possibility to assemble proteins of predesigned shapes, dimensions and structure, pave the way to generate tailored templates for nanostructures formation. Indeed, numerous recent studies have demonstrated the use of biological templates to yield conductive or semiconductive nanowires/nanotubes and their applications as electronic nanocircuitry or nanocontainer units were accomplished. Major breakthroughs were achieved by combining the nanowires with functional biomaterials to yield devices such as nanotransistors or nanotransporters. The three branches of nanobiotechnology that include the microscopy tools, the nanomaterial objects and the conjugated biomaterial/nanoobject hybrid systems provide a unique arsenal that could be the basis for new functional devices. The perspectives of nanobiotechnology and nanobioelectronics are at present questionable, and further research and creative ideas are essential to further develop the area. It is, however, certain that biomolecular/nanoobject hybrid systems will find applications in various disciplines such as intracellular drug delivery, gene therapy or *in vivo* imaging.

The advances in bioelectronics and the practical applications that originated from the research efforts in the field promise the continuous prosperity of this scientific topic. The incorporation of nanotechnology concepts and materials into the domain of bioelectronic systems paves the way to new challenges and highlights the long term and continuous interest in the field. It is expected that interdisciplinary efforts of chemists, biologists, physicists, material scientists and electronic engineers will highlight exciting scientific accomplishments in the coming years.

Subject Index

a

AC stimulation 359
– intracellular 353
acetylcholine esterase 332
acetylcholine esterase-CdS nanoparticles hybrid 233
ACHE *see* acetylcholine esterase-CdS nanoparticle hybrid
actin filaments 292 f., 298 f.
activity, neuronal 367 ff.
adders 448
adhesion, nerve cell 343
Adleman's first experiment 428 ff.
adsorbed biomolecules 194 f.
adsorptive transfer stripping 132, 167
ADTS *see* adsorptive transfer stripping
AFM *see* atomic force microscopy
AFM-cantilever tips, protein-functionalized 403
alcohol dehydrogenase electrodes 75
alignment, NAD^+-dependent enzymes 75 f.
allozymes 445
amalgam electrodes, DNA break detection 168 ff.
3-aminophenyl boronic acid 75
amperometric
– analysis 257
– electrodes 100
– immunosensors 336
– sensors 421
AND gate, DNA computing 448
anodic current 75
– electrocatalytic 75
antibodies, secondary 277
antibody-antigen detection 225 ff.
anticancer drugs 180
antigen, photoisomerizable 326 ff.
antigen-antibody affinity interactions 326
antigen-antibody complex, hydrophobic 325
apo-glucose oxidase 88

apo-horseradish peroxidase 69
apo-HRP *see* apo-horseradish peroxidase
apoenzymes 233
– reconstitution 91
apomyoglobin 38, 40
aptamer 42
architecture, biomolecule-nanoparticle 41
area-contact 343, 345
ArF *see* argon-fluorine
argon-fluorine 415
array, S-layer protein 398
artificial biomembranes 339
assays, ultrasensitive 247
assembly scheme 267
astrocytes 348
atomic force microscopy 6
ATP synthase 304
A-type spectrum 352
A-type stimulation 367
Au-nanoparticles 42, 57, 59, 61, 233
automaton, boolean 451
avidin 403
avidin-biotin-bond 291, 412
avidin-modified magnetic particles 241
azurin 195, 202

b

bacillus coagulans 406
bacillus sphaericus 403, 409, 416 ff.
bacillus stearothermophillus 400
bacteria cell walls 398
barrier model 16
bases, DNA 130
bath potential 351
bath stimulation 354
bimolecule-nanoparticle architecture 41
binding sites, S-layer protein 407
bioaffinity interactions, reversible 323 ff.
bioaffinity recognition events 330

bioanalytical sensors, optical 421
bioanalyzers 103
biocatalysis 147 f.
– activity 116
– charging 64
– oxidation 86
– processes 332
– reactions 310 ff., 330
biochemical reaction, molecular computing 435
biochips 102 ff.
bioelectrocatalysis 92, 253
– electrode 72
– function, cyclic 75
– oxidation 48, 56
– oxidation, glucose 256
– process 250
– reaction 250
– transformation 50
bioelectrocatalytic reduction, nitrate 83
bioelectrochemical systems 232 ff.
bioelectronic systems, hybrid 402 ff.
bioelectronics 1 ff., 287 ff., 395, 457
biofuel cells 1, 4, 35 ff., 49
– design 83 ff.
biological machinery 266
biological warfare, sensors 42
biomaterial 193 ff.
– light-controlled 331
– nanoparticle hybrid systems 231 ff.
biomimicry 395
biomolecules 197
– adsorbed 194 f.
– labeling 231
– manipulation 287 ff.
– mechanical properties 291 ff.
– nanoparticle hybrid systems 8
biorecognition
– assays 242
– event 235
– process 332
biosensing 235, 255
– dual 255
– strategy, label-free 135
biosensors 35 ff., 165, 180
– definition 102
– electroswitchable 88
– optical 232
– self-powered 86
biosynthetic polynucleotides 132
biotemplated circuits 309
biotemplates 416 ff.
biotin 277
biphasic reversible switch 330

bis-intercalators 146
blood 85
– fuel 91
– glucose meter 107
BNBIQ see dibutyl-naphthylamine-butylsulfonato-isoquinolinium
boolean automatons, DNA based 450
brain-computer interfaces 340
brain-silicon chips 383
brain slices 384 ff.
Brownian motion 300
Brownian ratchet model 303
B-type spectrum 352
B-type stimulation 368
burst stimulation 370

c

Ca1 region 386
cadmium sulfide superlattice 421
cantilever tips 403
capacitance, time-dependent 64
capacitive current 364
capacitive stimulation 367, 370
– cultured hippocampus slice 388
capacitors, circular 389
capping molecules 235
carbon electrodes 134, 167
carbon nanotubes 42, 53 ff., 59, 276
carbonic anhydrase containing liposomes 400
carcinogen 165, 178 f.
carrier bead 248
catalase 195
catalyst, nucleic acid 442 ff.
catalytic deposition 242
catalytic hydrogen evolution 133
catalytic P450 sensors 120 ff.
CCC see covalently closed circular DNA
CdS nanoparticle/ache hybrid systems 236
CdS nanoparticles, photoactivated 334
CdS semiconductors 235
cell-silicon junctions 343, 346 ff., 358 ff.
cell stimulation 352
channel proteins 396
channels, recombinant 361
charge transfer theory 3
charge transport, DNA-mediated 149
charged surfaces, glass 305
chemical
– energy conversion 84
– functionalization 213 ff.
– guidance 377
– immobilization 103

– modifiaction 400 f.
– photocurrent 334
– synapses 374
chemistry, electroanalytical 231
chips
– bioelectronic 459
– neuron-silicon 340
– stimulation 356
chromophores 37
chronopotentiograms 244
chronopotentiometric detection, DNA 249
circuits
– biotemplated 309
– DNA-templated 284
– electronic 3
– neuron-silicon 362 ff.
clauses, molecular computing 432
cleavage, DNA 170
cleft 346, 349 ff., 357
CNT see carbon nanotubes
coating, metal 273
cofactor, hemoproteins 67
cofactor-enzyme affinity 73
cofactor-modified electrodes 91
coherence parameter 26
comet assay 161
command interfaces 316
computing, nucleic acid 427 ff.
concomitant current 367
conductance 345, 349 ff., 357, 364
– leak 368
– ohmic 384
– single-channel 408 ff.
conduction band 250, 335
conductive polymers 4, 135 f.
– matrice 36
– semiconductive properties 92
– wires 53 ff.
conductive probe AFM 201
conductive redox-polymers 49
conductivity properties, metal nanoparticles 259
conductors 383 ff.
– planar 343 ff.
confocal microscopy 196
conformational changes, DNA 152
conformational distortions, DNA 172 f.
conjunctive normal form 430
connectivity, structural 267
contacted electrodes 65 ff.
contacting, electrical 73
conversion of chemical energy 84
core-coat conductors 372
– planar 343 ff.

coupling
– multipathway 24
– neurons 373
– pathways 21
coupling decay factors, proteins 23
covalent attachement
– biomolecule 213 ff.
– DNA 217 ff.
covalent DNA damage 163, 170 ff.
covalent linking, DNA 214
covalently bound labels 139
covalently bound photoisomerizable units 322
covalently closed circular DNA 133
CP see conductive probe AFM
cross-linking 73 f.
cryo electron microscopy 193
crystalline surfaces 416 ff.
crystals, dynamical closed 400
C-type stimulation 369
cumulative xor computation 438
current
– balance 355
– capacitive 364
– transient 404
cyclic bioelectrocatalytic functions 75
cyclic photoisomerization 315, 317
cyclic voltammograms 39, 48
cytochrome
– b5 29
– c 20 f., 70, 80, 82, 110 ff., 195, 317
– c' 115 f.
– P450 117 f., 194

d

damage detection 160 f.
daunomycin 137, 175
de novo hemoproteins 69 ff.
de novo proteins 70, 92
decay factor, pathway model 17
deep ultraviolet irradiation 414 ff.
deoxyribozyme 443
– based gate 447 f.
detection system, electrochemical 149
detector gate 445
dialyzers 102
diamond, biosensors 211 f.
diamond-surface, DNA-modified 225 ff.
diamond thin-film 226
dibutyl-naphthylamine-butylsulfonato-isoquinolinium 356
2,3-dichloro-1,4-naphthoquinone 251
differential pulse voltammetry 128

diffusional cofactors 80
diffusional electron mediators 232, 315, 321
diffusional redox mediators 73
dimyristoylphosphatidyletethanolamine 409
dinitrophenyl-antibody 326
dip-pen lithography 7, 198
direct electrical communication 117 ff.
direct electron transfer 68, 105
direct polarization 342
directly contacted proteins 109
displacement, microneedle 290, 300
distance-decay exponent 16
dithiol tunneling barrier 62
DME *see* dropping mercury electrode
DMPE *see* dimyristoylphosphatidyl-
    ethanolamine
DMSO 114
DNA 42
– adduct 171 f.
– adsorption 154
– algorithmic assembly 436
– base damage 173
– break sensor 166
– bridge 270
– chip 153
– cleavage 170
– computation 427, 430, 453
– computer 4, 11, 433
– covalent attachement 217 ff.
– covalent linking 214
– covalently closed circular 133
– detection, label-free 156
– detection scheme 239
– drug interaction 179 f.
– electrochemical detection 243
– electrochemistry 134
– hybridization 140 ff., 215 ff.
– hybridization sensor 150
– labels 136 ff., 147
– mediated charge transport 149
– metal coating 273
– metallization 266, 268 ff.
– modified diamond-surface 225 ff.
– modified electrode 128, 131 ff., 167 f.
– modified surface 219 ff.
– molecule, rotation 293
– noncovalently 173 ff.
– recognition 249, 334
– recognition layer 167
– repair 162
– scaffold 273
– structure 131, 133
DNA damage
– detection 160 ff.

– electrochemical detectors 162 ff.
– induced 176 f.
– sensor 159
DNA sensor 178
– electrochemical 127 ff.
DNA strand break, sensor 168 ff.
DNA template 56, 460
– circuit 284
– electronic 265 ff.
– field-effect transistor 280 f.
DNP-Ab *see* dinitrophenyl-antibody
docking, electron transfer 27 f.
donor-acceptor coupling 23
donor-acceptor interaction 19
doping, diamond 212
double-crossover tiles 438
double-stranded molecules 428
double-surface DNA hybridization sensors
    154
double-surface technique 153 ff., 155
DPL *see* dip-pen lithography
DPME-interface 408 ff.
dropping mercury electrodes 164
DST *see* double-surface technique
dual biosensing 255
DUV *see* deep ultraviolet irradiation
DX *see* double-crossover tiles
dyes, voltage sensitive 356
dynamic closed surface crystal 400
dynamical docking 27 f.
dynein 303

e

EB *see* electron beam lithography
ECHI 146
EIS *see* electrochemical impedance
    spectroscopy
electrical
– characterization 219 ff.
– coding 245
– contacts 49, 53 ff., 73
– – electrodes 65 ff.
– – enzymes 40, 99 ff.
– – hemoproteins 80
– – photochemical control 321
– DNA detection 237
– guidance 382
– polarization 341
– power generators, implantable 91
– wiring 235
electroactive
– bridges 121
– labels 136 ff.

– markers 153
– tags, encapsulation 248
electroactivity, natural 129 ff.
electroanalytical chemistry 231
electrocatalysis 133 ff.
electrocatalytic oxidation 151, 179
electrocatalytic process 250
electrochemical
– biosensors 102
– conduction 421
– contacting, multilayer arrays 324
– detection 143 ff., 162
– – DNA 243
– – systems 149
– DNA sensors 127 ff.
– driven epoxidation 121
– generators 86
– impedance spectroscopy 220
– interrogation 152
– method 128 f.
– microscopy, scanning 404
– response 171 f.
– sensors 177 ff.
– signals 131
– switching 87
electrodeposition, nanoparticles 399
electrodes
– cofactor-modified 91
– functionalized 35 ff., 43 ff.
– glucose-sensing 45
– surface 133, 143 f., 146 f., 193
electrodiffusion 346
electroless deposition 238
electrolyte oxide field-effect transistors 351
electron beam lithography 198
electron coupling matrix element 15
electron donor 117
– sacrificial 335
electron mediators 84
– ferrocene 40
electron microscopy 402 ff.
electron relays 4
electron transfer 15 ff., 35 ff.
– direct 105
– DNA 166
– heterogenous 120
– interproteins 27 ff.
– mediators 71, 73, 322
– – diffusional 315
– – photoswitchable 311
– photoinduced 332 ff.
– process 61
– proteins 3, 31
– rate 19 ff., 26

– resistance 45, 329
– theory 28
– vectorial 39
electron tunneling 15 ff.
electronic
– circuitry 3
– coupling 104
– – neurons 373
– devices 231 ff.
– interfaces 340 ff.
– transduction 310, 312 ff., 322
electronics
– DNA-templated 265 ff.
– molecular scale 266
electrooxidation 147
electrophoresis box, DNA computers 433
electrostatic control 320
electrostatic force 305
electroswitchable biosensors 88
ELISA *see* enzyme-linked immunosorbent assy technique
energy landscapes 15
ENFET *see* enzyme-based field-effect transistors
environmental monitoring 178
enzymatic networks 448
enzymatic probing 173
enzyme-active centers 35
enzyme-based field-effect transistors 80
enzyme-biocatalyzed reaction 332 ff.
enzyme electrodes 4, 36, 43, 50, 101, 315
– application 107
– NAD(P)$^+$-dependent 73 ff.
– reconstituted 83 ff.
enzyme family 117 ff.
enzyme-linked assay 157
enzyme-linked immunosorbent assy technique 414
enzyme multilayer array 323
enzyme-nanoparticle hybrid 59
enzyme reactors 102
enzyme test strips 100
enzymes 408 ff.
– electrically contacted 40, 99 ff.
– electrochemical process 316 ff.
– surface-reconstituted 320
EOSFET *see* electrolyte oxide field-effect transistors
epoxidation 121
EPSP *see* excitatory postsynaptic potential
equilibrium complex 73
ET *see* electron-transfer
excimer lasers 415
excitatory postsynaptic potential 375

– neuronal 385
extracellular potential 389
extracellular transient 371

f

FAD *see* flavin adenine dinucleotide
fad cofactor 51, 54, 313
– synthesis 314
faradaic impedance spectroscopy 327
fenton-type reaction 176
ferredoxin 204
ferrocene electron mediators 40
FET *see* field-effect transistors
FIB *see* focused ion beam
fibroblasts 350
field effect 223
– hybridization-induced 222
field-effect transistors 54, 276, 280f, 365
– electrolyte oxide 351
field potential, neurons 385 f.
filaments
– actin 298 f.
– myosin 305
films, lipids 405
flavin adenine dinucleotide 313
flavoenzymes 53 ff., 313
– electrical contacting 57 ff.
– electrodes 43 f.
FLIC *see* fluorescence interference contrast
FLIC microscopy 347
flickering interference 25
flow cells 220
fluorescence 195
– interference contrast 348
– internal reflection 288
– label 232
– lifetime 196
– molecule 300
– response 423
– silicon 346
focal contact, fibroblast cells 349
focused ion beam lithography 198
force
– direct measurement 304
– electrostatic 305
– microscopy 305
– noncovalent 401
formaldehyde dehydrogenase 334
four-helix bundles 70
fuel substrates 85
functionalized electrodes 35 ff.
functionalized nanoparticles 42, 231

g

gating 27
gating polarity 278
generators, electrochemical 86
genetic recombination, homologous 272
genotoxic substances 173 ff.
glas microneedles 289
glass, passivated 269
glassy carbon electrodes 119
glia cells 348
glucose
– determination 107
– fuel 84
– oxidation 256
glucose oxidase 4, 36, 46, 51, 55, 57, 59, 62, 85, 105, 195, 233, 255, 312, 331, 410, 422
– electrodes 50, 88
glucose sensing 53
– electrodes 45
– systems 86
glue, molecular 291
glutaric dialdehyde 83
glutathione reductase 323
gly-gly-cys units 70
glycoprotein subunits 398
gold metal tracers 237
GOx *see* glucose oxidase
gramicidin 411 f.
groove binder 137
guanine 151
– redox signal 171
guidance
– chemical 377
– electrical 382
guided outgrowth, neurons 378

h

half-adders 448 f.
α-hemolysin 408 ff.
hemoproteins 35, 37, 81, 317
– de novo 69 ff.
– electrically contacted 68 f., 80
heterogenous electron transfer 120
HIPIP *see* iron-sulfur proteins
Holliday junction 437
homologous genetic recombination 272
horse radish peroxidase 69, 147, 195
HRP *see* horseradish peroxidase
human immunoglobulin antigens 226, 238
hybrid systems 92, 457
– bioelectronic devices 395, 402 ff., 417
– biomaterial-nanoparticle 231 ff.
– biomolecule-nanoparticle 8

– CdS nanoparticle/ache 236
– neuroelectronic 10, 363
hybridization
– DNA 216 ff.
– efficiency 144
hybridization-induced field-effect 222
hydrogel 36
hydrophobic antigen-antibody complex 325
hyperpolarizing 368

i

IGg *see* human immunoglobulin antigen
immobilization 421, 143 f., 259
– chemical 103
– DNA 165
– neurons 380
immunoglobulin g 226
immunorecognition 310
immunosensing 237, 240
immunosensors, reversible 326 ff., 336
impedance spectroscopy 128, 225 ff.
in vivo detection 113
indicator electrodes 102, 128 f.
information storage 336
inorganic cations 137
input molecules 439
insulating template 395
integrated electrodes 80
integrated neuroelectronics 392
interaction, donor-acceptor 19
interaction force, direct measurement 304
intercalator 6, 137, 175
– redox-active 4
– threading 146
interfaces
– ionic-electronic 340 ff.
– lipid 408
– neuron semiconductor 339 ff.
– photoisomerizable 320
interfacing biological molecules 209 ff.
interfacing neurons 379
interferences, multiple pathways 24
interprotein electron transfer 27 f.
interrogation, electrochemical 152
intracellular
– AC stimulation 353
– dynamics 345
– voltage 369
ion channels 358 ff.
ionic-electronic interface 340 ff., 342
iontophoretic detection, glucose 108
iron-sulfur proteins 19

j

junction conductance 361
junctions
– cell-silicon 358 ff.
– membrane-silicon 355
– neuroelectronic 11

k

kinesin 296
– manipulation 303

l

label-free biosensing strategy 135
label-free DNA detection 156
labeling, biomolecule 231
– nucleic acid 129 ff.
labels, covalently bound 139
labile complex 73
lactate, oxidation 255
lactate dehydrogenase 74, 78, 254, 256
– electrode 75
laminin 348
Landweber-Lipton approach 434
laser traps 288 ff., 303
lateral patterning 408
LDH *see* lactate dehydrogenase
leak conductance 354, 368
leech neurons 354, 365, 369
ligase logic gate 452
light-controlled biomaterial 331
linking biomolecules 214
lipid core, cell membrane 349
lipid interface 405 ff., 408
lipid-supported lattices 411 f.
liposomes 249
lithography 198
– dip pen 7
– molecular 271 ff., 276
localized surface plasmons 61
logic gates 442
LSP *see* localized surface plasmons
lymnaea stagnalis 374
lysozome nanoarray 200

m

machinery, biological 266
magnetic beads 241
– streptavidin-coated 237
magnetic particles 231
– relay-functionalized 252
– rotation 257
magneto-controlled biosensing 255

470 | Subject Index

malate dehydrogenase 77
– electrode 75
MalD *see* malate dehydrogenase
mammalian brain 383
manipulation, biomolecule 287 ff.
manipulation technique, mechanical 297
– myosin 300
Marcus theory 3, 35
marker 237
– electroactive 153
matrix element, electron coupling 15
MCP *see* microcontact printing
mediated charge transport 149
mediatorless communication 99
mediators 35, 134, 151, 332
– diffusional 315, 321
– electron transfer 322
– photoswitchable 311
membrane-silicon junction 355
membranes, semifluid 405
mercury electrodes 130 f., 167
– DNA break detection 168 ff.
mercury-free electrodes 170
metal cluster arrays 420
metal coating, DNA 273
metal nanocircuitry 238
metallic
– electrodes 205
– nanoparticles 57 ff., 232, 259, 416 ff.
– wires 278
metallization, DNA 266, 268 ff.
microcontact printing 198, 200
microelectrodes 239
microfabrication techniques 199
microgravimetric quartz-crystal microbalance 249
microgravimetric transduction 329
micromolding in capillaries 408
microneedles 288, 298
– glass 289
microperoxidase-11 82
microscopy, scanning probe 6
MIMIC *see* micromolding in capillaries
miniaturization 102 ff.
molecular
– adresses 266
– beacons 152
– – stoichiometric 446
– devices, control 453
– dynamic trajectories 22
– glue 291
– lithography 276
– – sequence-specific 271 ff.
– motors 296 ff.

– movement, stepwise 301
– nanotechnology 414 ff.
– optobioelectronics 309
– relay systems 43 ff.
– self-assembly 435 ff.
– types 304
– wires 61, 66
molecules
– interacting 296
– interfacing 209 ff.
monolayers
– phospholipid 406 ff.
– protein 198
– thiolated 233
motors
– molecular 296 ff.
– rotary 304
MP-11 *see* microperoxidase-11
multilayer arrays
– electrochemical contacting 324
– enzymes 323
multiple amplification routes 247
multiple pathway coupling 24
multiple tunneling 23 ff.
mycobacterium smegmatis 405
myoglobin 29
myosin 292, 296, 298
– movement 299
– single 300
– unconventional 302 f.
myosin filemant 305

n

$NAD(P)^+$-dependent enzyme-electrode 73 ff.
NAD^+ *see* nicotinamide dinucleotide
NAD^+-cofactor 78
NAD^+-dependent enzymes 254
– alignment 75 f.
$NADP^+$ *see* nicotinamide dinucleotide phosphate
$NADP^+$-dependent 75
nanobiotechnology 3
nanocircuitry, metal 238
nanocrystal tracers 246
nanodevices 204
nanoengineering 91
nanogap electrodes 205 f.
nanogold 274
nanografting 201
nanometry
– displacement measurements 298
– resolution 290
nanoneedles 55

nanoparticle-enzyme hybrids 232 ff.
nanoparticle tags, semiconductive 245
nanoparticles 8
– DNA detection 155 f.
– functionalized 42, 231
– hybrid system 231 ff.
– metallic 57, 232
– semiconductive 232, 249
nanorods, semiconductor 40
nanotechnology 396, 459
nanotemplates 396
nanotransistors 9 f., 461
nanotubes, structural features 56
nanowires 9
natural electroactivity 129 ff.
near-resonance regime 27
nerve cells 341
nerve gases 42
networks, neuronal 3, 11
neuritic tree 381
neuroelectronic hybrids 363
– systems 10
neuroelectronic junctions 11
neuron-semiconductor interface 339 ff.
neuron-silicon chips 340
neuron-silicon circuits 362 ff.
neuronal
– activity 362, 365, 367 ff.
– excitation 366
– growth 377
– networks 3, 377 ff., 392
– stimulation 367 ff.
neurons 11
– chips 372 ff.
– displacement 379
– immobilzation 380
– presynaptic 376
– snail 370
neutron reflectometry 402
nicotinamide dinucleotide 73
nitrate reductase 71
nitric oxide sensors 115 f.
nitrite reductase 195
nitrogenase 30
nitromerocyanine 318
nitrospiropyran 312
– units 317
noncontact force microscopy 305
noncovalent force 401
noncovalent interaction 173
noncovalently bound DNA label 136 ff.
nonspecific adsorption, DNA 153
Norrish reaction 415
NOT gate 443

NOT logic, ribozymes 446
Np-complete problem 428
nucleases 173
nucleic acids 129 ff., 427 ff.
– analysis 131
– catalysts 442 ff.
– modification 139
– unlabeled 131 ff.

o

ohmic conductance 384
oligonucleotides 428
one-compartment model 344
one-electron transfer 117
open-circuit voltage 86
optical biosensors 232, 422
– analytical 421
optical recording 356
optobioelectronics, molecular 309
optodes 422
ordered structure, formation 435
organotypic slices 385
osmium-modified DNA 156 f.
osmium tetroxide complexes 138
outgrowth, guided 378
oxidation
– base 130
– electrocatalytic 151
– ethanol 75
– glucose 51, 61, 65
– lactate 255
oxygen-derived radicals 109 ff.

p

P450 see cytochrome P450
packing density 22, 145
PAN see polyaniline
parallel approach, computing 428 ff.
passivated glass 269
pathway
– coupling 28
– families 25
– model 17, 19 ff.
– tube 24
patterned surfaces 197
patterning 414 ff.
– lateral 408
PCR-amplified tDNA 158
PDMS see polydimethylsilane, see polydimethylsiloxane
PDR see putidaredoxin reductase
PDX see putidaredoxin
peptide nucleic acid 139

permeability barrier, S-layer protein 404
peroxidases 68
phosphodiesterases 445
phospholipid films 405
phospholipid monolayers 406 ff.
photoactivated CdS nanoparticles 334
photoactive reconstituted proteins 37
photochemical switching 310 ff.
photocurrent 235
– action spectra 250
– generation 332
photoelectrochemistry 260
photoinduced electron transfer 332 ff.
photoisomerizable units 312 ff.
– covalently bound 322
photoisomerization 310
– antigen 326 ff.
– fad cofactor, synthesis 314
– interface 320, 323 ff.
photolithography 198
photonic signals 310
photosensitizers 37
photoswitchable electron transfer mediators 311
photosynthetic reaction centers 38
photosystem I reaction centers 203
piezoelectric transduction 329
PLA2 see porcine pancreatic phospholipase a2
planar core-coat conductors 343 ff.
planar networks, neuronal 383
plastocyanin 204
platinum nanoclusters 418
PNA see peptide nucleic acid
point-contact model 344 f., 356
point mutation 148, 151
polarization
– direct 342
– electrical 341
polarography 164
polyaniline 49, 53, 78
polydimethylsilane 200
polydimethylsiloxane 220
polyester structure on chip 382
polymer films 66, 89
polymer matrices, conductive 36
polymer relay systems 43 ff.
polymer wires, conducting 53 ff.
polymers
– conducting 4, 135 f.
– protein 291
– redox-active 4
– semiconductive properties 92
porcine pancreatic phospholipase a2 409
porin 396

– structure 405
porosity, S-layer protein 398
positively charged surface 305
potassium channel 360
potential profile, Ca1 386
potentiometric electrodes 100
PQQ see pyrroloquinoline quinone
PQQ cofactor 65
(PQQ)-NAD$^+$ (cofactor) interface 80
premelting, DNA 127
presynaptic neuron 376
primer, molecule computing 430
protecting group 215
protection capacity 178 f.
protein chains, unfolding 294
protein-functionalized AFM-cantilever tips 403
protein-mediated tunneling 16
protein molecules 296
– unfolding 294
protein-protein
– binding 30
– complex 82
– docking 27
– interaction 305
proteins
– backbone 312
– coupling decay factor 23
– crystal 28
– directly contacted 109
– electrochemical process 316 ff.
– electrode 109 ff.
– electron transfer 15 ff., 31
– electron transport 3
– electronic tunneling 19
– folding 31
– photoactive reconstituted 37
– polymer 291
– reconstituted redox 43 ff.
– redox 73
– ruthenium modified 19
– S-layer 7
– structural engineering 4
– tube 419 ff.
proton, transport 304
PSI see photosystem I reaction center
putidaredoxin 201
putidaredoxin reductases 201
pyrroloquinoline quinone 105

q

quinoproteins 66
– reconstituted 65 ff.

Subject Index

r

radicals, oxygen-derived 109 ff.
rat neurons 358, 365
rational modular design 445
RDE *see* rotating disc electrodes
reaction center, photosynthetic 38
reactive oxygen species 177
readout, PCR-reactions 434
reagentless sensors 153
reca proteins 272
recognition layers, DNA 167
recombinant channels 361
recombination machinery 274
reconstituted
– enzymes 40, 83 ff.
– hemoproteins 68 f.
– quinoproteins 65 ff.
– redox proteins 43 ff.
reconstitution process, proteins 37
rectifying devices 204
redox-active 232
– intercalators 4
– polymers 4
– substrates 6
– transition metals 169
redox enzymes 4 f., 35 ff., 106, 231, 241, 312 ff., 322
– biocatalytic reaction 310 ff.
– interaction 316 ff.
redox-functionalized magnetic particles 250
redox markers 249
redox mediators 99
redox metalloproteins 202
redox proteins 35, 65, 68, 73
– reconstituted 43 ff.
redox reaction 28
redox relay 121
redox signal, guanine 171
reductive precipitation 416
reflectometry 402
relay, electron 4
relay-functionalized magnetic particles 252
relay systems 43 ff.
relay units 251
repetitive sequence 157
reporter probe 158
resistance 45, 278
– electron transfer 329
– gold wires 270
resists 273
response type, membrane 364
restriction enzymes 439
reusable immunosensors 326

reversible
– amperometric immunosensor electrodes 336
– bioaffinity interaction 323 ff.
– electron transfer 117
– immunosensors 326 ff.
– switch, biphasic 330
RNA 127
– detection, label-free 156
– hybridization 140 ff.
– polymerase 293, 304
rotary motor 304
rotating disc electrodes 258
rotaxane 48 f.
Rothemund-Shapiro paradigm 439 ff.
ruthenium complex 179
ruthenium modified azurin 26
ruthenium modified protein 19, 24
ruthenium wires 30

s

sacrificial electron donors 335
SAM *see* self-assembled monolayers
sandwich-assays 149
SAT *see* satisfiability problem
satisfiability problem 430, 432
scaffold, DNA 273
SCAL *see* carbonic anhydrase containing liposomes
scanning electrochemical microscopy 404
scanning probe lithography 198
scanning probe microscopy 6, 42, 195
SECM *see* scanning electrochemical microscopy
secondary antibodies 277
Seeman-Winfree paradigm 435 ff.
selective metallization 419 ff.
self-assembled monolayers 194
self-assembly 265, 276
– molecular 435 ff.
self-organization 283
self-powered biosensor devices 86
semiconductive properties, polymers 92
semiconductors 209 ff.
– interface 339 ff.
– nanoparticles 232, 245, 249, 416 ff.
– nano-rods 40
– substrates 210 ff.
semifluid membranes 405
sensing, bioelectronic 209 ff.
sensing devices 231 ff.
sensor calibration 112
sensorics 361

Subject Index

sensors 140
– bioelectronic 1
– biological warfare 42
– DNA 127 ff.
– DNA damage 159
– electrochemical 177 ff.
– types 100
sequence-specific molecular lithography 271 ff.
sheet-conductor model 384
sheet resistance 345
signal
– amplification 140
– approximation 362
– electrochemical 131
– ionic 341
– probe 147
signaling chip-neurons 380
silicon 219 f.
– single-crystal 210
silicon chips 372 ff.
silicon electrodes, impedance 221
silicon substrates 384
silicon support 408
silicon surfaces, DNA-modified 215 ff.
silver metal tracers 237
single
– action potential 374
– base mismatch 151
– biomolecule, adressing 201
– channel conductance 408 ff.
– crystal silicon 210
– fluorescent molecules 300
– molecule level 193 ff.
– molecule manipulation 287 ff.
– pore current 413
– protein molecules, electrochemistry 193
– surface technique 143 ff.
– wall carbon nanotubes 54, 276, 278
site-directed mutagenesis 194
S-layer proteins 7 f., 395
– array 398
– lattices 408 ff.
– lipid-supported 411 f.
– reconstitution 406 ff.
– tracks 410
– ultrafiltration membrane 412
Smith-Corn approach 434
snail neurons 370
– displacement 379
sodium montmorillonite 118
solid electrodes 170
solution library 431
soma-soma configuration 374

SP *see* nitrospiropyran
SPL *see* scanning probe lithographic
sporosarcina urea 418
SPR *see* surface plasmon resonance
spring force, laser traps 290
square barrier model 16
stem-loop structure 152
step stimulation 369
stimulating state macine 439 ff.
stimulus, ramp-shaped 389
stimulus-response relation 390
STM *see* scanning tunneling microscopy
STM lithography 198, 200
stoichiometric molecular beacons 446
streptavidin-coated magnetic beads 237
stripping, adsorptice transfer 132
stripping response 244
stripping voltammetry 235
structural connectivity 267
structureless barrier model 16
substrates 210 ff.
– redox-active 6
SUM *see* S-layer ultrafiltration membrane
superexchange 15
– charge transfer theory 3
superlattice 416 ff., 421
superoxide dismutase 112
superoxide sensors 110 ff.
supramolecular binding site 412
surface
– charged 305
– coverage 233
– cross-linking 91
– crystalline 416 ff.
– DNA-modified 219 ff.
– force 305
– plasmon resonance 51, 63
– plasmon resonance spectra 52
– reconstituted enzyme 320
– reconstitution 44
– S-layer protein 404 ff.
– techniques 42
surface crystal, dynamical closed 400
SWCNT *see* single wall carbon nanotubes
switches
– biphasic 330
– photobiocatalytic 315
switching, electrochemical 87
SWNT *see* single wall carbon nanotube
synapses, chemical 374
synaptic transmission 376, 388
synthetic cofactor 72

t

tags, semiconductive 245
test strips 107
tetraether lipids 412
– films 405
thermal fluctuation 290
thermoanaerobacter thermohydrosulfuricus 401
thin films 226
– configuration 86
thioacetylcholine 42
thiocholine 333
thiolated monolayers 233
thiolated oligonucleotide probes 246
threading intercalator 146
three-arm junctions 275 f.
three-terminal FETs 281
through bond model 31
through bond propagation 17f
through space propagation 17
tic-tac-toe automateon 452
tight-binding models 24
tips, cantilever 403
TIRFM *see* total internal reflection fluorescence microscopy
tissue-sheet conductors 383 ff.
titin 294
– unfolding 295
topaquinone 67
topographical guidance 381
total internal reflection fluorescence microscopy 196
toxic compounds 177
toxicity testing 178 f.
TPQ 67
transduction 310 ff., 329
– electronic 316 ff.
– photoelectrochemical 249, 334
transfer spectrum 352
transfer theory 28
transferase 15
transient current 404
transistors 276
– field-effect 351, 365
– recording 351 f., 354, 362, 385 ff.
transition metals 176
transition rules 441
transmission, synaptic 376, 388
triggering 250
triple-crossover tiles 437

tumor suppressor gene 148
tunneling 27 f.
– multiple 23 ff.
– protein-mediated 16
tunneling barrier 26
tunneling decay parameter 26
tunneling pathway 17 f., 21 ff.
tunneling propagation fluctuation 25
Turing machine 435
two neuron chip 372
Tx *see* triple-crossover tiles

u

ultrasensitive particle-based assays 247
unfolding, mechanically induced 294
unitary steps 300 ff.
UV lithography 198

v

vectorial electron transfer 39
visualization
– single molecule 288, 300
voltage, intracellular 369
voltage dependence, ion channel 359
voltage sensitive dyes 356
voltammetric signals 169
voltammetry 165 ff.
– differential pulse 128
– stripping 235
voltammograms, cyclic 39
volume conductors 383

w

Wang tiles 435
waste material 84
whole-cell patch-clamp technique 358
wires
– metallic 278
– polymer 53 ff.
wiring, electrical 231

x

XOR computation 438
X-ray reflectometry 402

y

YES gate 443

QH
509.5
.B56

2005